STARCH POLYMERS

From Genetic Engineering to Green Applications

STARCH POLYMERS

From Genetic Engineering to Green Applications

Edited by

Peter J. Halley

Luc Avérous

ELSEVIER

AMSTERDAM • BOSTON • HEIDELBERG • LONDON • NEW YORK • OXFORD
PARIS • SAN DIEGO • SAN FRANCISCO • SINGAPORE • SYDNEY • TOKYO

Elsevier
30 Corporate Drive, Suite 400, Burlington, MA 01803, USA
525 B Street, Suite 1800, San Diego, CA 92101-4495, USA

First edition **2014**

Library of Congress Cataloging-in-Publication Data
A catalog record for this book is available from the Library of Congress

British Library Cataloguing in Publication Data
A catalogue record for this book is available from the British Library

For information on all **Elsevier** publications
visit our web site at store.elsevier.com

Printed and bound in UK
14 15 16 17 18 10 9 8 7 6 5 4 3 2 1

ISBN: 978-0-444-53730-0

Dedication

To researchers, students, and engineers out there looking for the true sustainable triple bottom line solution

Contents

3. "Structure-Property" Relationships of Genetically Modified Starch

I. TAN, PETER J. HALLEY

III

NOVEL STARCH MODIFIED BY BIOTECHNOLOGY

4. Starch Modification by Biotechnology: State of Art and Perspectives

XUAN XU, RICHARD G.F. VISSER, LUISA M. TRINDADE

IV

NOVEL STARCH MODIFIED BY CHEMISTRY

5. Starch Modification to Develop Novel Starch-Biopolymer Blends: State of Art and Perspectives

ASHOK K. SHRESTHA, PETER J. HALLEY

6. Chemically Modified Starch; Allyl- and Epoxy-Starch Derivatives: Their Synthesis and Characterization

MAURICE C.R. FRANSSEN, CARMEN G. BOERIU

V

STRUCTURES AND PROPERTIES OF NOVEL STARCH-BASED MATERIALS

7. Mechanical Properties of Starch-Based Plastics

C. CHALÉAT, PETER J. HALLEY, R.W. TRUSS

8. Nanobiocomposites Based on Plasticized Starch

LUC AVÉROUS, ERIC POLLET

VI

NOVEL CHARACTERISATION METHODS

9. Starch NMR

MICHAEL J. GIDLEY

VII
NOVEL PROCESSING

10. Processing of Plasticized Starch-Based Materials: State of the Art and Perspectives

FENGWEI XIE, PENG LIU, LONG YU

11. Reactive Extrusion for Thermoplastic Starch-Polymer Blends

R.B. MALIGER, PETER J. HALLEY

VIII
DEGRADATION AND BIODEGRADATION

12. Degradation and Biodegradation Standards for Starch-Based and Other Polymeric Materials

JOSEPH GREENE

13. Effects of Formulation, Structure, and Processing on Biodegradation of Starches

PARVEEN SANGWAN, EUSTATHIOS PETINAKIS, KATHERINE DEAN

IX

APPLICATIONS

14. Starch Applications: State of Market and New Trends

BRONWYN G. LAYCOCK, PETER J. HALLEY

15. Starch Plastic Packaging and Agriculture Applications

GREGORY M. GLENN, WILLIAM ORTS, SYED IMAM, BOR-SEN CHIOU, DELILAH F. WOOD

Contributors

Luc Avérous BioTeam/ICPEES-ECPM, UMR 7515, Université de Strasbourg, Cedex 2, Strasbourg, France

Carmen G. Boeriu Wageningen UR Food and Biobased Research, Division Biobased Products, Wageningen, The Netherlands

C. Chaléat Australian Institute of Bioengineering and Nanotechnology, The University of Queensland, Brisbane, Queensland, Australia

Bor-Sen Chiou USDA-ARS, Western Regional Research Center, Bioproduct Chemistry and Engineering, Albany, California, USA

Katherine Dean CSIRO Materials Science and Engineering, Victoria, Australia

Maurice C.R. Franssen Wageningen University, Lab of Organic Chemistry, Wageningen, The Netherlands

Michael J. Gidley Centre for Nutrition and Food Sciences, ARC Centre of Excellence in Plant Cell Walls, Queensland Alliance for Agriculture and Food Innovation, The University of Queensland, Brisbane, Queensland, Australia

Gregory M. Glenn USDA-ARS, Western Regional Research Center, Bioproduct Chemistry and Engineering, Albany, California, USA

Joseph Greene California State University, Chico, California, USA

Peter J. Halley School of Chemical Engineering and Australian Institute of Bioengineering and Nanotechnology, The University of Queensland, Brisbane, Queensland, Australia and Centre for High Performance Polymers, Australian Institute for Bioengineering and Nanotechnology, The University of Queensland, Queensland, Australia

Syed Imam USDA-ARS, Western Regional Research Center, Bioproduct Chemistry and Engineering, Albany, California, USA

Stephen A. Jobling CSIRO Food Futures National Research Flagship, GPO Box 1600, Canberra, Australia and CSIRO Plant Industry, GPO Box 1600, Canberra, Australia

Bronwyn G. Laycock AIBN and the School of Chemical Engineering, The University of Queensland, St Lucia, Queensland, Australia

Zhongyi Li CSIRO Food Futures National Research Flagship, GPO Box 1600, Canberra, Australia and CSIRO Plant Industry, GPO Box 1600, Canberra, Australia

Peng Liu School of Chemistry and Chemical Engineering, Guangzhou University, Guangzhou, China

R.B. Maliger School of Chemical Engineering, The University of Queensland, Brisbane, Queensland, Australia

Matthew K. Morell CSIRO Food Futures National Research Flagship, GPO Box 1600, Canberra, Australia and CSIRO Plant Industry, GPO Box 1600, Canberra, Australia

William Orts USDA-ARS, Western Regional Research Center, Bioproduct Chemistry and Engineering, Albany, California, USA

Eustathios Petinakis CSIRO Materials Science and Engineering, Victoria, Australia

Eric Pollet BioTeam/ICPEES-ECPM, UMR 7515, Université de Strasbourg, 67087 Strasbourg, Cedex 2, Strasbourg, France

Ahmed Regina CSIRO Food Futures National Research Flagship, GPO Box 1600, Canberra, Australia and CSIRO Plant Industry, GPO Box 1600, Canberra, Australia

Parveen Sangwan CSIRO Materials Science and Engineering, Victoria, Australia

Ashok K. Shrestha School of Science and Health, University of Western Sydney, Hawkesbury Campus, Richmond, New South Wales, Australia

I. Tan Centre for High Performance Polymers, Australian Institute for Bioengineering and Nanotechnology, The University of Queensland, Queensland, Australia

Luisa M. Trindade Wageningen UR—Plant Breeding, Wageningen University and Research Center, P.O. Box 386, Wageningen, The Netherlands

R.W. Truss School of Mechanical and Mining Engineering, The University of Queensland, Brisbane, Queensland, Australia and School of Chemical Engineering, The University of Queensland, Brisbane, Queensland, Australia

Richard G.F. Visser Wageningen UR—Plant Breeding, Wageningen University and Research Center, P.O. Box 386, Wageningen, The Netherlands

Delilah F. Wood USDA-ARS, Western Regional Research Center, Bioproduct Chemistry and Engineering, Albany, California, USA

Fengwei Xie Australian Institute for Bioengineering and Nanotechnology, The University of Queensland, Brisbane, Queensland, Australia

Xuan Xu Wageningen UR—Plant Breeding, Wageningen University and Research Center, P.O. Box 386, Wageningen, The Netherlands and National Centre for Vegetable Improvement (Central China), Key Laboratory of Horticultural Plant Biology, Ministry of Education, Huazhong Agricultural University, Wuhan, China

Long Yu Centre for Polymer from Renewable Resources, ERCSPP, College of Light Industry and Food Sciences, South China University of Technology, Guangzhou, China and CSIRO Materials Science and Engineering, Melbourne, Victoria, Australia

Acknowledgments

Thanks to all the contributing authors for their time, patience, and skill in their chapters.

Thanks to Dr. Meir Bar for his technical editorial assistance.

Prof. Peter J. Halley thanks the funding provided from the Australian government and industry for starch-based plastics since 1995, and the efforts of the 60+ researchers who have worked in these projects.

Prof. Luc Avérous thanks the Masters, Ph.D. students, Postdocs, colleagues, and collaborators from all over the world, who have been together involved in this long (since 1996) and fruitful research story. We want also to associate with these acknowledgments, different French regions (Alsace and Champagne-Ardennes), the French government, and some companies for their funding and kind help.

Luc Avérous
Peter J. Halley

Preface

Over the past decades, international research has led to the strong development of materials from biomass, such as starch, for nonfood and polymer applications. These materials are biobased, but also they are often biodegradable. According to ASTM standard D-5488-94d or European norm EN 13432, biodegradable is defined as being capable of undergoing decomposition into carbon dioxide, methane, water, inorganic compounds, and biomass. The predominant mechanism is the enzymatic action of microorganisms. These materials are also often compostable. Compostability is defined as material biodegradability at set conditions using a compost medium.

Currently, biopolymers (recently defined by IUPAC as "Polymers which eminate from living organisms") from renewable source (biomass) represent enormous potential. This is mainly due to a growing societal demand for more friendly and environmental materials, where durable and long-lasting plastics are not always appropriate for short-term applications (packaging, leisure, agriculture, catering, hygiene, biomedical), since these polymers can easily be degraded and bioassimilated (biotic and abiotic degradations).

The worldwide demand for the biobased and biodegradable polymers has steadily grown over the last 10 years at an annual rate of between 10% and 20% per year. The market share, however, is very modest in terms of actual fraction of the total plastics market. According to the association European Bioplastics (EU), the worldwide production capacity of the biodegradable and biobased polymers for material applications was around 1161 kt in 2011 (the corresponding plastic production was 280 million tons), with around 250 kt of starch-based materials in 2011, growing to 500 kt by 2020. Up to now, the limited growth of the biopolymers can be explained by limited performance properties, high prices, limited legislative attention, the fact that biodegradability can be an added functional property not immediately perceived, and the lack of composting infrastructure. However, excitingly new environmental materials based on multiphase starch-based blends, modified starch, and other innovations in starch polymers have begun to address these major issues.

This book focuses on starch-based materials, from the plant to final objects and end-products. This book presents the most recent advances in science and technology in all related areas from genetically modified starch to end-user applications. These domains have not been covered before in a single book, and they include the major fields of biochemistry, chemistry, physics, and material science. These fields include specific areas in biosynthesis (biochemistry) and material synthesis (chemistry), multilevel (nano/micro/macro)structures, formulation, rheology and processing, characterization, and applications.

This book will benefit the scientific community as in 15 chapters; it covers almost all topics related to starch polymers. This book is divided into nine parts. For several parts, the first chapter presents a general state of-the-art overview of the corresponding subject and then the following chapter presents some very focused and specific research areas within the corresponding thematic. We feel this will maximize benefits for the readers.

This volume starts off with an introductory and general chapter, which briefly summarizes the importance of this subject.

Part 2 is based on starch genetics with two different chapters. This part presents the state of art of this fascinating topic.

Part 3 is focused on novel starch modified by biotechnology and bioproduction. A chapter presents the state of art of this subject.

Part 4 is related to novel starch modified by chemistry. This part synergizes with the previous one (also in the overreaching area of starch modification) and is divided into two chapters.

Part 5 is based on the structures and properties of novel starch-based materials at different scales. This major topic is presented in two different chapters.

Part 6 presents in a chapter, some novel characterization methods concerning the NMR analysis of these specific materials.

Part 7 presents some novel processing methods to produce these materials, in order to obtain multiphase and structured materials with high performance. This part is divided into two different chapters.

Part 8 is based on the study of the degradation and biodegradation of these starch-based materials, with two different chapters.

Part 9 is the last part and is mainly focused on the applications with two different chapters, which cover the different aspects related to the end-use of these challenging materials.

Given the global context, it is logical to us to consider the biobased systems discussed in this book as materials of the future. This book shows very important issues and topics related to these materials and should be attractive to the scientific community. Several books, book chapters, and other reviews have already been dedicated to the starch-based materials, and they show both the richness of this field and the huge potential for progress in our scientific understanding. However, this book fills a great void in presenting in a dedicated volume the global knowledge chain from genetic engineering to green applications. The remaining open questions that have to be addressed to improve their performances and applications further are just as challenging.

This book should be very useful for scientists, academicians, research scholars, polymer, agriculture and material engineers, and industries. This volume is also supportive for undergraduate and postgraduate students in institutes of agriculture, polymer, materials science, engineering and technology, and the corresponding labs. This is a special book with several important contributions from most of the best well-known experts from all over the world.

Both editors would like to express their great gratitude to all contributors of this book, who made excellent contributions. We would also like to thank all who helped us in the editorial work, especially Dr. Meir Bar who worked on technical editing during his sabbatical visit.

Luc Avérous
Peter J. Halley

Introduction

This book has evolved from over 500 researcher-years (combined) of research into starch-based plastics.

The target audience are undergraduate students, graduate students, and materials/chemical engineers or researchers who want to gain a broad understanding of starch-based plastics, or who want to focus on a specific area from genetics, characterization, processing, properties, and industrial applications.

This book is divided into themed sections where an overview is provided by an eminent scientist in the field and then two or three key research chapters are discussed by leading researchers. In this way, we aimed to provide the reader with both breadth and depth in the theme.

Also, we hope that this book is the first to attempt to showcase starch plastics research from starch genetics all the way through to industrial applications.

The idea of this book was born in Reims (France) in 2001. However, we cannot recall if it was before or after the tour of the Mumm tasting caves.

Luc Avérous
Peter J. Halley

PART I

INTRODUCTION

CHAPTER

1

Starch Polymers: From the Field to Industrial Products

Luc Avérous[1], *Peter J. Halley*[2]

[1]BioTeam/ICPEES-ECPM, UMR 7515, Université de Strasbourg, Cedex 2, Strasbourg, France

[2]School of Chemical Engineering and Australian Institute of Bioengineering and Nanotechnology, The University of Queensland, Brisbane, Queensland, Australia and Centre for High Performance Polymers, Australian Institute for Bioengineering and Nanotechnology, The University of Queensland, Queensland, Australia

1 INTRODUCTION

Starch is an attractive bio-based polymer because of its low-cost base and hence cost competitiveness compared to conventional plastics, its biodegradability, its abundance, and its renewable source. Starch-based biodegradable plastics were first developed in the 1970s (Westhoff, Otey, Mehltretter, & Russell, 1974) where thermoplastic starch (TPS) polymers were produced from starch and selected plasticizers and additives (e.g., glycerol, water, urea, and salts). However, due to poor properties and inherent water susceptibility, these materials have tended to focus on niche markets for low-performance applications. This book will focus on new research that is aimed to extend the capabilities of low-cost TPS polymers by focusing on research spanning plant physiology through to plastic product performance.

Additionally, this chapter will set the scene for the book and summarize some of the important technologies involved in producing starch polymers: from the field to industrial applications. This same breadth of technologies will then be further examined in more detail in the subsequent book sections with an overview chapter followed by specific state-of-the-art research chapters. The technologies of developing starch polymers are diverse and include genetic engineering, biotechnology, food science, polymer science and engineering, polymer structure-property relationships, product design and engineering, and chemical engineering scale-up.

2 STARCH SOURCES

Starch, the principal source of dietary calories to the world's human population, is the major polysaccharide reserve material present in photosynthetic tissues and in many types of plant storage organs such as seeds, tubers, roots, and fruits. Starch occurs in nature as water-insoluble granules, which are essentially composed of two main polysaccharides, amylose and amylopectin, with some minor components such as lipids and proteins, which can interact with the polysaccharides (e.g., Maillard reaction). Starches from different botanical origin have different biosynthesis mechanisms and may exhibit distinct molecular structure and characteristics and properties as well as diversity in shape, size, composition, and other macroscale constituents of the starch granules.

Roots and tubers, grains, cereals, and fruits have generally high starch content (40-80%, w/w, dry basis) (Hoover, 2001).

Root and tuber crops are plants yielding starchy roots, rhizomes, corns, stems, and tubers. They contain 70-80% water and 16-24% starch. A great number of the roots and tubers are grown in the tropics for edible purposes as yams, sweet potato, cassava, cocoyams, yam bean, canna (Queensland arrowroot), Indian arrowroot, West Indian arrowroot, tacca starch (from the savannah grassland), and ginger (Hoover, 2001).

Compared to the roots and tubers or fruits, grains and cereals (wheat, rice, corn, oats, barley, millet, sorghum, and quinoa) show higher starch content, due to low water content.

Among these different resources, starches obtained from some cereals (corn and wheat) or from some tubers and roots (particularly potato and cassava) are the main commercial products and largely dominate starch production.

3 STARCH PRODUCTION AND MAIN OUTPUTS

3.1 Starch Production and Main Outputs in the World

According to the European Starch Industry Association (AAF) (www.aaf-eu.org), in 2011 and worldwide, the global starch market was 45-50 billion € with a production of around 72 million tons of starch. Recently, production increases have been mainly due to the demand in emerging countries mainly China and Brazil (+10%/year), compared to the others countries, which the growths are of some 1-2% per year.

The principal industrial productions of starch are based only on four main resources with maize, cassava, wheat, and potatoes, which represent 76%, 12%, 7%, and 4%, respectively. The other botanical resources represent less than 1%. The main production areas are North America, China, Europe, Southeast Asia, and South America with 33%, 33%, 18%, 11%, and 5%, respectively.

The market is led by North America, China, and Europe, which represent in total 85% of the world production (and only around 1/3 the world population). This geographic predominance is also represented by the top 10 companies (data from 2010) and their corresponding levels of production:

1	Cargill (the United States)	8.0 million tons
2	Ingredion, formerly CPI (the United States)	5.9 million tons
3	ADM (the United States)	4.7 million tons
4	Tate & Lyle (Great Britain)	3.6 million tons
5	Roquette (France)	2.8 million tons
6	Zhucheng Xingmao (China)	2.0 million tons
7	Global Bio-Chem (China)	1.5 million tons
8	Tereo Syral (France)	1.4 million tons
9	COFCO (China)	1.3 million tons
10	Xiwang (China)	0.7 million tons

The main starch outputs are for food and nonfood applications. Food products represent 60% of the market with 31% and 29% for confectionary and drinks and processed food, respectively. Feed is around 1% (without taking into account the byproducts). Nonfood products represent 29%, with 19%, 6%, and 4% for corrugating and paper making, pharmaceuticals and chemicals, and others nonfood products, respectively.

3.2 Starch Production in Europe

In 2011, according to the AAF, the total European production of starch as processed raw materials was 22 million tons (30% of world production) with 34.6%, 34%, and 31.4% for wheat, maize, and potatoes, respectively.

In 2011, the production of starch in pure starch equivalent is 10 million tons with 47.4%, 38.6%, and 14% for maize, wheat, and potatoes, respectively. Since 2004, these ratios between each resource have been more or less the same. For the same period (2004-2011), the level of European production has increases of more or less of 10%.

In 2011, the European market for applications of starch and starch derivatives was 8.9 million tons with food and nonfood applications. Food products represent 32% and 29% for confectionary and drinks and processed food, respectively. Nonfood products represent 28%, 6%, and 4%

for corrugating and paper making, pharmaceuticals and chemicals, and others nonfood products, respectively. The percentages of these different outputs are closed to these previously given for the world.

4 STARCH MODIFICATION

4.1 From Native to Plasticized Starch

Native starch granules are generally not able to be processed directly into starch polymers and require some modification. The first type of process to consider is gelatinization that aids the breakdown of the inherent multilevel structures in the starch (Avérous and Halley, 2009). Starch is well known to have multilevel structures from starch granules (<1-100 μm), alternating amorphous and semicrystalline shells (growth rings) (100-400 nm), crystalline and amorphous lamellae (periodicity) (9-10 nm), and macromolecular chains (~nm) (Pérez, Baldwin, & Gallant et al., 2009). Native starch granules can also have a crystallinity varying from 15% to 45% depending on the particular plant species. In addition, the two major macromolecular components of starch are amylose and amylopectin polymers that are large macromolecules where amylose is a predominantly linear polysaccharide with α(1→4)-linked D-glucose units and can have a molecular mass as high as 10^6 and amylopectin is a highly branched polysaccharide with α (1→4)-linked D-glucose backbones and about 5% α-(1→6)-linked branches, which have a profound effect on the physical and biological properties and have a molecular weight ranging from 50 to 500×10^6: one of the largest natural polymers known (Pérez et al., 2009). Thus, it is important for processability and polymer production that this multilevel structure and associated interactions be broken down through gelatinization and destructuration. When native starch granules are heated in water, their semicrystalline nature is gradually eliminated, resulting in the phase transition from an ordered granular structure into a disordered state in water (destructuration of the 3D granule organization), which is known as "gelatinization" (Lelièvre and Liu, 1994). Gelatinization is an irreversible process that includes, in a broad sense and in time/temperature sequence, granular swelling, native crystalline melting (loss of birefringence), and molecular solubilization. Starch can also be gelatinized and destructured at lower moisture contents under thermomechanical input via, for example, extrusion processes. Water acts as a volatile plasticizer and destructuring agent. However, nonvolatile plasticizer such as polyols (glycerol, glycol, and sorbitol), compounds containing nitrogen (urea, ammonium-derived, and amines), sugars, and citric acid are also required to improve processability of the starch. High-temperature processing of starch polymers generally requires both gelatinization and plasticization processes, which are completed during the residence time in the extruder. Plasticized starch, the so-called thermoplastic starch (or TPS), is then obtained (Avérous, 2004; Xie, Halley, & Avérous, 2012).

4.1.1 *Biotechnological, Chemical, and Physical Modifications of Starch*

As unmodified starches are generally unsuitable for most food and industrial applications, starch is generally additionally modified biotechnologically, chemically, or physically to improve their processability or function. Biotechnologically modified starches are starches than have been altered in the plant by the control of growing conditions, seed variety selection, and understanding of genetic and biochemical pathways of the growing starch plant. Physically modified starches include pregelatinized, milled, cold-water-soluble, heat moisture treatment, and dry-heating of starch. Chemically modified starch includes acid hydrolyzed, oxidized starch, dextrinized starch, cross-linked starch, starch acetates, starch phosphates, starch succinates, cationic starch, and hydroxypropylated starches (wurzburg, 1982). These different modifications will be discussed in more detail later in Chapter 5.

5 NONFOOD STARCH POLYMER MATERIAL APPLICATIONS

5.1 Paper and Board Processing

Starch-based products are largely used in paper bags and others packaging papers, tissues, corrugating board, and stationery. Starch and modified starch (e.g., cationic starch) are used during the paper and board production, for example, for wet-end addition, size press, and surface coating. It is an indispensable ingredient for production of paper and board to elaborate a large range of varieties and qualities. For instance, starch modulates the wettability, the printability, and writing properties of paper. It is added at several stages of the production process, that is, while the paper is still a cellulose pulp, when it is passing through the drying rollers, and as a final coating to meet the many special requirements of printer. Unmodified starch is still being used via pulper addition or by "starch spraying." However, cationic starches ranging from D.S. about 0.020 to 0.040 are largely used to improve retention of (ligno)cellulose fibers and fillers during the process.

Typically, printing and writing papers contain, respectively, 4% starch, paper board 2% starch, and domestic and industrial papers 2% starch.

The growing use of recycled paper and recycled board requires more and more large quantities of starch-based products to avoid degradation of the quality and properties of the recycled fiber-based materials by retention, for instance, of the thinner and shorter fibers and microfibrils, which bring some mechanical properties.

5.2 Pharmaceuticals and Cosmetic Applications

Starches are a source for use in pharmaceuticals as a major ingredient. Starch acts as a binder in pharmaceutical tablet, as a disintegrating agent, and as sugarless sweetener to products as disparate as toothpaste, tablets, emulsions, lotions, liquid medicines, and creams.

Human well-being are also dependent on the role starch plays in the production of antibiotics, vitamins, penicillin, drip-feed systems, dialysis solutions, enteral nutrition, and even blood plasma substitutes.

5.3 Textile Applications

One of the earliest known industrial uses of starch was to "size" and stiffen textiles. This is still true, but today, starch does much more since it is added to individual yarns to increase mechanical strength and resistance to friction wear during the high-speed production and helps resist moisture penetration. It can serve as a stabilizer and filler for colored inks when fabrics are overprinted.

5.4 Chemicals Product Applications

The detergent industry uses starch products for the production of biodegradable, nontoxic, and skin-friendly detergents. Starch finds use as a redeposition inhibitor of "dirt" and fat components, once they have been released. Starch is also used in the construction industry for concrete mixtures. Starch finds use as a retarder in concrete and is also used for reducing set time in cement. It is also used for plasters and insulation as well as in glues, oil drilling, and mineral and metal processing.

Starch products are also used in an array of less obvious applications with strong developments in fermentation and white biotech, for the production of amino acids, organic acids, enzymes, and yeast. For example, starch is a common resource for the bioproduction of lactic acid that presents different outputs such as food, pharmaceutical, and cosmetic applications. Lactic acid is also the basic resource for the production in a multistep process of a biodegradable and bio-based polyester, the polylactic acid (Avérous, 2013).

Currently, the chemical and polymer industry used more and more starch for different applications such as the following:

- The production of surfactants or flocculants, in the water treatment plants for flocculation purposes
- Polyols for polyurethanes production or other polyol polymers
- For biodegradable materials, mainly in short-term applications (packaging, agriculture, leisure, catering, and hygiene) often in association with different biodegradable compounds (plasticizer, biodegradable polyesters, and cellulose fibers) to increase the performance and preserve the final biodegradability of the material

5.4.1 *Biodegradable Materials for Short-Term Applications*

Except, for instance, in the case of the association of polyolefins with starch such as the commercial thermoplastic named Gaialene© from Roquette (France) or Cerestech (Canada), starch is more and more used as an important ingredient to develop bio-based and biodegradable materials for short-term applications. Different companies such as Novamont (Italy), Plantic (Australia), Biotec-Sphere (France), Cereplast (the United States), and Limagrain (France) have developed different applications based on gelatinized or plasticized starches (Xie et al., 2012) or expanded starch. For instance, starch-based foams are good challengers to replace the environmentally unfriendly and very criticized expanded PS in a great number of packaging applications (loose fills and expanded trays).

Biodegradable materials for short-term applications are the subject of a crescent number of publications during the two last decades. This evolution is on agreement with the market progression, with an increasing number of real concretizations and product outputs. According to the Nova Institute (Germany) (http://www.nova-institut.de), the world market corresponding to the "destructured starch" was 95 ktons in 2011 and should be 156 ktons in 2020.

Different formulations (blends and composites) associated with multiprocessing steps (bulk, multilayer, and foam) allow developing a large range of end uses. The end of life of these materials can be biodegradation/compost or conventional recycling. The main utilizations for these starch-based materials are for short- or medium-term applications.

Examples include the following:

- Packaging (loose filler, foam packaging, bags, films, woven nets, fresh fruit and vegetable trays, boxes for transporting prepared food, and containers for dry cosmetics such as face powders, eye shadow, and lipstick). In Europe, for instance, these packaging should complied with European standard: UNI EN 13432: 2002—"Packaging—Requirements for packaging recoverable through composting and biodegradation—Test scheme and evaluation criteria for the final acceptance of packaging." In the United States, the "Standard Specification for Compostable Plastics" is the ASTM 6400: 04.

- Agriculture (mulching film, plant pots, bindings, clips, and devices). In France, mulching film should comply with the French standard NF U52-001 (Matériaux biodégradables pour l'agriculture et l'horticulture—Produits de paillage—Exigences et méthodes d'essai) specifically concerning their biodegradability.
- Leisure (children toys and fireworks).
- Hygiene (nappies, sanitary towels, and panty liners).
- Catering (plates, cutlery, cups, and trays).
- Others products (pens, pencil sharpeners, rulers, and bones and chewable items for pets).

6 CONCLUSIONS

The future for starch polymers is promising as the early restricted applicability to niche industrial applications is expanding into higher-performance, high-value applications as the science and technology of starch synthesis, starch modification, starch processing, and starch polymer structure and property relationships grows. The aim of this book is to highlight state-of-the-art understanding of the technologies in each stage of the processes from farm gate to industrial product in order to help facilitate even wider application areas for green starch polymers.

References

Avérous, L. (2004). Biodegradable multiphase systems based on plasticized starch: A review. *Journal of Macromolecular Science Part C Polymer Reviews, 44*, 231–274.

Avérous, L. (2013). Synthesis, properties, environmental and biomedical applications of polylactic acid. In S. Ebnesajjad (Ed.), *Handbook of biopolymers and biodegradable plastics*. Boston: Elsevier.

Avérous, L., & Halley, P. J. (2009). Biocomposites based on plasticized starch. *Biofuels, Bioproducts and Biorefining, 3*, 329–343.

Hoover, R. (2001). Composition, molecular structure, and physicochemical properties of tuber and root starches: A review. *Carbohydrate Polymers, 45*, 253–267.

Lelièvre, J., & Liu, H. (1994). A review of thermal analysis studies of starch gelatinization. *Thermochimica Acta, 246*, 309–315.

Pérez, S., Baldwin, P. M., & Gallant, D. J. (2009). Structural Features of Starch Granules I. In B. James, & W. Roy (Eds.), *Starch* (3rd ed.). San Diego: Academic Press, Chapter 5.

Westhoff, R. P., Otey, F. H., Mehltretter, C. L., & Russell, C. R. (1974). Starch-filled polyvinyl chloride plastics—Preparation and evaluation. *Industrial and Engineering Chemistry Product Research and Development, 13*, 123–125.

Wurzburg, O. B. (1982). Modified starches. *Food and Chemical Toxicology, 20*, 986–987.

Xie, F., Halley, P. J., & Avérous, L. (2012). Rheology to understand and optimize processibility, structures and properties of starch polymeric materials. *Progress in Polymer Science, 37*, 595–623.

PART II

STARCH GENETICS

CHAPTER

2

Genetically Modified Starch: State of Art and Perspectives

Ahmed Regina[1,2], *Zhongyi Li*[1,2], *Matthew K. Morell*[1,2], *Stephen A. Jobling*[1,2]

[1]CSIRO Food Futures National Research Flagship, GPO Box 1600, Canberra, Australia

[2]CSIRO Plant Industry, GPO Box 1600, Canberra, Australia

1 INTRODUCTION

Starch is commercially isolated from a wide range of sources including cereal grains such as corn, wheat, rice, and sorghum; roots and tubers such as potato, sweet potato, cassava, and arrowroot; and stem and pith such as sago. The composition of naturally occurring starch is more or less universal with a major component of amylopectin (~75%) and a minor component of amylose (~25%) irrespective of its source. Both amylose and amylopectin are polymers of α 1,4-linked glucan chains with varying proportions of α(1,6)-linked branch points. While amylose is predominantly linear with less than 1% α(1,6) linkages, amylopectin is a much larger and highly branched molecule with 4-5% α(1,6) linkages.

Starch modification implies efforts to both achieve increased starch production and modify the composition and component structure to impart specific properties to suit particular end uses. The most widely referred target for starch modification is the amylose-amylopectin ratio. Several plants have been modified in their starch biosynthetic pathway to yield both high-amylose and high-amylopectin starches. Both these type of starches have distinct end uses, for example, high-amylopectin starches are used in paper manufacture and in some specialized adhesive applications such as sticky labels on glass jars where the antiwetting properties of the starch prevent the label from peeling in damp conditions. Waxy starches are also widely used in food preparation as they make highly transparent gels that have improved freeze-thaw characteristics compared to normal amylose-containing starch gels. High-amylose starches are also used in adhesive manufacture as the long linear chains of amylose associate strongly together quickly on cooling, making them very useful in high speed applications such as corrugated cardboard production. These starches are also used in specialized food applications such as improving crispiness of coatings during frying, and they have general utility as thickeners and gelling agents (Jobling, 2004). More importantly, the nutritive value of high-amylose starches as a source of resistant starch, a special type of dietary fiber, is being increasingly recognized in the food industry, and this will be discussed in more detail later.

In this chapter, we will review recent advances in generating novel starch compositions and structures using new breeding technologies, focusing mainly on cereal starches but including other starches where appropriate advances have been made. We will first describe attempts to alter starch structure in order to improve starch functionality, then explain how increasing knowledge of the regulation of starch biosynthesis is being used to increase starch production, and finish by summarizing new methods for increasing the genetic diversity in crops as well as methods for fine-tuning gene expression in plants in order to bring improved starch-based products with value-added consumer benefits to the marketplace.

2 MODIFICATION OF STARCH QUALITY

Starch quality implies the net outcome of a combination of structural features such as the relative proportion of amylose and amylopectin, chain length distribution and branching frequency of these two components, molecular organization and higher-order structure, granule shape and size, lipid and protein content, and covalent modification such as phosphorylation. Our current knowledge suggests interdependence of these features to some extent. The physicochemical properties of starch are highly dependent on the structural features, which in turn influence its behavior while processing. The last few decades have seen successful attempts in altering the amylose content of starch in economically important crops, either toward lowering or toward elevating the amylose compared to the wild-type levels.

3 STARCH MODIFICATION TO LOWER AMYLOSE CONTENT

It is now commonplace knowledge that the absence or inactivation of the granule-bound starch synthase 1 gene (*GBSS1*) encoded by the waxy (*Wx*) locus in plants gives amylose-reduced or amylose-free phenotype in storage tissues with no apparent reduction in total starch content. The level of reduction in amylose content varies depending on the nature of mutation in the *GBSS1* gene. The first reported waxy hexaploid wheat had an amylose content of >1% (Nakamura, Yamamori, Hirano, Hidaka, & Nagamine, 1995) and was a combination of three null waxy alleles, the *Wx-A1* bearing a 23-bp deletion and a 4-bp insertion; the *Wx-B1* allele, in which the entire coding region is deleted; and the *Wx-D1* allele, which has a 588-bp deletion and a12-bp insertion (Vrinten, Nakamura, & Yamamori, 1999). Identification of variant alleles of *Wx* genes in each of the three waxy loci (six different alleles identified thus far each on *Wx-A1* and *Wx-B1* loci and seven alleles on *Wx-D1* locus) allowed the generation of grains with a range of phenotypes having 0-30% amylose content (Yasui, 2006; Yasui & Ashida, 2011). In diploid barley, low-amylose waxy cultivars appear to have a 413-bp deletion in the promoter and 5′-untranslated region (UTR) of the *GBSS1* gene, whereas an amylose-free barley gene carried an A to T substitution leading to the substitution of an aliphatic amino acid (Val) for an acidic amino acid (Asp) (Patron et al., 2002) In rice, a G/T polymorphism at the 5′-leader/first intron splicing site appeared to regulate the production of mature GBSS mRNA, which in turn influenced the amylose content. The presence of a G at the splicing site promotes normal splicing resulting in enhanced GBSS activity and high amylose content in indica rice compared to japonica rice, in which the presence of T at the junction results in cryptic splicing, leading to lesser GBSS efficiency and lower amylose levels (Cai, Wang, Xing, Zhang, & Hong, 1998; Ni et al., 2011; Wang et al., 1990). This in combination with two other single-nucleotide polymorphisms (SNPs) in *GBSS1*, an A/C SNP in exon 6 and a C/T SNP in exon 10, accounted for approximately 90% variation in amylose in two separate US and European germplasm collections (Dobo, Ayres, Walker, & Park, 2010).

Silencing of *GBSS1* using an RNA interference (RNAi) construct containing 3′-UTR region led to reduced amylose content ranging from 5.9% to 9.0% in transgenic rice lines compared to 17.7-18.0% in wild type (Park et al., 2010).

The most common waxy starch in use currently is from corn or maize, but the range of waxy starches on the market has increased recently with the introduction of amylose-free potato starches into Europe. AVEBE has developed a waxy potato variety Eliane™ using traditional mutation breeding techniques, while BASF has developed Amflora, a GM potato variety through silencing the *GBSS1* gene. The cultivation, processing, and industrial use of this GM potato were approved by the European Commission on 2 March 2010. Both of these potatoes are produced in a closed-loop system where seed production, growth, and supply to the processing factory are tightly controlled. Waxy potato starches have much improved functionality giving very clear gels with high freeze-thaw stability compared to normal potato starch. Even further increases in freeze-thaw stability are possible as was demonstrated in a triple *waxy/SSII/SSIII* downregulated transgenic potato line (Jobling, Westcott, Tayal, Jeffcoat, & Schwall, 2002). Starch from this line had extremely short-chain amylopectin and no amylose, and this reduced syneresis levels to virtually zero even after several freeze-thaw cycles.

4 STARCH MODIFICATION TO ELEVATE AMYLOSE CONTENT

Elevated amylose content is achieved through altering the expression of various enzymes involved in both starch elongation and starch branching (Figure 2.1). However, depending on the genetic mechanism involved, the extent of amylose increase and associated changes in starch properties can vary dramatically, even for the same gene among different species. The involvement of starch-branching enzyme in determining the amylose content was revealed first in rugosus pea in dicots and in amylose extender (*ae*) maize in monocots. The rugosus-wrinkled phenotype described by Mendel in pea was further demonstrated to contain an amylose content of >60% and found to be due to a lesion in starch-branching enzyme II

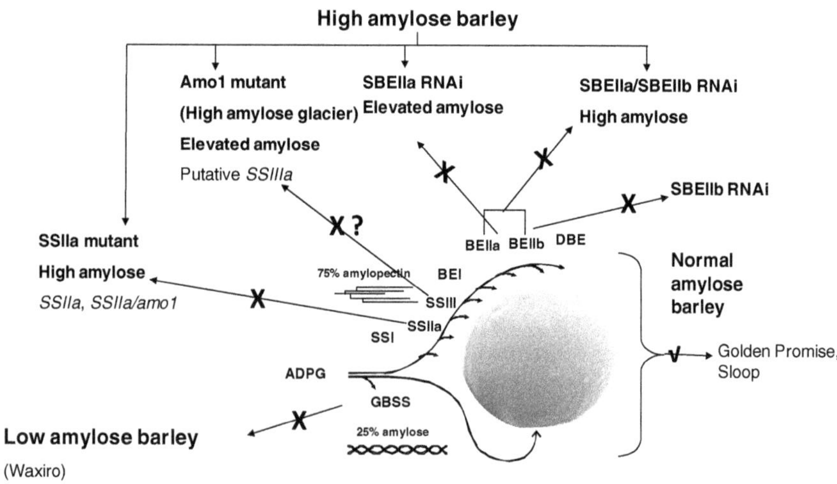

FIGURE 2.1 Genetically distinct barley starches with varying amylose content

(Bhattacharyya, Smith, Ellis, Hedley, & Martin, 1990). Three different isoforms of SBEs are commonly found in cereals, SBEI, SBEIIa, and SBEIIb although a fourth gene designated *SBEIII* with a slightly different structure has recently been described (Kang et al., 2013). In maize and rice endosperms, mutations eliminating the expression of SBEIIb lead to increased apparent amylose content (Garwood, Shannon, & Creech, 1976; Mizuno et al., 1993; Nishi, Nakamura, Tanaka, & Satoh, 2001). These starches are also characterized by longer amylopectin chain length proportions and higher gelatinization onset and peak temperatures. Likewise, the doubling of apparent amylose content in SBEIIb RNAi-silenced rice grain is not due to an increase in the relative proportion of true long linear amylose chains, but instead is due to significantly elevated levels of long and intermediate amylopectin chains (Butardo et al., 2011). In comparison, lesions in the *SBEIIa* gene in maize had no effect on endosperm starch but leaf starch was modified with elevated amylose levels (Blauth et al., 2001).

The first report of elevated amylose wheat came from the work of Yamamori, Fujita, Hayakawa, Matsuki, and Yasui (2000) who showed an increase in amylose content to ~37% from the wild-type level of 28-30% in a triple null mutant of starch synthase IIa (SSIIa). In the barley *sex6* mutant, a loss of function mutant of the *SSIIa* gene, a very high amylose content of more than 70% was observed, much higher than that observed in wheat (Morell et al., 2003). In both wheat and barley, the high-amylose phenotype is coupled with a reduction in starch content and grain weight due to a reduced rate of starch biosynthesis. Characteristic features of SSIIa mutant starch include amylopectin with increased short chains of DP 6-10, decreased long chains of DP 11-25, increased proportion of V-type crystallinity, and lowered gelatinization temperature (Konik-Rose et al., 2007; Morell et al., 2003).

The first targeted silencing of a *SBE* gene in wheat using RNAi demonstrated that the amylose content could be increased dramatically to a level of over 70% (Regina et al., 2006). In this high-amylose wheat, SBEIIa was specifically targeted for silencing in the endosperm; however, a concomitant reduction of SBEIIb also occurred, which was subsequently shown not to be due to cross silencing by the RNAi construct; rather, it indicated that the two enzymes work together to make the normal amylopectin structure. The starch properties of this wheat were similar to that of amylose extender maize mutants with long-chain amylopectin and higher gelatinization peak temperature. Silencing SBEIIb alone using RNAi in wheat did not reveal any significant change in starch properties including the amylose content. A similar study in barley also revealed that to get very high amylose contents of greater than 70%, it was necessary to significantly reduce the expression of both SBEIIa and SBEIIb (Regina et al., 2010).

In this same study, a reduction of SBEIIa alone leads to an elevation of amylose content of about ~50%, and an SBEIIb-alone knockout did not change the amylose content compared to wild type.

As indicated earlier, suppression of multiple starch-branching enzyme isoforms is often needed to achieve very high levels of amylose elevation particularly in cases where suppression of one SBE isoform gave either no increase or only a slight increase in amylose content. For example, potato has two SBE isoforms, SBEI and SBEII. SBEI is the major form of SBE in the potato tuber; however, antisense downregulation of SBEI did not change the amylose content, although small changes in the physical properties of the starch such as the gelatinization onset could be measured in differential scanning calorimetry (Safford et al., 1998). A second form of SBE (SBE II) was subsequently discovered, and when expression of this gene was reduced in the tuber, the amylose content was elevated to about 35% even though this SBE isoform made up less than 2% of the SBE activity of the tuber (Jobling et al., 1999). A combined downregulation of both these SBE isoforms led to a significant further elevation to 70% or higher (Schwall et al., 2000). In cereals also, SBEI mutants do not show any obvious change in the amylose level; however, detailed further analysis of an SBEI mutant of maize did show some slight changes to the fine structure of starch as well as altered physiological functions as the mutant showed less efficient starch utilization during germination due to resistance of the modified starch to enzyme digestion (Xia, Yandeau-Nelson, Thompson, & Guiltinan, 2011). In japonica rice, it was also recently shown that transgenic cosuppression of SBEI had no effect on amylose content and downregulation of SBEIIb produced only a small increase in amylose content from 15% to approximately 22% (Zhu, Gu, et al., 2012). It was necessary to reduce expression of both SBE I and SBEIIb to increase amylose content further up to only a maximum of 27%. Further increases in amylose were achieved by crossing the double-gene downregulated line to an indica rice with higher baseline levels of amylose (27%) and this eventually led to a line with more than 60% amylose (Zhu, Gu, et al., 2012).

Thus, over the last few years, very high-amylose starches have been produced for all the major cereals (wheat, barley, and rice). This is significant from a nutritional perspective as an important property of very high-amylose starches is that they are more resistant to digestion than wild-type starches, which are normally very rapidly digested. Resistant starch (RS) is a special type of dietary fiber that resists digestion in the upper gut and enters the large bowel where it is fermented by resident microbiota into short-chain fatty acids that contribute to optimal colonic function (Topping, 2007; Topping & Clifton, 2001). Foods high in RS generally have a low glycemic index and lower energy density and have many potential health benefits, protecting against lifestyle diseases such as obesity and diabetes, as well as the effects of high-protein diets, thus decreasing the risk of colorectal cancer. The physical properties of amylose such as slow gelatinization on heating and quicker retrogradation on cooling make it a good source of RS in processed food.

High-amylose maize starches with high levels of RS have been available on the market for some time and are readily incorporated as an ingredient in many food products, such as white bread to increase the dietary fiber levels. However, there is an additional nutritional benefit from getting RS in the diet in the form of whole grains rather than as an added ingredient from purified starch and the very high-amylose cereal grains such as wheat and barley described earlier will be very useful in this respect. Both these grains have been shown to improve the large bowel health indices in animal trials (Bird et al., 2004; Regina et al., 2006; Topping et al., 2003) and the barley grain has also been tested in human trials (Bird et al., 2008). For successful commercialization of these novel grains, it is necessary to produce consumer acceptable foods such as breads and other baked goods from these cereals, and further research and development has shown that this can be done (unpublished by authors). In fact, the high-fiber and high-RS barley has now been commercialized as BarleyMax™ and has been available in Australian supermarkets since 2009, first as a breakfast cereal and then as an increasing range of products such as flatbreads and high-fiber snack bars due to the high level of consumer demand for these healthy functional foods. There is high hope that high-amylose and high-RS wheat will also reach the market soon and this should have much larger socioeconomic benefits as wheat is used in a much larger number of food products consumed in much larger quantities than barley.

Nutritional evaluation of the high-amylose rice grain has also been tested in normal and diabetic rats where positive effects on bowel health and blood glucose response were observed (Zhu, Gu, et al., 2012) and also in humans, which showed that consumption of the high-RS rice decreased the postprandial glycemic and insulinemic responses as well as demonstrated fermentation in the large bowel by measurement of increased breath hydrogen compared to wild-type rice (Li et al., 2010). Although these rice lines demonstrated improved nutritional characteristics, the high-amylose phenotype severely affects the appearance of the rice grains that are completely opaque rather than translucent (Butardo et al., 2011; Zhu, Dogan, et al., 2012). As rice grains are normally eaten whole in a polished form exposing the central endosperm, this is a significant defect. It is also well known that high amylose content adversely affects the eating quality of rice and is likely to mean that these grains will only be used in a flour form as ingredients rather than as cooked whole grains.

5 MODIFYING STARCH QUANTITY IN STORAGE TISSUES

Many of the crops with modified starch structure also have effects on yield. This was particularly so with the very high-amylose potatoes where starch yield was only about half that of the controls (Schwall et al., 2000). This is a significant barrier to commercialization as farmers would have to be compensated for the reduced yield. Surprisingly, the very high-amylose wheats do not have much of a yield penalty, suggesting fundamental physiological differences perhaps exist between tuber and grain starch. Nevertheless, if novel starches are to be commercialized, this yield penalty has to be addressed, and to do this, we need a better understanding of what controls the level of starch synthesis in storage organs.

The initiation of starch biosynthesis in storage organs inevitably requires the mobilization of sucrose into glucose-6-phosphate (G6P), import of G6P into the amyloplast through inorganic phosphate (Pi) exchange, and subsequent conversion of G6P into glucose-1-phosphate (G1P) by plastidial phosphoglucomutase. The first committed step to starch synthesis is the formation of ADP glucose (ADPG) through ATP activation of G1P, catalyzed by ADPG pyrophosphorylase (AGPase). Metabolite compartmentation studies in barley revealed the localization of 80-90% of ADPGP in the cytosol as against potato tubers where AGPase is almost exclusively localized in the plastid. The AGPase from cytosol is transported to the plastid by *Brittle1*-encoded ADPG/ADP antiporter for starch biosynthesis to be initiated. AGPase has a heterotetrameric structure with two small subunits and two large subunits. Almost all known plant AGPases are allosterically regulated with 3-phosphoglyceric acid (3PGA) being the main activator and Pi the main inhibitor. The enzyme also exhibits redox modulation through oxidation/reduction of Cys12 residue of small subunit (Ballicora, Frueauf, Fu, Schurmann, & Preiss, 2000). While the small subunit performs catalytic and regulatory roles, the large subunit modulates the allosteric regulatory property of the small subunit in the heterotetrameric enzyme (Tuncel, Kavakli, & Keskin, 2008).

Mutations in AGPase, in both the small (*shrunken-2*) and the large (*brittle-2*) subunits, lead to dramatic reduction in starch content (Dickinson & Preiss, 1969; Hannah, Giroux, & Boyer, 1993; Smith, Bettey, & Bedford, 1989). Inactivation of the small subunit of cytosolic AGPase in barley *Riso 16* mutants (where the plastidial AGPase activity is unaffected) results in reduced accumulation of ADPG and endospermal starch levels with a concomitant reduction in storage protein buildup and seed size. A coordinated inhibition of glycolysis and amino acid and storage protein biosynthesis is observed in *Riso 16* mutant that is suggested to be useful in the prevention of osmotic imbalances and oxidative stress due to increased accumulation of sugars arising from decreased carbon fluxes into starch (Faix et al., 2012; Johnson et al., 2003).

Approaches so far undertaken for increasing starch production in plants included three target points: (1) precursor molecules for starch biosynthesis, (2) rate-limiting steps in starch biosynthesis involving AGPase, and (3) downstream starch biosynthetic enzymes. The majority of efforts to increase starch content have been focused on increasing the amount of ADPG available for starch biosynthesis. Allosteric properties of AGPase have been successfully altered in a number of species including potato, wheat, maize, rice, and cassava towards increasing starch accumulation (Ihemere, Arias-Garzon, Lawrence, & Sayre, 2006; Sakulsingharoj et al., 2004; Wang et al., 2007). Earlier work involved expressing a mutant bacterial AGPase with reduced allosteric dependence to activator fructose-6-phosphate in potato, resulting in a tuber starch increase to >35% (Mullerrober & Kossmann, 1994; Stark, Timmerman, Barry, Preiss, & Kishore, 1992). Transforming wheat with a modified form of maize shrunken 2 gene that encodes an altered AGP large subunit that has a reduced sensitivity to its negative allosteric effector, orthophosphate and more stable interactions with large and small subunits resulted in increased grain weight per plant up to 38% (Smidansky et al., 2002). Recently, it is shown that simultaneous overexpression of *shrunken 2* (encoding the large subunits of AGPase) and *brittle 2* (encoding small subunits of AGPase) in maize increased starch content to a level more than the increase when either of them individually was overexpressed (Li, Zhang, Zhao, Li, & Zhang, 2011). Sucrose synthase is another target that has been used to enhance starch levels in maize kernels by increasing the supply of ADPG to the endosperm (Li et al., 2013).

Increased starch production has also been demonstrated through exploiting alternate targets of precursor compounds and starch biosynthetic enzymes downstream of the initiation step. Heterologous expression of a potato sucrose transporter gene, *StSUT1*, in pea led to increased sucrose transport capacity of storage parenchyma cells from seed apoplasm (Rosche et al., 2002), demonstrating the potential of elevating the rate of starch synthesis through this approach. Downregulation of adenylate kinase yielding higher starch content through altering adenylate pool is another demonstration of targets prior to the initiation of starch synthesis (Regierer et al., 2002). *SRF1*, a transcription factor-encoding gene when overexpressed in sweet potato, yielded significant increases in starch content in storage roots with a marked decrease in glucose and fructose content. Simultaneous expression of a pea glucose-6-phosphate translocator and *Arabidopsis thaliana* adenylate transporter genes in potato resulted in increased starch content and tuber yield, suggesting that the synthesis of starch in the amyloplast is colimited by import of energy and carbon skeleton (Zhang, Hausler, et al., 2008).

Starch synthase IV as a target for increased starch production is more novel and recent. Gamez-Arjona et al. (2011) reported increased starch production in both autotrophic and heterotrophic organelles through enhanced expression of SSIV. Recently, exceptionally high levels of starch in leaf tissues were observed when plants were grown in close proximity to bacteria and fungi (Ezquer et al., 2010). There appear to be several underlying mechanisms behind this increase that centers on volatile microbial emissions causing changes to gene expression of starch biosynthetic enzymes and redox regulation and specifically involves *SSIII* and *SSIV* genes (Li, Ezquer, et al., 2011). Recently, overexpression of a particular thioredoxin gene in chloroplasts was also shown to lead to a very large increase in starch accumulation (Sanz-Barrio, Corral-Martinez, Ancin, Segui-Simarro, & Farran, 2013).

There is increasing evidence that starch biosynthetic enzymes interact, acting as complexes, to control starch structure (Liu et al., 2012, 2009; Zhang et al., 2011) and that combinations of starch mutants produce novel effects on starch structure and amounts (Szydlowski et al., 2011; Zhang et al., 2011). An unexpected starch content increase was observed in a barley double null mutant of *Sex6* (encoding SSIIa) and *amo1*, which is tightly linked to *SSIIIa* (Li, Ezquer, et al., 2011; Li, Zhang, et al., 2011; Li, Li, et al., 2011). In the *SSIIa/SSIIIa* double mutant of rice, starch synthase activity was doubled compared to the single *SSIIa* mutant (Zhang et al., 2011). These and other studies suggest that *SSIII* gene may be a negative regulator of other starch biosynthetic enzymes (Ral et al., 2006; Zhang, Szydlowski, et al., 2008) and that combining *SSIIIa* mutation with other mutants may be a good strategy to recover yield penalty due to starch reductions.

6 MODIFYING STARCH GRANULE SIZE

Starch granule size is an important feature that determines the suitability for many of the food and nonfood uses. Starch granule size is highly species-specific and starch granules can appear in many different forms. It could be large as in potato, small as in rice, large and small granules combined as in wheat and barley, or compound of many small starch granules as in oats (Zeeman, Kossmann, & Smith, 2010). The large size of potato starch granules is advantageous in applications that require high viscosity but limits its suitability for noodle making. Expression of multiple tandem starch-binding domains from *Bacillus circulans* cyclodextrin glycosyltransferase in potato resulted in a large number of small starch granules without affecting the starch yield (Ji et al., 2004; Nazarian-Firouzabadi, Trindade, & Visser, 2012). The two starch phosphorylases in plants, plastidic phosphorylase A (Pho 1a) and cytosolic phosphorylase

(Pho 2), catalyze reversible transfer of glucose from glucose-1-phosphate to α-glucan chain-releasing phosphorus (pi). In a recent mapping study in potato, *Pho 1a* emerged as a candidate gene linked to starch gelling and starch granule size (Werij, Furrer, van Eck, Visser, & Bachem, 2012) supporting previous finding in rice where Pho1 deficiency yielded grains containing smaller starch granules with lowered onset of starch gelling (Satoh et al., 2008). Antisense inhibition of isoamylase in potato induces massive numbers of small granules in tubers, suggesting that the debranching activity is necessary to prevent excessive granule initiation (Bustos et al., 2004). On the other hand, it has been observed that mutation of starch synthase IV in *Arabidopsis* increases the starch granule size in leaf (Roldan et al., 2007). However, despite these observations, we really have a poor understanding of what controls granule size and shape.

7 NEW METHODS FOR GENE MODIFICATION TO ALTER STARCH BIOSYNTHESIS

Many methods have been used to alter gene expression or create mutants in genes that control starch quantity and structure. These include natural mutations observed in germ pools, induced physical (e.g., using gamma ray and heavy ions) and chemical (ethyl methanesulfonate and sodium azide) mutagenesis, gene modification through gene silencing (hairpin RNA and artificial microRNA), overexpression, and heterologous gene expression, and there are more in development.

Mining for SNPs, either induced or naturally occurring, is a potent tool for identifying gene knockouts. The most recent exploitation of this technique for starch modification is the report of wheats with amylose contents of 47-55% through TILLING (targeted induced local lesions in genome) by Slade et al. (2012). Mutations in the form of SNPs in *SBEIIa* were identified in all three individual genomes of wheat, following which the new mutant alleles were combined through breeding to develop this wheat. A region of exon 8 of *SSIIa* gene in both rice and sorghum is strongly associated to starch gelatinization temperature (Hill, Lee, & Henry, 2012; Waters, Henry, Reinke, & Fitzgerald, 2006). Three simultaneous SNPs in this region led to a haplotype (designated haplotype 3) in sorghum that has increased amylose and fat contents and decreased starch content but with higher gelatinization temperature, contrary to *SSIIa* mutations in barley and wheat (Hill et al., 2012).

New tools to manipulate the genetic composition of crops and alter the expression of target genes continue to be developed, giving breeders more options to develop crops with altered starch characteristics. In wheat, we have developed a heavy-ion bombardment mutant population, which has the advantage over other irradiation methods (such as X-rays or gamma irradiation) of producing relatively small deletions—from single bases up to many thousands. We have screened this population for mutants of several starch biosynthetic genes and have recovered mutations in all three wheat genomes at a frequency of about one in 300 lines screened. We have also put considerable effort into generating a new wheat genetic resource called MAGIC (multiparent advanced generation intercross) that has great potential for uncovering genetic diversity and combining quantitative trait loci (Cavanagh, Morell, Mackay, & Powell, 2008; Huang et al., 2012). In rice, association analysis of diverse varieties has been applied to determine two major genes (*GBSSI* and *SSIIa*) and a number of minor genes (e.g., *SBEI, SBEII3, isoamylase,* and *SSIV*) affecting the eating and cooking quality indicators (Kharabian-Masouleh, Waters, Reinke, Ward, & Henry, 2012; Tiana et al., 2009). These studies provide directions for combining starch synthetic-related genes to improve the rice eating and cooking quality. In addition to the genetic modification knockdown and overexpression approaches described so far, new more sophisticated genome editing techniques are being developed that do not require insertion of heterologous genetic material. These techniques use DNA nucleases such as TALENS and CRISPRS targeted to very specific parts of the genome to create small changes to the DNA sequence in order to knock out (by creating a small deletion) or create a change in the amino acid sequence of a protein to produce a particular phenotype (Barrangou, 2013; Boch et al., 2009). It is currently in discussion whether these techniques will be treated as GM and regulated as such or whether they will be treated as other mutagenic techniques, which do not require such oversight and have been used to produce many of the crop varieties that we grow today.

8 CONCLUSIONS

In the recent years, there have been an increasing number of studies on starch biosynthetic enzyme mutants and combinations of mutants combined with studies of overexpressing putative regulatory genes, and these have helped increase our understanding of starch biosynthesis. Many modified starches have been produced using these techniques but few have reached the application stage and fewer have been successfully commercialized. Starch is a cheap and readily available raw material, so for any new starches to come to market, they have to be of sufficiently high value to justify the extra production costs. If the starch is produced from a crop by genetic modification, it faces another significant hurdle—that of huge regulatory costs. Only one starch has so far gone through that process and a non-GM alternative is available so it remains to be seen if this starch will be commercially successful. The most likely to be commercialized starches will be the very high-amylose starches that have a nutritional benefit, such as the high-RS wheat. These starches will not be sold as a purified ingredient, but rather, the whole grain high-amylose wheat will be used for food production. With sufficient market penetration, this will provide significant socioeconomic benefits at a population level by decreasing the incidence of lifestyle diseases and the huge health costs associated with treating these conditions.

References

Ballicora, M. A., Frueauf, J. B., Fu, Y. B., Schurmann, P., & Preiss, J. (2000). Activation of the potato tuber ADP-glucose pyrophosphorylase by thioredoxin. *Journal of Biological Chemistry*, *275*(2), 1315–1320.

Barrangou, R. (2013). CRISPR-Cas systems and RNA-guided interference. *Wiley Interdisciplinary Reviews-RNA*, *4*(3), 267–278.

Bhattacharyya, M. K., Smith, A. M., Ellis, T. H. N., Hedley, C., & Martin, C. (1990). The wrinkled-seed character of pea described by mendel is caused by a transposon-like insertion in a gene encoding starch-branching enzyme. *Cell*, *60*(1), 115–122.

Bird, A. R., Jackson, M., King, R. A., Davies, D. A., Usher, S., & Topping, D. L. (2004). A novel high-amylose barley cultivar (Hordeum vulgare var. Himalaya 292) lowers plasma cholesterol and alters indices of large-bowel fermentation in pigs. *British Journal of Nutrition*, *92*(4), 607–615.

Bird, A. R., Vuaran, M. S., King, R. A., Noakes, M., Keogh, J., Morell, M. K., et al. (2008). Wholegrain foods made from a novel high-amylose barley variety (Himalaya 292) improve indices of bowel health in human subjects. *British Journal of Nutrition*, *99*(5), 1032–1040.

Blauth, S. L., Yao, Y., Klucinec, J. D., Shannon, J. C., Thompson, D. B., & Guilitinan, M. J. (2001). Identification of Mutator insertional mutants of starch-branching enzyme 2a in corn. *Plant Physiology*, *125*(3), 1396–1405.

Boch, J., Scholze, H., Schornack, S., Landgraf, A., Hahn, S., Kay, S., et al. (2009). Breaking the code of DNA binding specificity of TAL-Type III effectors. *Science*, *326*(5959), 1509–1512.

Bustos, R., Fahy, B., Hylton, C. M., Seale, R., Nebane, N. M., Edwards, A., et al. (2004). Starch granule initiation is controlled by a heteromultimeric isoamylase in potato tubers. *Proceedings of the National Academy of Sciences of the United States of America*, *101*(7), 2215–2220.

Butardo, V. M., Fitzgerald, M. A., Bird, A. R., Gidley, M. J., Flanagan, B. M., Larroque, O., et al. (2011). Impact of down-regulation of starch branching enzyme IIb in rice by artificial microRNA- and hairpin RNA-mediated RNA silencing. *Journal of Experimental Botany*, *62*(14), 4927–4941.

Cai, X. L., Wang, Z. Y., Xing, Y. Y., Zhang, J. L., & Hong, M. M. (1998). Aberrant splicing of intron 1 leads to the heterogeneous 5′ UTR and decreased expression of waxy gene in rice cultivars of intermediate amylose content. *Plant Journal*, *14*(4), 459–465.

Cavanagh, C., Morell, M., Mackay, I., & Powell, W. (2008). From mutations to MAGIC: Resources for gene discovery, validation and delivery in crop plants. *Current Opinion in Plant Biology*, *11*(2), 215–221.

Dickinson, D. B., & Preiss, J. (1969). Presence of adp-glucose pyrophosphorylase in shrunken-2 and brittle-2 mutants of maize endosperm. *Plant Physiology*, *44*(7), 1058–1062.

Dobo, M., Ayres, N., Walker, G., & Park, W. D. (2010). Polymorphism in the GBSS gene affects amylose content in US and European rice germplasm. *Journal of Cereal Science*, *52*(3), 450–456.

Ezquer, I., Li, J., Ovecka, M., Baroja-Fernandez, E., Munoz, F. J., Montero, M., et al. (2010). Microbial volatile emissions promote accumulation of exceptionally high levels of starch in leaves in mono- and dicotyledonous plants. *Plant and Cell Physiology*, *51*(10), 1674–1693.

Faix, B., Radchuk, V., Nerlich, A., Hummer, C., Radchuk, R., Emery, R. J. N., et al. (2012). Barley grains, deficient in cytosolic small subunit of ADP-glucose pyrophosphorylase, reveal coordinate adjustment of C:N metabolism mediated by an overlapping metabolic-hormonal control. *Plant Journal*, *69*(6), 1077–1093.

Gamez-Arjona, F. M., Li, J., Raynaud, S., Baroja-Fernandez, E., Munoz, F. J., Ovecka, M., et al. (2011). Enhancing the expression of starch synthase class IV results in increased levels of both transitory and long-term storage starch. *Plant Biotechnology Journal*, *9*(9), 1049–1060.

Garwood, D. L., Shannon, J. C., & Creech, R. G. (1976). Starches of endosperms possessing different alleles at amylose-extender locus in zea-mays-l. *Cereal Chemistry*, *53*(3), 355–364.

Hannah, L. C., Giroux, M., & Boyer, C. (1993). Biotechnological modification of carbohydrates for sweet corn and maize improvement. *Scientia Horticulturae*, *55*(1–2), 177–197.

Hill, H., Lee, L. S., & Henry, R. J. (2012). Variation in sorghum starch synthesis genes associated with differences in starch phenotype. *Food Chemistry*, *131*(1), 175–183.

Huang, B. E., George, A. W., Forrest, K. L., Kilian, A., Hayden, M. J., Morell, M. K., et al. (2012). A multiparent advanced generation inter-cross population for genetic analysis in wheat. *Plant Biotechnology Journal*, *10*(7), 826–839.

Ihemere, U., Arias-Garzon, D., Lawrence, S., & Sayre, R. (2006). Genetic modification of cassava for enhanced starch production. *Plant Biotechnology Journal*, *4*(4), 453–465.

Ji, Q., Oomen, R. J. F., Vincken, J. P., Bolam, D. N., Gilbert, H. J., Suurs, L., et al. (2004). Reduction of starch granule size by expression of an engineered tandem starch-binding domain in potato plants. *Plant Biotechnology Journal*, *2*(3), 251–260.

Jobling, S. (2004). Improving starch for food and industrial applications. *Current Opinion in Plant Biology*, *7*(2), 210–218.

Jobling, S. A., Schwall, G. P., Westcott, R. J., Sidebottom, C. M., Debet, M., Gidley, M. J., et al. (1999). A minor form of starch branching enzyme in potato (Solanum tuberosum L.) tubers has a major effect on starch structure: cloning and characterisation of multiple forms of SBE A. *Plant Journal*, *18*(2), 163–171.

Jobling, S. A., Westcott, R. J., Tayal, A., Jeffcoat, R., & Schwall, G. P. (2002). Production of a freeze-thaw-stable potato starch by antisense inhibition of three starch synthase genes. *Nature Biotechnology*, *20*(3), 295–299.

Johnson, P. E., Patron, N. J., Bottrill, A. R., Dinges, J. R., Fahy, B. F., Parker, M. L., et al. (2003). A low-starch barley mutant, Riso 16, lacking the cytosolic small subunit of ADP-glucose pyrophosphorylase, reveals the importance of the cytosolic isoform and the identity of the plastidial small subunit. *Plant Physiology*, *131*(2), 684–696.

Kang, G. Z., Li, S. Y., Zhang, M. Q., Peng, H. F., Wang, C. Y., Zhu, Y. J., et al. (2013). Molecular cloning and expression analysis of the starch-branching enzyme III gene from common wheat (Triticum aestivum). *Biochemical Genetics*, *51*(5–6), 377–386.

Kharabian-Masouleh, A., Waters, D. L. E., Reinke, R. F., Ward, R., & Henry, R. J. (2012). SNP in starch biosynthesis genes associated with nutritional and functional properties of rice. *Nature*, *2*, 557.

Konik-Rose, C., Thistleton, J., Chanvrier, H., Tan, I., Halley, P., Gidley, M., et al. (2007). Effects of starch synthase IIa gene dosage on grain, protein and starch in endosperm of wheat. *Theoretical and Applied Genetics*, *115*(8), 1053–1065.

Li, J., Baroja-Fernandez, E., Bahaji, A., Munoz, F. J., Ovecka, M., Montero, M., et al. (2013). Enhancing sucrose synthase activity results in increased levels of starch and ADP-glucose in maize (Zea mays L.) seed endosperms. *Plant and Cell Physiology*, *54*(2), 282–294.

Li, J., Ezquer, I., Bahaji, A., Montero, M., Ovecka, M., Baroja-Fernandez, E., et al. (2011). Microbial volatile-induced accumulation of exceptionally high levels of starch in Arabidopsis leaves is a process involving NTRC and starch synthase classes III and IV. *Molecular Plant-Microbe Interactions, 24*(10), 1165–1178.

Li, Z., Li, D., Du, X., Wang, H., Larroque, O., Jenkins, C. L. D., et al. (2011). The barley amo1 locus is tightly linked to the starch synthase IIIa gene and negatively regulates expression of granule-bound starch synthetic genes. *Journal of Experimental Botany, 62*, 5217–5231.

Li, M., Piao, J. H., Tian, Y., Li, W. D., Li, K. J., & Yang, X. G. (2010). Postprandial glycaemic and insulinaemic responses to GM-resistant starch-enriched rice and the production of fermentation-related H-2 in healthy Chinese adults. *British Journal of Nutrition, 103*(7), 1029–1034.

Li, N., Zhang, S., Zhao, Y., Li, B., & Zhang, J. (2011). Over-expression of AGPase genes enhances seed weight and starch content in transgenic maize. *Planta, 233*(2), 241–250.

Liu, F. S., Ahmed, Z., Lee, E. A., Donner, E., Liu, Q., Ahmed, R., et al. (2012). Allelic variants of the amylose extender mutation of maize demonstrate phenotypic variation in starch structure resulting from modified protein-protein interactions. *Journal of Experimental Botany, 63*(3), 1167–1183.

Liu, F. S., Makhmoudova, A., Lee, E. A., Wait, R., Emes, M. J., & Tetlow, I. J. (2009). The amylose extender mutant of maize conditions novel protein-protein interactions between starch biosynthetic enzymes in amyloplasts. *Journal of Experimental Botany, 60*(15), 4423–4440.

Mizuno, K., Kawasaki, T., Shimada, H., Satoh, H., Kobayashi, E., Okumura, S., et al. (1993). Alteration of the structural-properties of starch components by the lack of an isoform of starch branching enzyme in rice seeds. *Journal of Biological Chemistry, 268*(25), 19084–19091.

Morell, M. K., Kosar-Hashemi, B., Cmiel, M., Samuel, M. S., Chandler, P., Rahman, S., et al. (2003). Barley sex6 mutants lack starch synthase IIa activity and contain a starch with novel properties. *Plant Journal, 34*(2), 173–185.

Mullerrober, B., & Kossmann, J. (1994). Approaches to influence starch quantity and starch quality in transgenic plants. *Plant Cell and Environment, 17*(5), 601–613.

Nakamura, T., Yamamori, M., Hirano, H., Hidaka, S., & Nagamine, T. (1995). Production of waxy (amylose-free) wheats. *Molecular & General Genetics, 248*(3), 253–259.

Nazarian-Firouzabadi, F., Trindade, L. M., & Visser, R. G. F. (2012). Production of small starch granules by expression of a tandem-repeat of a family 20 starch-binding domain (SBD3-SBD5) in an amylose-free potato genetic background. *Functional Plant Biology, 39*(2), 146–155.

Ni, D. H., Zhang, S. L., Chen, S., Xu, Y., Li, L., Li, H., et al. (2011). Improving cooking and eating quality of Xieyou57, an elite indica hybrid rice, by marker-assisted selection of the Wx locus. *Euphytica, 179*(2), 355–362.

Nishi, A., Nakamura, Y., Tanaka, N., & Satoh, H. (2001). Biochemical and genetic analysis of the effects of amylose-extender mutation in rice endosperm. *Plant Physiology, 127*(2), 459–472.

Park, H., Choi, M., Chun, A., Lee, J., Kim, M., Kim, Y., et al. (2010). Variation of amylose content using dsRNAi vector by targeting 3′-UTR region of GBSSI gene in rice. *Korean Journal of Breeding Science, 42*(5), 515–524.

Patron, N. J., Smith, A. M., Fahy, B. F., Hylton, C. M., Naldrett, M. J., Rossnagel, B. G., et al. (2002). The altered pattern of amylose accumulation in the endosperm of low-amylose barley cultivars is attributable to a single mutant allele of granule-bound starch synthase I with a deletion in the 5′-non-coding region. *Plant Physiology, 130*(1), 190–198.

Ral, J. P., Colleoni, C., Wattebled, F., Dauvillee, D., Nempont, C., Deschamps, P., et al. (2006). Circadian clock regulation of starch metabolism establishes GBSSI as a major contributor to amylopectin synthesis in Chlamydomonas reinhardtii. *Plant Physiology, 142*(1), 305–317.

Regierer, B., Fernie, A. R., Springer, F., Perez-Melis, A., Leisse, A., Koehl, K., et al. (2002). Starch content and yield increase as a result of altering adenylate pools in transgenic plants. *Nature Biotechnology, 20*(12), 1256–1260.

Regina, A., Bird, A., Topping, D., Bowden, S., Freeman, J., Barsby, T., et al. (2006). High-amylose wheat generated by RNA interference improves indices of large-bowel health in rats. *Proceedings of the National Academy of Sciences of the United States of America, 103*(10), 3546–3551.

Regina, A., Blazek, J., Gilbert, E., Flanagan, B. M., Gidley, M. J., Cavanagh, C., et al. (2012). Differential effects of genetically distinct mechanisms of elevating amylose on barley starch characteristics. *Carbohydrate Polymers, 89*, 979–991.

Regina, A., Kosar-Hashemi, B., Ling, S., Li, Z. Y., Rahman, S., & Morell, M. (2010). Control of starch branching in barley defined through differential RNAi suppression of starch branching enzyme IIa and IIb. *Journal of Experimental Botany, 61*(5), 1469–1482.

Roldan, I., Wattebled, F., Lucas, M. M., Delvalle, D., Planchot, V., Jimenez, S., et al. (2007). The phenotype of soluble starch synthase IV defective mutants of Arabidopsis thaliana suggests a novel function of elongation enzymes in the control of starch granule formation. *The Plant Journal, 49*, 492–504.

Rosche, E., Blackmore, D., Tegeder, M., Richardson, T., Schroeder, H., Higgins, T. J. V., et al. (2002). Seed-specific overexpression of a potato sucrose transporter increases sucrose uptake and growth rates of developing pea cotyledons. *Plant Journal, 30*(2), 165–175.

Safford, R., Jobling, S. A., Sidebottom, C. M., Westcott, R. J., Cooke, D., Tober, K. J., et al. (1998). Consequences of antisense RNA inhibition of starch branching enzyme activity on properties of potato starch. *Carbohydrate Polymers, 35*(3–4), 155–168.

Sakulsingharoj, C., Choi, S. B., Hwang, S. K., Edwards, G. E., Bork, J., Meyer, C. R., et al. (2004). Engineering starch biosynthesis for increasing rice seed weight: the role of the cytoplasmic ADP-glucose pyrophosphorylase. *Plant Science, 167*(6), 1323–1333.

Sanz-Barrio, R., Corral-Martinez, P., Ancin, M., Segui-Simarro, J. M., & Farran, I. (2013). Overexpression of plastidial thioredoxin f leads to enhanced starch accumulation in tobacco leaves. *Plant Biotechnology Journal, 11*(5), 618–627.

Satoh, H., Shibahara, K., Tokunaga, T., Nishi, A., Tasaki, M., Hwang, S. -K., et al. (2008). Mutation of the plastidial alpha-glucan phosphorylase gene in rice affects the synthesis and structure of starch in the endosperm. *Plant Cell, 20*(7), 1833–1849.

Schwall, G. P., Safford, R., Westcott, R. J., Jeffcoat, R., Tayal, A., Shi, Y. C., et al. (2000). Production of very-high-amylose potato starch by inhibition of SBE A and B. *Nature Biotechnology, 18*(5), 551–554.

Slade, A. J., McGuire, C., Loeffler, D., Mullenberg, J., Skinner, W., Fazio, G., et al. (2012). Development of high amylose wheat through TILLING. *BMC Plant Biology, 12*, 69.

Smidansky, E. D., Clancy, M., Meyer, F. D., Lanning, S. P., Blake, N. K., Talbert, L. E., et al. (2002). Enhanced ADP-glucose pyrophosphorylase activity in wheat endosperm increases seed yield. *Proceedings of the National Academy of Sciences of the United States of America, 99*(3), 1724–1729.

Smith, A. M., Bettey, M., & Bedford, I. D. (1989). Evidence that the rb locus alters the starch content of developing pea embryos through an effect on adp glucose pyrophosphorylase. *Plant Physiology, 89*(4), 1279–1284.

Stark, D. M., Timmerman, K. P., Barry, G. F., Preiss, J., & Kishore, G. M. (1992). Regulation of the amount of starch in plant-tissues by adp glucose pyrophosphorylase. *Science, 258*(5080), 287–292.

Szydlowski, N., Ragel, P., Hennen-Bierwagen, T. A., Planchot, V., Myers, A. M., Merida, A., et al. (2011). Integrated functions among multiple starch synthases determine both amylopectin chain length and branch linkage location in Arabidopsis leaf starch. *Journal of Experimental Botany, 62*(13), 4547–4559.

Tiana, Z., Qian Qian, Q., Liu, Q., Yan, M., Liu, X., Yan, C., et al. (2009). Allelic diversities in rice starch biosynthesis lead to a diverse array of rice eating and cooking qualities. *Proceedings of the National Academy of Sciences of the United States of America, 106*(51), 21760–21765.

Topping, D. (2007). Cereal complex carbohydrates and their contribution to human health. *Journal of Cereal Science, 46*(3), 220–229.

Topping, D. L., & Clifton, P. M. (2001). Short-chain fatty acids and human colonic function: Roles of resistant starch and nonstarch polysaccharides. *Physiological Reviews, 81*(3), 1031–1064.

Topping, D. L., Morell, M. K., King, R. A., Li, Z., Bird, A. R., & Noakes, M. (2003). Resistant starch and health—Himalaya 292, a novel barley cultivar to deliver benefits to consumers. *Starch - Stärke, 55*, 539–545.

Tuncel, A., Kavakli, I. H., & Keskin, O. (2008). Insights into subunit interactions in the heterotetrameric structure of potato ADP-glucose pyrophosphorylase. *Biophysical Journal, 95*(8), 3628–3639.

Vrinten, P., Nakamura, T., & Yamamori, M. (1999). Molecular characterization of waxy mutations in wheat. *Molecular and General Genetics, 261*(3), 463–471.

Wang, Z., Chen, X., Wang, J., Liu, T., Liu, Y., Zhao, L., et al. (2007). Increasing maize seed weight by enhancing the cytoplasmic ADP-glucose pyrophosphorylase activity in transgenic maize plants. *Plant Cell, Tissue and Organ Culture, 88*(1), 83–92.

Wang, Z. Y., Wu, Z. L., Xing, Y. Y., Zheng, F. G., Guo, X. L., Zhang, W. G., et al. (1990). Nucleotide-sequence of rice waxy gene. *Nucleic Acids Research, 18*(19), 5898.

Waters, D. L. E., Henry, R. J., Reinke, R. F., & Fitzgerald, M. A. (2006). Gelatinization temperature of rice explained by polymorphisms in starch synthase. *Plant Biotechnology Journal, 4*(1), 115–122.

Werij, J. S., Furrer, H., van Eck, H. J., Visser, R. G. F., & Bachem, C. W. B. (2012). A limited set of starch related genes explain several interrelated traits in potato. *Euphytica, 186*(2), 501–516.

Xia, H., Yandeau-Nelson, M., Thompson, D. B., & Guiltinan, M. J. (2011). Deficiency of maize starch-branching enzyme i results in altered starch fine structure, decreased digestibility and reduced coleoptile growth during germination. *BMC Plant Biology, 11*, 95.

Yamamori, M., Fujita, S., Hayakawa, K., Matsuki, J., & Yasui, T. (2000). Genetic elimination of a starch granule protein, SGP-1, of wheat generates an altered starch with apparent high amylose. *Theoretical and Applied Genetics, 101*(1–2), 21–29.

Yasui, T. (2006). Waxy and low-amylose mutants of bread wheat (Triticum aestivum L.) and their starch, flour and grain properties. *Japan Agricultural Research Quarterly, 40*(4), 327–331.

Yasui, T., & Ashida, K. (2011). Waxy endosperm accompanies increased fat and saccharide contents in bread wheat (Triticum aestivum L.) grain. *Journal of Cereal Science, 53*(1), 104–111.

Zeeman, S. C., Kossmann, J., & Smith, A. M. (2010). Starch: Its metabolism, evolution, and biotechnological modification in plants. In S. Merchant, W. R. Briggs & D. Ort (Eds.), *Annual review of plant biology: Vol. 61*(pp. 209–234). Palo Alto: Annual Reviews, USA.

Zhang, G. Y., Cheng, Z. J., Zhang, X., Guo, X. P., Su, N., Jiang, L., et al. (2011). Double repression of soluble starch synthase genes SSIIa and SSIIIa in rice (Oryza sativa L.) uncovers interactive effects on the physicochemical properties of starch. *Genome, 54*(6), 448–459.

Zhang, L., Hausler, R. E., Greiten, C., Hajirezaei, M. R., Haferkamp, I., Neuhaus, H. E., et al. (2008). Overriding the co-limiting import of carbon and energy into tuber amyloplasts increases the starch content and yield of transgenic potato plants. *Plant Biotechnology Journal, 6*(5), 453–464.

Zhang, X. L., Szydlowski, N., Delvalle, D., D'Hulst, C., James, M. G., & Myers, A. M. (2008). Overlapping functions of the starch synthases SSII and SSIII in amylopectin biosynthesis in Arabidopsis. *BMC Plant Biology, 8*, 96.

Zhu, L. J., Dogan, H., Gajula, H., Gu, M. H., Liu, Q. Q., & Shi, Y. C. (2012). Study of kernel structure of high-amylose and wild-type rice by X-ray microtomography and SEM. *Journal of Cereal Science, 55*(1), 1–5.

Zhu, L. J., Gu, M. H., Meng, X. L., Cheung, S. C. K., Yu, H. X., Huang, J., et al. (2012). High-amylose rice improves indices of animal health in normal and diabetic rats. *Plant Biotechnology Journal, 10*(3), 353–362.

CHAPTER

3

"Structure-Property" Relationships of Genetically Modified Starch

I. Tan, Peter J. Halley

Centre for High Performance Polymers, Australian Institute for Bioengineering and Nanotechnology, The University of Queensland, Queensland, Australia

1 INTRODUCTION

Starch, the major plant polysaccharide reserve material, is deposited as semicrystalline granules in the plastid compartment of plant cells such as in the amyloplast of plant storage organs, in leaf cell chloroplasts, or in other photosynthetically competent cells (Martin & Smith, 1995; Rahman et al., 2000; Slattery, Kavakli, & Okita, 2000). Despite the universality of starch distribution in plants, its major polysaccharide constituents, and the semicrystalline nature of its granules, there are substantial differences in their characteristics such as the polysaccharides composition, granules morphology, and granular ordered structures among different plant species and varieties. Such variations in starch structural attributes influence the functional properties of starch during processing (Fredriksson, Silverio, Andersson, Eliasson, & Aman, 1998; Hermansson & Svegmark, 1996; Morrison, Tester, Snape, Law, & Gidley, 1993; Tester & Morrison, 1990; Yuryev, Kalistratova, Van soest, & Nieman, 1998). These consequently restrict their suitability for a wider range of industrial applications.

The diversity in starch characteristics from different origins can be largely attributed to the variability in starch biosynthesis parameters between different plant species, namely, the differences in spatial and temporal regulation of genes encoding various isoforms of starch biosynthetic enzymes, the diversification in the activities of the participating enzymes, and modulation of the extent of their activities (Baga et al., 1999; Martin & Smith, 1995). Accordingly, starches from different plant species require different inhibitions of biosynthesis gene expressions to obtain a particular novel variability, which depends on the isoforms of starch biosynthesis enzymes that are present in the particular plant species. Hence, a thorough investigation on the properties of a range of novel starches derived from genetic modification of starches from the same species can provide useful insights into the interrelationship between starch synthesis and its properties, without the extra variability introduced by different species.

Therefore, this chapter considers a number of structural aspects that are inherent in most starch granules (e.g., the macromolecular properties, the extent of ordered structure, and the granule morphology) and relate them with the observed variation in their functional properties. The significance of each structural property and how changes in these properties correspond to the observed functional properties are explored. The variations in the structural and thermal properties of the novel starches are examined, aiming to establish the significance and the role of starch structural parameters (whether collectively or independently) on starch properties.

2 NOVEL STARCH MATERIALS

The structural and functional properties for a range of starches from important cereal crops, namely, maize, barley, and wheat (as listed in Table 3.1), are explored. The starch variants from these cereal crops were provided by the CSIRO Plant Industry. The properties of some commercial maize starches (Penford Australia Limited, Lane Cove, NSW 2066) and rice starches (Sunrice, Yanco Avenue, Leeton, and NSW 2705) are also considered.

3 MACROMOLECULAR PROPERTIES

A common approach to characterize the macromolecular properties of starch involves the measurement of the polysaccharide composition (amylose to amylopectin ratio) and their fine structures. The amylose to amylopectin ratio is an important measure of the constituting starch polysaccharides and is predictive of some starch granular and functional properties such as crystallinity, swelling power, and viscosity (Cheetham & Tao, 1998; Hermansson & Svegmark, 1996; Hoover, 2001; Morrison et al., 1993). The variation in the amylopectin fine structure is known to influence the granular

TABLE 3.1 List of Model Starch Materials Examined in This Chapter

Plant Origin	Starch Type	Description
Maize	W64A	Normal (nonmutant) maize
	Wx1M	Waxy maize—zero amylose; mutation of the *Waxy* locus, which encodes the GBSS protein (Jobling, 2004)
	Du1R	Mutation of *du1* gene, which reduces the activity of SSIII enzyme and also BEIIa enzyme (as secondary effect) (Boyer & Preiss, 1981; Cao et al., 1999; Gao, Wanat, Stinard, James, & Myers, 1998; Pan, 2000; Rahman et al., 2000)
	Su1R	*sugary1* mutant, associated with the deficiency of isoamylase-type debranching enzyme activity (Myers, Morell, James, & Ball, 2000)
	High amylose (HAM)	*amylose extender* mutant, known to lack the BEIIb isoform (Rahman et al., 2000; Takeda, Takeda, & Hizukuri, 1993)
Barley	Waxiro (WB)	Waxy barley—zero amylose; generated from barley mutant in which the GBSSI enzyme is lacking
	Himalaya (HB)	Normal (nonmutant) barley
	Tantangera (TB)	Normal (nonmutant) barley
	High-amylose glacier (HAG)	High-amylose barley phenotype; with an amylose content of 45-50% (Banks, Greenwood, & Walker, 1971; Rahman et al., 2000)
	292 Barley (292B)	Generated from barley *sex6* mutants, in which the SSII enzyme activity is deficient (Morell et al., 2003)
Wheat	Waxy (WW)	Waxy wheat phenotype—zero amylose
	Cadoux (CW)	Partial waxy wheat (GBSS-4A null); entailing a deficiency in the GBSSI B enzyme activity (Baga et al., 1999; Rahman et al., 2000; Zhao et al., 1998)
	Sunco (SW)	Normal (nonmutant) wheat
	Yamamori (YW)	Wheat carrying SSII null allele (Yamamori, Fujita, Hayakawa, Matsuki, & Yasui, 2000)
Wheat	Janz (JW)	Hard grain wheat

Continued

TABLE 3.1 List of Model Starch Materials Examined in This Chapter—cont'd

Plant Origin	Starch Type	Description
	Rosella (RW)	Soft grain wheat
Wheat carrying various SSII null allele genotypes	B68	Normal (nonmutant) wheat
	C20	SSII-A-null genotype (SSII-aBD)
	C3	SSII-B-null genotype (SSII-AbD)
	A59	SSII-D-null genotype (SSII-ABd)
	B10	SSII-AB-null genotype (SSII-abD)
	B21	SSII-AD-null genotype (SSII-aBd)
	B40	SSII-BD-null genotype (SSII-Abd)
	A63	SSII-ABD-triple-null genotype (SSII-abd)
Maize (Penford)	Waxy	Waxy maize—0% amylose
	Normal	Regular maize—28% amylose
	Gelose 50 (G50)	High-amylose maize—55% amylose
	Gelose 80 (G80)	High-amylose maize—85% amylose
Rice (Sunrice)	Indica rice (IR)	Long-grain rice
	Japonica rice (JR)	Short-grain rice

and functional properties of these starches (Fredriksson et al., 1998; Hizukuri, 1985, 1986; Hizukuri, Kaneko, & Takeda, 1983; Imberty, Buleon, Tran, & Perez, 1991). This can be partly attributed to the predominant role of amylopectin as the supporting framework of starch granular organization. Hence, the following discussions on the variation in starch macromolecular properties will primarily focus on the amylopectin fine structure (based on the analysis of chain length distribution) and the amylose to amylopectin ratio (amylose content). The characterization methods employed are described elsewhere (Tan, 2005).

3.1 Macromolecular Properties of Maize Starch Varieties

The amylose contents of the maize starch are 0.4%, 13.6%, 25.4%, 49.9%, and 75.3% for Wx1M, W64A, Du1R, Su1R, and high-amylose maize (HAM), respectively. Differences in the distribution of amylopectin chain

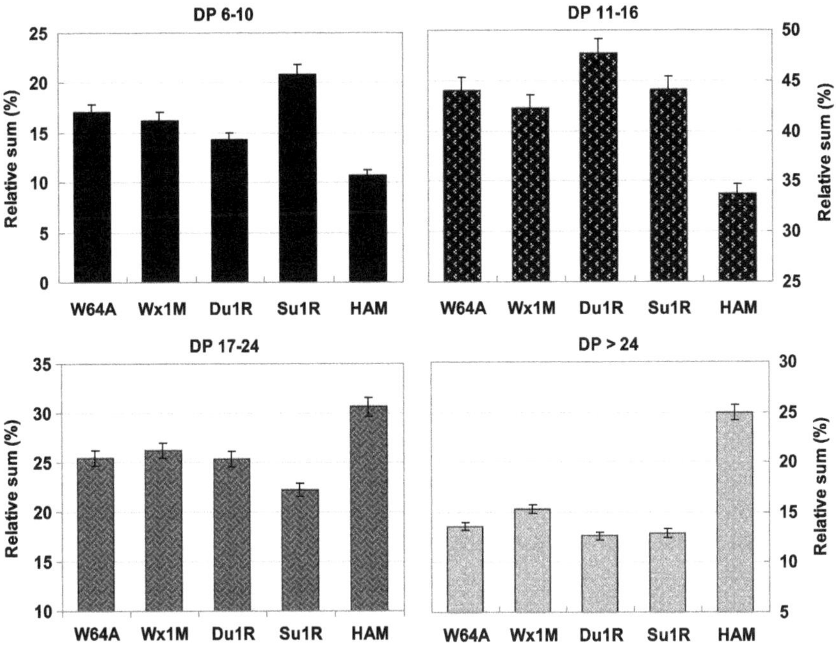

FIGURE 3.1 The sum of debranched amylopectin chain lengths relative proportions for maize starches, as fractionated into four different chain size ranges: DP 6-10, 11-16, 17-24, and >24.

lengths between the maize starch varieties are evident with HAM displaying the most distinct distribution (Figure 3.1). The amylopectin chain length distributions of waxy (Wx1M) maize starch and normal (W64A)-type starch are comparable. A noticeable increase in the proportion of amylopectin short chains with DP 10-12 is evident in Du1R and Su1R starches. The loss of debranching enzyme (DBE) activity in a Su1R mutant leads to an overall shift of the amylopectin chain distribution to shorter chain lengths (as evidenced by the increase in the fractions of short chains with DP 6-12 and a reduction in the longer-chain fractions of DP $\geq$17). The increase in short-chain fractions of Du1R amylopectin, on the other hand, is only observed in the proportion of DP 11-15 chains (Figure 3.1). The short-chain (DP 6-16) fraction of HAM amylopectin is reduced substantially to approximately half of those in W64A starch, while the proportion of longer chains (DP >16) is significantly higher compared to the others, showing an overall shift to longer-chain distribution.

These results demonstrate that the mutations in the key genes *waxy* (*wx*), *amylose extender* (*ae*), *dull1* (*du1*), and *sugary1* (*su1*) genes in maize cause changes in their amylopectin chain length distribution. The waxy mutant starch, with diminishing granule bound starch synthase enzyme I (GBSSI) enzyme activity, has reduced amylose content, but no alteration in its amylopectin structure is noted. The absence of BEIIb isoform results in starch with apparent amylose content of 75.3% (HAM), and its amylopectin acquires longer-chain-length characteristics compared to other maize varieties. Reduction in the SSIII enzyme activity does not result in substantial alteration in the starch macromolecular properties. Alternatively, the loss of the isoamylase-type DBE activity in Su1R maize mutants leads to altered amylopectin structure with enriched short-chain fractions.

3.2 Macromolecular Properties of Barley Starch Varieties

The amylose contents of the novel barley starches are 3.4%, 11.8%, 15.2%, 48.4%, and 76.5% for Waxiro, HB, TB, high-amylose glacier (HAG), and 292B, respectively (Tan, 2005). The amylopectin short chains of these barley starches exhibit peak maxima at DP 10-11 and a distinct shoulder at DP 17-18 (Tan, 2005). The amylopectin chains of TB starch are composed of a higher proportion of intermediate to long DP chains (DP $\geq$13) when compared to HB; refer to Figure 3.2. Waxiro and HAG

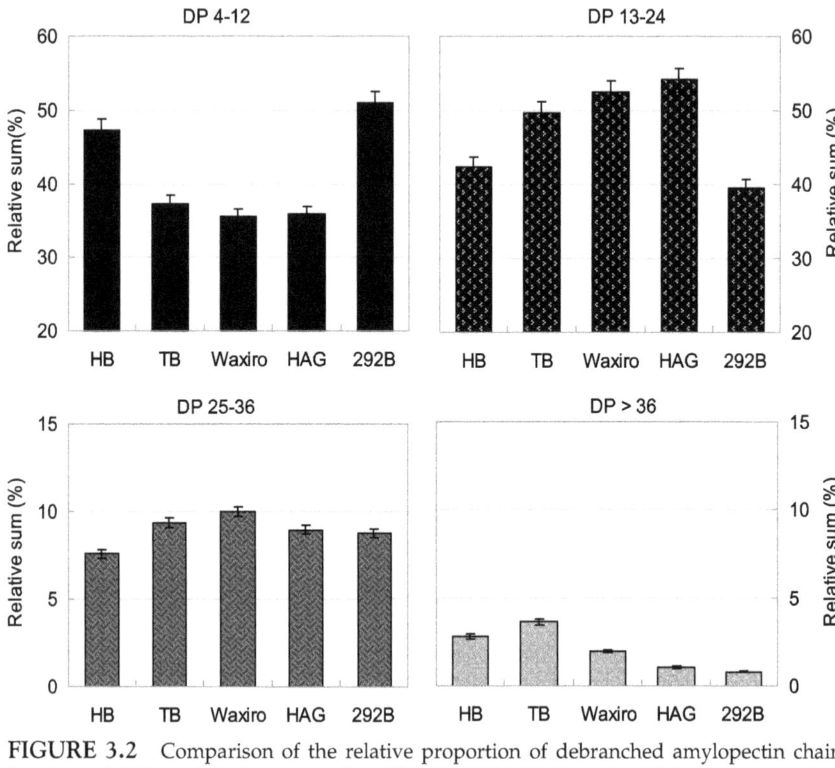

FIGURE 3.2 Comparison of the relative proportion of debranched amylopectin chain fractions of different DP size ranges (i.e., DP 4-12, 13-24, 25-36, and >36) for barley starches.

amylopectins consist of a higher proportion of chains with DP 13-36, showing a similarity in amylopectin chain distributions. The high-amylose 292 barley starch (292B) has significantly higher proportion of short chains with DP 4-12 and reduced amount of the DP 13-24 chain fractions, suggesting the possible role of SSII enzyme in the generation of amylopectin intermediate chains with DP 13-24. A considerably reduced amount of the amylopectin chains with DP >36 is noted in Waxiro, HAG, and 292B starches, and this supports the earlier deduction of the shorter amylopectin chain characteristics in these starches.

Unlike in HAM starches, the amylopectin of high-amylose barley starches have a higher proportion of short and intermediate chains. The high-amylose starch in barley (292B) has amylopectin with increased amount of short-chain fractions (DP 4-12). On the other hand, the amylopectin of HAG starch contains a higher proportion of the intermediate-size chains with DP13-24, which is similar to Du1R amylopectin.

3.3 Macromolecular Properties of Wheat Starch Varieties

The estimated values of amylose content for the novel wheat starch varieties ranged from 1.9% to 33.4% (Tan, 2005). Cadoux wheat starch has an amylose content of 26.6% amylose, which shows that the loss of GBSSI-4A isoform does not lead to significant amylose content reduction as compared to the waxy wheat starch (1.9% amylose), which contains all simultaneous GBSSI triple null alleles. Starches from wheat lines deficient in one or two isoforms of the SSII enzymes do not show considerable changes in the amylose content, and high amylose content is only observed in wheat lines with SSII triple null alleles, i.e., Yamamori and A63 (containing 31.9% and 33.4% amylose content, respectively).

The amylopectin chain length distributions of waxy wheat, Rosella, and Sunco (normal) wheat starches are comparable. These starches generally display short-chain peak maxima at DP 11-12 and a distinct shoulder at DP 18-19 (Tan, 2005). The Cadoux and Janz wheat starches display a general increase in the intermediate- to long-chain fractions (DP $\geq$ 12) and a considerable reduction in the fraction of amylopectin chains with DP 6-11 (cf. Figure 3.3). The observed increment in the relative proportion of DP 5-10 chains and reduction in the intermediate-chain fractions of DP 11-23 confirms the noted shift in the amylopectin chain distribution of Yamamori starch to the shorter chain lengths.

Figure 3.4 demonstrates the variation in amylopectin chain length distribution for wheat starches carrying different combinations of SSII null alleles. A general increase in the short-chain fractions of DP 4-12 is observed for most starches except for the SSII-ABd (single-D-null-type)

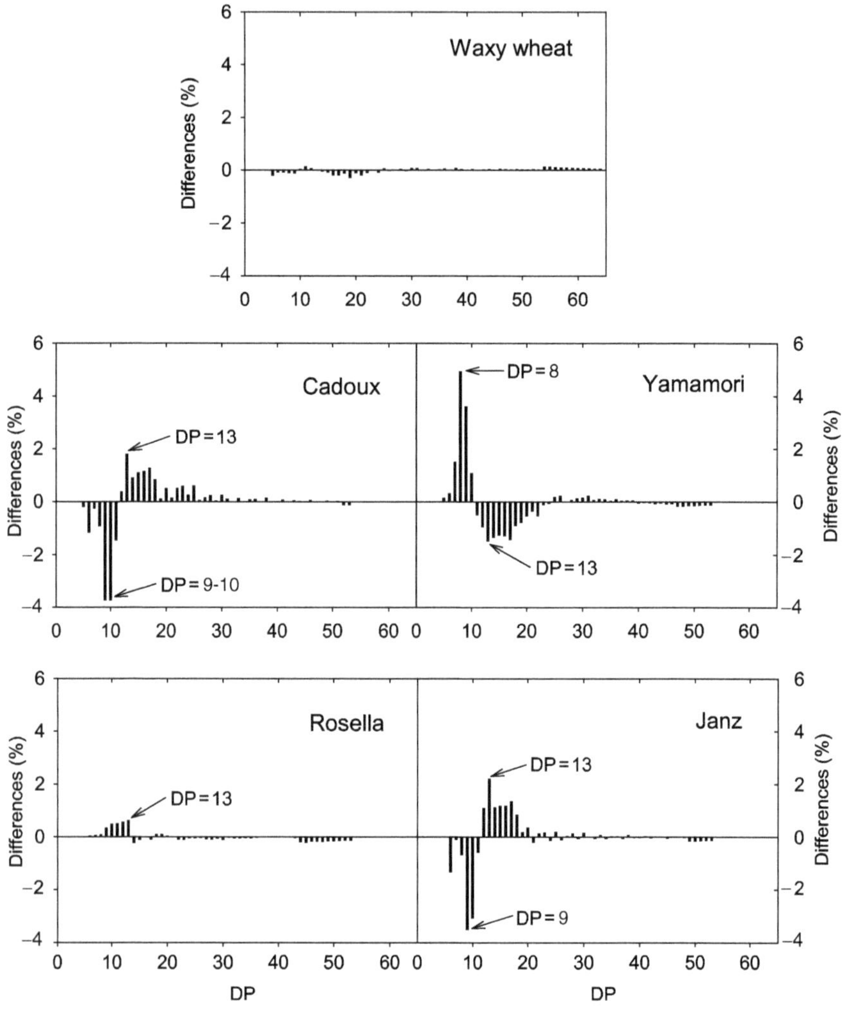

FIGURE 3.3 The differences in the relative distribution of debranched amylopectin chain lengths for some wheat starch varieties as compared to that of Sunco (normal type).

starch. The amylopectins of starches deficient in at least one of the A, B, and D genomes of SSII protein show a reduced amount of chain with DP of 13-24, with the most significant decrease noted for the triple-null starch (SSII-abd) type. The enrichment in the short-chain fraction (DP 4-12) of the triple-null starch type is accompanied by the depletion in the intermediate-chain fraction of DP 13-24, which indicates the possible role of the SSII enzyme on the generation of intermediate-chain fractions.

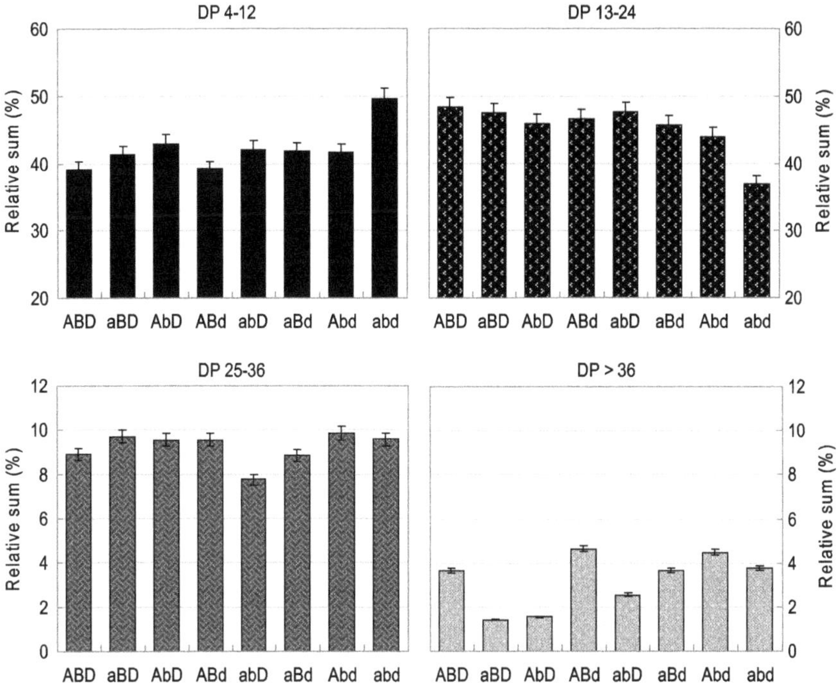

FIGURE 3.4 The relative proportion of isoamylase-debranched amylopectin chains, as fractionated into four different DP size ranges—DP 4-12, 13-24, 25-36, and >36—for wheat starches carrying different combinations of SSII null alleles.

3.4 Macromolecular Properties of Commercial Starches

The amylose contents of commercial maize starches are 3.4%, 24.4%, 56.3%, and 82.9% for PWM, PNM, G50, and G80, respectively (Tan, 2005). The amylose content of Indica rice starch (10.0%) is higher than that of the Japonica rice (5.6%). This is consistent with previous reports and is attributed to the presence of different alleles of the *Wx* gene (Wx^a and Wx^b) in the rice varieties (Mikami, Dung, Hirano, & Sano, 2000; Sano, 1984; Umemoto, Yano, Satoh, Shomura, & Nakamura, 2002).

The amylopectin short chains of the commercial HAM starches (G50 and G80) exhibit a peak maximum at DP 13 and a shoulder at DP 19 (Tan, 2005). Similar to HAM starch, an overall longer-chain distribution is noted for the G50 and G80 starches. Examination of the amylopectin chain length distributions for Indica and Japonica rice starch varieties showed that the amylopectin of Japonica rice starch has more short-chain fractions (DP 4-12) and reduced amount of intermediate-size chains with DP 13-24 and, when compared to those of Indica rice, is composed of more short-chain fractions (DP 4-12) than those of Indica rice (Tan, 2005).

4 GRANULAR PROPERTIES

Starches from different botanical origins have different granular characteristics such as the crystalline polymorph, degree of crystallinity, the amount of short-range molecular order (double-helix content), and the granule size distribution (Buleon, Colonna, Planchot, & Ball, 1998; Gernat, Radosta, Anger, & Damaschun, 1993; Gidley & Bociek, 1985; Gidley & Robinson, 1990; Zobel, 1988a,1988b). These granular properties can be characterized using a range of measurement techniques such as X-ray diffraction (crystallinity), ^{13}C CP/MAS solid-state NMR (short-range order), and laser light scattering (granules size distribution) as described elsewhere (Tan, Flanagan, Halley, Whittaker, & Gidley, 2007). The variable structural features of the novel starch varieties are reviewed to gain further knowledge on the manifested diversities in starch granular properties as a result of the mutation of a particular gene function and ultimately the structure-functionality interrelationship.

4.1 Granular Properties of Maize Starch Varieties

The granular properties of the maize starches, as obtained from the ^{13}C solid-state NMR and XRD studies, are described in Table 3.2. The characteristic triplet signal multiplicities (although weak) are apparent in Wx1M, W64A, and Du1R starches indicating the occurrence of the A-type polymorph crystallinity in these granules (Tan, 2005). The ^{13}C NMR spectrum for HAM starch demonstrates a closer resemblance in its spectral features to that of amorphous starch and plausibly the V-type polymorph. The waxy maize (Wx1M) starch contains the highest amount of double helices (50%), while HAM starch has the lowest proportion of double-helical conformation

TABLE 3.2 The Relative Amounts of the Amorphous, V-Type Single-Helix, and Double-Helix Conformations (^{13}C CP/MAS NMR) and the Degree of Crystallinity (X-Ray Diffraction) for Novel Maize Starch Varieties

Starch	Amylose Content (%)	Relative Proportion[a] (%) V-Type Polymorph	Double Helix	Amorphous	Relative Crystallinity[b] (%)	XRD Pattern
Wx1M	0.4	0	50	50	25	A
W64A	13.6	6	34	60	20	A
Du1R	25.4	5	31	64	17	A
HAM	75.3	11	25	64	8	B

[a] *The maximum standard deviation for the* ^{13}*C NMR analysis calculation was 3.6%.*
[b] *The maximum standard error for relative crystallinity calculation was 2.0%.*

(25%). The proportion of double helices and the degree of crystallinity decreases with increasing starch amylose content.

Analysis of the granule size distribution profiles showed that the W64A starch granules have the largest mean size diameter, i.e., 13.7 μm (Tan, 2005). Du1R and HAM starch granules have comparable mean granule sizes (10.8-10.9 μm), and their granules are smaller than those of W64A and Wx1M starches. The relative amounts of large granules are at least 10 times higher than the small granules and thus a unimodal granule size distribution is usually considered for maize starches (Tan, 2005). No trend is noted between the granule size distribution variation and macromolecular properties.

4.2 Granular Properties of Barley Starch Varieties

Both the ^{13}C NMR spectra and XRD patterns showed that all barley starches—except for the 292 Barley—adopt the A-type double-helical packing arrangement (Table 3.3). A decrease in the amount of double helices and the relative crystallinity are noted with increasing amylose content. The double helices in high-amylose (glacier) starch in barley adopt the same (A-type) packing arrangement as those in normal barley starches, possibly due to similarities in their amylopectin chain distributions. The 292 Barley starch exhibits the most different ^{13}C NMR spectral features, due to the presence of considerable amount of amylose-lipid inclusion complexes in 292B starch, as evidenced by the presence of

TABLE 3.3 The Relative Amounts of the Amorphous, V-Type Single-Helix, and Double-Helix Conformations (^{13}C CP/MAS NMR) and the Crystallinity Characteristics Along with Their Relative Amount (X-Ray Diffraction) for the Barley Starches

Starch	Amylose Content (%)	Relative Proportion[a] (%)			Relative Crystallinity[b] (%)	XRD Pattern
		V-Type Polymorph	Double Helix	Amorphous		
Waxiro	3.4	1	39	60	22	A
Himalaya	11.8	4	32	63	16	A
Tantangera	15.2	6	27	67	17	A
HAG	48.4	5	23	71	13	A
292 Barley	76.5	n.d.	n.d.	n.d.	14	B and V

[a] *The maximum standard deviation for the ^{13}C NMR analysis calculation was 1.3%.*
[b] *The maximum standard error for relative crystallinity calculation was 2.1%.*
n.d., not determined

a substantially higher resonance of the mid-chain methylene carbons of fatty acids in V_6 amylose complexes (Tan, 2005). Nevertheless, it is still unclear in this study as to whether the V-type polymorph structure present in the 292B granules was merely due to the amylose-lipid inclusion complexes or the adopted single-helical packing by the amylose molecules.

Characterization of the granule size distributions for barley starches revealed that barley starches have a generally higher proportion of the small (B- and C-) granules than maize starches (Tan, 2005). Deconvolution analyses showed that the high-amylose starches (HAG and 292B) have smaller A-type granules mean size than those of Waxiro and regular starches (cf. Figure 3.5) (Tan, 2005). The B-type granules of these high-amylose barley starches have larger mean diameter (3.5-4.0 μm) than those of waxy and regular barley starch varieties (2.3-2.6 μm). The relative ratios of A:B(+C) granules for HAG and 292B are much lower (3.5 and 1.7, respectively) than the other barley starches (7.2-8.7). This implies the occurrence of a higher proportion of small granule population in these starches compared to the regular barley starch varieties.

4.3 Granular Properties of Wheat Starch Varieties

Both the ^{13}C CP/MAS NMR and XRD studies showed that all wheat starches, excluding Yamamori and A63, adopt the A-type polymorphic structure (Table 3.4). The relative amounts of molecular orders (both double helices and crystallinity) are comparable for most starches, except

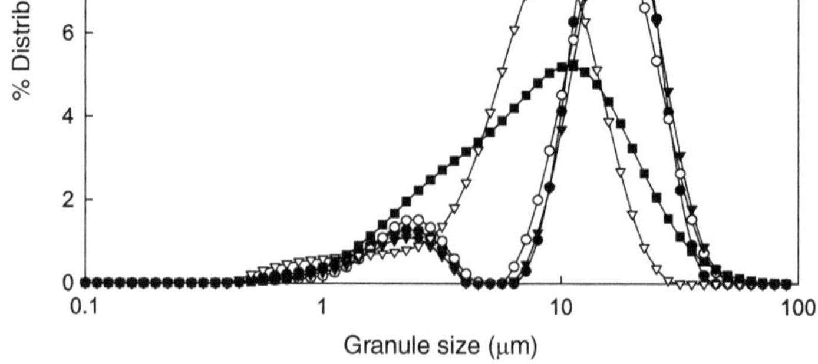

FIGURE 3.5 The granule size distribution profiles for barley starch varieties.

TABLE 3.4 The Estimated Relative Proportions of the Amorphous, Single-Helical, and Double-Helical Conformations for Wheat Starch Varieties and Their Crystallinity Properties

Starch	Amylose Content (%)	Relative Proportion[a] (%)			Degree of Crystallinity[b] (%)	XRD Pattern
		V-Type Polymorph	Double Helix	Amorphous		
Waxy	3.7	0	45	55	24	A
Cadoux	20.4	4	32	64	18	A
Sunco	22.2	4	33	63	18	A
Rosella	24.4	3	34	63	19	A
Janz	24.8	3	32	65	17	A
Yamamori	32.4	18	9	73	11	B and V
B68 (SSII-ABD)	23.9	0	37	63	17	A
C20 (SSII-aBD)	23.8	1	36	63	16	A
C3 (SSII-AbD)	21.5	7	31	62	18	A
A59 (SSII-ABd)	22.9	4	34	62	17	A
B10 (SSII-abD)	22.2	4	34	62	17	A
B21 (SSII-aBd)	24.7	1	38	61	18	A
B40 (SSII-Abd)	24.1	2	34	64	16	A
A63 (SSII-abd)	31.3	28	14	58	14	B and V

[a] *The maximum standard deviation for the* ^{13}C *NMR analysis calculation was 2.2%.*
[b] *The maximum standard error for relative crystallinity calculation was 2.6%.*

for the waxy, Yamamori, and A63 starches. Waxy wheat starch has the highest proportion of double-helical conformations, while Yamamori has the lowest amount of double helices. The ^{13}C NMR spectrum and X-ray diffraction pattern for the A63 triple-null wheat starch display similar characteristics to those of Yamamori starches (Tan, 2005). Both Yamamori and A63 starches have higher proportions of the amorphous and V-type component compared to the others. The presence of single or double null alleles of the *SSII* genes does not induce any significant variation in the packing arrangements of the double helices.

The ratios between the large A-type granules and small (B- and C-type) granules for wheat starches range between 2.7 and 6.2, suggesting that wheat starches have a generally higher proportion of the small B- and C-type granules as compared to barley and maize starches (Tan, 2005). Similar to Yamamori starch, the SSII triple-null starch (A63) shows a less prominent separation of the granule populations and has smaller A-type granules when compared to the other starches (Tan, 2005). A general reduction in the ratios of A to (B+C) granules is observed for most of the SSII null starches except for the SSII single-A-null starch type, C20. The A:(B+C) ratio decreases with an extent that is proportional to the number of null alleles, and thus, the most significant reduction is observed for the SSII triple-null (A63)-type starch (Tan, 2005).

4.4 Granular Properties of Commercial Starches

The extent of granular orders and structural characteristics for the commercial maize and rice starch varieties are summarized in Table 3.5. The amount of molecular orders (both short range and long range) for the commercial maize starches decreases with increasing amylose content, although both the HAM starches have comparable amount of molecular orders despite their differences in amylose content. Both of the commercial rice starches have comparable amount of molecular orders and

TABLE 3.5 The Relative Proportion of the Amorphous, Single-Helical, and Double-Helical Conformations and the Degree of Crystallinity Values for the Commercial Maize and Rice Starch Varieties

Starch	Amylose Content (%)	Relative Proportion[a] (%)			Degree of Crystallinity[b] (%)	XRD Pattern
		V-Type Polymorph	Double Helix	Amorphous		
Waxy maize	3.4	2	47	51	29	A
Regular maize	24.4	4	35	62	21	A
Gelose 50	56.3	7	20	73	13	B
Gelose 80	82.9	14	20	64	15	B
Japonica rice	11.6	4	37	59	21	A
Indica rice	16.5	3	36	61	25	A

[a] *The maximum standard deviation for the ^{13}C NMR analysis calculation was 2.4%.*
[b] *The maximum standard error for relative crystallinity calculation was 3.1%.*

packing arrangements. Therefore, the noted variation in their macromolecular properties in these commercial rice starches as detailed earlier is not significant enough to induce any substantial changes in their macromolecular organization within the granules.

Examination of the granule size distribution profiles showed that Gelose 50 and Gelose 80 starch granules have broader size distribution profiles than the waxy and regular starches (Tan, 2005). The commercial maize starches have larger granules (mean size diameters ranging from 12.9 to 184 μm) compared to the commercial rice starches (mean size diameters of 5.0 μm). The maize starches have much higher ratios of large to small granules (12.1-15.5), while the rice starches have relatively much higher proportions of the small granules population, with large to small granules ratio of ca. 6 (Tan, 2005).

5 THERMAL PROPERTIES

Most culinary and industrial starch processing typically involves the disruption of the structural orders within the starch granules, commonly known as gelatinization. Nevertheless, differences in the gelatinization characteristics such as the extent of granular swelling, the amount of heat or energy uptake (enthalpy), the transition temperatures of the gelatinization process, and the extent of viscosity increase are discernible between different types of starches (Bocharnikova et al., 2003; Fredriksson et al., 1998; Jane et al., 1999; Matveev et al., 2001; McPherson & Jane, 1999; Tester & Morrison, 1990; Yuryev et al., 1998; Zeng, Morris, Batey, & Wrigley, 1997). These gelatinization characteristics influence starch processing and hence determine the suitability of starch for particular end uses. This section discusses the variable thermal properties of novel starches, as obtained from a number of characterization techniques, namely, swelling power, MTDSC, RVA, and Rheoscope (Tan, 2005; Tan, Torley, & Halley, 2008; Tan, Wee, Sopade, & Halley, 2004). The contributions from different structural attributes (individually and/or collectively) are discussed to clarify the basis for differences in their thermal properties.

5.1 Thermal Properties of Maize Starch Varieties

The MTDSC thermograms for the gelatinization of maize starch varieties are presented in Figure 3.6. The heat capacity change accompanying gelatinization is discerned in the reversible heat-flow heat signal for all maize starches (although less distinct for HAM starch), and this step transition occurs within the temperatures of irreversible gelatinization

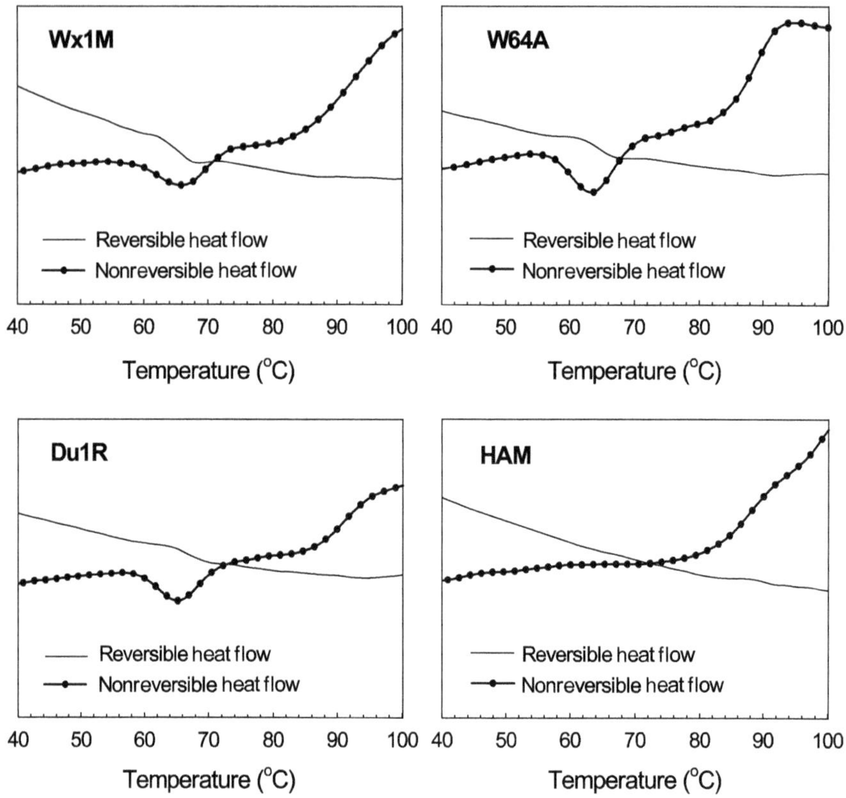

FIGURE 3.6 The reversible and nonreversible thermal events as measured by MTDSC during gelatinization of maize starch varieties (exotherm upward).

transition. The observation that the glass transition temperature overlaps with the gelatinization peak temperature (evident in all starches) suggests that this heat capacity change is possibly due to the change in state of starch macromolecules from being highly constrained within the granular packing to entangled macromolecules as the structural order collapse during the gelatinization process (Tan et al., 2004).

The gelatinization enthalpy value for Wx1M starch (which contains the highest double-helix content) is only slightly higher than those for W64A and Du1R starch (cf. Figure 3.7). The amylose-rich starch (HAM) exhibits the highest gelatinization enthalpy and the lowest heat capacity change, when compared to the other maize starches. In fact, the HAM starch demonstrates the most different thermal properties among all of the maize starches investigated. The presence of high amylose content seems to result in higher gelatinization temperatures, broader gelatinization endotherm, and reduction in the swelling ability of the starch granules.

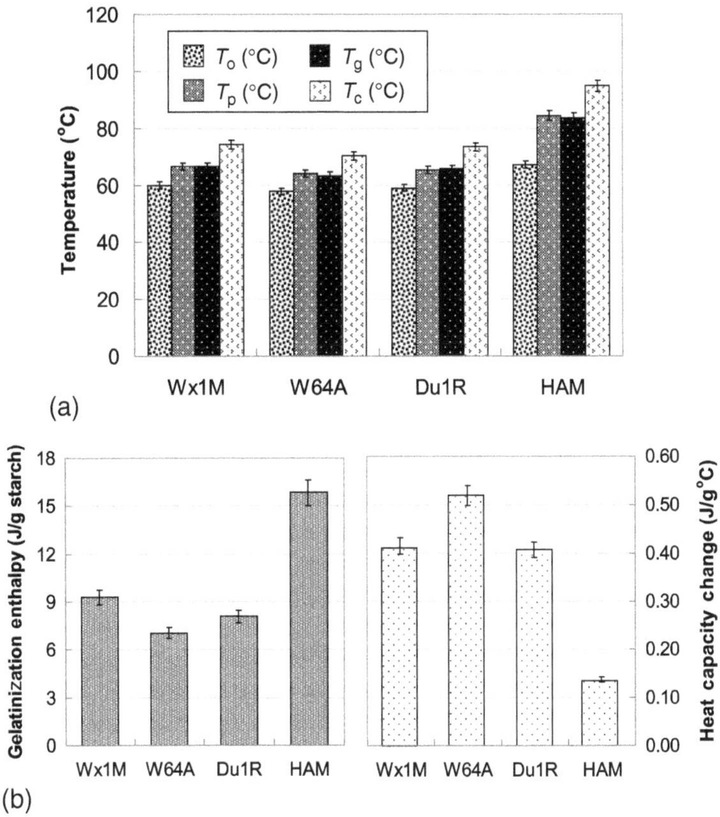

FIGURE 3.7 (a) The gelatinization temperatures (onset, T_o; peak, T_p; and end, T_c) and the glass transition temperature (T_g) for maize starch varieties. (b) The gelatinization enthalpy (ΔH) and heat capacity change (ΔC_p) for the maize starches.

The variation in gelatinization enthalpy and heat capacity change cannot be directly correlated with their amylose content and double-helix content (Tan, 2005). The anomalous amylopectin structure of HAM could also attribute to its unusual granular and thus functional properties.

Examination of the RVA pasting curves for the maize starch varieties reveals some differences in their rheological responses during gelatinization (Tan, 2005). The highest viscosity development is observed during the gelatinization of W64A starch, while the gelatinization of HAM starch does not show any significant viscosity development; refer to peak viscosity values in Table 3.6. No significant correlation could be deduced between the RVA parameters and amylose content for these maize starches. However, it seems that the swelling power is related to the starch amylose content, where the swelling power increases with a decrease in amylose content.

TABLE 3.6 The Swelling Power and RVA Pasting Parameters for Maize Starch Varieties

			RVA Pasting Parameters[b]					
Starch	**Amylose Content (%)**	**Swelling Power[a]**	**Peak Viscosity (cP)**	**Holding Strength (cP)**	**Breakdown (cP)**	**Setback (cP)**	**Final Viscosity (cP)**	**Pasting Temp (°C)**
Wx1M	0.4	25.8	2861	914	1947	648	1562	68.0
W64A	13.6	15.6	3779	1595	2184	2385	3980	69.5
Du1R	25.4	12.0	2418	982	1436	1064	2046	72.15
HAM	75.3	3.5	24	20	4	5	25	94.5

[a] *The maximum standard deviation for swelling power measurements was 1.8%.*
[b] *The maximum standard deviation for the analysis of RVA pasting parameters was 6.4%.*

The viscosity profiles and granules images for the gelatinization of maize starch varieties in 0.1% guar as measured by the Rheoscope are presented in Figure 3.8. Differences in the pasting profiles between RVA and Rheoscope can be attributed to differences in starch concentration used, geometries of the systems, and thus shear rate experienced during the measurements (Tan et al., 2008). Wx1M starch exhibits the largest and most rapid viscosity increase when compared to other maize starches. By contrast, the HAM starch does not show any significant viscosity development during the gelatinization process.

Figure 3.9 shows the average granule size (width and length) at different times of the gelatinization process, and it is evident that the viscosity increase for all maize starches generally follows the same rate as the increase in starch granule sizes. An approximately two- to threefold increase in the average granule size is observed for W64A and Du1R starches. W64A starch exhibits the highest viscosity increase when compared to Du1R and HAM starch, which can be attributed to its larger granules size and consequently the resulting swollen granules. The lack of viscosity increase during the gelatinization of HAM starch can be attributed to the lack of granular swelling during the temperature ramp.

5.2 Thermal Properties of Barley Starch Varieties

Figure 3.10 shows the gelatinization characteristics of the barley starch varieties. Unlike in maize, the gelatinization of high-amylose barley (292B) occurs at lower temperature when compared to the other barley starches. The gelatinization of HAG starch occurs at temperature similar to the regular (Himalaya and Tantangera) and waxy (Waxiro) barley starches. Hence, this suggests that the presence of high amylose does

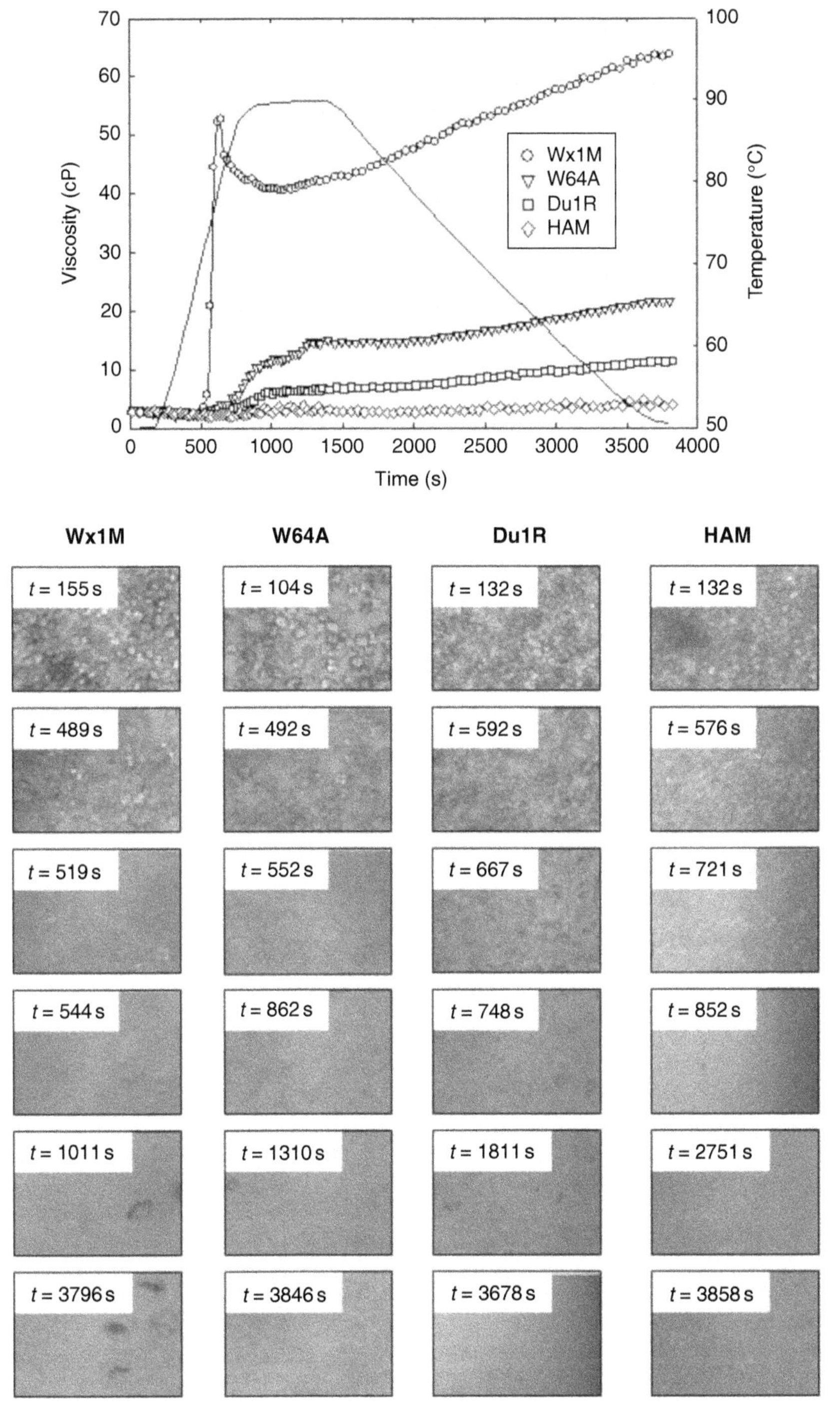

FIGURE 3.8 The viscosity profiles and granules images for the gelatinization of maize starches in 0.1% guar solution as measured by Rheoscope. Temperature profile (—).

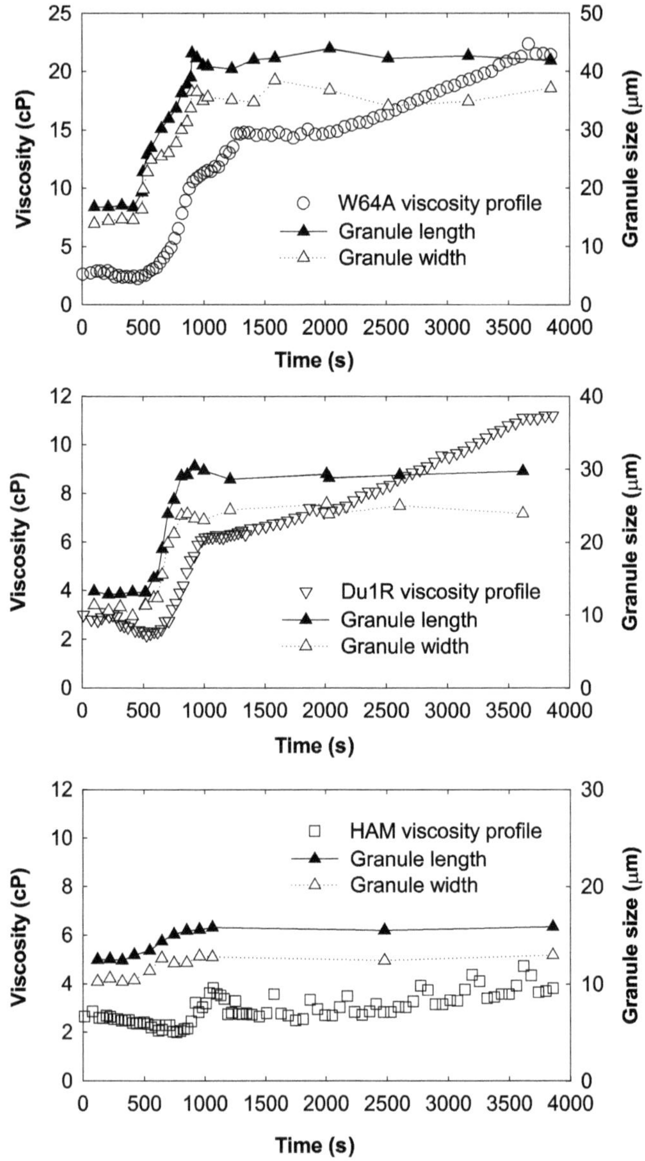

FIGURE 3.9 The average size of starch granules at different times of the gelatinization process for maize starch varieties. The average standard error was 10%.

not necessarily result in higher gelatinization temperature as observed in maize starches. Other factors such as shorter amylopectin chain lengths (as observed in 292B starch) could also attribute to differences in the gelatinization temperatures.

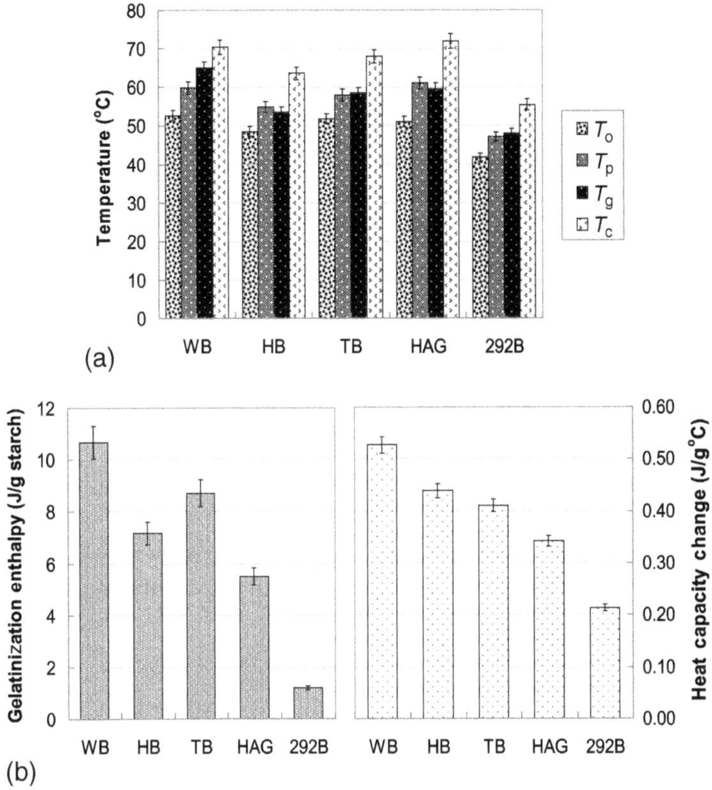

FIGURE 3.10 (a) The gelatinization (onset, peak, and end) temperatures and glass transition temperature for barley starches; (b) the enthalpy of gelatinization and the heat capacity change for barley starches.

Differences in the energy uptake during starch gelatinization (ΔH) and the heat capacity change (ΔC_p) values are also evident (Figure 3.10). Waxiro (WB), which has the highest double-helix content and crystallinity, exhibits the highest gelatinization enthalpy and heat capacity change. The gelatinization of high-amylose 292B starch requires the lowest energy and exhibits the lowest heat capacity change, possibly due to the small amount of molecular order in 292B (Section 4.2). Hence, for barley starches, the gelatinization enthalpy and heat capacity change seem to decrease with increasing amylose content and decreasing double-helix content. Consequently, it is evident that the variation of the gelatinization thermal properties with starch structural parameters between different cereal types. Such consequently requires a careful examination of the structural parameters variation within a particular cereal type in order to deduce a structure-functionality relationship that is applicable for starches from different origins. This will be explored further in

TABLE 3.7 The Swelling Power and the RVA Pasting Parameters for Barley Starch Varieties

			RVA Pasting Parameters[b]					
Starch	Amylose Content (%)	Swelling Power[a]	Peak Viscosity (cP)	Holding Strength (cP)	Breakdown (cP)	Setback (cP)	Final Viscosity (cP)	Pasting Temp (°C)
WB	3.4	26.7	3953	1057	2896	413	1470	62.5
HB	11.8	14.2	4471	3093	1378	2317	5409	76.9
TB	15.2	11.7	4350	2846	1504	2797	5643	73.4
HAG	48.4	5.5	299	304	0	413	716	91.5
292B	76.5	4.6	n.d.	n.d.	n.d.	n.d.	n.d.	n.d.

[a] *The maximum standard deviation for swelling power measurements was 2.6%.*
[b] *The maximum standard deviation for the analysis of RVA pasting parameters was 6.6%.*
n.d., the RVA measurement was not performed for 292B due to the limited availability of this starch.

Section 6, which aims to develop a hypothesis on structure-functionality relationship for starches in general by considering the significance of each starch structural parameters in influencing their functional properties.

The regular starches (Himalaya and Tantangera) demonstrate the highest RVA pasting viscosities, while no significant viscosity increase is noted for the HAG during the temperature ramp; refer to Table 3.7. Waxiro exhibits the most rapid increase in viscosity, which seems to be the typical characteristic of rheological responses of waxy starches during gelatinization. The pasting temperature for HAG starch is considerably higher than the other barley starches, and no breakdown is observed for HAG starch during the holding period (at 95 °C).

Rheoscope studies showed that Waxiro exhibited a much higher viscosity development compared to the other barley starches during the temperature ramp (Tan, 2005). An approximately twofold increase in the average granule size is noted for Waxiro and Himalaya starch granules as a result of swelling (Figure 3.11). The average size of HAG starch granules only increases from ca. 10 μm to 16 μm, indicating minimal swelling during the temperature ramp. Figure 3.11 shows the rate at which the viscosity does not follow the same rate as the granular swelling. The difference in the apparent rate of granular swelling and the rate of viscosity development for barley starches can be attributed to the higher proportions of smaller granule population (B-type granules). Therefore, although the starch granules in Waxiro and Himalaya swell as water penetrates into the granules, the viscosity increase is not observed immediately due to the packing of the smaller granules between the interstices of larger granules reducing the overall effect of granular swelling on the viscosity development.

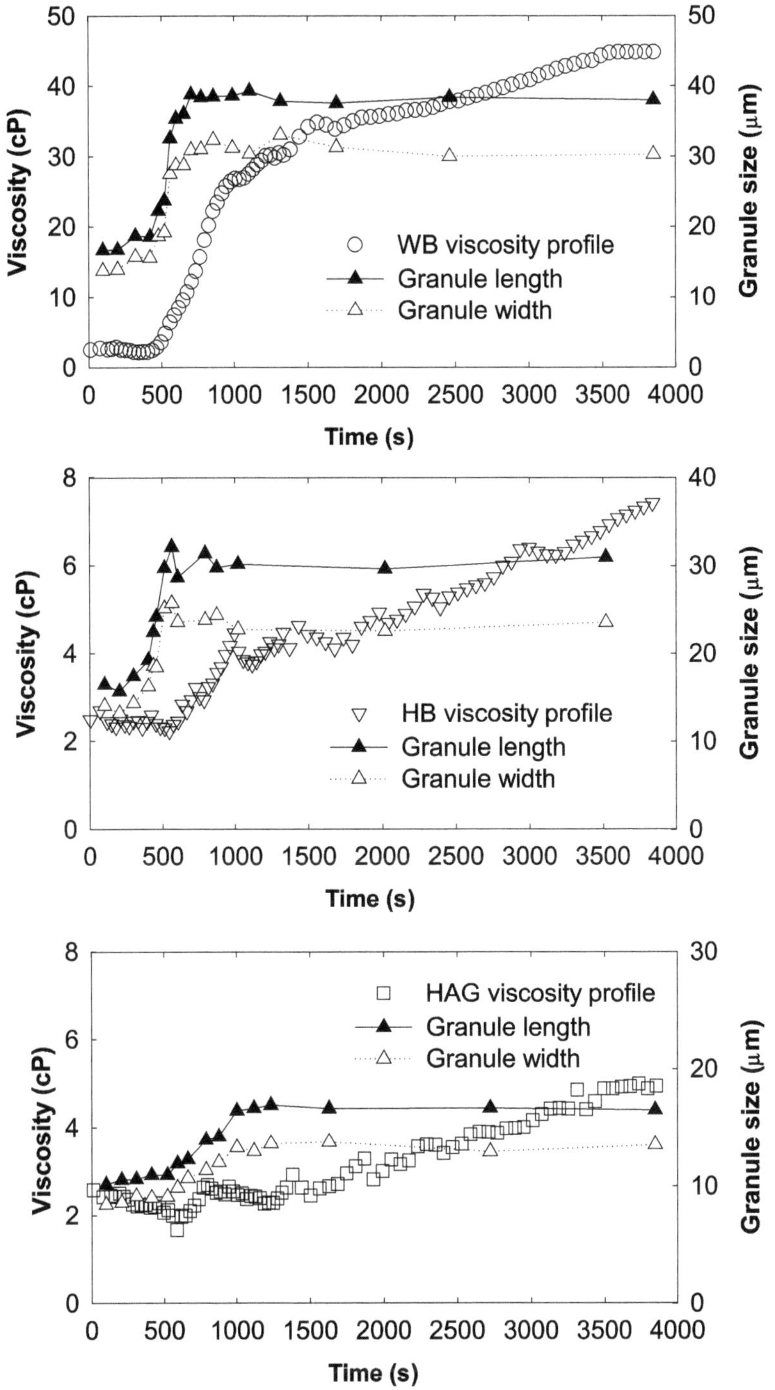

FIGURE 3.11 The average granule size for Waxiro (WB), Himalaya (HB), and HA-glacier (HAG) starches at different times of gelatinization as measured by Rheoscope. The average standard error was 10%.

5.3 Thermal Properties of Wheat Starch Varieties

MTDSC characterization study showed that the wheat starch varieties do not show any significant differences in their gelatinization characteristics, except for the Yamamori starch (Figure 3.12). The gelatinization of Yamamori starch occurs at a lower temperature and has lower enthalpy values than the other wheat starches.

The presence of SSII single or double null alleles in these wheat starches does not lead to significant variation in their gelatinization transition temperatures and enthalpy values (Tan, 2005). The gelatinization of wheat starches containing single and double null alleles occurs at slightly lower temperatures than the normal starch (B68). Significant alteration in the gelatinization properties is only observed in wheat starch containing SSII

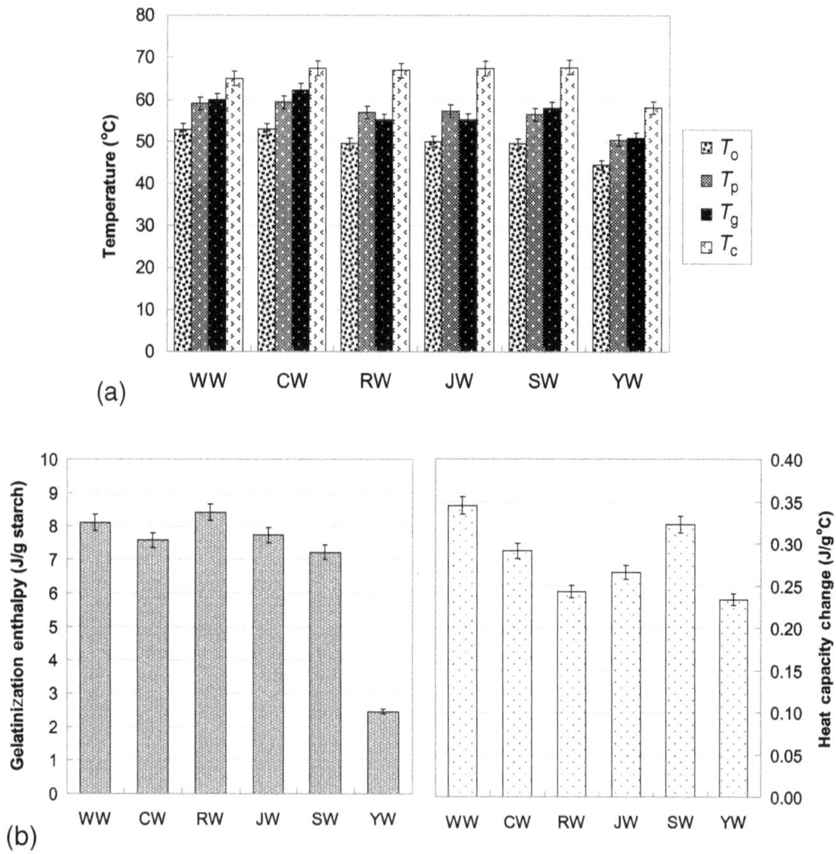

FIGURE 3.12 (a) The gelatinization (onset, peak, and end) temperatures and glass transition temperature for wheat starch varieties; (b) the gelatinization enthalpy (ΔH) and heat capacity change (ΔC_p) for wheat starches.

triple null alleles (A63), where its gelatinization occurs at a considerably lower temperature with a substantially lower gelatinization enthalpy and heat capacity change.

During RVA measurements, waxy wheat starch exhibits the highest peak viscosity, while Yamamori displays the lowest peak viscosity during RVA measurements; refer to Table 3.8. Additionally, waxy wheat starch exhibits the lowest holding strength value when compared to all wheat starches (apart from Yamamori starch). The Cadoux and Rosella wheat starches exhibit higher pasting viscosities than the regular wheat starch (Sunco). A general decrease in the RVA peak viscosity value is observed along with increasing amylose content. Similarly, the swelling power decreases with increasing amylose content. The pasting of waxy wheat starch proceeds at a lower temperature (ca. 62 °C) than the other wheat starches, while Yamamori starch only begins to paste at 91 °C. Consequently, an increase in starch amylose content apparently leads to an increase in pasting temperature.

For wheat starch varieties carrying different combinations of SSII null alleles, the starches with single and double null alleles have lower swelling power and breakdown values than the normal type (Tan, 2005). The most substantial reduction in the swelling power, however, is only observed in (triple null alleles) A63 starch. All the double-null-type starches have higher pasting temperature than the normal- and single-null-type starches.

The average granule size for the wheat starches at different times of gelatinization process obtained from the Rheoscope study is presented in Figure 3.13. An approximately two- to threefold increase in the average granule size is noted for all wheat starches except for Yamamori starch. Despite similarities in their granules size and swelling, differences in the extent of viscosity development are evident between waxy (WW) and Cadoux (CW) starch. This is possibly attributable to the presence of higher proportion of the smaller granule population (B-type granules) in Cadoux starch than in waxy wheat starch; refer to Section 4.3. Similarly, the rate at which the viscosity increases does not follow the same rate as the granular swelling during the gelatinization of Sunco wheat (SW) starch. Yamamori starch exhibits a slight increase in the average granule size, which is not accompanied by viscosity increase during the temperature ramp.

These studies show that the variation in the functional properties of wheat starch varieties is not as significant as those observed in maize and barley starches. Some slight variations in their gelatinization thermal characteristics, e.g., lower gelatinization temperature and enthalpy values for high-amylose starches (Yamamori/A63), are noted. An increase in the starch amylose content results in decreasing swelling power, and this correlation between the amylose content and swelling is consistently observed for starches from different plant origins.

TABLE 3.8 The Swelling Power and RVA Pasting Parameters for Wheat Starch Varieties

			RVA Pasting Parameters[b]					
Starch	**Amylose Content (%)**	**Swelling Power[a]**	**Peak Viscosity (cP)**	**Holding Strength (cP)**	**Breakdown (cP)**	**Setback (cP)**	**Final Viscosity (cP)**	**Pasting Temp (°C)**
Waxy	3.7	25.6	4771	1473	3298	489	1962	61.8
Cadoux	20.4	14.1	4008	2492	1516	2293	4785	65.5
Rosella	22.2	14.9	3749	2033	1716	1890	3923	64.5
Janz	24.4	11.2	2498	1688	810	1591	3279	65.0
Sunco	24.8	10.5	2123	1816	307	1644	3460	85.0
Yamamori	32.4	5.3	212	196	16	436	632	90.9

[a] *The maximum standard deviation for swelling power measurements was 2.5%.*
[b] *The maximum standard deviation for the analysis of RVA pasting parameters was 6.6%.*

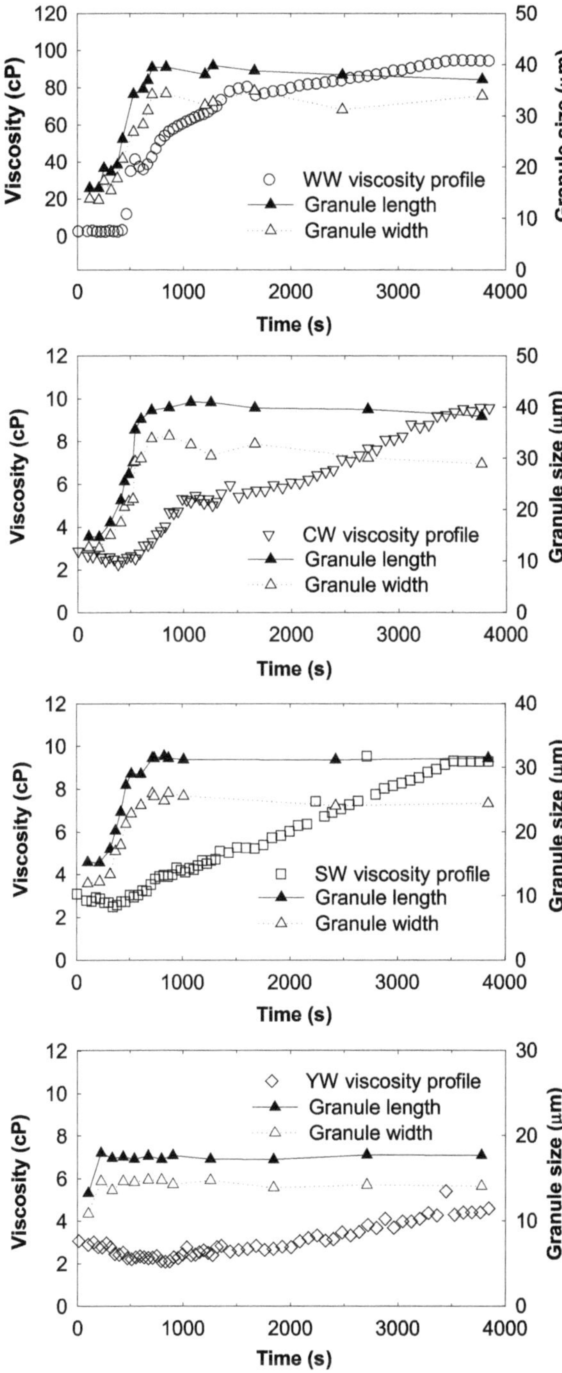

FIGURE 3.13 The approximate average size of waxy (WW), Cadoux (CW), Sunco (SW), and Yamamori (YW) starch granules at different times of gelatinization. The average standard error was 10%.

5.4 Thermal Properties of Commercial Starches

The gelatinization of regular maize starch (24% amylose) proceeds at a lower temperature than the waxy maize starch (3% amylose); refer to Table 3.9. The high-amylose starches, Gelose 50 and Gelose 80, have higher gelatinization temperature. The gelatinization enthalpy values for Gelose 50 and Gelose 80 are higher than those of waxy and regular maize starches, which can be attributed to the broad characteristics of their endotherm (Tan, 2005). The gelatinization of Indica rice proceeds at a higher temperature than the Japonica rice. On the other hand, the gelatinization enthalpy and heat capacity changes are similar for both the Indica and Japonica rice.

Reductions in RVA peak viscosity and breakdown values are observed with increasing amylose content for these starches (Table 3.10). Similarly, a decrease in the swelling power is also observed with an increase in starch amylose content. The commercial rice starches, on the other hand, do not show substantial variation in their RVA pasting properties (Tan, 2005). The Japonica rice starch begins to paste earlier than the Indica rice starch, and a higher pasting temperature is noted for Indica rice. This is consistent with the higher gelatinization temperatures observed for Indica rice as obtained from the MTDSC studies. Nevertheless, the Indica rice starch shows a more rapid increase in the viscosity when compared to Japonica rice starch.

TABLE 3.9 The Gelatinization Thermal Characteristics for Commercial Maize and Rice Starches

Starch	Amylose Content (%)	Gelatinization (Nonreversible Heat Flow)				Step Transition (Reversible Heat Flow)	
		T_o (°C)[a]	T_p (°C)[a]	T_c (°C)[a]	ΔH (J/g)[b]	T_g (°C)[c]	ΔC_p (J/g°C)[d]
Waxy maize	3.4	59.9	66.9	75.8	12.53	65.6	0.62
Regular maize	24.4	57.0	64.3	72.4	10.23	63.7	0.38
Gelose 50	56.3	64.9	81.7	96.1	22.50	69.6	0.30
Gelose 80	82.9	65.4	81.6	89.4	14.22	70.5	0.24
Japonica rice	5.6	51.4	58.8	69.1	10.58	58.9	0.43
Indica rice	10	59.1	67.0	75.2	10.61	66.1	0.44

[a] *The average standard error for the analyses of gelatinization temperatures (T_o, T_p, T_c) was 1.2%.*
[b] *The average standard deviation for analysis of gelatinization enthalpy (ΔH) was 1.6%.*
[c] *The average standard deviation for glass transition temperature (T_g) analysis was 1.1%.*
[d] *The average standard deviation for the analysis of ΔC_p was 3.2%.*

TABLE 3.10 The RVA Pasting Parameters and Swelling Power for Commercial Maize and Rice Starches

		RVA Pasting Parameters[b]					
Starch	**Swelling Power[a]**	**Peak Viscosity (cP)**	**Holding Strength (cP)**	**Breakdown (cP)**	**Setback (cP)**	**Final Viscosity (cP)**	**Pasting Temp (°C)**
Waxy maize	25.8	3919	1182	2737	522	1704	72.0
Regular maize	12.3	2828	1508	1320	1442	2950	75.9
Gelose 50	4.3	406	275	131	13	288	95.0
Gelose 80	3.5	39	33	6	7	40	95.0
Japonica rice	20.9	4881	1562	3319	1857	3419	68.9
Indica rice	19.0	4456	1589	2867	2494	4083	74.5

[a] *The average standard deviation for swelling power measurements was 1.7%.*
[b] *The average standard deviation for the analysis of RVA pasting parameters was 7%.*

6 OVERALL STRUCTURE-PROPERTY RELATIONSHIPS

Diversities in the structures and functionalities of the novel starches are evident from characterization studies of a range of novel starches from different species. The variation in their structural aspects and the significance of each structural aspect in dictating the observed functional properties are explored in this section. The alteration in macromolecular properties, namely, the amylose content and amylopectin chain length distribution, is first described in relation to the mutation of genes encoding the different enzymes involved in starch biosynthesis. Correlations between these macromolecular properties are also explored to further the understanding of starch structures and functionalities relationship.

6.1 Starch: From Synthesis to Macromolecules

The biosynthesis of starch in plants involves four main enzymatic steps, namely, the conversion of glucose-1-phosphate to ADP-glucose by ADP-glucose pyrophosphorylase, the elongation of α-1,4-glucan linear chains by the action of different isoforms of starch synthase (SS), the introduction of α-(1,6) branch points by branching enzymes (BE), and finally the cleavage of α-(1,6) glucosidic bonds by the action of DBEs

(Baga et al., 1999; James, Robertson, & Myers, 1995; Mouille et al., 1996; Rahman et al., 2000; Slattery et al., 2000; Smith, 1999). The novel starches were obtained from starch mutants, which lack the activity of one of these enzymes and lead to alteration in their properties. All of the zero amylose starches or waxy starches were obtained from starch mutants, which lack the GBSSI enzyme activity. The deficiency in GBSSI activity only inhibits the production of amylose, while the amylopectin structure remains similar to the amylopectin of their corresponding normal type. This is consistently observed for maize, barley, and wheat.

The high-amylose starches, on the other hand, were generated from different starch mutants depending on the cereal type. HAM starch was obtained from mutants lacking BEIIb enzymes, while the high-amylose barley (292B) and wheat (Yamamori or A63) starches were obtained from mutants deficient in the SSII enzyme activity. Since different isoforms of SS, BE, and DBEs are involved in the synthesis of amylopectin, the reduction in the amylopectin content (high amylose) can thus be obtained through inhibition of one of these enzymes (Baga et al., 1999; Mouille et al., 1996; Rahman et al., 2000; Slattery et al., 2000; Smith, 1999). It should be noted, however, that since these different enzymes have different roles in the amylopectin synthesis, the high-amylose starches obtained through different enzyme inhibitions might consequently have different amylopectin structures. The deficiency in BEIIb enzyme activity in HAM starch, which has a major role in defining the branching frequency in amylopectin in HAM starch (Rahman et al., 2000), results in the generation of amylopectin with a higher proportion of the intermediate and longer chain lengths (Figure 3.14). On the other hand, since the SSII enzyme is responsible for the elongation of amylopectin polymer in barley or wheat, the deficiency in SSII enzyme activity in Yamamori and 292B yields amylopectin with higher proportion of shorter chain lengths with DP 4-12 (Figure 3.14).

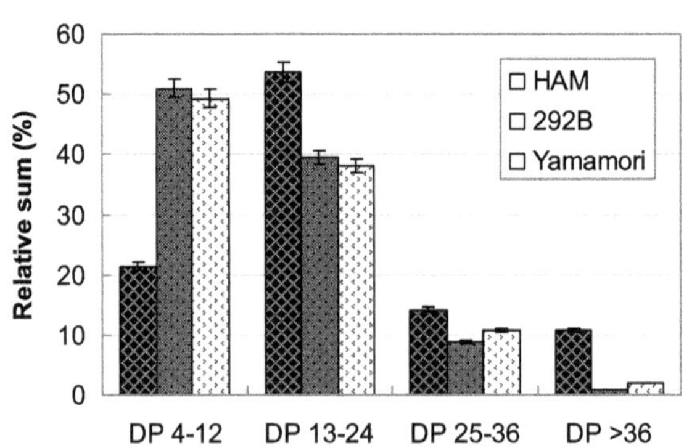

FIGURE 3.14 The amylopectin chain length distributions for HAM, 292B, and Yamamori starches, as fractionated into different DP size ranges: DP 4-12, 13-24, DP 25-36, and DP > 36.

6.2 From Starch Macromolecules to Their Granular Properties

The variations in the relative proportions of amorphous, single- and double-helical conformation in starch granules with amylose content are presented in Figure 3.15. A general increase in the proportion of single V-type conformation is noted with increasing amylose content. Yamamori and A63 starches, which are generated from wheat mutants carrying SSII triple null alleles, have a considerably high proportion of V-type polymorph. Similarly, the barley starch mutants lacking SSII enzymes (292B) have a substantial amount of V-type polymorph. A reduction in the molecular order (double-helix content) with increasing amylose content (or decreasing amylopectin content) is evident for all starch types. The r-squared value (r^2) for the correlation between double-helix content and amylose content ranged between 66% and 87% (Tan, 2005).

Similar trend is also observed in the variation of starch crystallinity with amylose content ($r^2 > 74\%$), where an increase in amylose content results in reduced crystallinity (cf. Figure 3.16). This consequently

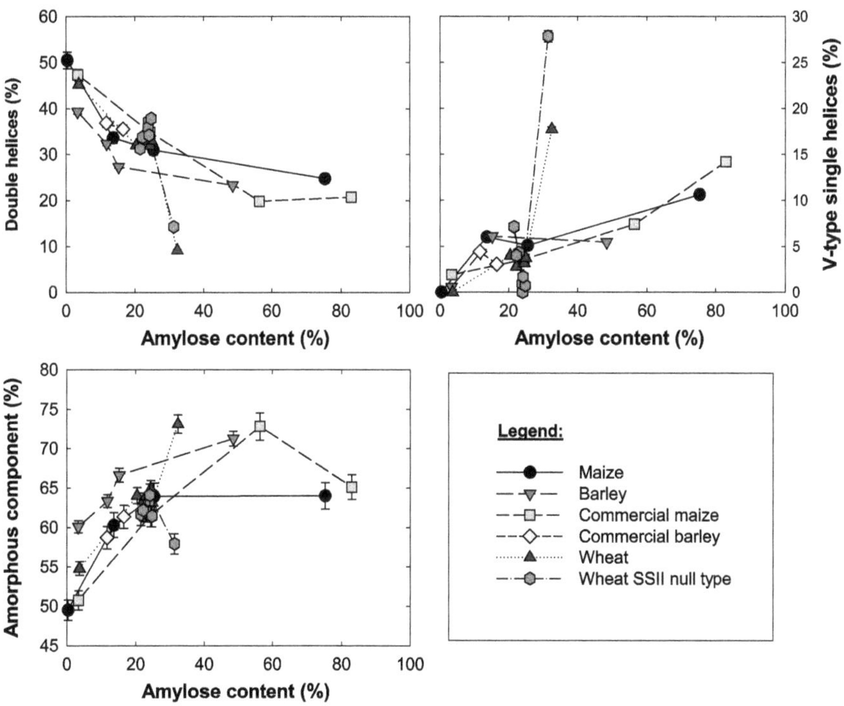

FIGURE 3.15 The variation in the relative proportions of amorphous, single-helical (V-type) and double-helical components in starch granules with amylose content for maize, barley, wheat, and rice starches. (For color version of this figure, the reader is referred to the online version of this chapter.)

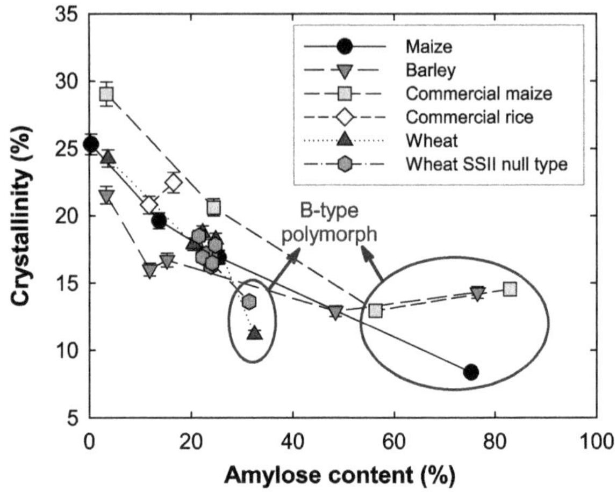

FIGURE 3.16 The variation in the degree of crystallinity with amylose content for maize, barley, wheat, and rice starches. (For color version of this figure, the reader is referred to the online version of this chapter.)

confirms the predominant role of amylopectin in supporting the crystalline framework in starch granules. In addition to decreasing crystallinity with increasing amylose content, a transition in the double-helix packing arrangement from A-type to B-type polymorph is noted at high amylose content for all starches, which can be attributed to the alteration in their amylopectin chain length distribution.

Despite the involvement of amylopectin side chains in forming double helices, which constitute the structural order in starch granules, the variation in starch double-helix content and crystallinity do not exhibit notable trend with variation in the average size of amylopectin chain lengths. However, changes in the helical packing conformation from A-type to B-type polymorph in HAM starches and starches carrying SSII null alleles can be attributed to alterations in their amylopectin chain lengths distribution. The high-amylose novel starches were obtained through different inhibitions of enzymes involved in amylopectin synthesis. Therefore, in addition to a reduction in the formation of amylopectin (thus increased amylose content), the amylopectin structures in these high-amylose starches are also altered (Figure 3.17). The high-amylose starches obtained from mutants lacking BEIIb enzymes (i.e., HAM, G50, and G80) have amylopectin with higher proportions of longer chains with DP $\geq$25. From considerations of entropy, the longer glucan chains in HAM starches experience larger (negative) entropy changes during crystallization and hence tend to favor the formation of polymorph with the highest entropy, i.e., the B-type polymorph (Gidley, 1987).

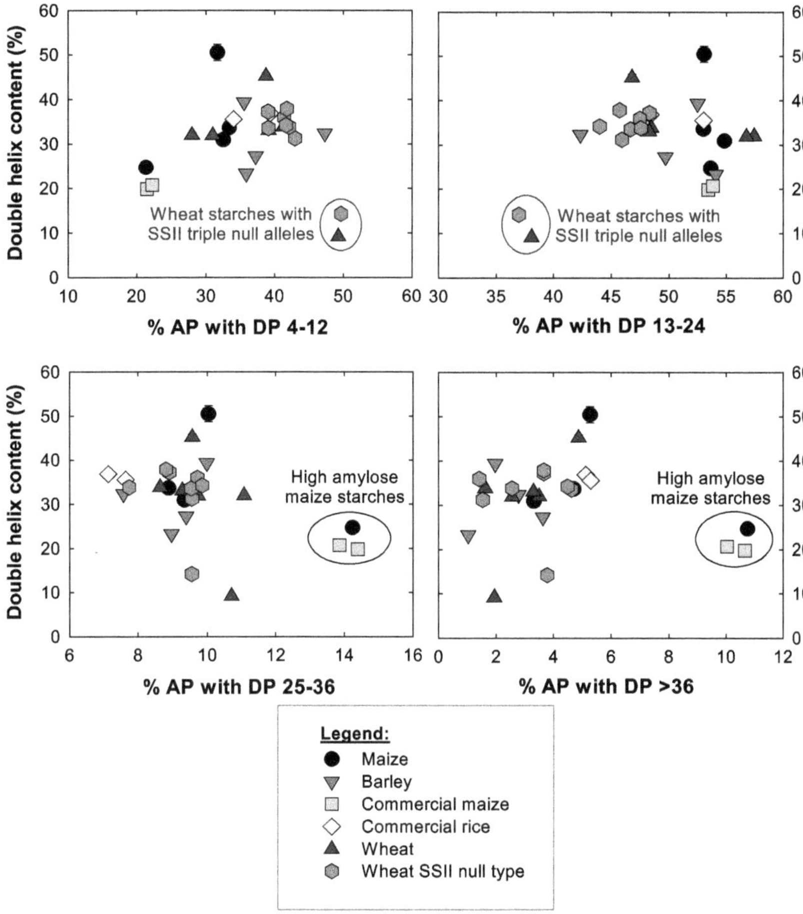

FIGURE 3.17 Variation in starch double-helix content with the relative proportions of amylopectin chains fractions (DP 4-12, DP 13-24, DP 25-36, and DP > 36), for maize, barley, wheat and rice starches. (For color version of this figure, the reader is referred to the online version of this chapter.)

On the other hand, the high-amylose starches obtained from mutants lacking SSII enzymes (i.e., Yamamori, A63, and 292B) have amylopectin with much higher proportion of short chains with DP 4-12 and lower fraction of chains with DP 13-24 (Figure 3.17). The B-type polymorph formation of their amylopectin double helices is possibly due to the lower proportion of amylopectin chains with DP 13-24. From investigations of the crystallization of maltooligosaccharides as models for starch crystallization, the minimum chain length for the formation of A-type polymorph is 10 (DP), and given that other factors are equal, the A-type form is only favored over B-type when higher concentrations of these oligosaccharides are present (Gidley, 1987; Gidley & Bulpin, 1987). For amylopectin on the other hand, studies with molecular models have shown that intertwined chains having DP 14-16 readily form double helices (Umeki & Kainuma, 1981). Hence, the presence of lower concentration of glucan chains with DP 13-24 during amylopectin side-chain crystallization in these starches leads to the tendency of the glucan chains to form the B-type polymorph. Consequently, a simple correlation between the average size of amylopectin chains and starch structural order cannot be deduced. This suggests that the involvement of amylopectin in supporting of starch crystalline framework is much more complicated than double-helix formation between their side chains.

The size distribution of starch granules is usually considered as a separate attribute of starch properties, which is important in determining the suitability of starch for particular end uses especially when the granular nature of starch is significant. An example is such the use of starch granules as fillers in biodegradable plastic films or incorporated into composites (Ellis et al., 1998). A slight reduction in the average size of A-type granules is noted with increasing amylose content for most starches except for the commercial maize starches (Tan, 2005). Nevertheless, the observed reduction in the average size of A-type granules is more likely due to the generation of anomalous amylopectin in these starches, which consequently prevents the normal and efficient packing mechanisms of these macromolecules. The ratio between large and small granules did not show any significant correlation with the amylose content and amylopectin chain distributions (Tan, 2005).

6.3 The Relationships Between Starch Structure and Thermal Properties

6.3.1 MTDSC Gelatinization Thermal Properties

The variation in gelatinization onset temperature with starch amylose content is presented in Figure 3.18a for all starch types. The gelatinization onset temperature increases with amylose content for maize starches. The high amylose content in wheat and barley starches, on the other hand, leads to lower gelatinization onset temperature when compared to their regular and waxy types. When the gelatinization onset temperature is plotted against the amylopectin chain length distribution, the variation in the gelatinization onset temperature is apparently related to the changes in the average size of their amylopectin chain length ($r^2 > 0.80$ except for wheat SSII-null type); refer to Figure 3.18b. For all starch types, an increase in the amylopectin chain lengths leads to higher gelatinization onset temperature, although the trend is less evident in wheat starches

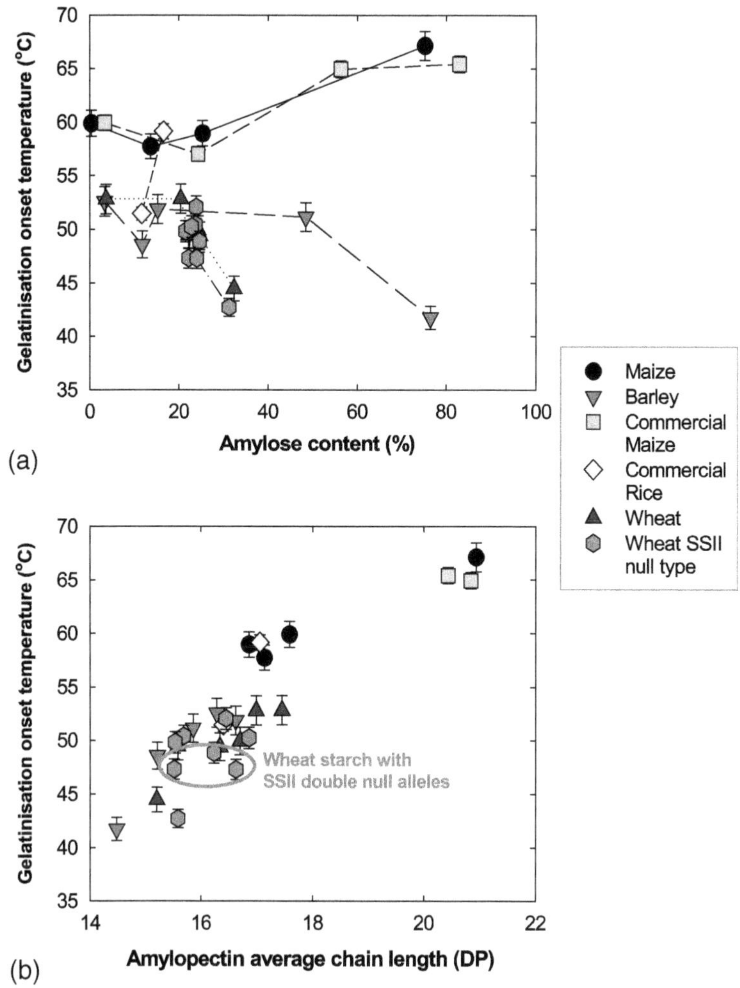

FIGURE 3.18 Variation in the gelatinization onset temperature with starch amylose content (a) and the average size of amylopectin chain length (b) for maize, barley, wheat, and rice starches. (For color version of this figure, the reader is referred to the online version of this chapter.)

carrying different combinations of SSII null alleles (due to the hexaploid nature of wheat starches).

The variation in starch gelatinization onset temperature is further explored in relation to the variation in the relative proportions of amylopectin chains as fractionated into different DP ranges, i.e., DP 4-12, DP 13-24, DP 25-36, and DP > 36. As demonstrated in Figure 3.19, the variation in starch gelatinization onset temperature can be correlated to the relative proportions of amylopectin chains with short (DP 4-12) and

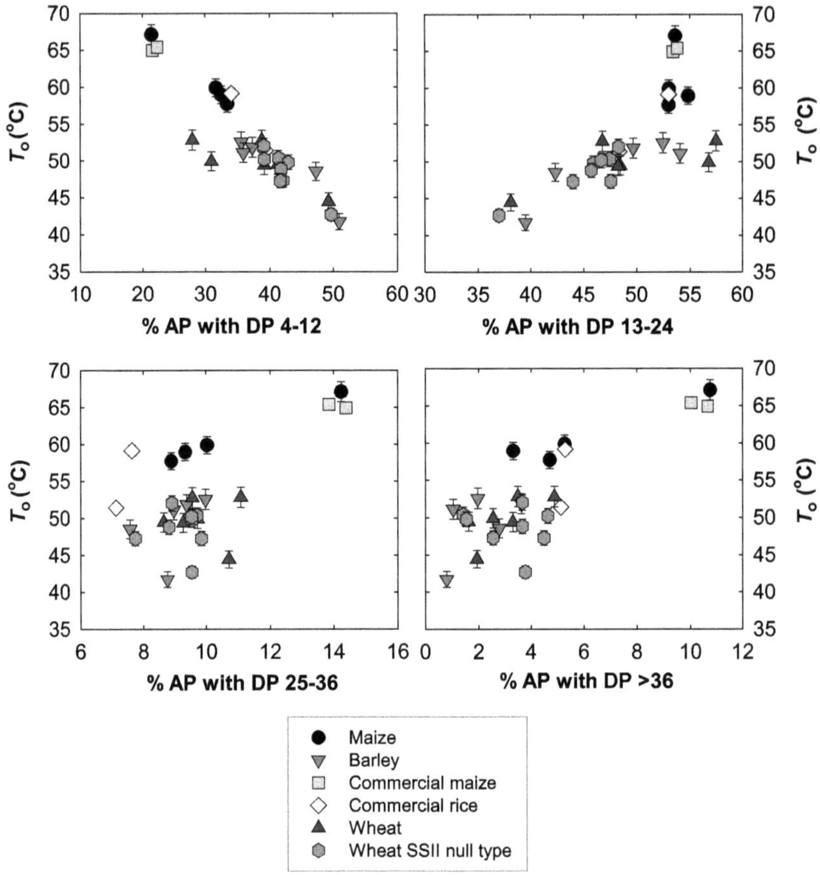

FIGURE 3.19 Variation in the gelatinization onset temperature (T_o) with the relative proportions of amylopectin chain lengths at different DP ranges (DP 4-12, DP 13-24, DP 25-36, and DP > 36) for maize, barley, wheat, and rice starches. (For color version of this figure, the reader is referred to the online version of this chapter.)

intermediate chain lengths (DP 13-24) for all starch types. An increase in the relative proportion of amylopectin short chains with DP 4-12 leads to a reduction in the gelatinization onset temperature ($r^2 > 0.61$). As discussed earlier, since the minimum chain lengths for intertwined chains to form double helices is ca. DP 14-16, these short chains cannot participate in the formation of double helices and may exist as single chains. Consequently, the presence of higher amylopectin short-chain fractions (DP 4-12) may act to destabilize the helix-helix association and hence causes the gelatinization to occur at lower temperature.

On the other hand, an increase in the relative proportion of intermediate amylopectin chains (DP 13-24) leads to an increase in the gelatinization onset temperature ($r^2 > 0.52$); refer to Figure 3.19. The observed trend between gelatinization onset temperatures with amylopectin intermediate-chain fractions may suggest the involvement of these chain fractions in the formation of a more stable helical structure and thus a higher thermal energy input is required to initiate the amylopectin helix-helix dissociation, which leads to a higher gelatinization onset temperature.

Figure 3.20 plots the variation in gelatinization enthalpy versus the amount of structural order (double-helix content) inside starch granules. Since gelatinization involves the disruption of structural order inside the granules, the amount of energy uptake during such process is therefore anticipated to correspond to the extent of order inside the granules. This is demonstrated in Figure 3.20, where an increase in gelatinization

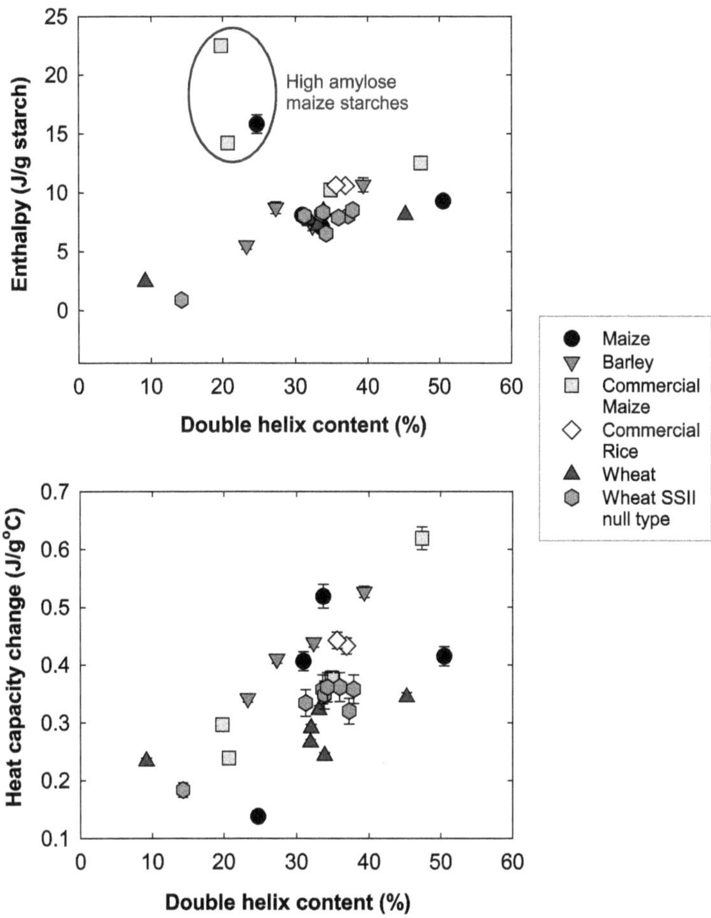

FIGURE 3.20 Variation in starch gelatinization enthalpy and heat capacity change with double-helix content for maize, barley, wheat, and rice starches. (For color version of this figure, the reader is referred to the online version of this chapter.)

enthalpy is noted with increasing double-helix content for all starches, except for the HAM starches ($r^2 > 0.68$ for barley and wheat starches). An increase in the values for heat capacity change accompanying starch gelatinization is also noted with increasing double-helix content in starch granules although the trend is not as obvious as that observed in gelatinization enthalpy (Figure 3.20). The lower heat capacity change values for starches with lower double-helix content are possibly attributable to the less significant state transition of starch macromolecules during the order-disorder transition in these starches.

6.3.2 Swelling Power

The variations in swelling power with variation in starch amylose content and the average size of amylopectin chain length are shown in Figure 3.21. The swelling power shows a notable decrease with increasing amylose content for all starches (the r^2values range from 0.67 to 0.98 for different starch types). On the other hand, the variation in starch swelling power is independent of the amylopectin average chain length. All waxy (zero amylose) starches exhibit comparable swelling power and have the highest swelling power. The high-amylose starches on the other hand show limited swelling. This may indicate the role of amylopectin in determining the swelling ability of starch granules as they are heated.

6.3.3 Starch Pasting Properties

Starch pasting properties are typically measured with RVA, which yield empirical results to describe starch rheological responses during heating and cooling down. There was no significant correlation observed between starch structures and most of the RVA parameters (Tan, 2005). A decrease in RVA peak viscosity and breakdown values is noted with increasing amylose content ($r^2 > 0.73$); refer to Figure 3.22.

Most waxy starches exhibit a substantial viscosity development, as indicated by the high peak viscosity values, while the high-amylose starches exhibit minimal viscosity development during the temperature ramp. Since the viscosity development during initial RVA temperature ramp can be attributed to the granular swelling, the decrease in peak viscosity with amylose content is thus consistent with the observed trend between amylose content and swelling power. Similarly, a decrease in RVA viscosity breakdown values is noted with increasing amylose content. This could be related to the variation in peak viscosity, where higher peak viscosity values imply generally larger swollen granules in these starches at peak viscosity. Accordingly, as these swollen granules are subjected to high temperature and shear (due to rigorous RVA impeller agitation), the larger swollen granules have higher tendency to collapse than the smaller swollen granules as they are subjected to continuous high temperature and shear. The collapses of these swollen granules consequently result in the breakdown, which is usually observed during the holding

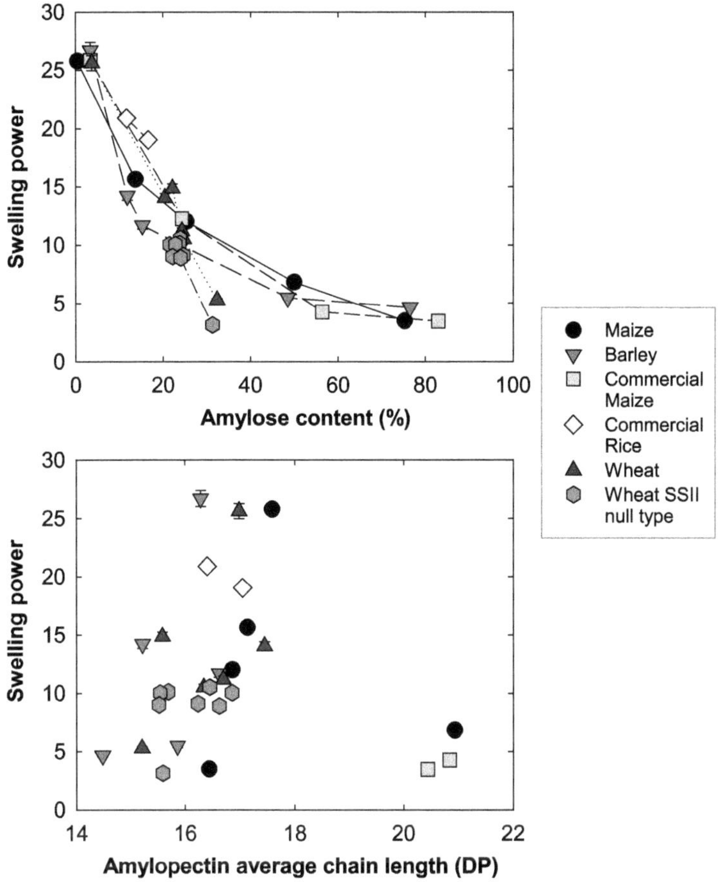

FIGURE 3.21 Variation in swelling power with amylose content and the average size of amylopectin chain length for maize, barley, wheat, and rice starches. (For color version of this figure, the reader is referred to the online version of this chapter.)

period at 95 °C at which the granules have swelled to their maximum size. Similar to RVA pasting profiles, an increase in starch amylose content results in an increase in the viscosity development in Rheoscope during the initial temperature ramp (Figure 3.23).

The differences in starch rheological responses (viscosity profiles) during gelatinization between different starch types can be attributed to the granule swelling ability and the proportion of small granules population (B-type granules). However, no significant trend could be deduced between the changes in Rheoscope viscosity and the granule size characteristics (Tan, 2005). This may indicate that the effect of variation in starch granules size distribution on starch rheological responses during gelatinization is not linear and their influences may not be as significant as the influence of the granules swelling ability.

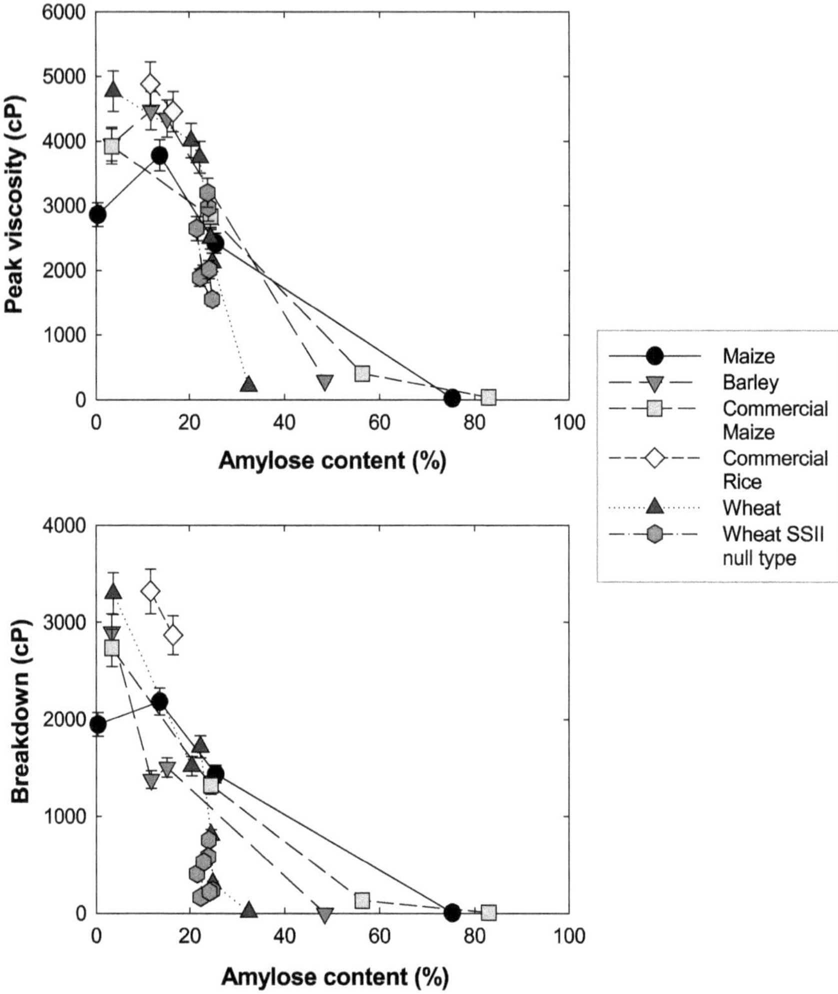

FIGURE 3.22 Variation in RVA peak viscosity and breakdown values with amylose content for maize, barley, wheat, and rice starches. (For color version of this figure, the reader is referred to the online version of this chapter.)

7 CONCLUSIONS

This chapter explores the variation in starch structural properties for a range of novel starches, aiming to investigate the correlations between these structural parameters and functionalities and the significance of each structural property in dictating the observed variation in starch functionalities.

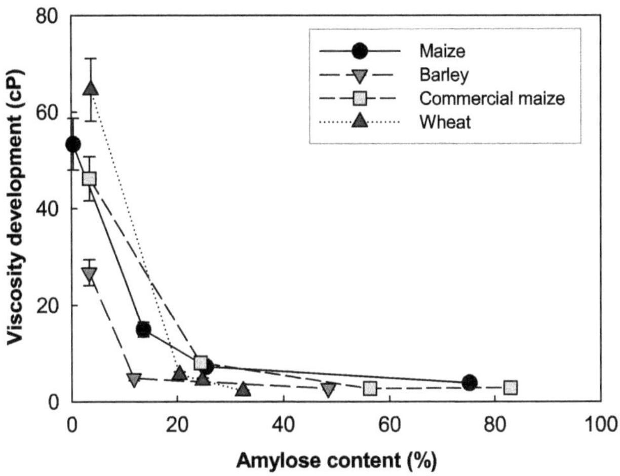

FIGURE 3.23 Variation in the maximum values for Rheoscope viscosity development during heating with the variation in amylose content for maize, barley, wheat, and rice starches. (For color version of this figure, the reader is referred to the online version of this chapter.)

Close examination of the variation in starch macromolecular properties reveals that the variation in variables describing amylopectin structures is independent of the variation in starch amylose content. While the occurrence of high amylose content results in substantial alteration in the amylopectin structures for all starch types, the observed changes in amylopectin structure are related to the mutation of genes expressing starch biosynthesis enzymes. The amylopectins of high-amylose wheat and barley starches (Yamamori and 292B) have enriched short-chain fractions with DP 4-12 due to the deficiency in SSII enzymes and thus result in the lack of amylopectin side-chain elongation. The amylopectins of HAM starches on the other hand are generated from starch mutants lacking BEIIb enzymes, and they have anomalous amylopectin with longer-chain-length distributions. The variation in amylose content and variables describing amylopectin structure is therefore considered as independent starch structural parameters dictating the functional properties of starch.

On the influence of starch macromolecular properties on their granular properties, the following deductions are obtained from this study. A general increase in the relative proportion of single-helix V-type polymorph is noted for high-amylose starches, which may indicate the possibility of single V-type helices as the adopted conformation of amylose. Nonetheless, a further analysis on the starch lipid content and subsequently the amylose-lipid complexes is required to confirm the origin of this V-type polymorph. For all starch types, an increase in the amylose content leads to a decrease in the amount of starch structural order, namely, the double-helix content and crystallinity. The variation in the amount of starch structural order (double-helix content and crystallinity) does not exhibit notable trend with variables describing amylopectin chain length distribution, which indicates that the involvement of amylopectin in supporting of starch crystalline framework is much more complicated than double-helix formation between their side chains. The noted transition in the packing arrangement of double helices in high-amylose starches to B-type polymorphic form can be attributed to the alteration in their amylopectin chain length distribution. The variation in starch granule size distribution is not related to the variation in starch macromolecular properties, and thus, the granule size distribution is considered as separate attribute of starch properties, which may dictate their functional properties. The noted changes in the average size of the large A-type granules in high-amylose starches can be attributed to the generation of anomalous amylopectin in these starches, which prevents the normal and efficient packing mechanisms of these macromolecules.

From examination of the variation in starch functional properties in relation to the variation in their macromolecular and granular properties, the following deductions are obtained. For all starch types, the variation in starch gelatinization onset temperature is related to the relative proportions of amylopectin chains with short (DP 4-12) and intermediate chain lengths (DP 13-24). Starches with higher proportion of short amylopectin chains have lower gelatinization temperature, possibly due to the inability of these short chains to participate in double-helix formation and thus result in higher destabilizing effect (by these short chains) on the amylopectin helix-helix association. On the contrary, an increase in the relative proportion of intermediate amylopectin chains (DP 13-24) leads to an increase in the gelatinization onset temperature. This may suggest that the occurrence of a higher proportion of amylopectin intermediate chains implies more stable helix-helix association and thus a higher thermal energy input is required to bring about the disruption of these structural orders. For most starches, an increase in the amount of starch structural order, namely, the double-helix content and crystallinity, leads to an increase in gelatinization enthalpy and heat capacity change (except for HAM starches). The unusual trend observed for HAM starches is possibly attributable to the unusual packing of starch macromolecules in these granules due to the occurrence of anomalous amylopectin with longer chain lengths. Therefore, it is possible that the gelatinization of these HAM starches proceeds differently from the other.

A strong inverse correlation is noted between starch swelling power and amylose content, which indicates that the ability of starch granules to swell is related to the presence of amylopectin. Starch granules with an essentially zero amylose content exhibit substantial swelling ability, while the high-amylose starches show a considerably lower swelling power. A reduction in RVA peak viscosity and breakdown values is noted with an increase in starch amylose content. Similarly, an increase in the viscosity development as measured by Rheoscope is noted with a decrease in starch amylose content. It is worth noting that the observed starch rheological responses during gelatinization can be mainly attributed to the swelling of starch granules. Therefore, these properties are strongly related to the swelling ability of granules, which is negatively correlated with the amylose content. A further examination on the variation in Rheoscope viscosity development with the variation in variables describing starch granules distribution indicates that although the granule size distribution has some effect on the Rheoscope viscosity development, their influences are not as significant as the influence of the granule swelling ability.

References

Baga, M., Repellin, A., Demeke, T., Caswell, K., Leung, N., Abdel-Aal, E. S., et al. (1999). Wheat starch modification through biotechnology. *Starch-Starke, 51*, 111–116.

Banks, W., Greenwood, C. T., & Walker, J. T. (1971). Studies on the starches of barley genotypes. A comparison of starches from normal and high-amylose barley. *Starch-Starke, 23*, 12–15.

Bocharnikova, I., Wasserman, L. A., Krivandin, A. V., Fornal, J., Blaszczak, W., Chernykh, V. Y., et al. (2003). Structure and thermodynamic melting parameters of wheat starches with different amylose content. *Journal of Thermal Analysis and Calorimetry, 74*, 681–695.

Boyer, C. D., & Preiss, J. (1981). Evidence for independent genetic control of the multiple forms of maize endosperm branching enzymes and starch synthases. *Plant Physiology, 67*, 1141–1145.

Buleon, A., Colonna, P., Planchot, V., & Ball, S. (1998). Starch granules: Structure and biosynthesis. *International Journal of Biological Macromolecules, 23*, 85–112.

Cao, H., Imparl-Radosevich, J., Guan, H., Keeling, P. L., James, M. G., & Myers, A. M. (1999). Identification of the soluble starch synthase activities of maize endosperm. *Plant Physiology, 120*, 205–215.

Cheetham, N. W. H., & Tao, L. (1998). Variation in crystalline type with amylose content in maize starch granules: An X-ray powder diffraction study. *Carbohydrate Polymers, 36*, 277–284.

Ellis, R. P., Cochrane, M. P., Dale, M. F. B., Duffus, C. M., Lynn, A., Morrison, I. M., et al. (1998). Starch production and industrial use. *Journal of the Science of Food and Agriculture, 77*, 289–311.

Fredriksson, H., Silverio, J., Andersson, R., Eliasson, A. C., & Aman, P. (1998). The influence of amylose and amylopectin characteristics on gelatinisation and retrogradation properties of different starches. *Carbohydrate Polymers, 35*, 119–134.

Gao, M., Wanat, J., Stinard, P. S., James, M. G., & Myers, A. (1998). Characterization of *dull1*, a maize gene coding for a novel starch synthase. *The Plant Cell, 10*, 399–412.

Gernat, C., Radosta, S., Anger, H., & Damaschun, G. (1993). Crystalline part of three different conformations detected in native and enzymatically degraded starches. *Starch-Starke, 45*, 309–314.

Gidley, M. J. (1987). Factors affecting the crystalline type (A-C) of native starches and model compounds: A rationalisation of observed effects in terms of polymorphic structures. *Carbohydrate Research, 161*, 301–304.

Gidley, M. J., & Bociek, S. M. (1985). Molecular organization in starches: A ^{13}C CP/MAS NMR study. *Journal of the American Chemical Society, 107*, 7040–7044.
Gidley, M. J., & Bulpin, P. V. (1987). Crystallisation of malto-oligosaccharides as models of the crystalline forms of starch: Minimum chain-length requirement for the formation of double helices. *Carbohydrate Research, 161*, 291–300.
Gidley, M. J., & Robinson, G. (1990). Techniques for studying interactions between polysaccharides. In P. M. Dey & J. B. Harborne (Eds.), *Methods in plant biochemistry*. London: Academic Press.
Hermansson, A. M., & Svegmark, K. (1996). Developments in the understanding of starch functionality. *Trends in Food Science and Technology, 7*, 345–353.
Hizukuri, S. (1985). Relationship between the distribution of the chain length of amylopectin and the crystalline structure of starch granules. *Carbohydrate Research, 141*, 295–306.
Hizukuri, S. (1986). Polymodal distribution of the chain lengths of amylopectins and its significance. *Carbohydrate Research, 147*, 342–347.
Hizukuri, S., Kaneko, T., & Takeda, Y. (1983). Measurement of the chain-length of amylopectin and its relevance to the origin of crystalline polymorphism of starch granules. *Biochimica et Biophysica Acta, 760*, 188–191.
Hoover, R. (2001). Composition, molecular structure, and physicochemical properties of tuber and root starches: A review. *Carbohydrate Polymers, 45*, 253–267.
Imberty, A., Buleon, A., Tran, V., & Perez, S. (1991). Recent advances in knowledge of starch structure. *Starch-Starke, 43*, 375–384.
James, M. G., Robertson, D. S., & Myers, A. M. (1995). Characterization of the maize gene *sugary1*, a determinant of starch composition in kernels. *The Plant Cell, 7*, 417–429.
Jane, J., Chen, Y. Y., Lee, L. F., McPherson, A. E., Wong, K. S., Radosavljevic, M., et al. (1999). Effects of amylopectin branch chain length and amylose content on the gelatinization and pasting properties of starch. *Cereal Chemistry, 76*, 629–637.
Jobling, S. (2004). Improving starch for food and industrial applications. *Current Opinion in Plant Biology, 7*, 210–218.
Martin, C., & Smith, A. M. (1995). Starch biosynthesis. *The Plant Cell, 7*, 971–985.
Matveev, Y. I., Van soest, J. J. G., Nieman, C., Wasserman, L. A., Prosterov, V. A., Ezernitskaja, M., et al. (2001). The relationship between thermodynamic and structural properties of low and high amylose maize starches. *Carbohydrate Polymers, 44*, 151–160.
McPherson, A. E., & Jane, J. (1999). Comparison of waxy potato with other root and tuber starches. *Carbohydrate Polymers, 40*, 57–70.
Mikami, I., Dung, L. -V., Hirano, H., & Sano, Y. (2000). Effects of the two most common *Wx* alleles on different genetic backgrounds in rice. *Plant Breeding, 115*, 505–508.
Morell, M. K., Kosar-Hashemi, B., Cmiel, M., Samuel, M. S., Chandler, P., Rahman, S., et al. (2003). Barley *sex6* mutants lack starch synthase IIa activity and contain a starch with novel properties. *The Plant Journal, 34*, 173–185.
Morrison, W. R., Tester, R. F., Snape, C. E., Law, R., & Gidley, M. J. (1993). Swelling and gelatinisation of cereal starches. IV. Some effects of lipid-complexed amylose and free amylose in waxy and normal barley starches. *Cereal Chemistry, 70*, 385–391.
Mouille, G., Maddelein, M., Libessart, N., Talaga, P., Decq, A., Delrue, B., et al. (1996). Pre-amylopectin processing: A mandatory step for starch biosynthesis in plants. *The Plant Cell, 8*, 1353–1366.
Myers, A. M., Morell, M. K., James, M. G., & Ball, S. G. (2000). Recent progress toward understanding biosynthesis of the amylopectin crystal. *Plant Physiology, 122*, 989–997.
Pan, D. (2000). Starch synthesis in maize. In A. K. Gupta & N. Kaur (Eds.), *Carbohydrate reserves in plants—Synthesis and regulation* (pp. 101–120). Amsterdam: Elsevier Science B.V.
Rahman, S., Li, Z., Batey, I., Cochrane, M. P., Appels, R., & Morell, M. (2000). Genetic alteration of starch functionality in wheat. *Journal of Cereal Science, 31*, 91–110.

Sano, Y. (1984). Differential regulation of waxy gene-expression in rice endosperm. *Theoretical and Applied Genetics*, *68*, 467–473.

Slattery, C. J., Kavakli, I. H., & Okita, T. W. (2000). Engineering starch for increased quantity and quality. *Trends in Plant Science*, *5*, 291–298.

Smith, A. M. (1999). Making starch. *Current Opinion in Plant Biology*, *2*, 223–229.

Takeda, C., Takeda, Y., & Hizukuri, S. (1993). Structure of the amylopectin fraction of amylomaize. *Carbohydrate Research*, *246*, 273–281.

Tan, I. (2005). Characterisation of novel starch materials: structure-functionality relationship (Doctor of Philosophy). Brisbane, Queensland: The University of Queensland.

Tan, I., Flanagan, B. M., Halley, P. J., Whittaker, A. K., & Gidley, M. J. (2007). A method for estimating the nature and relative proportions of amorphous, single, and double-helical components in starch granules by ^{13}C CP/MAS NMR. *Biomacromolecules*, *8*, 885–891.

Tan, I., Torley, P. J., & Halley, P. J. (2008). Combined rheological and optical investigation of maize, barley and wheat starch gelatinisation. *Carbohydrate Polymers*, *72*, 272–286.

Tan, I., Wee, C. C., Sopade, P. A., & Halley, P. J. (2004). Investigation of the starch gelatinisation phenomena in water-glycerol systems: Application of modulated temperature differential scanning calorimetry. *Carbohydrate Polymers*, *58*, 191–204.

Tester, R. F., & Morrison, W. R. (1990). Swelling and gelatinisation of cereal starches. II. Waxy rice starches. *Cereal Chemistry*, *67*, 558–563.

Umeki, K., & Kainuma, K. (1981). Fine structure of Nageli amylodextrin obtained by acid treatment of defatted waxy-maize starch-structural evidence to support the double helix hypothesis. *Carbohydrate Research*, *96*, 143–159.

Umemoto, T., Yano, M., Satoh, H., Shomura, A., & Nakamura, Y. (2002). Mapping of a gene responsible for the difference in amylopectin structure between *japonica*-type and *indica*-type rice varieties. *Theoretical and Applied Genetics*, *104*, 1–8.

Yamamori, M., Fujita, S., Hayakawa, K., Matsuki, J., & Yasui, T. (2000). Genetic elimination of a starch granule protein, SGP-1, of wheat generates an altered starch with apparent high amylose. *Theoretical and Applied Genetics*, *101*, 21–29.

Yuryev, V. P., Kalistratova, E. N., Van soest, J. J. G., & Nieman, C. (1998). Thermodynamic properties of barley starches with different amylose content. *Starch-Starke*, *50*, 463–466.

Zeng, M., Morris, C. F., Batey, I. L., & Wrigley, C. W. (1997). Sources of variation for starch gelatinization, pasting, and gelation properties in wheat. *Cereal Chemistry*, *74*, 63–71.

Zhao, X. C., Batey, I. L., Sharp, P. J., Crosbie, G., Barclay, I., Wilson, R., et al. (1998). A single genetic locus associated with starch granule properties and noodle quality in wheat. *Journal of Cereal Science*, *27*, 7–13.

Zobel, H. F. (1988a). Molecules to granules: A comprehensive starch review. *Starch-Starke*, *40*, 44–50.

Zobel, H. F. (1988b). Starch crystal transformations and their industrial importance. *Starch-Starke*, *40*, 1–7.

PART III

NOVEL STARCH MODIFIED BY BIOTECHNOLOGY

CHAPTER

4

Starch Modification by Biotechnology: State of Art and Perspectives

Xuan Xu[1,2], *Richard G.F. Visser*[1], *Luisa M. Trindade*[1]

[1]Wageningen UR—Plant Breeding, Wageningen University and Research Center, P.O. Box 386, Wageningen, The Netherlands

[2]National Centre for Vegetable Improvement (Central China), Key Laboratory of Horticultural Plant Biology, Ministry of Education, Huazhong Agricultural University, Wuhan, China

1 INTRODUCTION

Carbohydrates are the most abundant biocompounds on our planet. More than 100 billion metric tons of CO_2 and H_2O is converted into carbohydrates and other plant compounds by green plants during the process of photosynthesis every year (Nelson & Cox, 2004).

Starch, the major storage carbohydrate, is synthesized by most green plants and can be found in various types of tissues and organs. It is an insoluble glucan composed of two polymers: linear amylose (20-30%) and branched amylopectin (70-80%) (Kossmann & Lloyd, 2000). Although both are composed of glucosyl units linked by α-1,4-bonds and branched in α-1,6-bonds, the starch granule structure is more complex than expected from its components. The polymeric carbohydrates of the starch are organized in alternating amorphous and semicrystalline rings (Jenkins, Cameron, & Donald, 1993), and the forming of the semicrystalline organization is mainly dependent on amylopectin that has relatively short chains.

In plants, starch is synthesized in plastids and two types of starch can be distinguish, the transitory starch and the storage starch. The transitory starch is synthesized and accumulated in chloroplasts of green leaves during daytime photosynthesis and degraded throughout the night to provide substrates for leaf respiration and for continued sucrose synthesis to supply to sink tissues (Zeeman & Ap Rees, 1999). Mutants of *Arabidopsis thaliana* unable to synthesize transitory starch grow slower in the day-night regime relatively to the wild type, because they lack carbon supply in the dark (Caspar, Huber, & Somerville, 1985). Storage starch is synthesized in amyloplasts and typically associated with sink organs, such as stems, seeds, roots, and tubers. This type of starch is accumulated during different developmental stages and utilized for various periods of growth, dormancy, germination, or other specific processes. Apart from the different locations in the cell, the two types of starch have different characteristics. Storage starch is deposited in larger granules and contains a higher amylose content than that observed in transitory starch (Jobling, Westcott, Tayal, Jeffcoat, & Schwall, 2002; Matheson, 1996). Next to these differences within plants, there are also differences between plant species. For instance, tuber and root starches have lower protein and lipid contents than cereal starches, which have advantages in many applications. In a word, patterns of starch accumulation, starch granule size and shape, and physicochemical properties differ not only between plants of different botanic origins but also within accessions of the same species.

The importance of starch is not limited to the relevance of this polysaccharide in plant survival but, to a large degree, due to the application in food and nonfood industry. With the imminent depletion of fossil oil reserves, starch has gained increasingly interest as a sustainable and renewal feedstock for the production of biochemicals, biomaterials, and biofuels. With the marked development of industrial economy, starch, which is isolated from maize, rice, wheat, cassava, potato, pea, arrowroot (*Maranta arundinacea*), and sago palm (*Metroxylon sagu*), has become a cornerstone of a sustainable industry (Zeeman, Kossmann, & Smith, 2010). It has been shown that starch structure highly affects the physicochemical properties, such as gelatinization, texture, moisture retention, viscosity, and product homogeneity, which are determinant for its applications.

Native starches have shortcomings that restrict the end uses. Properties such as retrogradation and gelling tendencies of pastes are undesirable in many application fields (Kaur, Ariffin, Bhat, & Karim, 2012). Starch derivatization comprises the use of chemical, enzymatic, and/or physical modifications that require high amounts of energy and chemicals and produce chemical waste. Hence, modification of starch *in planta* could improve the functional properties with a broad range of specific applications and produce starches with better quality, using less energy and without pollutant chemical waste (Kaur et al., 2012). Up to now, modified starch has been widely used for many industrial applications such as dairy confectionery, meat products, oil drilling, paper and cardboard, adhesives, packing material, pharmaceuticals, and textile sand bioplastics (Ellis et al., 1998; Roper, 2002). It is worth noticing that genetic modification of starch *in planta* has received increased attention, due to not only the enormous potential for tailoring granules with new functionalities but also the reduction of the costs of the postharvest modifications and usage of other approaches that probably cause damage to the environment (Kok-Jacon, Ji, Vincken, & Visser, 2003). Understanding of the starch biosynthesis and degradation pathways will facilitate the progress of creating designer starches that have novel functionalities by using biotechnology.

This chapter focuses on the characterization of modified starches that have been generated using biotechnological approaches and the implications of these genetic modifications on the morphology of the starch granules and the subsequent starch physical-chemical properties.

2 STARCH BIOSYNTHESIS AND MUTANTS IN THE STARCH BIOSYNTHETIC PATHWAY

In plant's life cycle, the biosynthesis of starch works as a refined balance wheel between a continuous efficient assembly and the degradation of glucan chains. The synthesis of transitory and storage starch in the chloroplast and amyloplast involves several common enzymes as well as specific isoforms that are only involved in the synthesis of one of these two types of starch.

Current knowledge about starch biosynthesis in plants shows that the ADP-glucose pyrophosphorylase (AGPase, E.C.2.7.7.27), starch synthases (SS), granule-bound starch synthase (GBSSI), starch-branching enzymes (SBE), and possibly the starch-debranching enzymes (DBE) play a role in determining the final starch structure (Smith, Denyer, & Martin, 1997). These enzymes are present in one or more of the three major steps: (i) activation of glucose residues, (ii) chain elongation, and (iii) synthesis of side chains (Fig. 4.1).

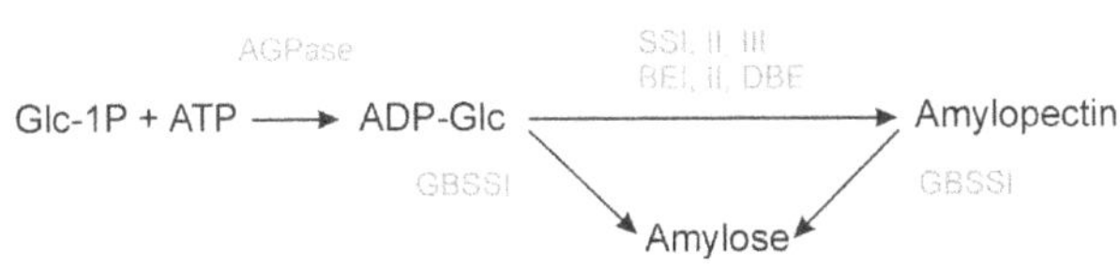

FIGURE 4.1 Amylose and amylopectin biosynthesis. ADP-glucose (ADP-Glc), the donor substrate for both amylose and amylopectin, is synthesized by the ADP-Glc pyrophosphorylase (AGPase). The combined action of different starch synthases (SSI, SSII, and SSIII), branching enzymes (BEI and BEII), and the debranching enzyme (DBE) is necessary for the synthesis of amylopectin. Granule-bound starch synthases (GBSSI and GBSSII) use amylopectin as the acceptor substrate to synthesize amylose, which is formed downstream of amylopectin.

2.1 Initiation of Starch Biosynthesis

The first step in starch biosynthesis, the synthesis of ADP-glucose, is catalyzed by ADP-glucose pyrophosphorylase (AGPase, E.C.2.7.7.27), first found in wheat flour (Espada, 1962). This enzyme converts glucose 1-phosphate (Glc-1-P) into adenosine diphosphoglucose (ADP-Glc, Preiss, Kumar, & Ghosh, 1988), which is the preferred substrate for starch formation. This enzyme also was found in many plant tissues and bacterial extract (Preiss & Sivak, 1998); however, the localization and regulation of ADP-Glc synthesis and import Glc-1-P are highly various between different species and in different organs within a species (Emes & Neuhaus, 1997; Neuhaus & Emes, 2000). As a key factor of this rate-controlling step, AGPase was extensively used, through biotechnology, to control the rate of ADP-Glc formation and, in turn, starch accumulation. The amylose content was severely reduced in transgenic potato that contained a lower expression of AGPase (Lloyd, Springer, et al., 1999). A mutation at the *rb* locus from pea (*Pisum sativum* L.), which reduced the activity of AGPase (Smith, Bettey, & Bedford, 1989), resulted in a decrease of starch in the embryo (Kooistra, 1962). Interestingly, inactivation of AGPase in a *Chlamydomonas* starch less mutant led to a 10-fold increase in TAG that can be used to produce biofuels (Li et al., 2010), but only to a slight decrease in the accumulation of starch (Ball et al., 1991). In addition, a mutant *Escherichia coli* AGPase gene (*glgc16*) has been overexpressed in some plants to enhance starch accumulation, such as potato (Stark, Timmerman, Barry, Preiss, & Kishore, 1992) and maize (Wang et al., 2007).

Some observations indicated that conversion of sucrose to starch is ATP-dependent. The ATP/ADP transporter (AATP), located on the inner-envelope membrane, imports ATP required for the AGPase reaction to the plastid (Heldt & Pfaff, 1969; Schunemann, Borchert, Flugge, & Heldt, 1993). Overexpression of an AATP, an energy supplier for the AGPase reaction within the plastid, leads to an increase of 16-36% in starch content; in contrast, antisense AATP tubers exhibited drastically reduced levels of starch (Tjaden, Schwoppe, Mohlmann, Quick, & Neuhaus, 1998). In chloroplasts, the catalytic activity of AGPase is allosterically activated by 3-phosphoglyceric acid (3-PGA) and inhibited by inorganic phosphate (Pi). The allosteric properties of AGPase are of potential importance in determining its role in controlling the rate of ADP-Glc formation and, in turn, starch biosynthesis (Smith et al., 1997). The small and large subunits' composition of AGPases mainly affects the allosteric properties. The small subunit alone can form a catalytically active enzyme and the large subunit is responsible for modulation of the sensitivity of the enzyme to effectors. The *sta1* mutant in the unicellular algae *Chlamydomonas reinhardtii*, containing lesions in the large subunit of AGPase, leads to a 95% reduction in starch biosynthesis (Ball et al., 1991; Vandenkoornhuyse et al., 1996). Maize mutants where the genes encoding the large subunit and small subunit of AGPase have been knocked out, respectively, *shrunken-2* and *brittle-2*, demonstrate that in the endosperm, the primary flux to starch is via the cytosolic enzyme. In seeds of the *shrunken-2* mutant, the amount of starch was significantly increase by 15%.

2.2 Chain Elongation

Synthesis of starch is catalyzed by starch synthases (SS, EC 2.4.1.21), which transfer the glucose residue from ADP-Glc to the nonreducing end of a growing α-1,4-linked glucan. Six isoforms of starch synthases have been characterized by multiple sequences alignment. SSI, SSII, SSIII, and SSIV are present in a diverse range of species as soluble forms or as both soluble and minor granule-bound isoforms, whereas GBSS, including GBSSI in storage organs and GBSSII in nonstorage organs, is involved in amylose biosynthesis. A novel high-amylose starch was discovered in the barley *sex6* mutant, which was lacking the activity of SSIIa (Morell et al., 2003). Compared to the high-amylose starches produced by the downregulation of SBE, the *sex6* starch has amylopectin with relatively short branches and gelatinized at lower temperatures. The synthesis of amylose-free starches has focused on engineering GBSS activity in different plant species. Amylose-free starches were reported in mutants of maize (*waxy*, Nelson & Rines, 1962), sorghum (*waxy*, Hseih, 1988), wheat (*waxy*, Nakamura, Yamamori, Hirano, Hidaka, & Nagamine, 1995), amaranth (*waxy*, Okuno & Sakaguchi, 1982), pea (*lam*, Denyer, Foster, & Smith, 1995), and unicellular green algae *C. reinhardtii* (*st-2-1* and *st-2-2*, Delrue et al., 1992). The activity of GBSS is inexistent in these mutants resulting in starches with better properties for industry; however, they show sometimes small yield penalties, which can be fixed by selection, irrespective of the method used to generate the mutants. Amylose-free starches also have been produced by downregulation of GBSS in potato (Visser et al., 1991) and sweet potato (Kimura et al., 2001; Noda et al., 2002), and no significant penalties on the starch content were observed. The natural GBSS mutant genotype producing amylose-free potato starch (Hovenkamphermelink et al., 1987) has been explored commercially for many years such as the ELIANE™ commercialized by AVEBE (www.avebe.nl). The amylose-free cassava starch derived from a mutant is opening the door to commercial applications in this crop as well (Ceballos et al., 2007). Jobling et al. (2002) observed a novel amylose-free starch with short-chain amylopectin via the simultaneous downregulation of GBSS, SSII, and SSIII.

2.3 Formation of Branch Points

The formation of the α-1,6 linkages as branching point of the α-1,4 glucan chain is of great importance in determining amylopectin structure. Comparing the SBE(EC 2.4.1.18) sequences from a wide range of species revealed two distinct classes of branching enzymes in plants: SBE A and SBE B. The analysis of mutants has provided direct evidence that expression levels of SBE affect the nature of synthesized starch. The mutation in the *r* locus in wrinkled peas, which is associated with complete absence of one isoform of SBE I, leads to an amylopectin content that is reduced from about 70% to about 30% of the total starch content (Bhattacharyya, Smith, Ellis, Hedley, & Martin, 1990). Antisense of the major isoform (SBE B) in potato reduces SBE activity in the tubers almost completely without affecting the amylose content in the starch (Kossmann et al., 1997; Safford et al., 1998). Schwall et al. (2000) created high-amylose potato starch (more than 60% of the granule) by simultaneously inhibiting two isoforms of SBE to below 1% of the wild-type activities.

Some results suggest that these enzymes are redundant. The antibody-mediated inhibition of SBEA activity is more efficient than using antisense technology and produces higher amylose levels (Jobling et al., 2002).

The action of DBE (EC 2.4.1.41) activities during biosynthesis is considered to be critical for the accurate assembly of the starch granule. Two types of DBE exist in plants, isoamylase (ISA) and limit dextrinase (LDA or pullulanase). The former one has a higher affinity for glycogen, while the latter one has a higher affinity for pullulan (Doehlert & Knutson, 1991). Both enzymes are responsible for hydrolyzing the α-1.6 branch points that are not correctly positioned in the molecule and for generating a branching structure appropriate for crystallization (Ball & Morell, 2003). Likewise SS and SBE, the role of DBE in the synthesis of amylopectin has been proposed by several studies with DBE-deficient mutants of various plant species, such as rice (Nakamura, 1996), maize (James, Robertson, & Myers, 1995; Pan & Nelson, 1984), *Arabidopsis* (Zeeman et al., 1998), and *Chlamydomonas* (Mouille et al., 1996). The results showed that wild-type amylopectin was replaced by glycogen-like α-glucan phytoglycogen and both morphological and physicochemical properties of starch were altered to a certain degree.

3 STARCH DEGRADATION AND MUTANTS IN THE STARCH BREAKDOWN

Starch degradation has been better studied in transitory starch in leaves, where starch is broken down during the night, than in storage starch where it is degraded over a relatively long time. Although starch degradation has been widely studied in cereal endosperm (Fincher, 1989; Ritchie, Swanson, & Gilroy, 2000), where starch is broken down over several days after seed germination, the mechanism and regulation of this process in other starch-containing organs such as roots, tubers, and non-cereal seeds are more fragmentary. This is most probably due to the fact that starch degradation pathways differ depending on the plant organ and different pathways may operate within the same organ (Zeeman et al., 2010). At least for *Arabidopsis*, the important genes have been identified and the major steps in the pathway are known. The initiation of starch degradation is the reversible phosphorylation of glucans at the surface of the starch granule by a class of enzymes called glucan water dekinases (GWDs, Ritte et al., 2002). Subsequently, hydrolases access to the glucan chains and starch is hydrolyzed to maltose and glucose, and both can be exported from the chloroplast and metabolized in the cytosol.

3.1 The Enzymatic Breakdown of Starch Granules

The initial event, seemingly required for starch degradation, must be catalyzed by an enzyme capable of metabolizing polymers on the surface of semicrystalline granules to release the soluble glucans. Three most likely candidates, α-amylase (AMY), β-amylase (BAM), and DBE, are capable of attacking the starch granule surface. Starch degradation occurs in germinating cereal endosperm and is thought to be initiated by AMY based on the characteristic visible pitting of the starch surface. It seems that AMY is in charge of attacking the insoluble granule, hydrolyzing α-1.4 linkages within polymers exposed on the surface or in channels within granules to release both linear and branched glucans for further degradation. However, surprisingly, the studies in *Arabidopsis* suggested that AMY is much less significant in starch breakdown. Three proteins predicted to be α-amylases (AMY1, 2, and 3) are encoded in the *Arabidopsis* genome. T-DNA insertion mutants lacking these isoforms, respectively, have normal rates of starch degradation in leaves at night, and starch metabolism also appears normal in *amy1/amy2/amy3* triple mutants (Yu et al., 2005). These results suggest that AMY is not required for starch degradation in *Arabidopsis*, whereas the studies described recently suggest that in some circumstances, AMY3 does participate in degradation of starch granule in *Arabidopsis* (Zeeman, Chen, Gatehouse, Smith, & Smith, 2007). It was reasonable to assume that there are other endoamylases involved in starch degradation that compensate for the deficiency of AMY3 in both *Arabidopsis* and potato (Zeeman, Smith, & Smith, 2007). The loss of a specific isoform of β-amylase (BAM3 in *Arabidopsis* and PCT-BMY1 in potato) and DBE ISA (ISA3 in *Arabidopsis* and potato) lead to the decreased rates of starch breakdown and excess accumulation of starch, indicated that these two enzymes affected the degradation of starch granules (Delatte et al., 2006; Kaplan & Guy, 2005; Scheidig, Frohlich, Schulze, Lloyd, & Kossmann, 2002; Wattebled et al., 2005). In addition, suppressed expression or overexpression of AMY in rice caused the correspondent changes of starch accumulation in leaves (Asatsuma et al., 2005).

3.2 The Importance of Glucan Phosphorylation

The GWD plays an important role in starch degradation and was first discovered in potato tubers by Lorberth, Ritte, Willmitzer, and Kossmann (1998). This enzyme transfers the β-phosphate group of ATP to a small proportion of the glucose residues of amylopectin chains, particularly at the 6-position (Mikkelsen, Baunsgaard, & Blennow, 2004; Ritte et al., 2002). The formed phosphate groups could affect the structure of the starch granule, and thus, the starch surface becomes more vulnerable to facilitate starch degradation (Blennow, Bay-Smidt, Olsen, & Moller, 2000). By antisense technology in potato, it was determined that downregulation of GWD leads to a reduction in the phosphate content of the starch and thereby impairs starch degradation in leaves and reduces cold sweetening in tubers (Lorberth et al., 1998). This result is similar with the data obtained from studies of *Arabidopsis* mutants of deficient GWD protein. Even the mature leaves of these mutants increase amounts of starch up to seven times greater than those in wild-type leaves (Caspar et al., 1991; Yu et al., 2001; Zeeman & Ap Rees, 1999). The picture of starch degradation has evolved to include the second GWD-like enzyme, phosphoglucan, water dikinase (PWD, Baunsgaard et al., 2005; Kotting et al., 2005). An *Arabidopsis* mutant with downregulated expression of this gene has increased amounts of starch in leaf (Baunsgaard et al., 2005). PWD will act on phosphorylated amylopectin rather than unphosphorylated glucans; therefore, it was presumed that the activity of PWD requires the presence of active GWD. With the discovery of GWD and PWD, it is now possible to create the starch of high phosphate content in plants and particularly in cereals, which are extremely low in phosphate content. Overexpression of GWD has been achieved in maize (Lanahan & Basu, 2005), rice (Frohberg, 2008), wheat (Schewe et al., 2002), and barley (Carciofi et al., 2011), resulting in increased starch phosphate content.

4 EXPRESSION OF HETEROLOGOUS ENZYMES IN PLANTS

Progress in understanding the high value of modified starches has encouraged researchers to produce modified starches using different strategies. Expression of heterologous genes from other organisms, which encode biosynthetic or modifying enzymes, has the potential to be one of the most economical systems for large-scale production of starches with novel properties. Bacterial genes provide an ideal source of variety because they may have properties that are slightly different from their plant counterparts and thus create different or novel phenotypes.

Several starch-modifying bacterial proteins have been introduced in different crops, including potato, maize, and cassava. These included a starch-binding domain (SBD) of *Bacillus circulans* cyclodextrin glycosyltransferase, an *E. coli* AGPase (GLGC16), an *E. coli* glycogen synthase (GLGA), an *E. coli* branching enzyme (GLGB), an *E. coli* maltose acetyltransferase (MAT), a *Neisseria polysaccharide* amylosucrase, a full-length (GTFI) and a truncated mutansucrase (GTFICAT) from *Streptococcus downei*, a *Leuconostoc mesenteroides* dextransucrase (DSRS), and an *L. mesenteroides* alternansucrase (ASR).

A SBD of cyclodextrin glycosyltransferase from *B. circulans* was expressed in potato tubers. SBDs accumulated in the starch granule during the biosynthesis process. The amalgamated starch granules consisted of many smaller granules have been achieved by the expression of multiple appended SBDs (SBD2-SBD5) in *amf* mutant potato plants (Ji et al., 2004; Nazarian-Firouzabadi, Trindade, & Visser, 2012).

A mutant variant of the *AGPase* gene from *E. coli* (*GLGC16*) was overexpressed in various plant species, such as cassava (Ihemere, Arias-Garzon, Lawrence, & Sayre, 2006), maize (Wang et al., 2007), and potato (Sweetlove, Burrell, & Aprees, 1996), and the starch content is slightly increased. Surprisingly, Stark et al. (1992) targeted this gene to plastids in transgenic potatoes resulting in significantly increased starch accumulation, and in some lines, there was almost a 60% increase in starch content relative to control tubers.

The potato tuber amyloplasts expressing the *E. coli* glycogen synthase gene (*GLGA*) were found to have a lowered specific gravity, about 50% reduction in the percentage of starch, and an increased branching degree of amylopectin (Shewmaker et al., 1994). For producing starches with higher branching degree of amylopectin and novel properties, the *E. coli* branching enzyme (GLGB) was introduced in amylose-containing background (Krohn, Stark, Barry, Preiss, & Kishore, 1994) and amylose-free potato mutant (Kortstee, Vermeesch, Devries, Jacobsen, & Visser, 1996). In the former, a slightly higher amylopectin percentage and a small difference in granule surface morphology were observed, while the branching degree of amylopectin was 25% higher in the latter as compared to the control.

Efforts have been made to introduce the maltose acetyltransferase gene (*MAT*) from *E. coli* into wild-type and mutant amylose-free (*amf*) potato plants in order to increase the acetylation of starch. A low concentration of acetyl groups was found in the starch granules of transgenic plants and also the starch granule morphology was altered (Firouzabadi, Vincken, Ji, Suurs, & Visser, 2007).

In order to alter the ratio of amylose and amylopectin, the amylosucrase gene from *N. polysaccharide* fused to a SBD was introduced in potato, which lead to a number of changes in the morphology of the starch granule and biochemical properties, but not to a change in the ratio of amylose and amylopectin (Huang et al., submitted for publication).

The production of bacterial extracellular polysaccharides *in planta* is another great way to generate material with unique nutritional or commercial added values (Moire, Rezzonico, & Poirier, 2003). Glucansucrases, produced by soil bacteria, *Streptococcus* species that form the oral flora and lactic bacteria *Lactococci*, catalyze synthesis of different α-glucans, such as mutan, dextran, alternan, and reuteran (Mooser, 1992; Sidebotham, 1974). Likewise, the production of these polysaccharides in the amyloplast could alter the properties of starch and be beneficial to replace postharvest chemical starch modifications by a more environmental-friendly bioprocessing.

4.1 Mutan

Mutansucrase (GTFI, E.C. 2.4.1.5) synthesizes the most adhesive and water-insoluble glucan, which is known as mutan (Hamada & Slade, 1980). In terms of basic research, two classes of chemical linkages were catalyzed by GTFI. One class is α-1,3 linkages (up to 90%) that are responsible for the backbone of mutan, and the other is α-1,6 linkages, which assemble single unit glucosyl residues to the backbone (Monchois, Vignon, & Russell, 1999). Expression of the *GTFI* gene, which was derived from oral cariogenic *S. downei* Mfe28 bacteria (Ferretti, Gilpin, & Russell, 1987), results in the accumulation of mutan polymers in the presence of sucrose. To develop the industrial applications of mutan polymers, different studies with genetically engineered GTFI have been carried out. One example is the investigation of the production of mutan polymers in amyloplasts of an amylose-containing potato by expressing a full-length (*GTFI*) and a glucan-binding domain-truncated gene (*GTFICAT*). By comparing transformants, GTFICAT plants showed a severely altered tuber phenotype and starch granule morphology in addition to starch structural changes (Fig. 4.2E and F). Next to that, the starch content was decreased in GTFICAT transformants, possibly because the expression of this gene interferes with starch biosynthesis by downregulating the *AGPase* gene (Kok-Jacon, Vincken, Suurs, & Visser, 2005). Further, in order to carry the GTFICAT enzyme close to or rather inside the growing starch granule, *GTFICAT* was fused to the N- or C-terminus of a SBD and introduced in amylose-containing and amylose-free potatoes, respectively (Firouzabadi, Kok-Jacon, et al., 2007). In these transgenic plants, mutan polymers were incorporated in the starch granule successfully and resulted in alterations of starch properties such as a higher melting temperature, granules with porous and spongy surfaces, and a more pronounced retrogradation behavior. However, these alterations, together with mutan polymers,

FIGURE 4.2 Scanning electron microscopy (SEM) analysis of starch granules from untransformed potato plants, cv Kardal (A) and amylose-free (*amf*, panel D), and transformed potato plants with the dextransucrase-encoding gene *dsrs* (B), with the alternansucrase-encoding gene *asr* (C), and with the full-length mutansucrase gene *gtfI* (E) and a truncated form of the latter *gtfI-CAT* (F).

seemed to be less strikingly than those from transformants with GTFICAT alone, implying that the appended SBD might interfere with the activity of GTFICAT (Firouzabadi, Kok-Jacon, et al., 2007).

4.2 Dextran

Dextran, an extracellular α-glucan containing a high proportion of α-1,6 glucosidic linkages and associated with variable proportions of α-1,2; α-1,4; or α-1,3 linkages, is synthesized by the DSRS (EC 2.4.1.5) of *Lactobacillus, Leuconostoc,* and *Streptococcus* bacteria in the presence of sucrose (Jeanes et al., 1954; Kralj et al., 2004; Monchois, Willemot, Remaudsimeon, Croux, & Monsan, 1996; Simpson, Cheetham, Giffard, & Jacques, 1995). One notable difference between *Leuconostoc* and *Streptococcus* bacteria is that in the former, the formation of DSRSs is induced and, in *Streptococcus* species, they are constitutively synthesized (Robyt, 1995b). The formation of dextran was observed as the result of bacterial transformation of sucrose solutions into viscous solutions, gels, and flocculent precipitates (Robyt, 1995a). *L. mesenteroides* B-512F formed only one dextran containing 95% α-1,6 linkages and 5% α-1,3 branch linkages, which is most widely studied and commercially produced (Groenwall & Ingelman, 1948). Kok-Jacon et al. (2005) introduced this gene into both an amylose-containing potato cultivar and an amylose-free (*amf*) potato mutant. In tuber juices, dextrans were detected in both the different genetic background series, and the dextran concentration was two times higher in amylose-containing lines than in the *amf* transgenic lines. In addition, the results revealed that despite altered granule morphology (Fig. 4.2B), the accumulation of dextran did not interfere with the physicochemical properties and starch content. This strategy could open a new avenue for commercialization of dextran production in plant systems.

4.3 Alternan

Alternan is a branched polymer with a unique backbone structure of alternating α-1,3 and α-1,6-D-glucosidic linkages, present for 46% and 54%, respectively (Cote & Robyt, 1982; Seymour & Knapp, 1980; Seymour, Knapp, Bishop, & Jeanes, 1979). It is branched by 3,6-disubstituted D-glucosyl residues, with approximately 7-11% branching (Seymour, Slodki, Plattner, & Jeanes, 1977). Therefore, alternan possesses the properties of high solubility and low viscosity as well as resistance to microbial and mammalian enzymes, which is suitable for the creation of material for functional foods (Cote, 1992, 2002). Leathers et al. (2002, 2003) isolated strains of *Penicillium* sp. that modify native, high-molecular-weight alternan in a novel bioconversion process to a lower-molecular-weight form with solution viscosity properties similar to those of commercial arabic gum. Alternan synthesized by *L. mesenteroides* NRRL B-1501 is catalyzed by the ASR (EC 2.4.1.140), which converts sucrose to alternan and fructose, and also has been cloned and sequenced (Arguello-Morales et al., 2000). Furthermore, ASR was transferred into potato tubers, and alterations in starch properties and morphology were analyzed (Kok-Jacon et al., 2007). The low accumulation of alternan (1.2 mg g^{-1} FW) resulted in a slight alteration of starch morphology (Fig. 4.2C). Although some of the key genes involved in starch biosynthesis seemed to be downregulated, particularly the *AGPase* and *GBSSI* genes, the physicochemical properties of starch remained unchanged (Kok-Jacon et al., 2007).

5 INDUSTRIAL APPLICATION OF IMPROVED STARCHES

5.1 Approaches for Biotechnological Modification of Starch

Along with the better understanding of starch structure and enzymes involved in starch biosynthesis, many of the genes that encode these enzymes have been cloned and transformed into plants using *Agrobacterium tumefaciens* to modify the starch metabolism. Transgenic plants have been generated by downregulation (antisense or cosuppression approaches) or overexpression of endogenous gene or expression of heterologous genes, where starch properties and morphology have been altered. The possibility to produce tailor-made starches *in planta* will broaden the functionality of starches in industrial applications (Zeeman et al., 2010). These *in planta* modified starches, such as the *amf* starch, are often of better quality relatively to the chemically derivative, and it precludes the use of hazardous chemicals and leads to energy savings in the production process (of up to 60% for, e.g., synthetic polymer replacers).

Ji, Vincken, Suurs, and Visser (2003) developed the so-called SBD technology to target effector proteins to starch granules during the biosynthesis process. Almost 10% of all microbial hydrolytic enzymes can bind to substrate mediated by a distinct domain referred to as carbohydrate-binding modules (CBMs) (Janecek, Svensson, & Macgregor, 2003; Machovic, Svensson, Ann Macgregor, & Janecek, 2005). SBDs from CBM family 20 have the affinity to bind to starch granules and enzymes can acquire affinity to starch granules by fusing to SBDs (Ohdan, Kuriki, Takata, Kaneko, & Okada, 2000). By using the SBD technology to generate potato starches with new or improved properties, some heterologous genes, including an *E. coli* MAT (Firouzabadi, Vincken, et al., 2007) and a truncated mutansucrase (GTFICAT) (Kok-Jacon et al., 2005), were successfully incorporated into starch granules and resulted in morphologically altered starch granules.

5.2 Application of Improved Starches

5.2.1 Waxy Starches

Waxy or *amf* (amylose-free) starch is a new species of starch composed only of amylopectin molecules. Within the last decade, there has been intense interest among researchers to create waxy starch in various species by mutating the *wx* locus that encodes the GBSS protein. Waxy maize containing 100% amylopectin, whereas normal maize starch contains 75% amylopectin and 25% amylose, was found in China in 1909. This mutant starch is mainly used due to its improved of uniformity, stability, and texture not only in food products but also in the adhesive industry due to its pasting characteristics.

The potato *amf* starch produces a clearer film and a stickier paste after gelatinization, and the resulting gel retrogradates less compared to regular potato starch (Visser, Suurs, Bruinenberg, Bleeker, & Jacobsen, 1997). Since 2005, the first natural potato variety ELIANE™, obtained through traditional well-accepted breeding techniques, is being cultivated and marketed by the starch company AVEBE. Recently, the European Food Safety Authority (EFSA) allowed the company BASF to commercialize potato variety Amflora™ within the EU. Although it was produced using genetic modification, Amflora™ is just as safe for humans, animals, and the environment as conventional potatoes via the evidence of extremely rigorous studies (Schmidt, 2009). These starches have much better performance in different industrial processes including the manufacturing of paper, adhesive, and textiles. In food industry, another much desired property is freeze-thaw stability of gelatinized starch. Normally, the use of unmodified starches in frozen foods is achieved by means of chemical modification to prevent the retrogradation of glucan chains. Jobling et al. (2002) created a freeze-thaw-stable potato starch by simultaneous downregulation of three starch synthase genes (GBSS, SSII, and SSIII) *in planta* (Jobling et al., 2002). From an environmental point of view and a consumer perspective, this starch may have a great potential for application in the food industry.

Waxy wheat is more suitable for different food applications. Pastas made with durum wheat containing lower amylose content are of better quality (Kim, Johnson, Graybosch, & Gaines, 2003; Sharma, Sissons, Rathjen, & Jenner, 2002).

5.2.2 High Amylose

High-amylose starches have the ability to form strong gels and films. These particular properties are especially exploited in various industry, such as textiles, candies, and adhesives. For example, the high gelling ability of these starches is exploited in jelly gum candies (William, 2009). Moreover high-amylose starches are used in extruded and fried snack products to obtain crisps, because they can expand to tender structure when heated or dried (Jobling, 2004). The first high-amylose maize hybrid was developed in the 1950s with the discovery of the *ae* (SBE, "amylose extender") gene by the Bear Hybrid Corn Company. Amylomaize is the generic name for maize that has amylose content higher than 50% and is grown exclusively for wet milling. Likewise, the high-amylose phenotype in cereals or potato is caused by the loss or downregulation of the corresponding *ae* gene, respectively (Jobling, 2004; Jobling et al., 1999). Unlike the normal starch, high-amylose starch escapes digestion in the small intestine of healthy individuals and reaches the large intestine where it is fermented by the gut bacteria. The fermentation of starch in the large intestine has health benefits by promoting a better colonic function. This starch, known as resistant starch (RS), is considered the third type of dietary fiber and well suited for food applications (Bird et al., 2000). In Australia, foods enriched in RS as a high-amylose maize starch have gained consumer acceptance. A company, Ascentia Pty Ltd., has been established to develop a RS barley cultivar (*Hordeum vulgare* var. *himalaya 292*), which was proven to have potential health benefits through the reduction of plasma cholesterol and the production of increased large bowel short-chain fatty acids. The starch of this cultivar is a promising vehicle to deliver health benefits to consumers (Topping et al., 2003). A high-amylose transgenic rice line modified by downregulation of two SBEs in transgenic rice endosperm resulted in high content of resistant starch with high potentialities in food and nonfood industry (Wei et al., 2010). High-amylose wheat generated by RNA interference also has a potential to deliver significant health benefits to consumers given its RS content (Regina et al., 2006).

5.2.3 *Phosphate*

Native starches contain small amounts of covalently bound phosphate. Potato starch shows a naturally high degree of phosphorylation compared to starches from other crops (Hizukuri, Tabata, & Nikuni, 1970). The presence of phosphate in potato starch results in the stable-paste properties and transparent gels. Hence, potato starch is preferred for use in fish-paste products and as an ingredient in noodles. It is universally accepted that starch with longer polymer chains tends to contain higher levels of phosphate because the longer chains provide a better substrate for the phosphorylating enzyme. The phosphate content of high-amylose starches had levels up to more than fivefold compared to the wild-type starch and was generated by simultaneous inhibition of SBE A and B (Schwall et al., 2000). A crucial enzyme (GWD) responsible for phosphorylating starch has been identified in potatoes, and regulating the expression of this gene could change the phosphate content and viscosity of potato starch (Donath et al., 1998).

5.2.4 *Altered Amylopectin Structure*

The synthesis of amylopectin requires the concerted action of several enzymes, including starch synthases, branching enzymes, and debranching enzymes. The modification of amylopectin structure *in planta* by the modulation of endogenous genes is complex and involves the targeting of multiple genes. Amylopectin containing more very long chains and more chains with less than 15 glucose units in length (Lloyd, Landschutze, & Kossmann, 1999) was obtained by knocking down the activity of SSII and SSIII simultaneously. As a result of the increase in the number of short chains in amylopectin, the starch can gelatinize at a low temperature (lower than 50 °C) and have valuable uses in the food industry. *Japonica*-type rice varieties have better eating quality and their starch gelatinizes at a lower temperature than *indica*-type rice varieties. The SSIIa protein is responsible for the difference in amylopectin structure between these two varieties, which makes that the amylopectin of *japonica* varieties is enriched in short chains and depleted in intermediate-sized chains (Umemoto, Yano, Satoh, Shomura, & Nakamura, 2002).

5.2.5 Granule Size and Number

Starch granule size and number vary considerably across plant sources. Rice starch is relatively small (about 2 μm), while potato starches have larger granules (up to 100 μm). Starch granule size affects starch digestibility properties (Dhital, Shrestha, & Gidley, 2010; Noda et al., 2008; Parada & Aguilera, 2009), food quality such as noodle quality (Chen, Schols, & Voragen, 2003), and potentially crop productivity (Hofvander, Andersson, Larsson, & Larsson, 2004). Some studies of modified potato starches via inhibition of starch biosynthetic genes have altered granule morphology and number (Edwards et al., 1995; Jobling et al., 2002; Marshall et al., 1996). In the barley mutants (*sugary 1*), loss of ISA gene causes a profound effect on the number and timing of initiation of starch granules (Burton et al., 2002). And similar results were observed in potato tubers by downregulation of different ISA isoforms (Bustos et al., 2004). Expression of a tandem repeat of multiple starch-binding domains (SBD2-SBD5) from cyclodextrin glycosyltransferase of *B. circulans* in an amylose-free potato genetic background resulted in a large number of smaller starch granules (Ji et al., 2004; Nazarian-Firouzabadi et al., 2012).

5.2.6 Other Uses

Along with the traditional bulk utilization of starch in various fields such as paper and board industry, a range of new research directions and applications for starch and its derivatives has been developed during recent years. For instance, due to the biodegradability of starch-based materials, commercial trials with starch-based packagings for pasta (Italy) and PLA-based pots for yogurt (Germany) have already been performed (Weber, Haugaard, Festersen, & Bertelsen, 2002). Besides, potato starch granules seem, based on their ability to swell, the best source for natural microencapsulation of small molecules (Korus, Tomasik, & Lii, 2003; Lii, Tomasik, Hung, Yen, & Lai, 2003).

5.2.7 Perspectives of GMO Starches

In an era towards a biobased economy, the knowledge on how to improve complex carbohydrates such as starch is essential. A deeper understanding of the starch biosynthetic pathway, how storage starch granules are formed and how the composition, size, and shape can be changed and optimized for different bioproducts, is of great importance for food and nonfood applications.

In spite of its great importance, the development and commercialization of crops with altered starch properties using biotechnological approaches is being hampered by regulatory hurdles. The very high costs and the great deal of time needed, associated with the regulation of genetically modified crops (GMOs), are major problems. Although there is currently one GMO potato variety in the market, the Amflora, the commercialization of this variety has been challenged by farmers and environmental organizations. The development of new methods in plant breeding that would circumvent these regulatory problems would be of greatly stimulated the development of novel starches (Lusser, Parisi, Plan, & Rodriguez-Cerezo, 2012). The identification of genetic marker associated with starch properties and the exploitation of new mutations in tilling populations are other tools with great potential for uncovering key genes determining starch properties (Werij, Furrer, Van Eck, Visser, & Bachem, 2012).

Another bottleneck for the production of improved starches is associated with the difficulties in predicting beforehand the effect of a (trans) gene. The understanding of mechanism by which starch granules are made in the form of dense granules would be a great step forward in the synthesis of tailored starches for different biobased applications.

References

Arguello-Morales, M. A., Remaud-Simeon, M., Pizzut, S., Sarcabal, P., Willemot, R. M., & Monsan, P. (2000). Sequence analysis of the gene encoding alternansucrase, a sucrose glucosyltransferase from Leuconostoc mesenteroides NRRL B-1355. *FEMS Microbiology Letters*, *182*, 81–85.

Asatsuma, S., Sawada, C., Itoh, K., Okito, M., Kitajima, A., & Mitsui, T. (2005). Involvement of alpha-amylase I-1 in starch degradation in rice chloroplasts. *Plant and Cell Physiology*, *46*, 858–869.

Ball, S., Marianne, T., Dirick, L., Fresnoy, M., Delrue, B., & Decq, A. (1991). A Chlamydomonas-Reinhardtii low-starch mutant is defective for 3-phosphoglycerate activation and orthophosphate inhibition of Adp-Glucose pyrophosphorylase. *Planta*, *185*, 17–26.

Ball, S. G., & Morell, M. K. (2003). From bacterial glycogen to starch: Understanding the biogenesis of the plant starch granule. *Annual Review of Plant Biology*, *54*, 207–233.

Baunsgaard, L., Lutken, H., Mikkelsen, R., Glaring, M. A., Pham, T. T., & Blennow, A. (2005). A novel isoform of glucan, water dikinase phosphorylates pre-phosphorylated alpha-glucans and is involved in starch degradation in Arabidopsis. *Plant Journal*, *41*, 595–605.

Bhattacharyya, M. K., Smith, A. M., Ellis, T. H. N., Hedley, C., & Martin, C. (1990). The wrinkled-seed character of pea described by Mendel is caused by a transposon-like insertion in a gene encoding starch-branching enzyme. *Cell*, *60*, 115–122.

Bird, A. R., Hayakawa, T., Marsono, Y., Gooden, J. M., Record, I. R., Correll, R. L., et al. (2000). Coarse brown rice increases fecal and large bowel short-chain fatty acids and starch but lowers calcium in the large bowel of pigs. *Journal of Nutrition*, *130*, 1780–1787.

Blennow, A., Bay-Smidt, A. M., Olsen, C. E., & Moller, B. L. (2000). The distribution of covalently bound phosphate in the starch granule in relation to starch crystallinity. *International Journal of Biological Macromolecules*, *27*, 211–218.

Burton, R. A., Jenner, H., Carrangis, L., Fahy, B., Fincher, G. B., Hylton, C., et al. (2002). Starch granule initiation and growth are altered in barley mutants that lack isoamylase activity. *Plant Journal*, *31*, 97–112.

Bustos, R., Fahy, B., Hylton, C. M., Seale, R., Nebane, N. M., Edwards, A., et al. (2004). Starch granule initiation is controlled by a heteromultimeric isoamylase in potato tubers. *Proceedings of the National Academy of Sciences of the United States of America*, *101*, 2215–2220.

Carciofi, M., Shaif, S. S., Jensen, S. L., Blennow, A., Svensson, J. T., Vincze, E., et al. (2011). Hyperphosphorylation of cereal starch. *Journal of Cereal Science, 54*, 339–346.

Caspar, T., Huber, S. C., & Somerville, C. (1985). Alterations in growth, photosynthesis, and respiration in a starchless mutant of Arabidopsis-Thaliana (L) deficient in chloroplast phosphoglucomutase activity. *Plant Physiology, 79*, 11–17.

Caspar, T., Lin, T. P., Kakefuda, G., Benbow, L., Preiss, J., & Somerville, C. (1991). Mutants of Arabidopsis with altered regulation of starch degradation. *Plant Physiology, 95*, 1181–1188.

Ceballos, H., Sanchez, T., Morante, N., Fregene, M., Dufour, D., Smith, A. M., et al. (2007). Discovery of an amylose-free starch mutant in cassava (Manihot esculenta Crantz). *Journal of Agricultural and Food Chemistry, 55*, 7469–7476.

Chen, Z., Schols, H. A., & Voragen, A. G. J. (2003). Starch granule size strongly determines starch noodle processing and noodle quality. *Journal of Food Science, 68*, 1584–1589.

Cote, G. L. (1992). Low-viscosity alpha-D-glucan fractions derived from sucrose which are resistant to enzymatic digestion. *Carbohydrate Polymers, 19*, 249–252.

Cote, G. L. (2002). Subcutaneous fluorescence glucose sensing using a competitive binding assay. *Abstracts of Papers of the American Chemical Society, 224*, U116.

Cote, G. L., & Robyt, J. F. (1982). Isolation and partial characterization of an extracellular glucansucrase from Leuconostoc mesenteroides NRRL B-1355 that synthesizes an alternating (1 goes to 6), (1 goes to 3)-alpha-D-glucan. *Carbohydrate Research, 101*, 57–74.

Delatte, T., Umhang, M., Trevisan, M., Eicke, S., Thorneycroft, D., Smith, S. M., et al. (2006). Evidence for distinct mechanisms of starch granule breakdown in plants. *Journal of Biological Chemistry, 281*, 12050–12059.

Delrue, B., Fontaine, T., Routier, F., Decq, A., Wieruszeski, J. M., Van Den Koornhuyse, N., et al. (1992). Waxy Chlamydomonas reinhardtii: Monocellular algal mutants defective in amylose biosynthesis and granule-bound starch synthase activity accumulate a structurally modified amylopectin. *Journal of Bacteriology, 174*, 3612–3620.

Denyer, K., Foster, J., & Smith, A. M. (1995). The contributions of adenosine 5′-diphosphoglucose pyrophosphorylase and starch-branching enzyme to the control of starch synthesis in developing pea embryos. *Planta, 197*, 57–62.

Dhital, S., Shrestha, A. K., & Gidley, M. J. (2010). Relationship between granule size and in vitro digestibility of maize and potato starches. *Carbohydrate Polymers, 82*, 480–488.

Doehlert, D. C., & Knutson, C. A. (1991). 2 Classes of starch debranching enzymes from developing maize kernels. *Journal of Plant Physiology, 138*, 566–572.

Donath, H., Avtomonov, E. V., Sarraje, I., Von Dahlen, K. H., El-Essawi, M., Lorberth, J., et al. (1998). Organoplatinum compounds VII—Trimethylplatinum fluoride [(CH3)(3)PtF](4), the missing link in organoplatinum cluster chemistry: Its synthesis, crystal structure and a comparison to the crystal structure of [(CH3)(3)PtOH](4). *Journal of Organometallic Chemistry, 559*, 191–196.

Edwards, A., Marshall, J., Sidebottom, C., Visser, R. G. F., Smith, A. M., & Martin, C. (1995). Biochemical and molecular characterization of a novel starch synthase from potato-tubers. *Plant Journal, 8*, 283–294.

Ellis, R. P., Cochrane, M. P., Dale, M. F. B., Duffus, C. M., Lynn, A., Morrison, I. M., et al. (1998). Starch production and industrial use. *Journal of the Science of Food and Agriculture, 77*, 289–311.

Emes, M. J., & Neuhaus, H. E. (1997). Metabolism and transport in non-photosynthetic plastids. *Journal of Experimental Botany, 48*, 1995–2005.

Espada, J. (1962). Enzymic synthesis of adenosine diphosphate glucose from glucose 1-phosphate and adenosine triphosphate. *Journal of Biological Chemistry, 237*, 3577.

Ferretti, J. J., Gilpin, M. L., & Russell, R. R. (1987). Nucleotide sequence of a glucosyltransferase gene from Streptococcus sobrinus MFe28. *Journal of Bacteriology, 169*, 4271–4278.

Fincher, G. B. (1989). Molecular and cellular biology associated with endosperm mobilization in germinating cereal-grains. *Annual Review of Plant Physiology and Plant Molecular Biology, 40*, 305–346.

Firouzabadi, F. N., Kok-Jacon, G. A., Vincken, J. P., Ji, Q., Suurs, L. C. J. M., & Visser, R. G. F. (2007). Fusion proteins comprising the catalytic domain of mutansucrase and a starch-binding domain can alter the morphology of amylose-free potato starch granules during biosynthesis. *Transgenic Research, 16*, 645–656.

Firouzabadi, F. N., Vincken, J. P., Ji, Q., Suurs, L. C. J. M., & Visser, R. G. F. (2007). Expression of an engineered granule-bound Escherichia coli maltose acetyltransferase in wild-type and amf potato plants. *Plant Biotechnology Journal, 5*, 134–145.

Frohberg, C. (2008). Genetically modified plants which synthesize a starch having increased swelling power. Int Patent WO 08/017518

Groenwall, A. J., & Ingelman, B. G. A. (1948). Manufacture of infusion and injection fluids, U.S. Patent 2, 437–518

Hamada, S., & Slade, H. D. (1980). Biology, immunology, and cariogenicity of Streptococcus mutans. *Microbiological Reviews, 44*, 331–384.

Heldt, H. W., & Pfaff, E. (1969). Adenine nucleotide translocation in mitochondria—Quantitative evaluation of correlation between phosphorylation of endogenous and exogenous Adp in mitochondria. *European Journal of Biochemistry, 10*, 494–500.

Hizukuri, S., Tabata, S., & Nikuni, Z. (1970). Studies on starch phosphate. 1. Estimation of glucose-6-phosphate residues in starch and presence of other bound phosphate(S). *Starke, 22*, 338.

Hofvander, P., Andersson, M., Larsson, C. T., & Larsson, H. (2004). Field performance and starch characteristics of high-amylose potatoes obtained by antisense gene targeting of two branching enzymes. *Plant Biotechnology Journal, 2*, 311–320.

Hovenkamphermelink, J. H. M., Jacobsen, E., Ponstein, A. S., Visser, R. G. F., Vosscheperkeuter, G. H., Bijmolt, E. W., et al. (1987). Isolation of an amylose-free starch mutant of the potato (Solanum-Tuberosum-L). *Theoretical and Applied Genetics, 75*, 217–221.

Hseih, J. S. (1988). Genetic studies on the *Wx* gene of sorghum (*Sorghum bicolor* (E.) Moench.) I. Examination of the protein product of the waxy locus. *Botanical Bulletin of Academia Sinica, 29*, 293–299.

Ihemere, U., Arias-Garzon, D., Lawrence, S., & Sayre, R. (2006). Genetic modification of cassava for enhanced starch production. *Plant Biotechnology Journal, 4*, 453–465.

James, M. G., Robertson, D. S., & Myers, A. M. (1995). Characterization of the maize gene sugary1, a determinant of starch composition in kernels. *Plant Cell, 7*, 417–429.

Janecek, S., Svensson, B., & Macgregor, E. A. (2003). Relation between domain evolution, specificity, and taxonomy of the alpha-amylase family members containing a C-terminal starch-binding domain. *European Journal of Biochemistry, 270*, 635–645.

Jeanes, A., Haynes, W. C., Wilham, C. A., Rankin, J. C., Melvin, E. H., Austin, M. J., et al. (1954). Characterization and classification of dextrans from 96 strains of bacteria. *Journal of the American Chemical Society, 76*, 5041–5052.

Jenkins, J. P. J., Cameron, R. E., & Donald, A. M. (1993). A universal feature in the structure of starch granules from different botanical sources. *Starch-Starke, 45*, 417–420.

Ji, Q., Oomen, R. J. F., Vincken, J. P., Bolam, D. N., Gilbert, H. J., Suurs, L. C. J. M., et al. (2004). Reduction of starch granule size by expression of an engineered tandem starch-binding domain in potato plants. *Plant Biotechnology Journal, 2*, 251–260.

Ji, Q., Vincken, J. P., Suurs, L. C. J. M., & Visser, R. G. F. (2003). Microbial starch-binding domains as a tool for targeting proteins to granules during starch biosynthesis. *Plant Molecular Biology, 51*, 789–801.

Jobling, S. (2004). Improving starch for food and industrial applications. *Current Opinion in Plant Biology, 7*, 210–218.

Jobling, S. A., Schwall, G. P., Westcott, R. J., Sidebottom, C. M., Debet, M., Gidley, M. J., et al. (1999). A minor form of starch branching enzyme in potato (Solanum tuberosum L.)

tubers has a major effect on starch structure: Cloning and characterisation of multiple forms of SBE A. *Plant Journal, 18,* 163–171.

Jobling, S. A., Westcott, R. J., Tayal, A., Jeffcoat, R., & Schwall, G. P. (2002). Production of a freeze-thaw-stable potato starch by antisense inhibition of three starch synthase genes. *Nature Biotechnology, 20,* 295–299.

Kaplan, F., & Guy, C. L. (2005). RNA interference of Arabidopsis beta-amylase8 prevents maltose accumulation upon cold shock and increases sensitivity of PSII photochemical efficiency to freezing stress. *Plant Journal, 44,* 730–743.

Kaur, B., Ariffin, F., Bhat, R., & Karim, A. A. (2012). Progress in starch modification in the last decade. *Food Hydrocolloids, 26,* 398–404.

Kim, W., Johnson, J. W., Graybosch, R. A., & Gaines, C. S. (2003). Physicochemical properties and end-use quality of wheat starch as a function of waxy protein alleles. *Journal of Cereal Science, 37,* 195–204.

Kimura, T., Otani, M., Noda, T., Ideta, O., Shimada, T., & Saito, A. (2001). Absence of amylose in sweet potato [Ipomoea batatas (L.) Lam.] following the introduction of granule-bound starch synthase I cDNA. *Plant Cell Reports, 20,* 663–666.

Kok-Jacon, G. A., Ji, Q., Vincken, J. P., & Visser, R. G. F. (2003). Towards a more versatile alpha-glucan biosynthesis in plants. *Journal of Plant Physiology, 160,* 765–777.

Kok-Jacon, G. A., Vincken, J. P., Suurs, L. C. J. M., & Visser, R. G. F. (2005). Mutan produced in potato amyloplasts adheres to starch granules. *Plant Biotechnology Journal, 3,* 341–351.

Kok-Jacon, G. A., Vincken, J. P., Suurs, L. C. J. M., Wang, D. N., Liu, S., & Visser, R. G. F. (2007). Expression of alternansucrase in potato plants. *Biotechnology Letters, 29,* 1135–1142.

Kooistra, E. (1962). On differences between smooth and 3 types of wrinkled peas. *Euphytica, 11,* 357–373.

Kortstee, A. J., Vermeesch, A. M. S., Devries, B. J., Jacobsen, E., & Visser, R. G. F. (1996). Expression of Escherichia coli branching enzyme in tubers of amylose-free transgenic potato leads to an increased branching degree of the amylopectin. *Plant Journal, 10,* 83–90.

Korus, J., Tomasik, P., & Lii, C. Y. (2003). Microcapsules from starch granules. *Journal of Microencapsulation, 20,* 47–56.

Kossmann, J., Buttcher, V., Abel, G. J. W., Duwenig, E., Emmermann, M., Frohberg, C., et al. (1997). Starch biosynthesis and modification of starch structure in transgenic plants. *Macromolecular Symposia, 120,* 29–38.

Kossmann, J., & Lloyd, J. (2000). Understanding and influencing starch biochemistry. *Critical Reviews in Biochemistry and Molecular Biology, 35,* 141–196.

Kotting, O., Pusch, K., Tiessen, A., Geigenberger, P., Steup, M., & Ritte, G. (2005). Identification of a novel enzyme required for starch metabolism in Arabidopsis leaves. The phosphoglucan, water dikinase. *Plant Physiology, 137,* 242–252.

Kralj, S., Van Geel-Schutten, G. H., Dondorff, M. M. G., Kirsanovs, S., Van Der Maarel, M. J. E. C., & Dijkhuizen, L. (2004). Glucan synthesis in the genus Lactobacillus: Isolation and characterization of glucansucrase genes, enzymes and glucan products from six different strains. *Microbiology, 150,* 3681–3690.

Krohn, B. M., Stark, D. M., Barry, G. F., Preiss, J., & Kishore, G. M. (1994). Modification of starch structure in transgenic potato. *Plant Physiology, 105,* 37.

Lanahan, M. B. & Basu, S. S. (2005). Modified starch, uses methods for production thereof, Int Patent WO 05/002359.

Leathers, T. D., Nunnally, M. S., & Cote, G. L. (2002). Modification of alternan by novel Penicillium spp. *Journal of Industrial Microbiology & Biotechnology, 29,* 177–180.

Leathers, T. D., Nunnally, M. S., Ahlgren, J. A., & Cote, G. L. (2003). Characterization of a novel modified alternan. *Carbohydrate Polymers, 54,* 107–113.

Li, Y. T., Han, D. X., Hu, G. R., Dauvillee, D., Sommerfeld, M., Ball, S., et al. (2010). Chlamydomonas starchless mutant defective in ADP-glucose pyrophosphorylase hyperaccumulates triacylglycerol. *Metabolic Engineering, 12,* 387–391.

Lii, C. Y., Tomasik, P., Hung, W. L., Yen, M. T., & Lai, V. M. F. (2003). Granular starches as dietary fibre and natural microcapsules. *International Journal of Food Science and Technology, 38*, 677–685.
Lloyd, J. R., Landschutze, V., & Kossmann, J. (1999). Simultaneous antisense inhibition of two starch-synthase isoforms in potato tubers leads to accumulation of grossly modified amylopectin. *Biochemical Journal, 338*, 515–521.
Lloyd, J. R., Springer, F., Buleon, A., Muller-Rober, B., Willmitzer, L., & Kossmann, J. (1999). The influence of alterations in ADP-glucose pyrophosphorylase activities on starch structure and composition in potato tubers. *Planta, 209*, 230–238.
Lorberth, R., Ritte, G., Willmitzer, L., & Kossmann, J. (1998). Inhibition of a starch-granule-bound protein leads to modified starch and repression of cold sweetening. *Nature Biotechnology, 16*, 473–477.
Lusser, M., Parisi, C., Plan, D., & Rodriguez-Cerezo, E. (2012). Deployment of new biotechnologies in plant breeding. *Nature Biotechnology, 30*, 231–239.
Machovic, M., Svensson, B., Ann Macgregor, E., & Janecek, S. (2005). A new clan of CBM families based on bioinformatics of starch-binding domains from families CBM20 and CBM21. *FEBS Journal, 272*, 5497–5513.
Marshall, J., Sidebottom, C., Debet, M., Martin, C., Smith, A. M., & Edwards, A. (1996). Identification of the major starch synthase in the soluble fraction of potato tubers. *Plant Cell, 8*, 1121–1135.
Matheson, N. K. (1996). The chemical structure of amylose and amylopectin fractions of starch from tobacco leaves during development and diurnally nocturnally. *Carbohydrate Research, 282*, 247–262.
Mikkelsen, R., Baunsgaard, L., & Blennow, A. (2004). Functional characterization of alpha-glucan, water dikinase, the starch phosphorylating enzyme. *The Biochemical Journal, 377*, 525–532.
Moire, L., Rezzonico, E., & Poirier, Y. (2003). Synthesis of novel biomaterials in plants. *Journal of Plant Physiology, 160*, 831–839.
Monchois, V., Vignon, M., & Russell, R. R. B. (1999). Isolation of key amino acid residues at the N-terminal end of the core region Streptococcus downei glucansucrase, GTF-I. *Applied Microbiology and Biotechnology, 52*, 660–665.
Monchois, V., Willemot, R. M., Remaudsimeon, M., Croux, C., & Monsan, P. (1996). Cloning and sequencing of a gene coding for a novel dextransucrase from Leuconostoc mesenteroides NRRL B-1299 synthesizing only alpha(1–6) and alpha(1–3) linkages. *Gene, 182*, 23–32.
Mooser, G. (1992). Glycosidases and glycosyltransferases. *The Enzymes, 20*, 187–233.
Morell, M. K., Kosar-Hashemi, B., Cmiel, M., Samuel, M. S., Chandler, P., Rahman, S., et al. (2003). Barley sex6 mutants lack starch synthase IIa activity and contain a starch with novel properties. *Plant Journal, 34*, 172–184.
Mouille, G., Maddelein, M. L., Libessart, N., Talaga, P., Decq, A., Delrue, B., et al. (1996). Pre-amylopectin processing: A mandatory step for starch biosynthesis in plants. *Plant Cell, 8*, 1353–1366.
Nakamura, Y. (1996). Some properties of starch debranching enzymes and their possible role in amylopectin biosynthesis. *Plant Science, 121*, 1–18.
Nakamura, T., Yamamori, M., Hirano, H., Hidaka, S., & Nagamine, T. (1995). Production of waxy (amylose-free) wheats. *Molecular and General Genetics, 248*, 253–259.
Nazarian-Firouzabadi, F., Trindade, L. M., & Visser, R. G. F. (2012). Production of small starch granules by expression of a tandem-repeat of a family 20 starch-binding domain (SBD3-SBD5) in an amylose-free potato genetic background. *Functional Plant Biology, 39*, 146–155.
Nelson, D. L., & Cox, M. M. (2004). Chapter 7: Carbohydrates and glycobiology. In D. L. Nelson & M. M. Cox (Eds.), *Lehninge principles of biochemistry* (4th ed., pp. 238–272). NewYork, USA: W.H. Freeman.

Nelson, O. E., & Rines, H. W. (1962). Enzymatic deficiency in waxy mutant of maize. *Biochemical and Biophysical Research Communications, 9*, 297–300.
Neuhaus, H. E., & Emes, M. J. (2000). Nonphotosynthetic metabolism in plastids. *Annual Review of Plant Physiology and Plant Molecular Biology, 51*, 111–140.
Noda, T., Kimura, T., Otani, M., Ideta, O., Shimada, T., Saito, A., et al. (2002). Physicochemical properties of amylose-free starch from transgenic sweet potato. *Carbohydrate Polymers, 49*, 253–260.
Noda, T., Takigawa, S., Matsuura-Endo, C., Suzuki, T., Hashimoto, N., Kottearachchi, N. S., et al. (2008). Factors affecting the digestibility of raw and gelatinized potato starches. *Food Chemistry, 110*, 465–470.
Ohdan, K., Kuriki, T., Takata, H., Kaneko, H., & Okada, S. (2000). Introduction of raw starch-binding domains into Bacillus subtilis alpha-amylase by fusion with the starch-binding domain of Bacillus cyclomaltodextrin glucanotransferase. *Applied and Environmental Microbiology, 66*, 3058–3064.
Okuno, K., & Sakaguchi, S. (1982). Inheritance of starch characteristics in perisperm of Amaranthus-Hypochondriacus. *Journal of Heredity, 73*, 467.
Pan, D., & Nelson, O. E. (1984). A debranching enzyme deficiency in endosperms of the sugary-1 mutants of maize. *Plant Physiology, 74*, 324–328.
Parada, J., & Aguilera, J. M. (2009). In vitro digestibility and glycemic response of potato starch is related to granule size and degree of gelatinization. *Journal of Food Science, 74*, E34–E38.
Preiss, J., Kumar, A., & Ghosh, P. (1988). A study of the catalytic and regulatory functions of a sugar nucleotide synthetase using invitro and invivo mutagenesis techniques. *Glycoconjugate Journal, 5*, 365.
Preiss, J., & Sivak, M. N. (1998). Biochemistry, molecular biology and regulation of starch synthesis. *Genetic Engineering, 20*, 177–223.
Regina, A., Bird, A., Topping, D., Bowden, S., Freeman, J., Barsby, T., et al. (2006). High-amylose wheat generated by RNA interference improves indices of large-bowel health in rats. *Proceedings of the National Academy of Sciences of the United States of America, 103*, 3546–3551.
Ritchie, S., Swanson, S. J., & Gilroy, S. (2000). Physiology of the aleurone layer and starchy endosperm during grain development and early seedling growth: new insights from cell and molecular biology. *Seed Science Research, 10*, 193–212.
Ritte, G., Lloyd, J. R., Eckermann, N., Rottmann, A., Kossmann, J., & Steup, M. (2002). The starch-related R1 protein is an alpha-glucan, water dikinase. *Proceedings of the National Academy of Sciences of the United States of America, 99*, 7166–7171.
Robyt, J. F. (1995a). Mechanisms in the glucansucrase synthesis of polysaccharides and oligosaccharides from sucrose. *Advances in Carbohydrate Chemistry and Biochemistry, 51*, 133–168.
Robyt, J. F. (1995b). Properties and uses of dextrans and related glucans. *Abstracts of Papers of the American Chemical Society, 209*, 86-Pmse.
Roper, H. (2002). Renewable raw materials in Europe—Industrial utilisation of starch and sugar. *Starch-Starke, 54*, 89–99.
Safford, R., Jobling, S. A., Sidebottom, C. M., Westcott, R. J., Cooke, D., Tober, K. J., et al. (1998). Consequences of antisense RNA inhibition of starch branching enzyme activity on properties of potato starch. *Carbohydrate Polymers, 35*, 155–168.
Scheidig, A., Frohlich, A., Schulze, S., Lloyd, J. R., & Kossmann, J. (2002). Downregulation of a chloroplast-targeted beta-amylase leads to a starch-excess phenotype in leaves. *Plant Journal, 30*, 581–591.
Schewe, G., Knies, P., Amati, S. F., Lörz, H., Becker, D., Uwer, U., Landschütze, V., Pilling, J., & Frohberg, K. (2002). Monocotyledon plant cells and plants which synthesise modified starch. Int Patent WO 02/34923.

Schmidt, R. M. (2009). *Amflora facts. http://www.basf.com/biotechnology.*
Schunemann, D., Borchert, S., Flugge, U. I., & Heldt, H. W. (1993). Adp/Atp translocator from pea root plastids—Comparison with translocators from spinach-chloroplasts and pea leaf mitochondria. *Plant Physiology, 103*, 131–137.
Schwall, G. P., Safford, R., Westcott, R. J., Jeffcoat, R., Tayal, A., Shi, Y. C., et al. (2000). Production of very-high-amylose potato starch by inhibition of SBE A and B. *Nature Biotechnology, 18*, 551–554.
Seymour, F. R., & Knapp, R. D. (1980). Structural analysis of dextrans, from strains of leuconostoc and related genera, that contain 3-*O*-α-d-glucosylated α-d-glucopyranosyl residues at the branch points, or in consecutive, linear positions. *Carbohydrate Research, 81*, 105–129.
Seymour, F. R., Knapp, R. D., Bishop, S. H., & Jeanes, A. (1979). Structural analysis of leuconostoc dextrans containing 3-*O*-α-d-glucosylated α-d-glucosyl residues in both linear-chain and branch-point positions, or only in branch-point positions, by methylation and by ^{13}C-N.M.R. spectroscopy. *Carbohydrate Research, 74*, 41–62.
Seymour, F. R., Slodki, M. E., Plattner, R. D., & Jeanes, A. (1977). Six unusual dextrans: Methylation structural analysis by combined g.l.c.—m.s. of per-O-acetyl-aldononitriles. *Carbohydrate Research, 53*, 153–166.
Sharma, R., Sissons, M. J., Rathjen, A. J., & Jenner, C. F. (2002). The null-4A allele at the wax locus in durum wheat affects pasta cooking quality. *Journal of Cereal Science, 35*, 287–297.
Shewmaker, C. K., Boyer, C. D., Wiesenborn, D. P., Thompson, D. B., Boersig, M. R., Oakes, J. V., et al. (1994). Expression of Escherichia-coli glycogen-synthase in the tubers of transgenic potatoes (Solanum-tuberosum) results in a highly branched starch. *Plant Physiology, 104*, 1159–1166.
Sidebotham, R. L. (1974). Dextrans. In R. S. Tipson & D. Harton (Eds.), *Advances in Carbohydrate Chemistry and Biochemistry: 30.* (pp. 371–444). San Diego: Academic Press.
Simpson, C. L., Cheetham, N. W. H., Giffard, P. M., & Jacques, N. A. (1995). 4 Glucosyltransferases, Gtfj, Gtfk, Gtfl and Gtfm, from Streptococcus-Salivarius Atcc-25975. *Microbiology, 141*, 1451–1460.
Smith, A. M., Bettey, M., & Bedford, I. D. (1989). Evidence that the Rb locus alters the starch content of developing pea embryos through an effect on Adp glucose pyrophosphorylase. *Plant Physiology, 89*, 1279–1284.
Smith, A. M., Denyer, K., & Martin, C. (1997). The synthesis of the starch granule. *Annual Review of Plant Physiology and Plant Molecular Biology, 48*, 67–87.
Stark, D. M., Timmerman, K. P., Barry, G. F., Preiss, J., & Kishore, G. M. (1992). Regulation of the amount of starch in plant-tissues by Adp glucose pyrophosphorylase. *Science, 258*, 287–292.
Sweetlove, L. J., Burrell, M. M., & Aprees, T. (1996). Characterization of transgenic potato (Solanum tuberosum) tubers with increased ADPglucose pyrophosphorylase. *Biochemical Journal, 320*, 487–492.
Tjaden, J., Schwoppe, C., Mohlmann, T., Quick, P. W., & Neuhaus, H. E. (1998). Expression of a plastidic ATP/ADP transporter gene in Escherichia coli leads to a functional adenine nucleotide transport system in the bacterial cytoplasmic membrane. *Journal of Biological Chemistry, 273*, 9630–9636.
Topping, D. L., Morell, M. K., King, R. A., Li, Z. Y., Bird, A. R., & Noakes, M. (2003). Resistant starch and health—Himalaya 292, a novel barley cultivar to deliver benefits to consumers. *Starch-Starke, 55*, 539–545.
Umemoto, T., Yano, M., Satoh, H., Shomura, A., & Nakamura, Y. (2002). Mapping of a gene responsible for the difference in amylopectin structure between japonica-type and indica-type rice varieties. *Theoretical and Applied Genetics, 104*, 1–8.
Vandenkoornhuyse, N., Libessart, N., Delrue, B., Zabawinski, C., Decq, A., Iglesias, A., et al. (1996). Control of starch composition and structure through substrate supply in the monocellular alga Chlamydomonas reinhardtii. *Journal of Biological Chemistry, 271*, 16281–16287.

Visser, R. G. F., Somhorst, I., Kuipers, G. J., Ruys, N. J., Feenstra, W. J., & Jacobsen, E. (1991). Inhibition of the expression of the gene for granule-bound starch synthase in potato by antisense constructs. *Molecular and General Genetics, 225*, 289–296.

Visser, R. G. F., Suurs, L. C. J. M., Bruinenberg, P. M., Bleeker, I., & Jacobsen, E. (1997). Comparison between amylose-free and amylose containing potato starches. *Starch-Starke, 49*, 438–443.

Wang, Z. Y., Chen, X. P., Wang, J. H., Liu, T. S., Liu, Y., Zhao, L., et al. (2007). Increasing maize seed weight by enhancing the cytoplasmic ADP-glucose pyrophosphorylase activity in transgenic maize plants. *Plant Cell, Tissue and Organ Culture, 88*, 83–92.

Wattebled, F., Dong, Y., Dumez, S., Delvalle, D., Planchot, V., Berbezy, P., et al. (2005). Mutants of Arabidopsis lacking a chloroplastic isoamylase accumulate phytoglycogen and an abnormal form of amylopectin. *Plant Physiology, 138*, 184–195.

Weber, C. J., Haugaard, V., Festersen, R., & Bertelsen, G. (2002). Production and applications of biobased packaging materials for the food industry. *Food Additives and Contaminants, 19*, 172–177.

Wei, C. X., Qin, F. L., Zhu, L. J., Zhou, W. D., Chen, Y. F., Wang, Y. P., et al. (2010). Microstructure and ultrastructure of high-amylose rice resistant starch granules modified by antisense RNA inhibition of starch branching enzyme. *Journal of Agricultural and Food Chemistry, 58*, 1224–1232.

Werij, J. S., Furrer, H., Van Eck, H. J., Visser, R. G. F., & Bachem, C. W. B. (2012). A limited set of starch related genes explain several interrelated traits in potato. *Euphytica, 186*, 501–516.

William, R. M. (2009). Starch use in foods. In J. BeMiller & R. Whistler (Eds.), *Starch: Chemistry and technology* (3rd ed., pp. 746–788). Oxford, UK: Elsevier.

Yu, T. S., Kofler, H., Hausler, R. E., Hille, D., Flugge, U. I., Zeeman, S. C., et al. (2001). The Arabidopsis sex1 mutant is defective in the R1 protein, a general regulator of starch degradation in plants, and not in the chloroplast hexose transporter. *Plant Cell, 13*, 1907–1918.

Yu, T. S., Zeeman, S. C., Thorneycroft, D., Fulton, D. C., Dunstan, H., Lue, W. L., et al. (2005). alpha-Amylase is not required for breakdown of transitory starch in Arabidopsis leaves. *Journal of Biological Chemistry, 280*, 9773–9779.

Zeeman, S. C., & Ap Rees, T. (1999). Changes in carbohydrate metabolism and assimilate export in starch-excess mutants of Arabidopsis. *Plant Cell and Environment, 22*, 1445–1453.

Zeeman, S., Chen, J., Gatehouse, J., Smith, A., & Smith, S. (2007). The pathways and control of starch breakdown in Arabidopsis thaliana. *Photosynthesis Research, 91*, 237.

Zeeman, S. C., Kossmann, J., & Smith, A. M. (2010). Starch: its metabolism, evolution, and biotechnological modification in plants. *Annual Review of Plant Biology, 61*, 209–234.

Zeeman, S. C., Smith, S. M., & Smith, A. M. (2007). The diurnal metabolism of leaf starch. *Biochemical Journal, 401*, 13–28.

Zeeman, S. C., Umemoto, T., Lue, W. L., Au-Yeung, P., Martin, C., Smith, A. M., et al. (1998). A mutant of Arabidopsis lacking a chloroplastic isoamylase accumulates both starch and phytoglycogen. *The Plant Cell, 10*, 1699–1712.

PART IV

NOVEL STARCH MODIFIED BY CHEMISTRY

CHAPTER

5

Starch Modification to Develop Novel Starch-Biopolymer Blends: State of Art and Perspectives

Ashok K. Shrestha[1], *Peter J. Halley*[2]

[1]School of Science and Health, University of Western Sydney, Hawkesbury Campus, Richmond, New South Wales, Australia

[2]Centre of High Performance Polymers, School of Chemical Engineering/ AIBN, The University of Queensland, St. Lucia, Queensland, Australia

1 STARCH STRUCTURE

Starch is one of the most abundant biomolecule on earth and a major part of plant foods, providing energy to humans. The structure of starch in plant science is quite intriguing as it is formed for the purpose of storage of molecular glucose structures in the form of macroscopic granules. The starch structure dictates the release of glucose into digestive tract during digestion as well as determining various functional properties of starch in other applications.

Starch molecules are composed of hundreds or thousands of D-glucose units, which are joined together by the α-D (1→4) glucosidic linkages to form linear polymer chains (amylose) and α-D (1→4) linkages with α-D (1→6) glucosidic side chains to form branched chains (amylopectin) (Figure 5.1). However, starch in plant is stored in the starch granules, which consists of a complex hierarchical structure (Cameron & Donald, 1992), which is given as the molecules (~nm), the growth ring (~100 s nm), and the whole grain (~μm) (Figure 5.2). The starch molecules may have small amount of lipids, proteins, phosphorous, and some minerals (Galliard & Bowler, 1987; Tester, Karkalas, & Qi, 2004; Tester, Karkalas, & Xin, 2004). The starches from most plants are composed of around 15-30% amylose

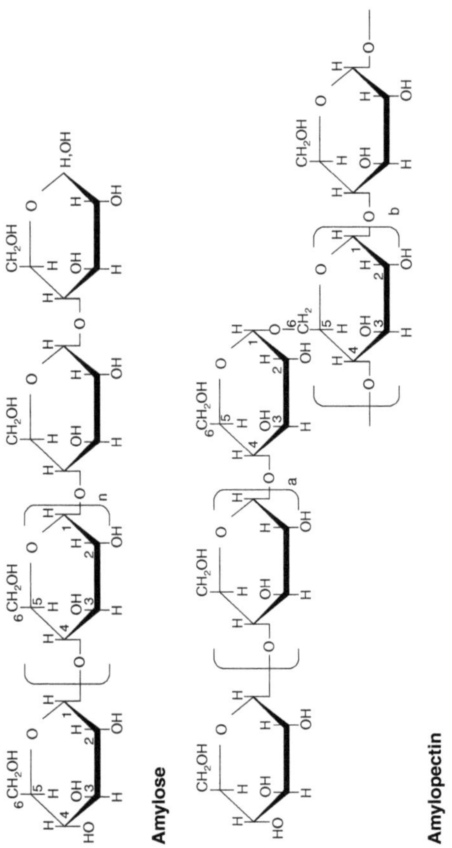

FIGURE 5.1 Structure of amylose and amylopectin (Tester, Karkalas, & Qi, 2004).

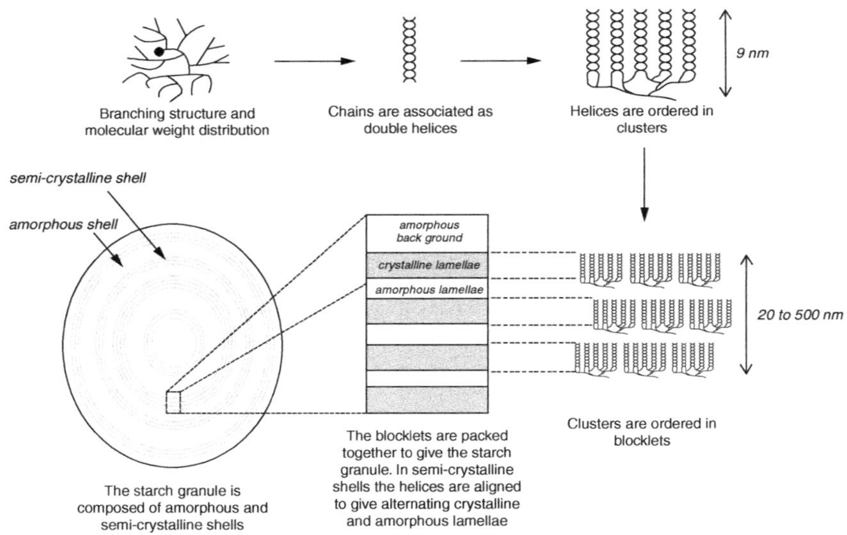

FIGURE 5.2 Hierarchy of starch structure, from the molecular to the granular level. (For color version of this figure, the reader is referred to the online version of this chapter.)

and 70-85% amylopectin, although, mutations affecting starch biosynthesis can dramatically affect the amount of both molecules in starch granules (Davis, Supatcharee, Khandelwal, & Chibbar, 2003; Galliard & Bowler, 1987).

Amylose is a long-chain molecule with a typical molecular weight of about 10^5-10^6 and degree of polymerization of 500-5000 (Yoshimoto, Tashiro, Takenouchi, & Takeda, 2000). Approximately 25-55% (molecular basis) of the amylose have secondary chains connected through intermittent α $(1 \rightarrow 6)$ branch points ranging from 2 to 11 depending on the botanical origin (Morrison & Karkalas, 1990). Compared to amylose, amylopectin is a very large and more frequently branched molecule with 95% α-D $(1 \rightarrow 4)$ and 5% α-D $(1 \rightarrow 6)$ glucosidic linkages and molecular weight of 10^7-10^9 (Tester, Karkalas, & Qi, 2004; Tester, Karkalas, & Xin, 2004). Although the degree of polymerization of amylopectin molecules ranged from 700 to 26,500, the individual chains are relatively shorter and more complex compared to those of amylose molecules (Tester, Karkalas, & Qi, 2004; Tester, Karkalas, & Xin, 2004). These short chains in amylopectin are not distributed randomly throughout and instead form a highly organized three-dimensional structure in the form of clusters (Hizukuri, 1986).

Starch structure is also known as "the other double helix" (Hancock & Tarbet, 2000) due to its helical structure, similar to DNA strand (Figure 5.2). The maltose structure indicates the torsion angles across the α-D $(1 \rightarrow 4)$ glucosidic bond between two glucose molecules are slighted twisted. If a series of glucose units is connected maintaining these torsion angles across the glycosidic bonds, a left-hand helix is obtained. Individual helices are intertwined to give the double helix (Figure 5.2) that forms the basis of crystalline structure of A, B, and C, as shown by recent crystallographic studies (Imberty, Buleon, Tran, & Perez, 1991). It has been proposed that at least 10 glucose units are required to form double helices within the ordered arrays of amylopectin (Imberty et al., 1991). The "A" type is commonly found in cereals such as maize, wheat, and rice; "B" type in tubers such as potatoes; and "C" type, which is a combination of A and B types, in legumes such as smooth peas, e.g., the center part exhibiting the B form and the outermost regions the A type (Gidley, 1987). A less prominent starch polymorph in starch granule is V type, which is characterized by the presence of internally absorbed small molecules such as iodine, DMSO, *n*-butanol, or lipids in the single helical structure of amylose (Tester, Karkalas, & Qi, 2004; Tester, Karkalas, & Xin, 2004).

It is now well established that amylopectin, a highly branched molecule with its side chain branches, intertwined to form double helices forming the basis for crystals. The amylose molecules are known to exist inside the granules as amylose-lipid inclusion complexes or free amylose (Morrison, 1988). The role of amylose in starch granules is not well understood.

According to "cluster model," the amylopectin molecule made up of three broad classes of glucose chains, A, B, and C. The outer chains (A) are glycosidically linked at their potential reducing group through C-6 position of glucose residue to an inner chain (B). B chains bind to other B chains or to a C chain that has a single reducing terminal end (Hizukuri, 1986). The amylopectin side chains are incorporated into alternating crystalline and amorphous lamellae, 9-10 nm in size (Jenkins & Donald, 1996). The side chain clusters with double helices form the crystalline lamellae, whereas the branch points form the amorphous lamellae. Figure 5.2 showed these alternating structures form a blocklet structure, representing one segment of lamellar growth ring in the starch granule. A number of these lamellae combine to form crystalline growth rings, separated by amorphous growth rings in concentric, onion-like granule architecture of ~100 nm size. These growth rings are in fact the result of multiple concentric shells (or lamellae) of increasing diameter extending from hilum toward the surface of the granule. Every variety of starch granules has its own characteristics size and shapes. Appearance of birefringence or Maltese cross when granules are observed under the polarized light gives some assumption about the correlation between molecular axis and principal axis that the underlying packing of the molecule is radial (Banks, Geddes, Greenwood, & Jones, 1972).

2 STARCH FUNCTIONALITY

Starch has ability to impart a broad range of functional properties in food and equally in industrial products (BeMiller, 2007). The functionalities of the starch are largely influenced by the properties of the granular as well as supramolecular assembly of the starch, processing conditions, and modification of structure (Donald, 2004). Starch structure-functionality relationships have been interlinked in numerous characterizations of different starches (Cooke & Gildley, 1992; Jenkins & Donald, 1998). Development of various starch-based products needs useful insights on the possible structural factors governing the variations in starch functionalities. The major determinants of starch functionality are listed as granule size and shape, starch crystallinity, amylose and amylopectin ratio, packing density, presence of fat, encapsulated starch granules, swelling power and solubility, gelatinization, retrogradation, rheological properties, and many more. This consequently highlights the need for a more robust starch structure-functionality relationship, which considers each of the structural aspects inherent in most starch granules and relates them with the observed variation in their functional properties (Tan, 2005).

2.1 Gelatinization and Retrogradation

Due to its complex architecture, the processing of starch granules typically involves the disruption of the structural order within the granules when it is heated in water. This leads to disruption of the molecular organization within starch granules, an irreversible event. The chain of events that leads to the formation of "starch paste" or "gel" is known as *gelatinization*. The gelatinization temperature (GT) varies with the origin of starch and the amylose and amylopectin ratio and other factors (Cooke & Gildley, 1992; Lang, 2004). Gelatinization is a key functional property of starch granules that often determines its end use. Gelatinization process is investigated by the physical events such as granular swelling, loss of crystallinity (birefringence and its X-ray diffraction pattern), loss of double helical order (X-ray diffraction, 13C-CP/MAS-NMR), uptake of heat (differential scanning calorimeter), increased viscosity (Rapid Visco Analyser (RVA), thermomechanical analysis), and solubilization of amylose (Cooke & Gildley, 1992; Jenkins & Donald, 1998; Ring, Gee, Whittam, Orford, & Johnson, 1988; Shrestha et al., 2012).

When cooled, the dispersed "amorphous" starch molecules undergo slow reassociation forming a tightly packed structure, the process commonly called *retrogradation* (Miles, Morris, & Ring, 1985). The chain of events leading retrogradation of starch pastes or solutions may have the following effects (Swinkels, 1985): increase in viscosity, development of opacity and turbidity, formation of insoluble "skins" on hot pastes, precipitation of insoluble starch particles, formation of gels, and syneresis of water from the paste (weeping). Retrograded starch is thermally stable, B-type crystallinity, high melting temperature, and resists mammalian amylolytic enzymes (Bird et al., 2009). Retrogradation is linked to increased firmness or rigidity of processed starch. In industrial starch, products retrogradation may lead to change in the properties of starch, such as shrinkage of starch film used as edible film, and biodegradable thermoplastic starch (TPS).

2.2 Rheological and Pasting Properties

Many starch-based materials exhibit time-dependent behavior-like deform under a constant pressure and recover when stress is released (Hookean material) and also deform immediately under stress but do not recover immediately when stress is released (Newtonian material) (Cornell, 2004).

Analysis of pasting properties of starch granules using amylographs, such as Brabender Amylograph (Viscograph) and RVA, is a key method for characterization of (cooked) starch. Figure 5.3 demonstrates the pasting profile measurement of starch at various temperatures using RVA,

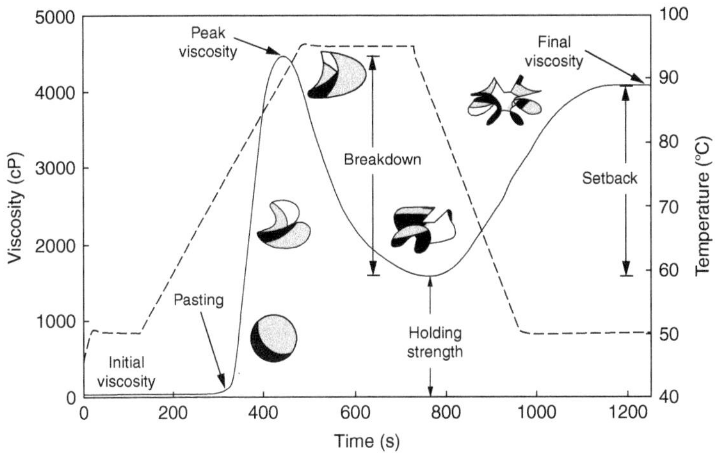

FIGURE 5.3 RVA Viscograph showing an example of starch pasting curve and temperature profile. Also showing the swelling of wheat starch granules and bursting of granules and subsequent change in viscosity (Bowler et al., 1980; Tan, 2005). (For color version of this figure, the reader is referred to the online version of this chapter.)

Newport Scientific, Australia. Viscographs, however, do not fully characterize a starch in relation to its utilization in a certain food products or in certain physical environment. In such cases, starch gel is characterized with certain moduli, such as shear modulus (*G*), storage modulus (*G*′), loss modulus (*G*″), or complex modulus (*G**). The phase angle (δ) is used to indicate the degree of elasticity (Eliasson & Gudmundsson, 2006).

Amylose is insoluble in water, gives turbidity mixture, and eventually precipitates (from dilute solution) or forms gel (concentrated solutions). It is reported to form a gel above a certain value (Eliasson & Gudmundsson, 2006). Generally, waxy maize has low or almost absent gel strength, gels of normal maize starches (~27% amylose) are firm, whereas high-amylose starches (>50% amylose) form high-strength gels in addition to gelling rapidly (Fleche, 1985). Amylose gels are thermally stable that do not melt even after being heated to 120 °C (Ellis & Ring, 1985). The peak temperatures measured for amylose gels by DSC at concentration 2.73-7.94% were 117 and 125 °C (Leloup, Colonna, Ring, Roberts, & Wells, 1992).

3 MODIFICATION OF STARCH

It is estimated global production of dry starch during 2009 was 68 million ton and the projected production for 2012 is 72 million ton (Patil, 2010). Wanous (2004) reported that starch, native or modified, contributes toward more than 85% of all hydrocolloids used in food systems worldwide. Native starches are inherently unsuitable for most of the processed foods as well as industrial applications, mainly due to viscosity and rigidity of gel of the cooked starch (Wurzburg, 1995).

Therefore, the starch molecules must be modified chemically or physically to improve their positive attributes and/or to minimize their defects (Chiu & Solarek, 2009) or blend with other materials to achieve a more appropriate property balance. According to Taggart (2004), starch modifications can best be described as a mean of altering the structure and affecting the hydrogen bonding in a controllable manner to enhance and extend their application. Modification of starch brings about a number of specific functional properties needed for a particular starch-based product. Modified starches are broadly categorized as physically modified and chemically modified starches. In general, physically modified starches include pregelatinized, milled, and cold-water-soluble starches prepared by annealing, heat moisture treatment, and dry heating of starch. On the other hand, chemically modified starch includes acid hydrolyzed, oxidized starch; dextrinized starch; cross-linked starch; stabilized starches, such as starch acetates, starch phosphates, and starch sodium octenyl succinate (SSOS); and hydroxypropylated starches (HPSs) (Wurzburg, 1995). Table 5.1 briefly presents various modified starches, their properties, advantages, and applications in foods products.

TABLE 5.1 Modified Starches: Properties, Uses, and Applications

Modifications	Purpose/Property	Advantages	Applications
Pregelatinization	Heat to gelatinize starch, dry and cool to give cold-water-thickening properties	A precook starch; avoid cooking as it thicken in cold water, convenient and energy saving	Instant soups, sauces, dressings, desserts, bakery mixes
Heat moisture treatment	Strengthen starch granule through higher temperature and low moisture Delay viscosity development, starch solubility, swelling power, and amylose leaching Minor alteration in granule's supramolecular structure	Unique functional native starch Improve process tolerance to heat, acid, and shear Production efficiency: increased heat penetration allowing shorter process time	Ambient stable products Bottles sauces Sterilized soups and sauces

TABLE 5.1 Modified Starches: Properties, Uses, and Applications—cont'd

Modifications	Purpose/Property	Advantages	Applications
Milling	Alter granule morphology, crystallinity, solubility, and swelling Depolymerization of starch	Decreased pasting temperature and increased water absorption, swelling capacity, and swelling Lower pasting temperature and susceptible to enzymatic hydrolysis	Thickeners, extenders, emulsion stabilizers, and ingredient carriers in cakes, noodles, pie fillings, soups, sauces Gum candies, formulated liquid foods
Acid hydrolysis/ thinning	Low viscosity and increase gel strength	Enhances textural properties at higher usage concentrations of starch, e.g., disperse at high concentration and gives clear gel	Gums, pastilles, jellies, formulated liquid foods
Pyroconversion/ dextrinization	Breakdown and rearrange starch molecule providing lower viscosity, increased solubility, and encapsulation Range of viscosity, stability from liquid to gel	Achieve low viscosity at higher concentration than parent starch Create film-forming properties	Fat replacer Bakery glaze, protecting coatings in confectionery
Enzyme conversion	Produce varied viscosity, gel strength, with thermoreversibility and sweetness	Contribute texture and rheology Economic dispersant	Fat mimetics Flavor carrier, dry mix fillers
Oxidation	Depolymerize and introduce carbonyl and carboxyl groups that increase clarity and reduce retrogradation of cooked starch pastes and stabilization Provide low viscosity, high solid dispersions, and resistance to viscosity increase upon gelling in aqueous dispersion and low-temperature stability	Improve adhesion of coating, gelling, and clarification Create soft stable gels at higher dosage than parent native starch	Battered meat, poultry, and fish Gum confectionery

Continued

TABLE 5.1 Modified Starches: Properties, Uses, and Applications—cont'd

Modifications	Purpose/Property	Advantages	Applications
Stabilization/ dual modification	Prevent shrinkage of starch granules and retrogradation and give stability at low temperature Lower gelatinization temperature	Excellent chill and freeze-thaw stability to extend shelf life Easy to cook in high solid system	Chilled and frozen processed foods High brix fillings and toppings
Lipophilic substitution	Introduce lipophilic groups	Emulsion stabilizer that improves quality of any fat-/oil-containing product Reduce rancidity by preventing oxidation	Beverage, salad dressing Flavor-encapsulating agents
Cross-linking	Introduce intermolecular bridges by multifunctional reagent Strengthen starch granules, restricted swelling of the granule during gelatinization, resistance shear, high temperature, and low pH Thickening, stabilizing, delay viscosity development by retarding granule swelling	More tolerant to physical, chemical, and enzymatic attack More efficient: increase heat penetration, shorter process time, desired product quality	Ambient stable product, foods to bioplastics Pie fillings, breads, frozen bakery products, puddings, infant foods, soup, gravies, and salad dressing
Esterification	Introducing acetate group Hydrophobic, cationic, or anionic character; prevent or minimize association with other amylopectin branches	Prevent cloudiness and syneresis, promote viscosity stability and clarity in low temperatures Plasticity, hydrophobicity, and moldability	Canned, frozen, baked, and dry foods Thermoplastic starch and foam
Hydroxyalkyl starches	Introducing hydroxyalkyl or hydroxypropyl	Low-temperature stability	Low-temperature food storage conditions

TABLE 5.1 Modified Starches: Properties, Uses, and Applications—cont'd

Modifications	Purpose/Property	Advantages	Applications
	groups Dispersion stability, nonionic character, decreased glass transition temperature		
Starch phosphate monoesters	Introduction of phosphate groups Clarity, high viscosity, long and cohesive texture	Stability against retrogradation	Emulsifiers
Cationic starches	Introduction of a positive charge through tertiary amino or quaternary ammonium groups Decreased gelatinization temperature; improved stability and clarity of dispersions	Clarity and dispersibility	Papermaking as wet-end additives for dry strength Surface sizing agents, emulsion stabilizer

References: Taggart (2004), Fleche (1985), Huber and BeMiller (1997), Wurzburg (1995), and Eliasson and Gudmundsson (2006)

4 PHYSICALLY MODIFIED STARCHES

A number of traditional methods of heating and drying such as drum drying and spray drying and more energy intensive methods such as extrusion are traditionally used to prepare starch pastes that swell on agitation with cold water (or liquid foods), forming a dispersion. These modified starches in the form of pregelatinized starches are some of the early modified starches. Water absorption capacity of starch granules depends on a number of factors, e.g., source of starch, amylose/amylopectin ratio, and granule morphology to name few. High-amylose starches are some of the most hydrophobic starches that need hydrothermal treatment as well as high shear force to disperse. Under the combination of heat and agitation, swollen granules are disintegrated and form of large molecule aggregates (Elder & Schoch, 1959). Therefore, the methods that use violent agitation and high temperature, such as jet cooker, extrusion, spray drying, and drum drying, are more effective in producing homogenous modified starches.

Drum drying, probably, is the most widely used technique to make pregelatinized or cold-water-soluble modified starch due to simplicity of the method. This method involves passing of partially or fully cooked starch slurry through the heated cylinders to dehydrate/cook, giving a dried starch film that is pulverized into pregelatinized starch.

Spray drying is a better method of producing high-quality pregelatinized starch. The method employs injection of starch slurry through an atomization orifice in the nozzle assembly to form a fine spray. The atomized slurry when comes in contact with the high-pressure steam in chamber gelatinize the starch (Chiu & Solarek, 2009). Interestingly, the gelatinized starch is recovered essentially as granules. Current innovation in spray drying technology has made it possible to produce flavor-free pregelatinized starch that has viscosity-building capability and more uniform granules that produce a uniformly pregelatinized cold-water-soluble starch with desirable qualities. Spray drying has also been used to improve the dispersibility of pregelatinized starches by agglomeration of the fines by exposing to a fine mist of water (Katcher, Mahon, & Matherly, 1989).

Alcohol-treated physically modified cold-water swellable starch has been reported by Eastman and Moore (2009) where a slurry of starch granules is soaked in 70-80% alcohol, heated to 157-177 °C for 4-5 min. Slurry of this in a sugar or corn syrup sets to a sliceable gel without cooking or chilling. This modified starch is useful in pie fillings, confectioneries, demoldable desserts, and instant puddings.

Annealing is a process that improves crystallinity. If the water content is high enough (~35%) for gelatinization but the temperature is too low, the condition is suitable for annealing (Knutson, 1990). Jacobs and Delcour (1998) reported low-moisture heating as the true annealing. Annealing generally lead to both an increase in GT and a sharpening and increase in the magnitude of gelatinization. It has been associated with better local packing of the amylopectin side chain branches, facilitating an improvement in the local order (Donald, 2004). Many commercial starches might undergo an annealing treatment during the production process.

Extrusion combines several unit operations such as mixing, kneading, shearing, heating, cooling, shaping, and forming. The thermomechanical treatment, in presence of sufficient water, triggers a number of physicochemical and functionality changes in starch granules, such as the loss of granular structure associated with melting of crystallites and underlying helices, and generating an amorphous structure (Shrestha et al., 2010). The product inside the extruder exposed up to 200 °C, shorter residence time and high pressure that generally pregelatinized the starch. The shearing effect on the starch granules inside the extruder tends to breakdown longer amylose chains, increase *in vitro* starch digestibility, and lower the molecular order and crystallinity (Shrestha et al., 2010). Some other physicochemical properties acquired by extruded starch are higher uncooked paste viscosity, reduced paste consistency, increased water solubility, decreased water absorption capacity, and rapid absorption of water to form gel in room temperature. Extrusion is extensively used in food manufacturing such as ready-to-eat cereals, texturized vegetable protein, noodles, and pet foods. The high pressure and temperature exerted by extruder means it is effectively used as a chemical reactor for chemical modification of starch, such as carboxymethyl cellulose, cationic potato starch, starch phosphates, anionic starch, oxidized starches, and starch fatty acid esters (Xie et al., 2005).

5 CHEMICAL MODIFICATION

Chemical modifications of starch by various reagents add chemical moieties on the linear chains of the α-D-glucopyranosyl units of starch chain by molecular scission, oxidation, or molecular arrangements, etc., which are common industrial means of enhancing starch functionalities (White & Tziotis, 2004). In starch modifications, some common substituents introduced into starch molecules are hydroxypropyl, hydroxyethyl, methyl, and acetyl groups, whereas oxidized starches largely form depolymerized group of modified starch (Wurzburg, 1986a, 1986b). The extent of modification for reactions that introduce new chemical substituent group is most commonly quantified as degree of substitution (DS; Huber & BeMiller, 2009). The DS value is the average number of hydroxyl groups on the D-glucosyl units that have been substituted. DS value ranges from 0 to 3. For example, if DS value is 3, that means all three possible hydroxyl groups on the D-glucosyl units have been substituted. Molar substitution (MS) is the average number of moles of the substituent per D-glucosyl residue. In general, MS equals DS.

Generally, DS of commercial starches, for foods, is less than 0.2 (Chiu & Solarek, 2009). Chemical modifications are influenced by two factors: intrinsic, factors involving the inherent properties of starch, e.g., granule composition and structure, and extrinsic, related to the conditions imposed by the reaction system, e.g., reagent parameters and reagent medium conditions (Huber & BeMiller, 2009). In general, increasing substitution eventually make starch cold-water-soluble unless some insolublizing, cross-linking treatment or a hydrophobic substitution is carried out. Chemical events such as disruption of hydrogen bonding and weakening of granular structure by breaking of glycosidic bonds such as thinning under acidic conditions or oxidation under alkaline conditions are few examples (Wootton & Bamunuarachchi, 1979). The substituent groups in modified starch are almost exclusively confined to the granule amorphous regions and to the outer lamellae of crystalline regions (Chen, Schols, & Voragen, 2004; Manelius, Nurmi, & Bertoft, 2000).

One of the major extrinsic factors identified for chemical modification is pH of the medium. The lower the pH, the faster is the hydrolysis of the α-1,4 linkage (Fleche, 1985). Extremely high pH values cause cold gelatinization of starch. In the majority of chemical modification, starch hydroxyl ionization is essential to activate starch for reaction (Gray & BeMiller, 2005). In general, increased pH leads to increase ionization of starch hydroxyl group as its pK_a is 12.6 (Huber & BeMiller, 2009).

Starch modification speed and pattern are also affected by the reagent types. Generally, slow-reacting reagents act slowly but produce more homogeneous reaction, acting throughout the granule matrix, whereas fast-reacting reagents act mostly on the surface and produce heterogeneous reaction patterns (Huber & BeMiller, 2009). Chemical modification of native starch lowers the GTs as well as enthalpy (White, Abbas, & Johnson, 1989; Wootton & Bamunuarachchi, 1979). Rheological properties are another extrinsic properties affected by chemical modification of starches.

With the aforementioned background, the goals of chemically modifying starch are listed as follows: to modify the gelatinization and cooking characteristics; to decrease the retrogradation and gelling tendencies of amylose-containing starch; to minimize syneresis; to enhance hydrophilic character; to impart the hydrophobic properties; to introduce the ionic substances; to improve binding and adhesion; to impart film-forming functionality and low viscosity at high content of solids; and to maintain the viscosity on exposure to high-temperature cooking, high shear, or acid (Eliasson & Gudmundsson, 2006).

Chemically modified starches are generally grouped based on their reaction characteristics or active functional groups. Wurzburg (1995) and Huber and BeMiller (2009) had classified chemically modified starches as converted or acid hydrolyzed starch, oxidized starch, dextrinized starch, cross-linked starch, and stabilized starches. Stabilized group includes some major modified starches, e.g., starch acetate, starch phosphate, starch sodium octenyl succinate, and HPS. A brief review of these modified starches is presented in forthcoming chapters.

5.1 Converted Starches/Acid or Enzyme Hydrolyzed Starches

The swelled and viscous cooked starches are difficult to handle in food and industrial applications due to their unpredictable behavior. Converted starches are partially depolymerized starch to produce products that generates low viscosity so that the granules no longer maintain its integrity on swelling on water. It makes possible to produce modified starches with lower viscosity that can be dispersed at increasingly higher concentration of starch (Wurzburg, 1995). The method primarily attacks at α-D-(1→4)-glucosidase linkages and in some extent α-D-(1→6)-glucosidase linkages in the starch molecules. Three major basic processes for making converted starches are the use of acid, the use of oxidants and alkaline system, and the application of heat, e.g., pyroconversion or dextrinization (Huber & BeMiller, 2009).

Also known as "acid modified" or "thin boiled" starches, converted starches are prepared by suspending starch granules (36-40%) in aqueous solution (or alcoholic solution) of mineral acids such as hydrochloric (HCl) or sulfuric acid (H_2SO_4) between ambient to just below the pasting temperature. Converted starch is neutralized and starch recovered by filtration (Ma & Robyt, 1987).

Starch hydrolyzing enzymes such as α-, β-, and glucoamylases, pullulanase, and isoamylase are commonly used to breakdown α-D-(1→4) and α-D-(1→6) linkages. The first stage of starch hydrolysis is liquefaction, where α-amylase hydrolyze α-D-(1→6)-glucosidase linkages in an endoaction that rapidly reduce the size of the starch molecule as well as its viscosity. The semihydrolyzed starch chains are act upon by enzymes like glucoamylase that saccharify the starch into glucose. Both, acid and enzyme, hydrolysates contain a mixture of low-molecular-weight starch molecules, oligosaccharides, and simple sugars. Enzyme hydrolysis has more control over the conversion and better control over the process and few by-products (Xie, Liu, & Cui, 2005). The degree of starch conversion is determined by fluidity or thinned starch (based on cooked dispersion, scaled from 0 to 90) (Wurzburg, 1995). Acid and enzyme hydrolysis of starch results in the decreased molecular weights that correlate to lower intrinsic viscosity but increased GT, hot water solubility and increased percentage crystallinity (Xie et al., 2005).

Starch hydrolyzates such as syrups, maltodextrins, and cyclodextrins are important ingredients for many processed foods. Sugar syrups are used in baked foods, beverages, sweets, etc., for sweetness, taste, texture, and color. Acid-converted cornstarch is widely used in gum drops, jams, jelly, ice cream, and similar products. Glucose syrups of low degree of esterification are also used to carry calcium and other nutrients into the lower digestive tract. Besides, hydrolyzed starches have many industrial applications where sugars are fermented into ethanol, butanol, acetone, organic acids, and many other useful industrial chemicals. Maltodextrins with a range of degree of esterification or DE (3-20, higher the DE, shorter glucose chains) have a number of industrial uses such as a filler, sweetener, and bulking agent in foods. Cyclodextrins are good encapsulating agents and often used to carry flavor, aroma, and heat-sensitive compounds during various processing operations, e.g., spray drying and extrusion.

5.2 Oxidized Starch

Oxidized starches are the group of converted starches where a β-alkoxycarbonyl system is formed when a carbonyl unit is introduced at C-2, C-3, or C-4 of a 4-substituted D-glucopyranosyl unit that finally leads to chain cleavage in an alkaline system (BeMiller, 2007) (Figure 5.4). A number of oxidizing agents such as hydrogen peroxide, hypochlorite, periodate, permanganate, dichromate, persulfate, and chlorites have been

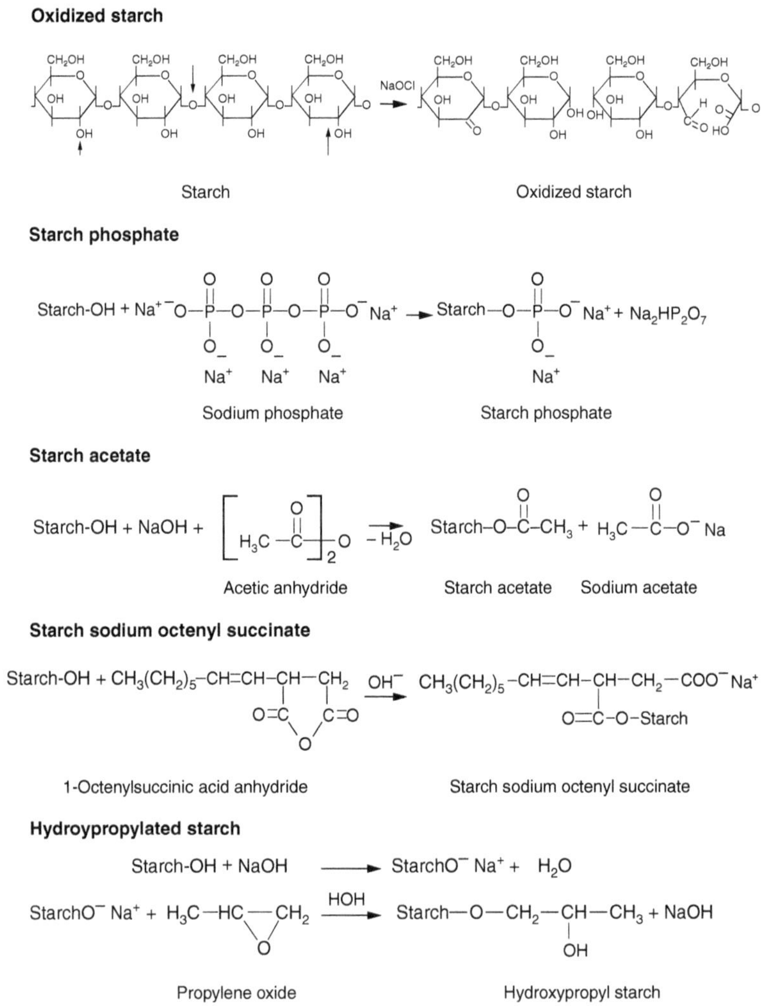

FIGURE 5.4 Chemical synthesis and structure of various chemically modified starches from starch.

used for starch oxidation. Oxidation reaction takes place in milk phase. The reaction is exothermic and consequently requires keeping temperature in the range of 30-35 °C by adjusting the rate at which the reagents are added. This reaction is sensitive to parameters such as concentration of oxidizing agent, pH, and temperature (Fleche, 1985). The most commonly used oxidizing agent for starch oxidation is sodium hypochlorite. Huber and BeMiller (2009) reported that although there are many similarities between oxidized and acid hydrolyzed starch pastes, the former are generally clearer and more stable than those made from the latter method.

Carboxyl groups in the oxidized starch allow improved solvation and provide relatively stable dispersions, with better film properties. The main applications of oxidized starches are in paper industries as a binder for coating and surface sizing compounds; textile industries as a sizing compound; and food industries as low-viscosity filler, providing stability, clarity, and binding properties (Fleche, 1985; Xie et al., 2005).

5.3 Dextrinization or Pyroconversion

"Dextrin" is a broad term assigned to any starch degradation product from conversion. Pyroconversion involves the heat treatment of a dried starch with or without adding acid. It involves two aspects of structural modification of starch: partial depolymerization by hydrolysis and repolymerization. The cleavage of glycosidic bond can be achieved by dry roasting of starch either alone using natural moisture (10-20%) or in the presence of catalytic quantities of acid (Taggart, 2004). Hydrolysis of starch produces a range of starch fractions with varying chain length (low conversion). The second aspect involves a recombination of these fragments that can be described as transglycosidation and repolymerization. Transglycosidation involves the transfer of portions of glucan chains to hydroxyl groups on the same or different chains to create a branch point. This is a highly branched polymer and more soluble (high conversion). Repolymerization or reversion occurs with catalytic amounts of acid at high temperatures and low-moisture content. This further creates high molecular weight and more branched structure (Fleche, 1985; Huber & BeMiller, 2009; Taggart, 2004).

Three common pyrodextrins are white dextrins, yellow dextrins, and British gums. White dextrins are prepared by using an acid catalyst, relatively low temperatures, and short time; yellow dextrins are products of higher temperature and longer times; British gums are made with higher temperature and long times but few or no acids. Yellow dextrins are widely used as adhesive for various purposes. White dextrins are used by pharmaceutical industries both for coating capsules and as a nutritive source in antibiotic fermentation. Few other common uses of these dextrins are binders in foundries, in textile printing, and in paper industries (Fleche, 1985). Highly soluble British gums and yellow dextrins are also used as a diluents or carriers for active food ingredients such as flavorings, spices, and colorants. Yellow corn dextrin is also used in the encapsulation of water-insoluble flavorings and oils (Wurzburg, 1995).

5.4 Cross-Linked Starches

Cross-linked starches are the starches that have hydroxyl groups (—OH) reacted with the multifunctional reagents resulting in chemical bonds responsible for granule integrity. Besides strengthening granules, it enhances the resistance to viscosity breakdown as a result of mechanical shear, acid conditions, or high temperature (Huber & BeMiller, 2009; Taggart, 2004). In a way, the starch granule is "spot-welded" at random locations to reinforce it, which restricts swelling of the starch under cooking conditions or prevents gelatinization. FDA has permitted few reagents for cross-linking of food grade starches, such as phosphoryl chloride ($POCl_3$), sodium tripolyphosphate, sodium trimetaphosphate (STMP), adipic acetic mixed anhydride, epichlorohydrin (EPI), and mixtures of STMP and tripolyphosphates (Sherry, 2005; Huber & BeMiller, 2009).

Huber and BeMiller (2009) reported the functionality of the crossed-linked starches depend on the degree of cross-linking with the reagents. For example, very low level of cross-linking generally stabilizes granular structure to allow the modified starch to attain higher degree of swelling during heating, similar to native starch, whereas the relatively higher level of cross-linking usually causes reduced granule swelling, solubility, extent of amylose leaching, paste clarity, and paste viscosity. Higher degree of cross-linking results in increased pasting temperatures, stability to shear, and tolerance to acid pH conditions.

Only a small DS is needed to impart the beneficial effects, typically per 100-3000 anhydroglucose units of the starch (Taggart, 2004). It has been shown that even as low as 0.1% cross-linking increases the viscosity of unmodified starch, whereas slightly higher levels modification tends to decrease the viscosity. Increasing cross-linking levels in starch eventually prevent the granule from swelling, and the starch cannot be gelatinized in boiling water even under autoclave condition (Srivastava & Patel, 1973). Starches such as waxy corn, tuber, or root starch are quite fragile and tend to be fragmented during prolonged heating or agitation. In presence of acids, these starches result in a rapid breakdown in viscosity. Cross-linking helps to maintain the integrity and loss of viscosity under acidic conditions.

Cross-linked starches are extensively used in food industries as a thickener, stabilizer, and texturizer in the food systems. They are used in salad dressing to provide thickening with stable viscosity at low pH and high shear during homogenization process. Some common uses of this group of modified starches are in the preparation of pie fillings, bread, puddings, baby foods, soups and sauces, gravies, and salad dressings; in retorted foods; and in aseptic processing (Wurzburg, 1995). Cassava starch paste tends to undergo breakdown when subject to heat and shear. The use of cross-linking strengthens the starch molecules, preventing easy disintegration of starch (Moorthy, 2004).

5.4.1 Distarch Phosphate

Distarch phosphates and distract adipates are the most common cross-linked starches that contain a phosphate or adipate bridge. Distarch phosphate is produced by cross-linking of starch with phosphorous oxychloride. The reaction is effected by high pH (11) and sodium sulfate, as shown in Figure 5.4 (Furin, 1973; Seib, 1996). The nature of cross-linking in the granule is often determined by the reactivity of the reagents. For example, the highly reactive reagents, such as $POCl_3$, are reported to react predominantly at external granule surfaces, including those in channels and pores, where slowly reactive reagents such as STMP and EPI are reported to yield more uniform reaction patterns throughout the starch granules (Hirsch & Kokini, 2002; Huber & BeMiller, 1997).

5.5 Stabilized Starches

This group of modified starches includes those modifications made by reacting some of the hydroxyl groups on the starch molecules with monofunctional reagents to introduce substitutional groups (Wurzburg, 1995). The major objective of stabilization is to stabilize the amylose against retrogradation and thereby enhance shelf life through tolerance to temperature fluctuation such as freeze-thaw cycles (Taggart, 2004; Wurzburg, 1995). It is usually used in conjunction with cross-linking.

Stabilization involves substitution of the bulky groups onto the starch in order to take up space and hinder any tendency for dispersed (cooked), linear fragments to realign and retrograde. The effectiveness of the stabilization depends on replacement of hydroxyl groups of starch molecules into larger ester or ether groups, interchain associations blocked, resulting in more stable pastes and gels and less tendency to retrograde (Huber & BeMiller, 2009; Wurzburg, 1995). The nature and type of substituent used for modification affect the effectiveness of stabilization. There are four major types of stabilized starches prepared by reacting four monofunctional reagents with starches: *acetylated, hydroxypropylated (and HPS ether), monophosphates, and sodium octenyl succinate*. Low DS starch, particularly below 0.2 DS, is the most important commercially. Generally, the higher the DS value, the lower the starch-starch granules interaction, and consequently, hydration and gelatinization by cooking are achieved at lower temperature.

5.5.1 Starch Acetates

Starch acetates or acetylated starches are one of the starch esters in which some hydroxyl groups have been replaced by ester (acetyl) group. The first starch acetates were reported as early as 1865 (Schuetzenberger, 1865). Some of the physicochemical changes acetylation reaction brings to starch are low GT, good cooking, and storage stability; stability of product texture and appearance (prevent retrogradation); reinforced hydrophobicity and thermal stability; film-forming ability; and improved thermoplasticity (Moorthy, 1985).

Figure 5.4 shows the possible reactions when starch suspension is reacted with acetic anhydride in presence of sodium hydroxide as a catalyst and heat. There are three free —OH groups in glucose structure that have different reactivities. The primary —OH on C6 is more reactive and is acetylated more readily than the secondary ones at C2 and C3 due to steric hindrance. The primary —OH located at the exterior surface of the starch molecules reacts readily with the acetic groups, while the two secondary ones located within the interior surface of starch form hydrogen bonds with the —OH groups on the neighboring glucose unit (Xu, Miladinov, & Hanna, 2004).

The preparation procedure for the synthesis of acetic anhydride for food purpose is relatively mild. Initially, starch slurry is prepared at pH 8-9.5 at about 30 °C, and acetic anhydride is slowly added with continuous mixing. Recommended use of sodium hydroxide solution is 3% as stronger solution tend to gelatinize the starch. The modification occurs within about 30 min and the solution is neutralized at about pH 7. The modified starch is washed and dried. The modified starches destined for food ingredient purpose cannot have more than 8% weight of acetic anhydride in the final product and no more than 2.5% acetyl groups on an anhydrous basis (equivalent to approximately 0.1% DS) (Rutenberg & Solarek, 1984; Wurzburg, 1995). Sodium acetate is the by-product of the acetylation reaction that must be completely removed by a series of washings. It tends to cause typical acetyl odors in puddings made from these starch acetates (Katcher & Ackilli, 1980). Among the modified starches, acetate starches (weak substitute ~0.1 DS) are the most sold due to its use in food ingredients as well as industrial applications. The reaction produces acetaldehyde as a by-product that can be removed by washing (Jarowenko, 1986; Tuschoff & Smith, 1962). Vinyl acetate is used as catalyst mostly in countries where acetic anhydride is banned.

It has been reported that starch granules with DS up to 0.5 showed no changes from native starch under the light microscope. X-ray diffraction study of native starch and its acetylated counterparts showed the individual signature diffraction patterns/peaks of native start to disappear as acetylation progresses and eventually turn into troughs and valleys (Xu et al., 2004). Fringant, Desbrieres, and Rinaudo (1996) synthesized acetylated high-amylose starches with DS ranging from 0 to 3.0. Incorporation of acetate groups in the starch structure stabilizes the starch against the retrogradation and intermolecular association; acetylated groups weaken the native granule structure slightly so that swelling and gelatinization can take place at relatively lower temperature (Kruger & Rutenburg, 1967). As a consequence of this, pasting properties of native starch improve with acetylation. Liu, Ramsden, and Corke (1997) reported that acetylation of native normal- and high-amylose starches results in lower GT as well as slightly higher peak viscosity.

Acetylation of high-amylose starch to DS of 0.1-0.2 lowers GT so that this modified starch can be dispersed in a boiling water bath. High-amylose starch (66% amylose) does not show pasting property in RVA, but upon acetylation (DS at 0.130), it clearly exhibits pasting property (Figure 5.5). Cornstarch containing 1.8% acetyl paste at a temperature 6 °C lower than its unmodified counterpart reached its peak viscosity at temperature 10 °C lower than its unmodified counterpart (Jarowenko, 1986; Tuschoff & Smith, 1962). Acetylated starch holds water at low temperature or freezing, particularly in low-amylose starches such as tapioca,

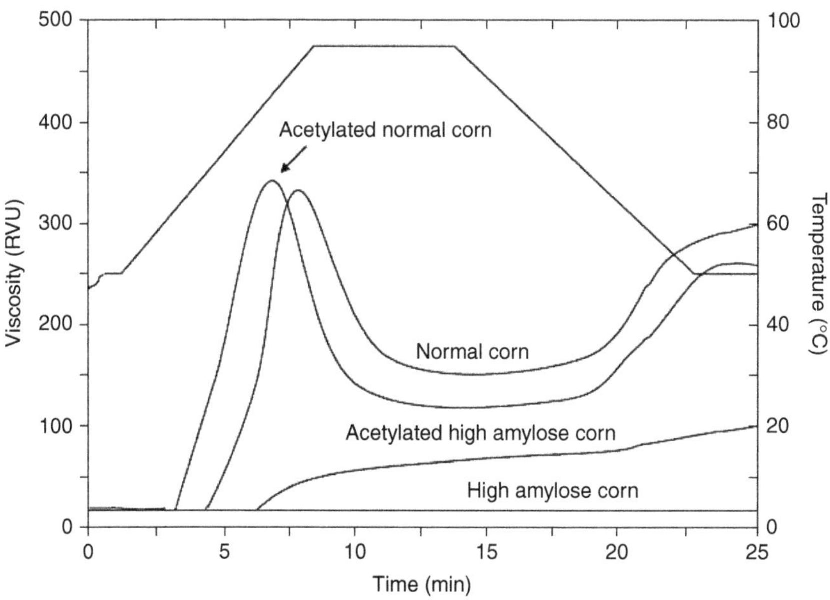

FIGURE 5.5 RVA pasting curves of normal and high-amylose starches (66% amylose) and their acetylated starch (Liu et al., 1997).

potato, and cornstarch. This prevents the opacity, gel formation, and syneresis in these starches. Therefore, acetylated starch generally improves the stability and clarity of sols by increasing the degree of swelling and dispersion of starch granules as well as prevents retrogradation (Rutenberg & Solarek, 1984).

The major industrial application of acetylated starch is sizing in paper manufacture or textiles as adhesives. It is commonly used in formulated foods is as antigelling agent. High-amylose starch acetates are heavily used in gummed candies. It is also used as acid pH-resistant binders in food industries. As acetate starches normally lack resistance to mechanical shear and acids, they are generally cross-linked to achieve viscosity stability, clarity, and low-temperature stability, particularly waxy corn, tapioca, and potato (Jarowenko, 1986; Seib, 1996). Cross-linked acetylated starches are commonly used in food industries in a large variety of processed foods that include baked foods, canned pie fillings, canned cream-type corn, sauces, retorted soups, foods processed by continuous sterilization instead of retorting, frozen foods, baby foods, salad dressings, and snack foods (Tuschoff & Smith, 1962; Wurzburg, 1995). It is also used as biodegradable TPS, ERS, and encapsulating agents and in control release of biochemicals.

5.5.2 Starch Phosphates (Monoesters)

This group of modified starch has only one starch hydroxyl group esterified by phosphoric acid to form the monoester. Esterification with phosphate salts affects the colloidal properties of the starch dramatically. The reaction between granular starch with monosodium orthophosphate and sodium polyphosphates can be used for modification. Initially, starch is mixed with phosphate salts in an aqueous suspension; pH is adjusted from 4 to 8.5, mixed, filtered, and dried at 100-160 °C (Bergthaller, 2004). These starch phosphates can result in DS up to 0.2. The esterification occurs mainly at the primary positions at C-6, with lower level substitution at C-2 and C-3 (Lim & Seib, 1993).

Starch phosphate monoesters yield higher viscosity, clarity, very stable dispersions, even under freezing conditions including freeze-thaw cycles. Substitution of monophosphates in starch lowers their GT significantly. Monophosphates with 0.07 DS lower GT in such a low that it swells in cold water (Solarek, 1986).

With greater solubility in low temperature, phosphorylated starches can be used as thickeners and stabilizers for food system. They are attributed to fairly clear, stable dispersions suitable for use in thickening food systems having improved freeze-thaw stability. One of the limitations of this group of modified starches is their anionic character that tends to react with polyvalent cations but works best in salt-free or acidic conditions (Wurzburg, 1995). Starch phosphate along with guar gum and propylene glycol has been used as an emulsion stabilizer for vinegar and vegetable oil in water (Solarek, 1986). Cold-water-soluble starch phosphate is dry mixed with sugar and flavoring agent that can be added to cold milk to form puddings with a soft, smooth, even-textured superior eating quality of puddings (Neucom, 1958).

5.5.3 Starch Sodium Octenyl Succinate

Octenyl succinic anhydride (OSA) starch is a relatively new modified starch that was first synthesized by Caldwell and Wurzburg (1953). It is a lightly substituted half-ester of sodium octenyl succinate prepared by reacting OSA to a suspension of granular starch at neutral pH under agitation. The chemical structure of SSOS is given in Figure 5.4. The maximum DS for SSOS is about 0.02, i.e., one octenyl monosuccinate substitution group for about every 50 anhydroglucose units (Wurzburg, 1995).

Starch octenyl succinate half-ester may be present as either the sodium salt (or SSOS) or the acid, depending on the medium pH. As with other stabilized starches, octenyl succinate half-ester substituent interrupts the linearity of the starch molecule, stabilizing dispersions against the gelling or pasting (Wurzburg, 1995). Compared to native starch, OSA starch contains both hydrophilic groups (carboxylic or sodium carboxylate group) and hydrophobic group (octenyl) in a stable ratio of 1:1. This special structure of OSA starch is advantageous for its use in food, thereby approved as a food additive in the United States since 1972 (Wang, Lianghua, & Wang, 2009).

Succinylation of starch is highly desirable in the food industries as this modification improves the physicochemical properties, such as decreasing GT; increasing freeze-thaw stability, thickening power, viscosity stability, paste clarity, ability to swell in cold water, stability in acid, and salt; reducing the tendency of retrogradation; and modifying texture of starch (White & Tziotis, 2004). It has been used in a number of food formulations such as emulsifiers or stabilizers in beverages and salad dressing, flavor-encapsulating agent, and clouding agent for preparation of beverages (Trubiano, 1986; Whistler & BeMiller, 1997). OSA starch is also useful in a number of applications such as processing aids, body powders and lotions, emulsion stabilizers in coffee whiteners, hypoallergenic infant food formulas, water-repellent dusting powder, and dehydrating agent in water-insoluble vitamins (Wurzburg, 1995). This group of modified starches are amphiphilic in nature with a wide range of viscosities. This property can be used in a variety of oil-in-water emulsion in food, pharmaceutical, and industrial areas.

5.5.4 Hydroxypropylated Starch

Low DS hydroxyprolated starch in the granular form is prepared by partial etherification of starch with propylene oxide in alkaline condition. The preparation involves the reaction between concentrated aqueous suspension of starch granules and propylene oxide with high agitation (BP 35 °C) in alkaline solution (Figure 5.4). Generally, sodium sulfate is added to the reaction mixture to minimize starch gelatinization. About 10% of propylene oxide is added to the dry weight of starch for the preparation of HPS for food use (Wurzburg, 1995). Most HPS used in the food industry are sold as hydroxypropylated distract phosphate. As obvious, these products require a subsequent treatment to introduce phosphate cross-links after the hydroxypropylation reactions (Wurzburg, 1995).

Introduction of hydroxypropyl group reduces bond strength between starch molecules that facilitate gelatinization, lowers molecular association or retrogradation, increases water holding and gelling property, improves freeze-thaw and cold stability, improves textural property of products, and enhances shelf life of the food products (Goff, 1994; Tuschoff, 1986). Unlike acetylated and phosphorylated starches, the hydroxypropylated groups are linked to starch through ether linkages, which are chemically more stable than ester linkages, particularly at pH >6.0 and also at high temperature (Wurzburg, 1995). Hydroxypropylation is also reported to reduce tensile strength of high-amylose starch films and increase elongation and burst strength (Bae & Lim, 1998; Roth & Mehltretter, 1967). It is chemically more stable at higher temperature and during long retorting times of high-pH foods such as cream soups and retorted milk-based puddings (Wurzburg, 1995). HPS can also be used as a raw material in the production of starch hydrolysates with low caloric values than sucrose or regular starch hydrolysate and also as noncariogenic substitute (Wurzburg, 1995).

6 STARCH MODIFICATION AND RESISTANT STARCH

Starch molecules contain only α-glycosidic linkages and are potentially digestible by the amylolytic enzymes secreted by the human digestive tract. However, the structural conformation of starches, native and modified, and other processing factors can influence the rate of starch hydrolysis, *in vitro* and *in vivo*, and absorption in humans and animals (Englyst, Kingman, & Cummings, 1992). On the basis of how starch is acted upon by the digestive enzymes (*in vitro*), it is classified as rapidly digestible starch (RDS), slowly digestible starch (SDS), and resistant starch (RS). RDS is the amount of glucose released in during digestion of starch in 20 min. Generally, RDS consists mainly of amorphous and dispersed starch and richly present in starchy foods cooked with moist heat, e.g., bread, potato, and rice. SDS is the amount of glucose released between 20 and 120 min digestion. Compared to RDS, it has controlled and slowly release of glucose in the alimentary canal (low glycemic). The classical definition of RS is the total starch minus amount of glucose released within 120 min of digestion (RS = TS − (RDS + SDS). However, physiologically, RS is defined as the sum of starch and products of starch degradation not absorbed in the small intestine of healthy individuals (Asp, 1992).

In current perspective, high amount of RDS in human diet is a global health issue as high blood glucose is linked to obesity, diabetes, coronary heart diseases, colon cancer, and related diseases and disorders. Food scientists and *nutritionists* are working toward lowering RDS in diet and increasing SDS and RS. SDS and in some cases RS offer the advantage of sustained postprandial blood glucose over an extended period of time, compared to RDS. In addition to hormonal and metabolic responses, postprandial glycemic response and generation of secondary metabolites (short chain fatty acids) contribute toward the maintenance of the colonic health. This can have implications for physical and mental performance, satiety, and diabetes management. Another benefit of products rich in SDS is their moderate impact on the glycemic index.

6.1 Resistant Starch

RS constitutes a large proportion of the fermentable carbohydrates entering the colon and provides an energy source to colonic microflora. RS is first digested by bacterial amylases then glucose is metabolized into organic acids (mostly short chain fatty acids, SCFA, and lactic acid) and gases likes CO_2, H_2, and CH_4 (Topping & Clifton, 2001). The presence of fermentable substrate is reported to prevent inflammatory bowel disease and maintains the metabolic requirements of the mucous and subsequent restoration of apoptosis (programmed cell death) (Jacobasch, Schmiedl, Kruscheski, & Schmehl, 1999; Moreau et al., 2003). These factors are believed to reduce the incidence of colorectal cancer. Some other health benefits of consuming high-RS diets are low-energy intake leading to weight management and obesity, control over glycemic release and diabetes, system benefits such as improving blood lipid profile (lowering plasma cholesterol/triglycerides), and improved bowel health (Bird, Shrestha, Lopez-Rubio, & Gidley, 2009). RS is classified into four major groups, namely, RS_1, RS_2, RS_3, and RS_4. The classification is based on origin and physical characteristics of the starches usually considered "as eaten" as most technological treatments, as including reheating, can affect the RS content in foods (Champ, 2004).

6.1.1 RS_1 (Encapsulated Starch)

RS_1 arises from physically inaccessible starches, e.g., within plant tissue structures or dense food matrices. Pulses are the major source of RS_1 as their thick cell walls resist breakdown during preparation of the meal, mastication, and stomach and intestinal digestion. The complex matrices of legume seeds where dietary fiber and proteins encapsulate starch render them unavailable to digestive enzymes in intestine. Coarse ground cereal grains are also only partially digested in gut. Unlike RS_1, other RSs are either native in endosperm (RS_2) or physically or chemically modified (RS_3 and RS_4) to acquire resistant to digestive enzymes.

6.1.2 RS_2 (Native Starch Granules)

Native starch granules present in plant sources such as in unripe banana and raw potatoes. As these granules become digestible if (thermally) processed, RS_2 level in human diet is very low. There are a number of "new" high-amylose maize starches (HAMS) such as Gelose 80®, Hylon VII®, Novelose 240®, and Hi-Maize®, which are relatively resistant to digestion due to their granular architecture. In general, most of these starches have B-type crystallinity (Htoon et al., 2009). These HAMS are resistant to gelatinization (>120 °C) and, therefore, need relatively extreme conditions to cook, e.g., extrusion and high-pressure cooking. Huber and BeMiller (2009) reviewed few potential methods to increase their RS_2 level. For example, annealing or heating granular starch in water (>40%) at a temperature above T_g (but below GT) increases the formation of new double helices in high-amylose starches but not in normal and waxy starches. Heat moisture treatment or HMT (<35%) and high temperature (80-140 °C) bring about alteration within amorphous regime such as development of new order and crystallinity, formation or enhancement of AM-lipid complexes, and conversion of amylase from unordered to a partial helical form. These events lead to newly acquired resistance to enzyme digestion.

6.1.3 RS_3 (Retrograded Starches)

RS_3 or retrograded starches is the starch of the interest as it is the form starch most often eaten or used as an ingredient in a wide range of conventional foods. Retrograded starch resists enzyme digestion due to reorganization of linear chains after decrystallization. It is difficult to predict the digestibility of starch after processing (and storage) as a large number of factors such as processing methods, e.g., extrusion, pressure cooking, drying, acid/enzyme hydrolysis, and baking; amylase content; moisture; food matrix; and starch source can affect the orientation of starch molecules.

Boiling and high-pressure cooking affect the gelatinization and retrogradation of starch. Several investigations have reported an increase in RS value of cooked starches (Parchure & Kulkarni, 1997; Siljestrom & Bjorck, 1990); however, others reported a decrease in RS value (Sagum & Arcot, 2000). Berry (1986) reported autoclave of high-amylose starches causes significant RS formation (20-34%) but no effect on waxy or high amylopectin starch unless they are debranched and cooked. The number of autoclaving-cooling cycles found to enhance the degree of RS formation significantly. In comparison to cooked foods stored at ambient condition, cooked food kept under frozen condition (−20 °C) (for 30 days) considerably increases the RS content in various cereals and cereal products and legumes (Rosin, Lajolo, & Mendezes, 2002). Baking is reported to increase of RS level. Addition of soluble sugars such as glucose, maltose, sucrose, and ribose has been found to reduce the level of crystallization and subsequently reduce the RS yield (Eerlingen, Bjorck, Delcour, Slade, & Levine, 1994). It has been shown that after debranching (acid/enzyme hydrolysis), the linear chains can contribute to a high RS content (Berry, 1986; Gonzalez-Soto, Escobedo, Hernandez-Sanchez, Sanchez-Rivera, & Bello-Perez, 2006). Microwave cooking reported to redistribute insoluble nonstarch polysaccharides to soluble fraction, thereby lowering the RS level of legumes and cereals significantly (Geng, Zongdao, & Toledo, 2003; Marconi, Ruggeri, Cappelloni, Leonardi, & Carnovale, 2000). Extrusion of high-amylose maize starch (80% amylose) lowers the RS level from 45.4% to 22.6%. On the other hand, extrusion slightly increase the RS level in low amylose maize starch (27% amylose) from 0.5% to 1.8% (Shrestha & Khadka, 2007).

6.1.4 RS_4 (Chemical Modified Starches)

This is a group of chemically tailored RS for a controlled and predicted digestibility. Acetylated starches are most commonly studied RS_4. Hoover and Sosulski (1985) reported the *in vitro* digestibility of acetylated legume starch with DS 0.05 was reduced by up to 10%, as compared to native starch. The *in vitro* digestibility with amyloglucosidase of acetylated cornstarches containing 1.5% and 2.41% acetyl levels was 84% and 69%, respectively, as compared to unmodified starch (Federation of America Societies for Experimental Biology, 1979). Besides, an *in vivo* study showed the intake of a meal from acetylated potato starch reported to improve the glycemic, insulinemic, and satiating properties (Raben, Andersen, Karberg, Holst, & Astrup, 1997).

In hydroxypropylated wheat starch, *in vitro* digestion showed the increased rate of substitution restricting the ability of enzymes to hydrolyze the substituted groups in starch molecule (Hoover, Hannouz, & Sosulski, 1988). Heacock, Hertzier, and Wolf (2004) found that esterification of starch with OSA might impair the binding of α-amylase, thus decreasing the extent of starch digestion. Their study indicated that OSA starch appeared to be a RS, which could be used as a functional fiber for the treatment of certain human diseases.

However, all chemical modifications do not involve increase in resistance to enzyme hydrolysis. Researchers have reported that the cross-linking with phosphorylation may reduce the rate of digestion (Woo & Seib, 2002), whereas others have not observed any change (Hoover & Sosulski, 1985). From the *in vitro* test, when these modified starches were in a granular (ungelatinized) state, cross-linking or acetylation did not affect the digestibility. Etherification with hydroxypropyl groups, however, increased the amylolysis because the enzymes had better access to starch as hydroxypropylation helped the starch granules to swell (Björck, Gunnarsson, & Östergård, 1989). Phosphorylation of rice starch has impact on the physicochemical properties, and reduces the degree of amylase hydrolysis (Sitohy & Ramadan, 2001).

7 STARCH MODIFICATION: BIODEGRADABLE TPS

With annual global consumption of more than 200 million tons, plastic industries are the major source of nonbiodegradable wastes (Siracusa, Rocculi, Romani, & Rosa, 2008). Biodegradable materials are eco-friendly and user-friendly as packaging film (bioplastic) and also considered to be safer to use food applications. Starch is a popular bioplastic material but not easy to use as a synthetic plastic. The natural polymers have diverse stereochemistry and architecture that pose difficulty in tailoring them into desirable packaging materials. Their vulnerability to factors such as temperature, RH, UV exposure, and microbial contamination affects product quality and the product shelf life. The ability of TPS to provide all the required end-product performance characteristics, such as ease of molding, thermal stability during molding, mechanical properties, and dimensional stability, is always under scrutiny (Chaudhary, Torley, Halley, McCaffery, & Chaudhary, 2009).

Chemically modified starches such as derivatized starches involve substitution of free —OH group from starch polymer by various reagents, improving water sensitivity and mechanical performance disrupting linearity and acquiring desirable functionality and properties. High-amylose starches are preferred as a starting material for biodegradable material due to their superiority over the low-amylose starches, such as comparatively higher hydrophobicity, less water absorption, and greater barrier properties. The heated and cooled amylose-rich starch forms strong gel and has better film-forming properties. Considering a large number of biodegradable polymers being used as packaging materials, a comprehensive review of these materials is beyond the scope of the current paper. Instead, this chapter will briefly cover the major chemically modified starches used in biodegradable TPS.

7.1 Modified Starch as Biodegradable TPS

One of the prerequisites of application of starch into bioplastic is transforming starch granules from semicrystalline into a homogeneous amorphous matrix that enhance its processability. Heating of moist starch in close volume above the melting and glass transition temperatures renders endothermic changes as a consequence of which oxidative and thermal degradation of starch polymers occurs (Thire, 2010). Under the melting conditions, water tends to evaporate, and addition of compatible plasticizers such as glycerol, sorbitol, and triacetin yields TPS melt after thermomechanical transformation. Breakdown or melting of starch granular structure can be achieved by thermal and mechanical input such as extrusion, kneading, and injection molding. Extrusion is common in starch processing that operates up to melting temperature, e.g., 220-240 °C of starch granule, higher than degradation temperature, e.g., 200-220 °C; therefore, a plasticizer is needed to process the starch (Shogren, 1996). Modified starches alone cannot form films with satisfactory mechanical properties unless it is plasticized or blended with other materials. Plasticizers are low-molecular-weight compounds that form secondary bonds to polymer chains and spread them apart. Plasticizers reduce modulus, tensile strength, hardness, density, melt viscosity, glass transition temperature, electrostatic chargeability, and volume resistivity of a polymer and at the same time increase flexibility, elongation at break, toughness, dielectric constant, and power factor (Matthews, 1996). Therefore, the properties of TPS mainly rely on hydrogen bond-forming ability between starch molecules and plasticizers. Glycerol is the most commonly used plasticizer but there are a range of plasticizers such as triacetin, dibutyl sebacate, triethyl citrate, and propylene glycol used in starch processing. Some common chemically modified starches used as a TPS are briefly mentioned in the succeeding text.

7.1.1 Starch Acetate as TPS

When acetyl groups are introduced on starch molecule, they reinforce hydrophobicity, improve thermoplasticity, and enhance thermal stability and film-forming ability, clarity, and low tendency to syneresis when held at low temperature (Tarvainen, 2004; Xu et al., 2004). The acetate groups act as pegs on the amylopectin chains to reduce or prevent retrogradation. Increasing degree of acetylation lowers the T_g of amylose-rich SA that demonstrates increased mobility of the polymer chains, hydrophobicity, and compatibility with other additives (Shogren, 1996; Tessler & Billmers, 1996).

Acetylated amylopectin-based films have poor elongation due to the densely packed branched structure of amylopectin (Fringant, Rinaudo, Foray, & Bardet, 1998). Elasticity of SA films can be improved by external plasticizers such as glycerol but internal plasticization with some fatty acid esters is not effective (Bonacucina et al., 2006; Fringant et al., 1998). Better rheological properties are achieved with DS value higher than 1.7 (Fringant et al., 1996). Starch acetate with a higher DS is a possible alternative to plastic and water-soluble and biodegradable loose fill packaging materials (Guan, Fang, & Hanna, 2004; Xu & Hanna, 2005). Takagi, Koyama, Kameyama, and Tokiwa (1994) showed blends of SA/polycaprolactone can maintain their biodegradability and have stable viscosities with increasing acetylation. Frische and Best (1994) showed that the strength properties and water resistance are improved by extruding blends of starch acetate and low-molecular-weight esters (e.g., triethyl citrate).

7.1.2 HPS as TPS

Hydroxypropylation of starch disrupts the internal bonds that change the thermal behavior and physicochemical properties, making it suitable for biodegradable TPS. Kaur, Singh, and Singh (2004) demonstrated that hydroxypropylation of potato starch decreases GT and increases heterogeneity in the granule and widens the GT range with increasing MS. The decreased GT and enthalpy were more profound for the high-amylose starches than others (Hoover et al., 1988). Chaudhary et al. (2009) investigated the effect of starch type, screw speed, and aging after molding on final product properties such as mechanical properties, moisture absorption, morphology, and retrogradation. They concluded that waxy maize and hydroxypropylated 80% amylose TPS have mechanical properties comparable to that of low-density polyethylene (LDPE) and high-density polyethylene.

HPS recommended of having the potential to be an environmentally friendly alternative to current polymer resins. Lafargue, Pontoire, Buleon, Doublier, and Lourdin (2007) analyzed the properties of solution casted hydroxypropylated pea starch with varying molecular weights. This study showed the solvent casting conditions such as room temperature and humidity (40%) increase the crystallinity for the shorter starch chains. The casting temperature also enabled the control of the crystallinity of films from 16% at 25 °C to almost none at 65 °C. They concluded that the control of molecular weights and the casting temperature of modified samples enable to obtain desired structural properties. Hydroxypropylation of normal and high-amylase starches causes an increase in ease of extrusion, decrease in water absorption, and increase in flexibility and strength.

7.1.3 Other Biodegradable Polymers Derived from Starch

Bioplastics synthesized by microbial fermentation of polysaccharides, e.g., starch and sugar, are polylactides (PLAs), polyhydroxyalkanoates (PHAs), and many others. Polylactide or polylactic acid is a biodegradable, thermoplastic and aliphatic polyester that is derived directly from lactic acid. This thermoplastic film material has excellent barrier properties and strong enough to withstand the injection blowing and vacuum forming processes. PHAs are linear polyesters produced via bacterial fermentation of sugars and lipids. They have low-moisture-vapor permeability, comparable to LDPE (Halley, 2005).

Starches from various sources are widely blended with biodegradable polymers, such as polycaprolactone, polylactic acid, polyhydroxybutyrate, and polyhydroxyester ether (as reviewed by Thire, 2010). TPS blended with polyesters, such as polycaprolactone, is the most popular base film (Bastioli, 2000). Starch can be associated with other polymers to obtain multilayer structure. Starch-based material coated with water-resistant biodegradable polyesters has been used food preparation, packaging, and consumption (Shogren & Lawton, 1998). The method involves application of natural resin to the surface of a starch-based article and subsequent application of water-resistant polyester, e.g., PLA. Extrusion of thermoplastic cornstarch-glycerol-polyvinyl alcohol (PVOH) blends showed that PVOHs improve the mechanical properties and also slow biodegradation (Mao, Imam, Gordon, Cinelli, & Chiellini, 2000). It has been suggested that molecular weight and type of PVOH are likely to affect biodegradability of starch.

8 MODIFIED STARCH AS ENCAPSULATION AND CONTROLLED RELEASE

Modified starches have good film-forming properties and commonly used in a variety of industry applications such as in pharmaceutical coatings, protecting coatings on fruits, and generally packaging materials. Film-formation from the polymer solution takes place as evaporation of the solvent proceeds. During the drying process, viscosity and concentration of polymer increases that bring the polymer chains in proximity eventually form a gel-like phase (Tarvainen, 2004). Eventual loss of solvent from the gelled phase forms a continuous film of a dense, three-dimensional polymeric network with tightly intertwined network (Banker, 1966). Some of the key requirements of edible films and coatings are easier handling and processing, increased shelf life, reduced volatility, mask odor, color or taste, good elasticity and flexibility, low brittleness, and high toughness to prevent cracking during handling and storage and control release of the encapsulant (Barreto, Pires, & Soldi, 2003; Forssell, 2004). Plasticization generally lowers the tensile strength and softening temperature, and decreases the glass transition temperature (Tg) of the plasticized film (Banker, 1966). Therefore, plasticizers are generally required for polysaccharide-based edible films, from 10% and 60%. The most commonly used plasticizers are polyols (propylene glycol), glycerol, sorbitol, polyethylene glycol, oligosaccharides, and water (Gupta, Brennan, & Tiwari, 2012). The gel-forming ability, biodegradability, and biocompatibility are few properties that put starch and their modified counterparts as potential materials for a sustained release agent. A number of techniques such as spray drying, extrusion, precipitation, and chelation are used in encapsulation and controlled release of active ingredients.

Tarvainen, Sutinen, Peltonen, Tiihonen, and Paronen (2000) investigated starch acetates (DS 2.8) as novel, multifunctional excipients for the direct compression tableting process, as compared to ethylcellulose (EC). Study showed SA films are tougher and stronger than EC films at the same plasticizer concentration. DSC thermograms supported the findings of tensile test that showed plasticizers such as triacetin and triethyl citrate were the most compatible with SA. It also appeared to be promising film former for pharmaceutical coating. SA with varying DS have been demonstrated to be multifunctional excipients for direct compression tablet process, as they form more hydrophobic and mechanically stronger tablets than unsubstituted starch (Korhonen et al., 2000). Lopez, Garcia, and Zaritzky (2008) studied the effect of plasticizers on the film properties of solvent casted SA. They showed 5% concentration is optimum for film formation.

Rofarma (Italy) has a popular commercial starch acetate product, AMPRAC01, marketed as a water-based polymer for controlled drug delivery of solid dosage form. It is reported to have a wide range of additives and excipients for improving storage stability, taste and odor masking, pH-dependent solubility, protection of sensitive actives from gastric fluid, protection of gastric mucosa from aggressive drugs, and clear or colored coating (AMPRAC01, undated). On the other hand, a significant commercial success for encapsulation has been noted for utilization of OSA starch. They are good stabilizer of oil-in-water emulsion and good film former, providing an effective matrix for encapsulating volatile flavors and fragrances. Wurzburg, Herbust, and Cole (1963) and Wurzburg, Trubiano, and Herbsi (1970) patented encapsulation of flavor compounds in the presence of moisture in OSA starch and also the controlled release of encapsulated oils such as vitamin A in the presence of moisture. It was also reported that the inclusion of polyvalent metal ions, e.g., magnesium and calcium, in the reaction with OSA-treated starch produces a water-repellent starch (Zallie, 1999).

Onofre and Wang (2010) demonstrated that hydroxypropylation is more effective in Hylon VII as compared to medium- and lower-amylose starches as a controlled released agent. Introduction of hydroxypropyl group increases the water holding capacity of starch, however, decreases the tablet porosity and storage modulus of swollen tablets. Microscopic analyses revealed smoother and less porous tablet structure. However, hydroxypropylation had a detrimental impact on drug release from waxy cornstarch matrices. Another study has shown that tensile strength of amylomaize film is reduced by hydroxypropylation, but elongation and burst strength is increased (Roth & Mehltretter, 1967).

Hydroxypropylation is also used to improve mechanical properties of starch-based films and expanded loose fill packaging (foams) because of its structural linearity (Altieri & Lacourse, 1991).

References

Altieri, P. A., & Lacourse, N. L. (1991). Starch-based protective loose fill material. In: *Proceedings of corn utilization conference IV*, St. Louis, MO: National Corn Grower's Association.

AMPRAC01 (undated). http://www.rofarma.com/default.aspx?menuItem=26.

Asp, N. G. (1992). Resistant starch. Proceedings from the second plenary meeting of EURESTA: European FLAIR Concerted Action No. 11 on physiological implications of the consumption of resistant starch in man. *European Journal of Clinical Nutrition*, *46*(Suppl. 2), S1.

Bae, S. O., & Lim, S. T. (1998). Physical properties of extruded strands of normal and high amylose hydroxypropylated amylomaize corn starch. *Cereal Chemistry*, *75*, 449–454.

Banker, S. (1966). Film coating theory and practice. *Journal of Pharmaceutical Science*, *55*, 81–89.

Banks, W., Geddes, R., Greenwood, C. T., & Jones, I. G. (1972). Physicochemical studies of starches. *Starch*, *24*, 245–280.

Barreto, P. L. M., Pires, A. T. N., & Soldi, V. (2003). Thermal degradation of edible films based on milk proteins and gelatin in inert atmosphere. *Polymer Degradability and Stability*, *79*, 147–152.

Bastioli, C. (2000). Global status of the production of biobased packaging materials. In: J. W. Claus (Ed.), *Proceedings of the food biopack conference, Copenhagen*, pp. 2–7.

BeMiller, J. N. (2007). *Carbohydrate chemistry for food scientists.* St. Paul, MN: AACC International, Inc.

Bergthaller, W. (2004). Development of potato starch. In A. C. Eliasson (Ed.), *Starch in foods: Structure, function and applications* (pp. 241–257). Cambridge, London: Woodhead Publishing.

Berry, C. S. (1986). Resistant starch: Formation and measurement of starch that survives exhaustive digestion with amylolytic enzymes during the determination of dietary fibre. *Journal of Cereal Science*, *4*, 301–314.

Bird, A. R., Shrestha, A. R., Lopez-Rubio, A., & Gidley, M. J. (2009). Resistant starch *in vitro* and *in vivo*: Factors determining yield, structure, and physiological relevance. In I. T. Norton, J. C. Ubbink, & S. Kasapis (Eds.), *Modern biopolymer science: Bridging the divide between fundamental treatise and industrial application* (pp. 449–512). London: Academic Press.

Björck, I., Gunnarsson, A., & Östergård, K. (1989). A study of native and chemically modified potato starch. Part II. Digestibility in the rat intestinal tract. *Starch*, *41*, 128–134.

Bonacucina, G., Martino, P. D., Piombetti, M., Colombo, A., Roversi, F., & Palmieri, G. F. (2006). Effect of plasticizers on properties of pregelatinised starch acetate (Amprac 01) free films. *International Journal of Pharmaceutics*, *313*, 72–77.

Bowler, P., Williams, M. R., & Angold, R. E. (1980). A hypothesis for the morphological changes which occur on heating lenticular wheat starch in water. *Starch Chemistry*, *32*, 186–189.

Caldwell, C. G., & Wurzburg, O. B. (1953). Polysaccharide derivatives of substituted dicarboxylic acids. U.S. Patent 2,661,349.

Cameron, R. E., & Donald, A. M. (1992). A small angle X-ray scattering study of the annealing and gelatinization of starch. *Polymer*, *33*, 2628–2635.

Castro, J. C. (2005). *The molecular weight distribution of debranched starch.* Sydney, Australia: University of Sydney.

Champ, M. (2004). Resistant starch. In A. C. Eliasson (Ed.), *Starch in food structure, function and applications* (pp. 560–574). Cambridge, London: Woodhead Publishing.

Chaudhary, A. L., Torley, P. J., Halley, P. J., McCaffery, N., & Chaudhary, D. S. (2009). Amylose content and chemical modification effects on TPS from maize—Processing and characterization using conventional polymer equipment. *Carbohydrate Polymers, 78*, 917–925.

Chen, Z., Schols, H. A., & Voragen, A. G. J. (2004). Differentially sized granules from acetylated potato and sweet potato starches differ in the acetyl substitution pattern of their amylose populations. *Carbohydrate Polymers, 56*, 219–226.

Chiu, C. -W., & Solarek, D. (2009). Modification of starches. In J. BeMiller & R. Whistler (Eds.), *Starch chemistry and technology* (pp. 629–655). (3rd ed.): Oxford, UK: Elsevier, Inc.

Cooke, D., & Gildley, M. J. (1992). Loss of crystalline and molecular order during starch gelatinization: Origin of the enthalpic transition. *Carbohydrate Research, 227*, 103–112.

Cornell, H. (2004). The functionality of wheat starch. In A. C. Eliasson (Ed.), *Starch in foods: Structure, function and applications* (pp. 211–240). Cambridge, London: Woodhead Publishing.

Davis, J. P., Supatcharee, N., Khandelwal, R. L., & Chibbar, R. N. (2003). Synthesis of novel starches in plants: Opportunities and challenges. *Starch-Starke, 55*, 107–120.

Donald, A. M. (2004). Understanding starch structure and functionality. In A. C. Eliasson (Ed.), *Starch in foods: Structure, function and applications* (pp. 156–184). Cambridge, London: Woodhead Publishing.

Eastman, J. E., & Moore, O. (2009). U.S. Patent 4,465,702 (as referred by Chiu and Solarek, 2009).

Eerlingen, R. C., Bjorck, I. V. D., Delcour, J. A., Slade, L., & Levine, H. (1994). Enzyme-resistant starch. VI. Influence of sugars on resistant starch formation. *Cereal Chemistry, 71*, 472–476.

Elder, A. E., & Schoch, T. J. (1959). Measuring the useful properties of starch. *Cereal Science Today, 4*, 202.

Eliasson, A. -N., & Gudmundsson, M. (2006). Starch: Physicochemical and functional aspects. In A. C. Eliasson (Ed.), *Carbohydrates in food* (pp. 391–470). (2nd ed.). Boca Raton, FL: Taylor and Francis.

Ellis, S. H., & Ring, S. G. (1985). A study of some factors influencing amylose gelation. *Carbohydrate Polymers, 5*, 201.

Englyst, H. N., Kingman, S. M., & Cummings, J. H. (1992). Classification and measurement of nutritionally important starch fractions. *European Journal of Clinical Nutrition, 46*, 33–50.

Federation of America Societies for Experimental Biology, FASEB (1979). Evaluation of the health aspects of starch and modified starches as food ingredients, SCOGS 115. Bethesda, MD: Life Sciences Research Office.

Fleche, G. (1985). Chemical modification and degradation of starch. In G. M. A. Van Beyunm & J. A. Roels (Eds.), *Starch conversion technology* (pp. 15–45). New York: Marcel Dekker, Inc.

Forssell, P. (2004). Starch-based microencapsulation. In A. C. Eliasson (Ed.), *Starch in food structure, function and applications* (pp. 462–473). Cambridge, England: Woodhead Publishing Limited.

Fringant, C., Desbrieres, J., & Rinaudo, M. (1996). Physical properties of acetylated starch based materials: Relation with their molecular characteristics. *Polymer, 37*, 2663–2673.

Fringant, C., Rinaudo, M., Foray, M. F., & Bardet, M. (1998). Preparation of mixed esters of starch or use of an external plasticizer: Two different ways to change the properties of starch acetate films. *Carbohydrate Polymers, 35*, 97–106.

Frische, R., & Best, B. (1994). European Patent Application. EP 579197.

Furin, J. C. (1973). Batter starch. U.S. Patent No. 3,767,826.

Galliard, T., & Bowler, P. (1987). Morphology and composition of starch. In T. Galliard (Ed.), *Starch: Properties and potential* (pp. 55–77). Great Britain: John Wiley and Sons.

Geng, Z., Zongdao, C., & Toledo, R. (2003). Effects of different processing methods on the glycemic index of kudzu starch. *Journal of Chinese Cereals Oils Association, 18*, 5–10.

Gidley, M. (1987). Factors affecting the crystalline type (A-C) of native starches and model compounds: A rationalization of observed effects in terms of polymorphic structures. *Carbohydrate Research, 161*, 301–304.

Goff, H. D. (1994). Modified starches and the stability of frozen foods. In A. C. Eliasson (Ed.), *Starch in foods: Structure, function and applications* (pp. 425–440). Cambridge, London: Woodhead Publishing.

Gonzalez-Soto, R. A., Escobedo, R., Hernandez-Sanchez, H., Sanchez-Rivera, M., & Bello-Perez, L. A. (2006). The influence of time and storage temperature on resistant starch formation from autoclaved debranched banana starch. *Food Research International, 40*, 304–310.

Gray, J. A., & BeMiller, J. N. (2005). Influence of reaction conditions on the location of reactions in waxy maize starch granules reacted with a propylene oxide analogue at low substitution levels. *Carbohydrate Polymers, 60*, 147–162.

Guan, J., Fang, Q., & Hanna, M. A. (2004). Selected functional properties of extruded starch acetate and natural fibers foams. *Cereal Chemistry, 81*, 199–206.

Gupta, M., Brennan, C., & Tiwari, B. K. (2012). Starch-based edible films and coatings. In J. Ahmad, B. K. Tiwari, S. H. Imam, & M. A. Rao (Eds.), *Starch-based polymeric materials and nanocomposites: Chemistry, processing, and applications* (pp. 239–254): Boca Raton, FL: CRC Press.

Halley, P. J. (2005). TPS biodegradable polymers. In R. Smith (Ed.), *Biodegradable polymers for industrial applications* (pp. 140–163). Boca Raton, FL: Woodhead Publishing Limited, CRC Press.

Hancock, H. D., & Tarbet, B. J. (2000). The other double helix-the fascinating chemistry of starch. *Chemistry for Everyone, 77*, 988–992.

Heacock, P. M., Hertzier, S. R., & Wolf, B. (2004). The glycemic, insulinemic, and breath hydrogen responses in humans to a food starch esterified by 1-octenyl succinic anhydride. *Nutrition Research, 24*, 581–592.

Hirsch, J. B., & Kokini, J. L. (2002). Understanding the mechanism of cross-linking agents (POCl3, STMP, and EPI) through swelling behaviour and pasting properties of cross-linked waxy maize starches. *Cereal Chemistry, 79*, 102–107.

Hizukuri, S. (1986). Polymodal distribution of the chain lengths of amylopectin and its significance. *Carbohydrate Research, 147*, 342–347.

Hoover, R., Hannouz, D., & Sosulski, F. W. (1988). Effect of hydroxypropylation on thermal properties, starch digestibility and freeze-thaw stability of field pea (*Pisum sativum cv trapper*) starch. *Starch-Starke, 40*, 383–387.

Hoover, R., & Sosulski, F. (1985). A comparative study of the effect of acetylation on starches of *Phaseolus vulgaris* biotypes. *Starch, 37*, 397–404.

Htoon, A., Shrestha, A. K., Flanagan, B. M., Lopez-Rubio, A., Bird, A. R., Gilbert, E. P., et al. (2009). Effects of processing high amylose maize starches under controlled conditions on structural organization and amylase digestibility. *Carbohydrate Polymers, 75*, 236–245.

Huber, K. C., & BeMiller, J. N. (1997). Visualization of channels and cavities of corn and sorghum starch granules. *Cereal Chemistry, 74*, 537–541.

Huber, K. C., & BeMiller, J. N. (2009). Modified starch: Chemistry and properties. In A. C. Bertolini (Ed.), *Starches: Characterization, properties, and application* (pp. 145–203): Boca Raton, FL: CRC Press.

Imberty, A., Buleon, A., Tran, V., & Perez, S. (1991). Recent advances in knowledge of starch structure. *Starch, 43*, 375–384.

Jacobasch, G., Schmiedl, D., Kruscheski, M., & Schmehl, K. (1999). Dietary resistant starch and chronic inflammatory bowel diseases. *International Journal of Colorectal Diseases, 14*, 201–211.

Jacobs, H., & Delcour, J. A. (1998). Hydrothermal modifications of granular starch, with retention of the granular structure: A review. *Journal of Agricultural and Food Chemistry, 46*, 2895–2905.

Jarowenko, W. (1986). Acetylated starch and miscellaneous organic esters. In O. B. Wurzburg (Ed.), *Modified starches: Properties and uses* (pp. 55–78). Boca Raton, FL: CRC Press.

Jenkins, P. J., & Donald, A. M. (1996). Application of small angle neutron scattering to the study of the structure of starch granules. *Polymer, 37*, 5559–5568.

Jenkins, P. J., & Donald, A. M. (1998). Gelatinization of starch: A combined SAX/WAXS/DSC and SANS study. *Carbohydrate Research, 308*, 133–147.

Katcher, J. H., & Ackilli, J. A. (1980). Process for preparing an odor-free acetylated starch. U.S. Patent 4,238,604.

Katcher, J. H., Mahon, J. T., Matherly, J. W., McCarty, L. T., & Schara, T. E. (1989). U.S. Patent 4,871,398 (From Wurzburg, 1995).

Kaur, L., Singh, N., & Singh, J. (2004). Factors influencing the properties of hydroxypropylated potato starches. *Carbohydrate Polymers, 55*, 211–223.

Knutson, C. A. (1990). Annealing of maize starches as elevated temperatures. *Cereal Chemistry, 67*, 376.

Korhonen, O., Raatikainen, P., Harjunen, P., Nakari, J., Suihko, E., Peltonen, S., et al. (2000). Starch acetates multifunctional direct compression excipients. *Pharmaceutical Research, 17*, 1138–1143.

Kruger, L. H., & Rutenburg, M. W. (1967). Production and uses of starch acetates. In R. L. Whistler & E. F. Paschall (Eds.), *Starch: Chemistry and technology, II. Industrial aspects* (p. 367). New York: Academic Press.

Lafargue, D., Pontoire, B., Buleon, A., Doublier, L., & Lourdin, D. (2007). Structure and mechanical properties of hydroxypropylated starch films. *Biomacromolecules, 8*, 3950–3958.

Lang, G. (2004). Development of a range of industrialised cereal-based food stuffs, high in slowly digestible starch. In A. C. Eliasson (Ed.), *Starch in foods: Structure, function and applications* (pp. 477–504). Cambridge, London: Woodhead Publishing.

Leloup, V. M., Colonna, P., Ring, S. G., Roberts, K., & Wells, B. (1992). Microstructure of amylose gels. *Carbohydrate Polymers, 18*, 189–197.

Lim, S., & Seib, A. (1993). Location of phosphate esters in a wheat starch phosphate by phosphorun-31 nuclear magnetic resonance spectroscopy. *Cereal Chemistry, 70*, 145.

Liu, H., Ramsden, L., & Corke, H. (1997). Physical properties and enzymatic digestibility of aceylated *ae*, *wx*, and normal maize starch. *Carbohydrate Polymers, 34*, 283–289.

Lopez, O. V., Garcia, M., & Zaritzky, N. E. (2008). Film forming capacity of chemically modified corn starches. *Carbohydrate Polymers, 73*, 573–581.

Ma, W., & Robyt, J. F. (1987). Preparation and characterization of soluble starches having different molecular sizes and composition, by acid hydrolysis in different alcohols. *Carbohydrate Research, 166*, 283–297.

Manelius, R., Nurmi, K., & Bertoft, E. (2000). Enzymatic and acid hydrolysis of cationized waxy corn starch granules. *Cereal Chemistry, 77*, 345–353.

Mao, L. J., Imam, S., Gordon, S., Cinelli, P., & Chiellini, E. (2000). Extruded cornstarch-glycerol-polyvinyl alcohol blends: Mechanical properties, morphology and biodegradability. *Journal of Polymer Environment, 8*, 205–211.

Marconi, E., Ruggeri, S., Cappelloni, M., Leonardi, D., & Carnovale, E. (2000). Related physicochemical, nutritional, and microstructural characteristics of chickpeas (*Cicer arietinum, L.*) and common beans (*Phaseolus vulgaris, L.*). *Journal of Agricultural and Food Chemistry, 48*, 5986–5994.

Matthews, G. (1996). *PVC: Production, properties and uses.* London, England: Institute of Metals.

Miles, M. J., Morris, V. J., & Ring, S. G. (1985). The roles of amylose and amyl pectin in the gelatinization and retrogradation of starch. *Carbohydrate Research, 135*, 257–269.

Moorthy, S. N. (1985). Acetylation of cassava starch using perchloric acid catalysis. *Starch, 37*, 307–308.

Moorthy, S. N. (2004). Tropical sources of starch. In A. C. Eliasson (Ed.), *Starch in foods: Structure, function and applications* (pp. 363–392). Cambridge, London: Woodhead Publishing.

Moreau, N. M., Martin, L. J., Tooquet, C. S., Laboisse, C. L., Nguyen, P. G., Siliart, B. S., et al. (2003). Restoration of integrity for rat caeco-colonic mucosa by resistant starch, but not by fructo-oligosaccharides, in dextran sulfate sodium-induced experimental colitis. *British Journal of Nutrition, 90*, 75–85.

Morrison, W. R. (1988). Lipids in cereal starches: A review. *Journal of Cereal Science, 8*, 1–15.

Morrison, W. R., & Karkalas, J. (1990). Starch. In P. M. Dey (Ed.), *Methods in plant biochemistry: Vol. 2.* (pp. 323–352). London: Academic Press.

Neucom, H. (1958). Pudding mix. U.S. Patent 2,865,762.

Onofre, F. O., & Wang, Y. J. (2010). Hydroxypropylated starches of varying amylose contents as sustained release matrices in tablets. *International Journal of Pharmaceutics, 385*, 104–112.

Parchure, A. A., & Kulkarni, P. R. (1997). Effect of processing treatments on generation of resistance starch. *International Journal of Food Science and Nutrition, 48*, 257–260.

Patil, S. R. P. (2010). Global modified starch products and carbohydrates functional foods derivatives and markets—A strategic review, August. http://www.skpatilassociates.com/reports/jan11/Report%202%20Abstract%20Table%20of%20Content%20Sample%20pages%20Global%20Mod%20St%20Derivatives%20Strategic%20Rev%20Aug%202010.pdf.

Raben, A., Andersen, K., Karberg, M. A., Holst, J. J., & Astrup, A. (1997). Acetylation of or β-cyclodextrin addition to potato starch: Beneficial effect on glucose metabolism and appetite sensations. *American Journal of Clinical Nutrition, 66*, 304–314.

Ring, S. G., Gee, J. M., Whittam, M., Orford, P., & Johnson, I. T. (1988). Resistant Starch: Its chemical form in foodstuffs and effect on digestibility *in vitro*. *Food Chemistry, 28*, 97–109.

Rosin, P. M., Lajolo, F. M., & Mendezes, E. W. (2002). Measurement and characterization of dietary starches. *Journal of Food Composition and Analysis, 15*, 367–377.

Roth, W. B., & Mehltretter, C. L. (1967). Some properties of hydroxypropylated amylomaize starch films. *Food Technology, 21*, 72.

Rutenberg, M. W., & Solarek, D. (1984). Starch derivatives: Production and uses. In R. Whistler, J. N. BeMiller, & E. F. Paschall (Eds.), *Starch: Chemistry and technology* (pp. 314–388). New York: Academic Press, Inc.

Sagum, R., & Arcot, J. (2000). Effect of domestic processing methods on the starch, non-starch polysaccharides and *in vitro* starch and protein digestibility of three varieties of rice with varying levels of amylose. *Food Chemistry, 70*, 107–111.

Schuetzenberger, P. (1865). *Seances Academic Science, 61*, 465.

Seib, P. A. (1996). *Starch chemistry and technology, syllabus* (p. 38). Manhattan, KS: Kansas State University.

Shogren, R. L. (1996). Preparation, thermal properties, and extrusion of high-amylose starch acetates. *Carbohydrate Polymers, 29*, 57–62.

Shogren, R. L., & Lawton, J. W. (1998). Enhanced water resistance of starch-based materials. U.S. Patent 5,756,194.

Shrestha, A. K., Blazek, J., Flanagan, B. M., Dhital, S., Larroque, O., Morell, M. K., et al. (2012). Molecular, mesoscopic and microscopic structure evolution during amylase digestion of maize starch granules. *Carbohydrate Polymers, 90*, 23–33.

Shrestha, A. K., & Khadka, R. (2007). Effects of extrusion on resistant starch content and structure of low and high amylose starches. *Journal of Food Science and Technology Nepal, 3*, 22–28.

Shrestha, A. K., Ng, C. S., Lopez-Rubio, A., Blazek, J., Gilbert, E. P., & Gidley, M. J. (2010). Enzyme resistance and structural organization in extruded high amylose maize starch. *Carbohydrate Polymers, 80*, 699–710.

Siljestrom, M., & Bjorck, I. (1990). Digestible and undigestible carbohydrates in autoclaved legumes, potatoes and corn. *Food Chemistry, 38*, 145–152.

Siracusa, V., Rocculi, P., Romani, S. And, & Rosa, M. D. (2008). Biodegradable polymers for food packaging: A review. *Trends in Food Science and Technology, 19*, 634–643.

Sitohy, M. Z., & Ramadan, M. F. (2001). Degradability of different phosphorylated starches and thermoplastic films prepared from corn starch phosphomonoesters. *Starch, 53*, 317–322.
Solarek, D. B. (1986). Phosphorylated starches and miscellaneous inorganic starches. In O. B. Wurzburg (Ed.), *Modified starches: Properties and uses* (pp. 97–112). Boca Raton, FL: CRC Press.
Srivastava, H. C., & Patel, M. M. (1973). Viscosity stabilization of tapioca starch. *Starch, 25*, 17–21.
Swinkels, J. J. M. (1985). Sources of starch, its chemistry and physics. In G. M. V. Beyunm & J. A. Roels (Eds.), *Starch conversion technology* (pp. 15–45). New York: Marcel Dekker, Inc.
Taggart, P. (2004). Starch as an ingredient: Manufacture and applications. In A. C. Eliasson (Ed.), *Starch in foods: Structure, function and applications* (pp. 363–392). Cambridge, London: Woodhead Publishing.
Takagi, S., Koyama, M., Kameyama, H., & Tokiwa, Y. (1994). Development of polycaprolactone/gelatinized starch blends and their enzymatic degradation. *Studies in Polymer Science, 12*, 437.
Tan, I. (2005). Characterization of novel starch materials: structure-functionality relationship (Ph.D. thesis). Brisbane, Australia: The University of Queensland.
Tarvainen, M. (2004). Starch acetate as a coating polymer for oral extended released products: film-forming ability, permeability and plasticization (Ph.D. dissertation). Kuopio, Finland: University of Kuopio.
Tarvainen, M., Sutinen, R., Peltonen, S., Tiihonen, P., & Paronen, P. (2000). Starch acetate-a novel film-forming polymer for pharmaceutical coatings. *Journal of Pharmaceutical Sciences, 91*, 282–288.
Tessler, M. M., & Billmers, R. L. (1996). Preparation of starch esters. *Journal of Environment Polymer Degradation, 4*, 85–89.
Tester, R. F., Karkalas, J., & Qi, X. (2004). Starch-composition, fine structure and architecture. *Journal of Cereal Science, 399*, 151–165.
Tester, R. F., Karkalas, J., & Xin, Q. (2004). Structure of amylose and amylopectin. *Journal of Cereal Science, 39*, 151–165.
Thire, R. M. S. M. (2010). Starch based plastics. In A. Bartolin (Ed.), *Starch characterization, properties, and applications* (pp. 103–128). Boca Raton, FL: Taylor and Francis, LLC.
Topping, D. L., & Clifton, P. M. (2001). Short-chain fatty acids and human colonic function: Roles of resistant starch and non starch polysaccharides. *Physiological Review, 81*, 1031–1064.
Trubiano, P. C. (1986). Succinate and substituted succinate derivatives of starch. In O. B. Wurzburg (Ed.), *Modified starches: Properties and uses* (pp. 131–147). Boca Raton, FL: CRC Press.
Tuschoff, J. V. (1986). Hydroxypropylated starch. In O. B. Wurzburg (Ed.), *Modified starches: Properties and uses* (pp. 89–96). Boca Raton, FL: CRC Press.
Tuschoff, J. V., & Smith, C. E. (1962). U. S. Patent 3,022,089.
Wang, J., Lianghua, S., & Wang, S. (2009). Physicochemical properties of octenyl succinic anhydride-modified potato starch with different degree of substitution. *Journal of Science Food and Agriculture, 90*, 424–429.
Wanous, M. P. (2004). Texturizing and stabilizing, by gum. *Prepared Foods, 173*, 108–118.
Whistler, R. L., & BeMiller, J. N. (1997). Starch. In Jame N. BeMiller (Ed.), *Carbohydrate chemistry for food scientists* (pp. 117–152). St. Paul, MN: Eagan Press.
White, P. J., Abbas, I. R., & Johnson, L. A. (1989). Freeze-thaw stability and refrigerated-storage retrogradation of starches. *Starch-Starke, 41*, 176.
White, P. J., & Tziotis, A. (2004). New corn starches. In A. C. Eliasson (Ed.), *Starch in foods: Structure, function and applications* (pp. 295–320). Cambridge, London: Woodhead Publishing.

Woo, K. S., & Seib, P. A. (2002). Cross-linked resistant starch: Preparation and properties. *Cereal Chemistry, 79*, 819–825.

Wootton, M., & Bamunuarachchi, A. (1979). Applications of differential scanning calorimetry to starch gelatinization. I. Commercial native and modified starches. *Starch-Starke, 31*, 116–121.

Wurzburg, O. B. (1986a). Converted starches. In O. B. Wurzburg (Ed.), *Modified starches: Properties and uses* (pp. 17–40). Boca Raton, FL: CRC Press.

Wurzburg, O. B. (1986b). Nutritional aspects and safety of modified food starches. *Nutrition Review, 44*, 74–79.

Wurzburg, O. B. (1995). Modified starch. In A. M. Stephen (Ed.), *Food polysaccharides and their applications* (pp. 67–97). New York: Marcel Dekker, Inc.

Wurzburg, O. B., Herbust, W., & Cole, H. M. (1963). U. S. Patent 3,091,507.

Wurzburg, O. B., Trubiano, P., & Herbsi, W. (1970). U.S. Patent 3,499,962.

Xie, S. X., Liu, Q., & Cui, S. W. (2005). Starch modification and application. In S. W. Cui (Ed.), *Food carbohydrates: Chemistry, physical properties and applications* (pp. 357–407). Boca Raton, FL: Taylor and Francis Group.

Xu, Y., & Hanna, M. A. (2005). Physical, mechanical, and morphological characteristics of extruded starch acetate foams. *Journal of Polymers and the Environment, 13*, 221–230.

Xu, Y., Miladinov, V., & Hanna, M. (2004). Synthesis and characterization of starch acetates with high substitution. *Cereal Chemistry, 81*, 735–740.

Yoshimoto, Y., Tashiro, J., Takenouchi, T., & Takeda, Y. (2000). Molecular structure and some physicochemical properties of high-amylose barley starch. *Cereal Chemistry, 77*, 279–285.

Zallie, J. (1999). Function of speciality starches in the confection industry. In http://management.ebooks6.com/The-role-and-function-of-specialty-starches-in-the-confection-industry-download-w12860.pdf. National Starch and Chemical Corporation.

CHAPTER

6

Chemically Modified Starch; Allyl- and Epoxy-Starch Derivatives: Their Synthesis and Characterization

Maurice C.R. Franssen[1], *Carmen G. Boeriu*[2]

[1]Wageningen University, Lab of Organic Chemistry, Wageningen, The Netherlands

[2]Wageningen UR Food and Biobased Research, Division Biobased Products, Wageningen, The Netherlands

1 INTRODUCTION

1.1 Chemical Modification of Starch

Starch is very often modified to produce derivatives with improved properties for specific applications and industrial use. Commonly, starch is modified by chemical reactions and/or enzymatic treatment. Also, the starch granule can be modified by physical treatment such as extrusion, gelatinization, blending, and drying (Bao & Bergman, 2004; Bemiller, 1997; Richardson & Gorton, 2003; Van Soest, Hulleman, Dewit, & Vliegenthart, 1996). Chemical modification is based on the reaction of the free hydroxyl groups of the anhydrous glucose unit (AGU) monomers with a functional group, resulting in starch derivatives. Usually, the modification involves esterification or etherification of the hydroxyl groups. In general, these modifications of the hydroxyl groups in starch in the presence of small amounts of base favor 2-O-substitution above 3-O- and 6-O-substitution. The hydroxyl group on C-2 exhibits the highest acidity and is therefore the most reactive. However, the other groups may also react because the reactivity of the hydroxyl groups depends on electronic and conformational factors and their availability for reagents in general (Richardson & Gorton, 2003; Tomasik & Schilling, 2004).

The behavior and properties achieved after chemical modification are depending upon the number, the distribution, and the nature of the substituents as well as the starch source and reaction conditions. Those starch derivatives with a degree of substitution (DS) up to 0.20 are of commercial interest (Bemiller, 1997; Richardson & Gorton, 2003; Roper, 2002; Rutenberg & Solarek, 1984; Singh, Kaur, & McCarthy, 2007; Swinkels, 1990; Tomasik & Schilling, 2004). Commercially available starch derivatives display a variety of properties, which are suitable for industrial applications (Table 6.1).

Generally, these starch derivatives can be manufactured in an aqueous solution at a temperature below 60 °C. NaOH and Na_2SO_4 can be added, respectively, to enhance the reactivity of the hydroxyl groups in starch and to prevent gelatinization of the granular starch in alkaline condition (Bemiller, 1997; Richardson & Gorton, 2003). Under these conditions, the granular structure of starch can be maintained, and subsequently, granular starch derivatives can easily be recovered by simple filtration and drying. Starch modification generally leads to a reduction in gelatinization temperature. This implies that only a limited level of substitution can be obtained when one wants to retain the starch in granular form (Rutenberg & Solarek, 1984).

The DS is defined as the average number of hydroxyl groups on the AGU that have been substituted (Rutenberg & Solarek, 1984; Wurzburg, 1986c). DS ranges from 0 to 3. If DS is 3, all possible hydroxyl groups are substituted per AGU, and if DS is less than 1, the average number of substituents per AGU is <1. Another term (molar substitution) is preferred when the substitution is performed by chemical compounds that generate new free hydroxyl groups for further substitution. Molar substitution is defined as the average number of moles of the functional group per AGU. Commonly, molar substitution equals DS, (Moser, 1986) but if more than one substituent per AGU can be attached at the same position on the AGU, molar substitution can exceed three in contrast to DS (Bertoft, 2004; Richardson & Gorton, 2003; Rutenberg & Solarek, 1984; Singh et al., 2007; Wurzburg, 1986c). The determination of DS is often performed chemically (Bhuniya, Rahman, Satyanand, Gharia, & Dave, 2003; Lin, Lee, & Chang, 2003; Manelius, Buleon, Nurmi, & Bertoft, 2000) or by NMR (Bien et al., 2001; Radosta et al., 2004), dependent on the type of substituent.

1.2 Starch Derivatives

Chemical modifications can be accomplished by either controlled acid conversion, oxidation, substitution, grafting, or cross-linking (Bemiller, 1997; Richardson & Gorton, 2003; Roper, 2002; Rutenberg & Solarek, 1984; Singh et al., 2007; Swinkels, 1990; Tomasik & Schilling, 2004).

TABLE 6.1 Properties and Application of the Main Commercial Available Starch Derivatives

Modified Starch	Reagents	Properties	Application	References
Acetylated	Acetic anhydride Vinyl acetate	High viscosity, decreased gelatinization and retrogradation, improved storage stability, good film properties	Food: thickeners, stabilizers; paper: surface sizing agents; textile: warp sizing agents; gummed tapes	Elomaa et al. (2004), Jarowenko (1986), Lawal (2004), Singh, Chawla, and Singh (2004), Tomasik and Schilling (2004), Wang and Wang (2002)
Phosphorylated	Phosphoric acid	Decreased gelatinization, high viscosity, stable dispersion, cohesive texture, polyelectrolyte	Food: emulsifiers, thickening agents, adhesives; paper: wet-end additives and binders; textile: thickeners, warp sizing agents, stiffening; pharmaceuticals: detergents, encapsulation, flocculants	Deetae et al. (2008), Solarek (1986), Tomasik and Schilling (2004), Zhu (2003)
Succinylated	Succinic anhydride Octenyl succinic anhydride	High swelling, viscous pastes, reduced retrogradation, improved freeze-thaw stability	Food: additives to beverages, binders, and thickening agents; paper: surface sizing agents, coating binders; films; pharmaceuticals: encapsulation, drug delivery systems, tablets decomposer	Massicotte, Baille, and Mateescu (2008), Moorty (2004), Tomasik and Schilling (2004), Trubiano (1986), Zhang, Ju, Zhang, and Yang (2008)
Alkylated	Propylene oxide Ethylene oxide Acrylonitrile	Decreased gelatinization, lower pasting temperature, reduced retrogradation, improved storage stability, increased swelling, film properties	Food: thickening agents, coatings; paper: surface sizing, coatings, and binders; adhesives for plastic shaping; films; pharmaceuticals: drugs and protein carriers	Bien, Wiege, and Warwel (2001), Funke and Lindhauer (2001), Gonera, Goclik, Baum, and Mischnick (2002), Moser (1986), Tuschhoff (1986)

Continued

TABLE 6.1 Properties and Application of the Main Commercial Available Starch Derivatives—cont'd

Modified Starch	Reagents	Properties	Application	References
Carboxymethylated	Monochloroacetic acid	Enhanced solubility, improved clarity, polyelectrolyte	Paper: hydrophilic sizing agents, coatings; textile: sizing agents; pharmaceuticals: antitumor, drug delivery system, chelation ion exchange or polyanion flocculations, mostly cross-linked (absorbent product)	Heinze and Koschella (2005), Taylor (1979), Tijsen, Voncken, and Beenackers (2001), Tomasik and Schilling (2004)
Cationic	(3-Chloro-2-hydroxypropyl) trimethylammonium chloride (2,3-Epoxypropyl) trimethylammonium chloride	Better dispersity and solubility, polyelectrolyte	Paper: wet-end additives and binders, coatings, surface sizing agents; textile: warp sizing agents, flocculants, detergents	Khalil and Farag (1998), Radosta et al. (2004), Solarek (1986), Tomasik and Schilling (2004), Tuting, Wagemann, and Mischnick (2004)
Cross-linked	Phosphorous oxychloride Sodium trimetaphosphate Epichlorohydrin	Decreased gelatinization, reduced swelling and solubility, reinforced granules, improved storage stability	Food: thickener agents; paper: wet-rub-resistant coating and adhesives; other: emulsions, surgical dusting powder, antiperspirants, personal sanitary applications	Anwar, Khotimah, and Yamar (2006), Hamdi and Ponchel (1999), Kovats (1971), Le Bail, Morin, and Marchessault (1999)

Acid hydrolysis (Lawal, Adebowale, Ogunsanwo, Barba, & Ilo, 2005; Singh et al., 2007; Swinkels, 1990; Wang, Truong, & Wang, 2003; White & Tziotis, 2004; Wurzburg, 1986a) is performed by suspending starch in dilute acid solution at a temperature between ambient and just below pasting temperature. The reduced molecular weight results in reduced viscosity, less swelling, and a more soluble product.

Dextrinization (Campechano-Carrera, Corona-Cruz, Chel-Guerrero, & Betancur-Ancona, 2007; Singh et al., 2007; Swinkels, 1990; Taggart, 2004; Wurzburg, 1986a) also known as *pyroconversion* is the partial depolymerization in the presence of small quantities of acid. Alternatively, there is partial hydrolysis of the molecules and recombining of starch fragments, which is known as transglucosidation. These fragments or (pyro)dextrins exhibit low viscosity, reduced sugar content, and varying solubility, depending on the conversion.

Oxidation (Besemer & Van Bekkum, 1994; Manelius et al., 2000; Singh et al., 2007; Swinkels, 1990; White & Tziotis, 2004; Wurzburg, 1986a) of starch is achieved by a variety of oxidizing reagents such as hydrogen peroxide, alkaline hypochlorite, peracetic acid, persulfate, or permanganate. During the oxidation process, the starch is depolymerized, while AGUs become functionalized with carboxyl and carbonyl groups depending on the conditions used (Besemer & Van Bekkum, 1994; Manelius et al., 2000). *Bleaching*, which introduces $\leq$0.1% carboxyl groups per AGU, is considered to be a very light oxidization. Oxidized starch has a low viscosity and low temperature stability.

Cross-linked (Goff, 2004; Rutenberg & Solarek, 1984; Singh et al., 2007; Swinkels, 1990; Thomas & Atwell, 1999; Tomasik & Schilling, 2004; Wurzburg, 1986b) starch is achieved by the reaction of two or more hydroxyl groups in starch with each other, through the action of a bifunctional compound. A small number of the polymer chains are chemically linked to form covalent inter- and intramolecular bridges. The introduced bridges reinforce the granule to withstand chemical and physical treatment. Cross-linked starch exhibits high viscosity. Gelatinization and swelling of the granule is inhibited, depending on the type of cross-linker. Commercial cross-linking is often performed by the reaction of bi- or polyfunctional reagents, e.g., phosphorus oxychloride, sodium trimetaphosphate, epichlorohydrin, and mixtures of adipic anhydride and acetic acid (Figure 6.1) (Bhuniya et al., 2003; Hamdi & Ponchel, 1999; Kovats, 1971; Le Bail et al., 1999). Cross-linking is often performed in combination with esterification or etherification, to provide appropriate gelatinization, viscosity, and texture properties. Multiple treated starches are often used as thermoplastics (Duanmu, Gamstedt, & Rosling, 2007; Wach, Mitomo, Nagasawa, & Yoshii, 2003), hydrogels (Wang & Wang, 2000), and ingredients for food products such as bread (Tomasik & Schilling, 2004; Wu & Seib, 1990; Wurzburg, 1986a).

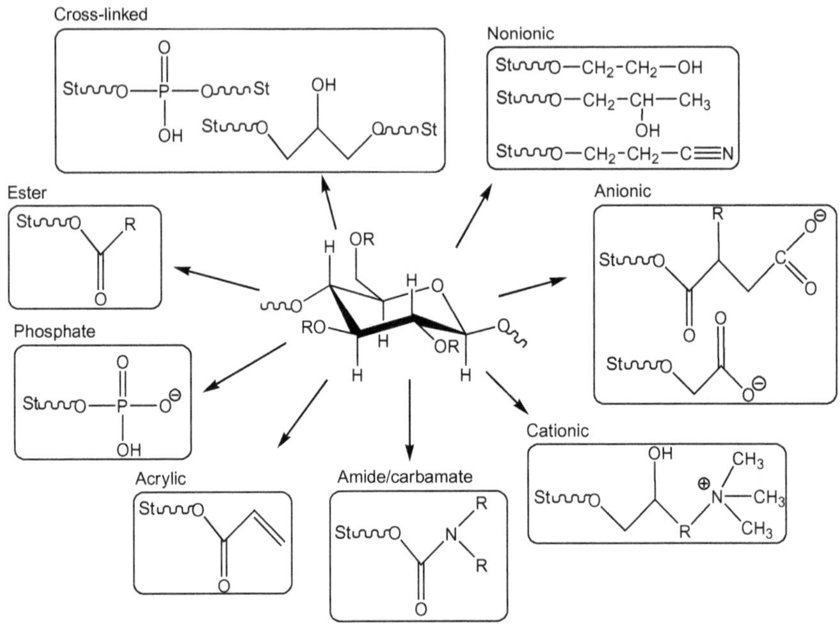

FIGURE 6.1 Starch modification for the production of commercial starch derivatives; St: starch; R: alkyl group.

Grafting (Athawale & Rathi, 1997; Fanta & Doane, 1986; Tomasik & Schilling, 2004; Zhang, Yang, & Yan, 2005) of starches is obtained by generating free radicals on the biopolymer for reaction with vinylic or acrylic monomers. Free radicals are generally formed by chemical initiation (often using ceric salts) or irradiation. Grafted starches can be used as hydrogels (Bhuniya et al., 2003), bioadhesive drug carriers (Ameye et al., 2002), thermoplastics, and paper additives (Nud'ga, Petrova, & Lebedeva, 2003; Tsai & Meier, 1990; Wilham, Mcguire, Rudolphi, & Mehltretter, 1963).

Substitution (Rutenberg & Solarek, 1984; Singh et al., 2007; Swinkels, 1990; Tomasik & Schilling, 2004; White & Tziotis, 2004) of starch is commonly accomplished through esterification and etherification. The reaction of starch with an etherifying or esterifying group in an alkaline medium introduces side chains in the starch structure, i.e., adding irregularities to amylopectin and especially to amylose, leading to lower gelatinization temperature. Furthermore, substitution is often performed to increase the pasting consistency and to limit retrogradation. Therefore, these starch derivatives are described as "stabilized starches." A broad variety of starch esters and ethers have been produced with nonionic, anionic, or cationic groups (Figure 6.1). Ionic starches are applied to improve interactions with oppositely charged surfaces, e.g., succinylated or cationic starches, which can be applied as coating binders in paper industry (Table 6.1). Esterification is commonly performed with anhydrides such as acetic, vinyl, or phosphoric anhydride (Deetae et al., 2008; Elomaa et al., 2004; Jarowenko, 1986; Lawal, 2004; Singh et al., 2004; Wang & Wang, 2002). Esterified starches are used in food applications as emulsion stabilizers, in frozen food, and for encapsulation. The ether groups are mostly introduced upon reaction with alkyl halides (Khalil & Farag, 1998; Laakso & Sjoholm, 1987; Radosta et al., 2004; Tuting et al., 2004), acrylonitrile (Gonera et al., 2002; Tomasik & Schilling, 2004), or alkylene oxides (Bien et al., 2001; Burton & Harding, 1997; Funke & Lindhauer, 2001; Hjermstad & Kesler, 1962; Kesler & Hjermstad, 1950a, 1950b; Moser, 1986; Tuschhoff, 1986; Wu, Pittman, & Gardner, 1995). Mostly, etherified starches are used in food applications as thickeners, in paper as additives, and in textile as sizing agents.

1.3 Aim of This Chapter

Functionalization of starch is widely used to modify the physical and chemical characteristics of the polysaccharide to obtain desirable properties for industrial applications. There is a strong interest in the development of new granular multifunctional starch derivatives for applications as biodegradable delivery systems, stabilizers, and coatings. This requires new relevant starch-based derivatives, which possess chemically active functional groups and allow easy binding of reagents with nucleophilic groups such as hydroxyl and amino groups. Among these new starch derivatives are allyl- and epoxy-starch derivatives. When epoxy groups are introduced into starch, they are able to react with compounds having a coupling, complexing, or cross-linking function or with compounds having anionic or cationic groups. These starch-based products can serve various purposes, e.g., flocculants, complexants, delivery systems, and pharmaceuticals.

This chapter focuses on epoxy-derivatized starch and its direct precursor allyl starch. Preparation and physicochemical properties are described for both compounds.

2 SYNTHESIS AND CHARACTERIZATION OF ALLYL STARCH

2.1 Synthesis

1-Allyloxy-2-hydroxypropyl (AHP) derivatives of waxy maize starch (WMS) and amylose-enriched maize starch (AEMS) were synthesized through the reaction of allyl glycidyl ether (AGE) with granular starch in an alkaline suspension in the presence of Na_2SO_4 as starch granule stabilizer (Scheme 6.1 and Table 6.2) (Fang, Fowler, Sayers, & Williams, 2004;

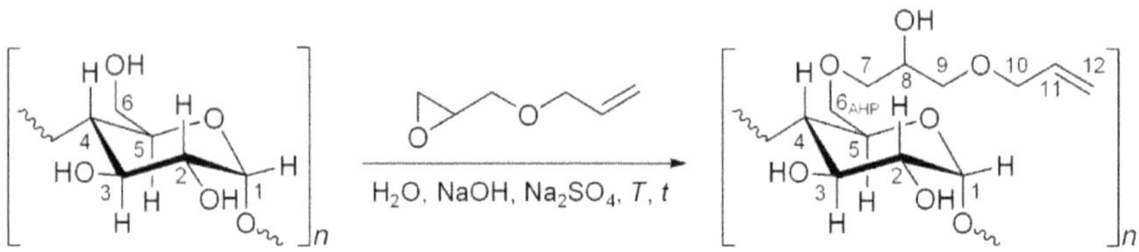

SCHEME 6.1 Synthesis of 1-allyloxy-2-hydroxy-propyl (AHP) starch.

TABLE 6.2 DS Values of AHP-AEMS and AHP-WMS (Huijbrechts et al., 2007)

Compound	NaOH	Na_2SO_4	Ratio AGE/AGU	DS_{NMR}	$DS_{titration}$
AEMS	0.12 M	40%, w/w	0.6	0.20 ± 0.01	0.24 ± 0.06
WMS	0.12 M	20%, w/w	0.3	0.19 ± 0.01	0.26 ± 0.06

Gotlieb & Cappelle, 2005; Richardson & Gorton, 2003). To prevent swelling of starch under the alkaline reaction conditions, a 0.20 or 0.48 equivalent of sodium sulfate was added for WMS and AEMS, respectively. Sodium hydroxide (9% mol/mol) was added to enhance the reactivity of the starch hydroxyl groups in the S_N2 reaction through the formation of alkoxide anions (Chen, Schols, & Voragen, 2004; Richardson & Gorton, 2003; Tsai & Meier, 1990; Van Der Burgt et al., 2000; Whistler & Bemiller, 1997). The reaction temperature was kept below the gelatinization temperature of the substrate starches to facilitate the production of a starch derivative in a dry, granular form (Fang et al., 2004; Richardson & Gorton, 2003; Whistler & Bemiller, 1997). After the penetration of the AGE into the starch granule, the reaction took place at both amylose and amylopectin fractions. The etherification of AEMS and WMS, respectively, with a 0.6 equivalent and 0.3 equivalent of AGE, proceeded smoothly and resulted in products with DSs of 0.20 and 0.19 per AGU, correspondingly (Table 6.2). Under these reaction conditions, this is a relatively high DS (Richardson & Gorton, 2003; Whistler & Bemiller, 1997) for granular starch derivatives. It suggests a good availability of the AGE reagent inside the granule for AEMS (70% amylose) and WMS (0% amylose) under the heterogeneous conditions.

We have optimized the etherification of granular starch with AGE to achieve a high DS (Huijbrechts et al., 2009). The synthesis of the AHP-starch derivatives of maize starch (MS) and WMS was investigated with the aim to find the optimum process conditions. A two-step approach was followed: first, the process parameters were identified (screening phase), and second, the processes were optimized based on these parameters (improving phase). In the screening phase, the starch concentration, the reaction time, the temperature, the amount of NaOH, the amount of Na_2SO_4, and the AGE were evaluated as reaction variables. Based on our experience and reported literature (Huijbrechts et al., 2007; Kesler & Hjermstad, 1950a; Radosta et al., 2004; Tsai & Meier, 1990; Vanwarners, Stamhuis, & Beenackers, 1989), six process factors, which were expected to have a significant influence on the DS, were selected for the screening phase. The effect of the following factors was investigated: the starch concentration (g kg^{-1} starch suspension), the reaction time (h), the temperature (°C), the amount of NaOH (as weight percentage related to the amount of dry starch, % ds), the amount of Na_2SO_4 (% ds), and the amount of AGE (% ds). The different reaction factors, inclusive of boundary values, are presented in Table 6.3. All treatments were performed in a random order, and data were analyzed using a response surface regression procedure. The responses of the different samples were analyzed using the statistical module of the experimental design program. The responses are statistically described with second-order interactions (2^m), which are characteristic for a full factorial design. This approach allows to estimate the main effects and all interactions up to m (m is six variables in this study). The generalized regression model was used, as shown in Equation (6.1):

$$Y = b_0 + \sum b_i X_i + \sum b_{ii} X_i^2 + \sum_{i<j}\sum b_{ij} X_i X_j \quad \text{with} \quad i = 1-6 \quad \text{and} \quad j = 2-6 \tag{6.1}$$

where Y = response, X_1 = starch concentration, X_2 = reaction time, X_3 = temperature, X_4 = the amount of NaOH, X_5 = the amount of Na_2SO_4, X_6 = the amount of AGE, b_0 = intercept, and b_i = corresponding regression coefficients.

TABLE 6.3 Experimental Design: Six Variables at Three Levels[a] (Huijbrechts et al., 2008)

	Min-Max Levels		
Variables	**−1**	**0**	**1**
C_{starch} (g kg^{-1} slurry)	200	300	400
Reaction time (h)	4	10	16
Temperature (°C)	20	34	48
NaOH (% ds)[b]	0.4	0.7	1.0
Na_2SO_4 (% ds)[b]	5	20	35
AGE (% ds)[b]	0.0016	0.0091	0.0166

[a] *The levels are indicated with 1, 0, and −1.*
[b] *Variables are calculated as % (w/w) on dry starch (ds); molar ratio of AGE, NaOH, and Na_2SO_4 per anhydrous glucose units are, respectively, 0.24-2.36, 0.02-0.04, and 0.06-0.40 mol mol^{-1}.*

TABLE 6.4 Effect of the Critical Process Factors in the Screening Phase for DS_{WMS} and DS_{MS}. (Huijbrechts et al., 2009)

	DS_{WMS}		DS_{MS}	
Factor	***F*-Value**	***p*-Value**[a]	***F*-Value**	***p*-Value**[a]
C_{starch}[b]	0.63	0.4340	0.36	0.5520
Reaction time	1.29	0.2674	77.81	**<0.0001**
Temperature	19.65	**0.0002**	120.02	**<0.0001**
NaOH[c]	4.07	0.0544	6.71	**0.0164**
Na_2SO_4[c]	0.81	0.3762	0.58	0.4559
AGE[c,d]	4.05	0.0550	1.05	0.3153
$C_{starch} \times C_{starch}$	15.48	**0.0006**	—	—
$C_{starch} \times$ NaOH	—	—	4.35	**0.0483**
Reaction time × Temperature	—	—	68.65	**<0.0001**

[a] *Factors in bold are significant ($p < 0.05$).*
[b] *Starch suspension.*
[c] *Process factors are in amounts.*
[d] *Allyl glycidyl ether.*

Significant factors were selected based on their *F* and *p* values in the statistical analysis. An overview of the effect of the critical process factors is shown in Table 6.4. In general, the temperature has the highest effect on AGE substitution for both starches. Based on this screening, it was not possible to present a mathematical model (=equation) to describe the DS as a function of factor settings, since variables were only set at their min-max levels plus center points. However, this initial screening design allowed a primary selection of the reaction parameters that are significant for the DS. Figure 6.2 shows the effect of the interactions between the process factors on the DS of WMS and MS.

Waxy Maize starch: as shown in the response surface plots for WMS (Figure 6.2a), if the temperature is increased to around 34 °C, the highest substitution is obtained. A higher temperature reduces the DS value for the used amounts of AGE or the used reaction time. Furthermore, when the NaOH concentration or the amount of AGE is increased, the DS increases in the temperature range of 20-48 °C (not shown).

Maize starch: the response surface plots for MS (Figure 6.2b and c) show that when the NaOH concentration is increased at the optimum temperature of 37 °C, the highest DS value is obtained. Similarly, the amount of AGE induces the best AGE substitution at 0.010%, and with an increasing NaOH concentration, an increasing DS value is generated (not shown).

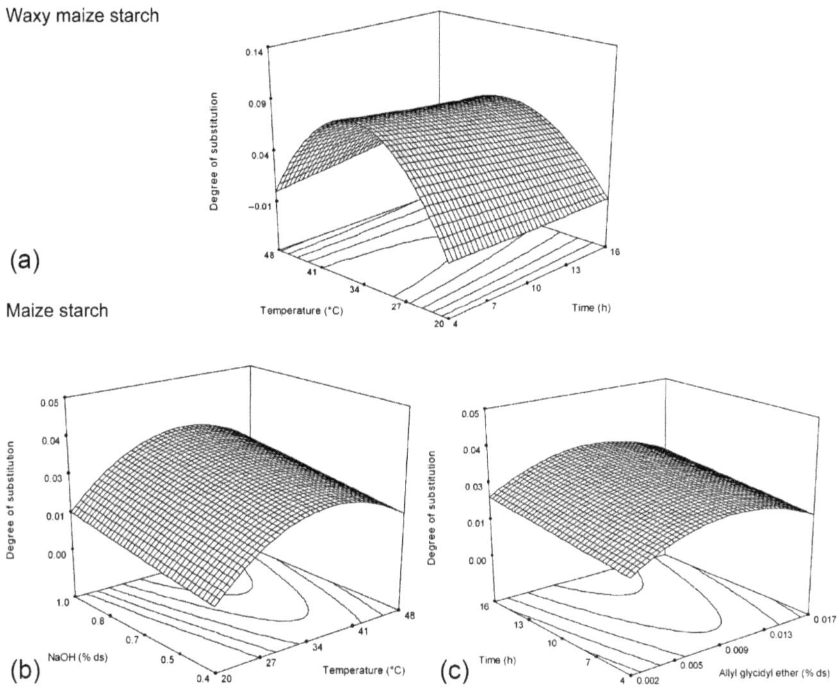

FIGURE 6.2 Effects of temperature, the amount of AGE, the amount of NaOH, and reaction time on DS for WMS (a) and MS (b and c) (Huijbrechts et al., 2009).

In the improving phase, the most pronounced process factors were obtained. Based on these parameters, the optimized conditions were generated using the optimization module of the experimental design program (Table 6.5). The aim was to have either a maximum DS, or a maximum DS at low cost (i.e., lowest amount of AGE), or the shortest reaction time for both starches. As can be observed from Table 6.5, the optimal conditions for the best conversion of WMS and MS are different for all optimizations. The optimized conditions generated with a maximum DS with no constraints of the other parameters show that the maximum DS that can be achieved for WMS and MS differs significantly, i.e., 0.102 for WMS and 0.039 for MS. It suggests that WMS is easier to substitute than MS with the generated process conditions. The most important differences in the process conditions between WMS and MS are the temperature and AGE concentration, 34.2 °C and 0.0166% ds and 37.0 °C and 0.0098% ds, respectively. However, even at these most optimal conditions, the conversion efficiency is very low.

Subsequently, reaction conditions were generated, which would give the maximum DS at the lowest amount of AGE and no constraints of

TABLE 6.5 Optimum Conditions for the Best Conversion of WMS and MS with a Maximum DS, the Lowest Amount of AGE and the Shortest Reaction Time (Huijbrechts et al., 2009)

	WMS			MS		
Process Factors	**Max. DS[a]**	**Lowest AGE[b]**	**Shortest Time[c]**	**Max. DS[a]**	**Lowest AGE[b]**	**Shortest Time[c]**
Reaction time (h)	4.0	4.0	4.0	16.0	16.0	4.0
Temperature (°C)	34.2	34.2	34.3	37.0	37.0	37.0
NaOH[d]	0.99	0.99	0.99	0.99	0.99	0.99
AGE[d]	0.0166	0.0016	0.0016	0.0099	0.0031	0.0038
DS	0.102	0.096	0.096	0.039	0.031	0.027
Conversion efficiency (%)	4.3	42.1	42.0	2.8	7.1	4.9

[a] *No constraints of the process factors.*
[b] *Max. DS and no constraints of the other factors.*
[c] *Max. DS and no constraints for temperature and the amount of NaOH.*
[d] *Variables are calculated as % (w/w) on dry starch.*

the other conditions. The best substitution of WMS and MS is again at a different temperature, respectively, at 34.2 °C for WMS and at 37.0 °C in case of MS. The maximum DS of AHP-WMS is generated with the lowest possible amount of AGE (0.0016% ds) with a remarkably efficient conversion. Oppositely, the synthesis of AHP-MS needs a higher amount of AGE (0.0031% ds) to generate the maximum DS. A slight decrease in the substitution of both starches is revealed at these optimized conditions.

In the third optimization, the conditions were optimized for a maximum DS at the lowest amount of AGE in the shortest reaction time and no constraints for the temperature and the amount of NaOH. The best results for both starches are obtained after 4 h. Even no differences in process conditions are obtained for a maximum DS of WMS because the shortest reaction time has already been achieved in the previous optimization. Consequently, the maximum DS for WMS is the same as in the optimized analysis. The maximum DS for MS (0.027) is lower than the DS value of the previous optimization. The optimum temperature for the synthesis of AHP-MS is 37.0 °C. Clearly, in all optimized analyses, the best substitution of WMS is at a moderate temperature, around 34 °C in all cases. Furthermore, the highest NaOH concentration is needed for the best conversion of WMS and MS in the generated optimizations. The AHP-MS synthesis needs a higher amount of AGE (0.0038%) than AHP-WMS synthesis in this optimized analysis. Additionally, the conversion efficiency of WMS is similar to the previous optimization, although the conversion for MS is less efficient due to high amount of AGE and very low DS.

2.2 NMR Characterization of AHP Starches

Various methods are available to determine DS and the position of substituents in starch derivatives. Especially, NMR is considered a very powerful technique. Figure 6.3a shows the ^{1}H NMR spectrum of AHP-AEMS. The peaks were assigned to the assumed structure of AHP-AEMS starting with the peak at 3.95 ppm, which belongs to the protons H-8 (1H) and H-10 (2H). The peaks at 5.90 and 5.20 ppm were assigned to the ethylenic protons H-11 (1H) and H-12 (2H), and the peak at 4.93 ppm was attributed to the alpha-anomeric proton of glucose residues (H-1, 1H). The characteristic peaks for the H-6 (2H) protons in AHP-substituted and unsubstituted glucose were identified at 3.76 and 3.66 ppm. The broader peak between 3.50 and 2.95 ppm was assigned to the magnetically similar protons H-2 to H-5, H-7, and H-9. In the ^{13}C NMR, the peaks characteristic for the allylic and ethylenic carbon atoms were identified at 69.9 ppm (C-10), 136.2 ppm (C-11), and 117.3 ppm (C-12), respectively (spectra not shown). The peaks characteristic for the α-D-glucopyranosyl moiety (C-1 to C-6) and the 2-hydroxypropyl ether chain (C-7 to C-9) were identified in the ^{13}C NMR spectrum.

The peak at 5.90 ppm in the ^{1}H NMR spectrum of the AHP starches assigned to the H-11 of the ethylenic bond in the side chain was used to quantify the DS. Maleic acid (MA) was used as an internal standard (Figure 6.3b). With this method, DS values of 0.20 and 0.19 were determined for AHP-AEMS and AHP-WMS, respectively. These values for DS, based on ^{1}H NMR, were compared to the DS determined from adapted double-bond titration measurements using reported methods

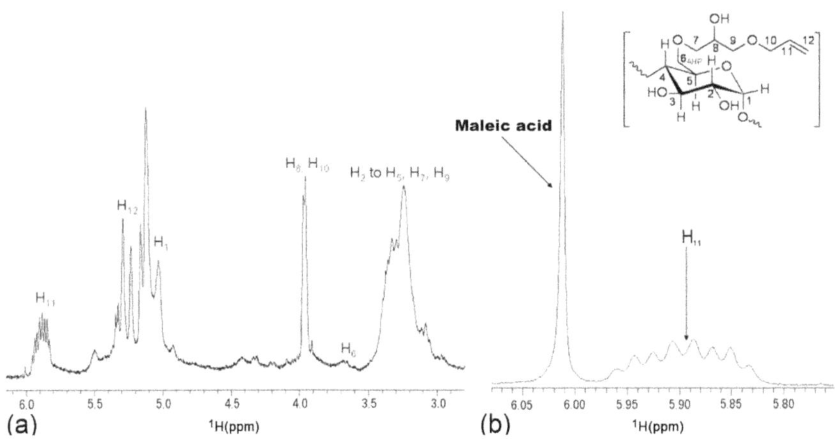

FIGURE 6.3 (a) ^{1}H NMR spectrum of AHP-AEMS with water suppression at 3.3 ppm. (b) ^{1}H NMR spectrum containing the peaks of maleic acid and H-11 of AHP starch for the determination of the degree of substitution (Huijbrechts et al., 2007).

TABLE 6.6 DS Values of AHP-AEMS and AHP-WMS, Determined by Two Different Methods (Huijbrechts et al., 2007)

Compound	NaOH	Na_2SO_4	Ratio AGE/AGU	DS_{NMR}	$DS_{titration}$
AEMS	0.12 M	40% w/w	0.6	0.20 ± 0.01	0.24 ± 0.06
WMS	0.12 M	20% w/w	0.3	0.19 ± 0.01	0.26 ± 0.06

(Martin, 1949) (Table 6.6). The NMR method appeared to be more accurate than the titration method. Furthermore, the method was faster in performance. Therefore, the ^{1}H NMR approach is a more preferable method for DS determination compared to the titration method.

2.3 Regioselectivity of the AHP Substitution

One important question addressed in this study is the location of the AHP substitution on glucose residues. According to the literature, a relatively high reactivity of O-2 and O-3 compared to O-6 substitution for numerous starches and cellulose has been reported (Richardson & Gorton, 2003). In highly acetylated starches, it was shown that the glucose residues are equally substituted in O-2 and O-3 position, whereas for hydroxypropyl starches, a predominant O-2 monosubstitution is proposed (Kavitha & Bemiller, 1998; Richardson & Gorton, 2003; Rutenberg & Solarek, 1984; Xu & Seib, 1997). In the case of AHP starches with a low degree of etherification, using the method described in this study, the ^{1}H NMR, ^{13}C NMR, and 2D NMR analyses suggest that the glucose moiety is preferentially substituted at the O-6 position (Huijbrechts et al., 2007). To validate the NMR information for O-6 substitution, AHP-MG was synthesized and characterized. MG (methyl-α-glucopyranoside and methyl-α-glucoside) was chosen as a model structure for the glucose residues in starch. The synthesis of AHP-MG has been performed with a low-molar-ratio MG and AGE to favor a specific substitution. The reaction product with an overall DS of 0.11 contained 15.8% AHP-MG and was therefore a good monomer model of the AHP starch with a DS of 0.20 (Huijbrechts et al., 2007).

Apart from ^{1}H NMR, ^{13}C NMR, DEPT90, and DEPT135 spectra, the structure of AHP-MG was studied using ^{1}H-^{13}C *het*eronuclear chemical shift *cor*relation (HETCOR, one-bond correlation), ^{1}H-^{13}C heteronuclear multiple bond correlation (HMBC), and ^{1}H-^{1}H *co*rrelation NMR *s*pectroscop*y* (COSY). Via 1D NMR, the chemical shift for both the carbon atoms and the hydrogen bonds was obtained (Table 6.7). The matching hydrogen atoms were elucidated with ^{1}H NMR and HETCOR. The sequence of the carbon atoms was obtained via COSY and HMBC spectra. Using HMBC, a specific correlation showed an O-6 substitution. This substitution was

TABLE 6.7 Chemical Shift (ppm) for the ^{1}H- and ^{13}C-Signals of the Mixture AHP-MG[a] and MG (Huijbrechts et al., 2007)

Number[b]	Group	^{13}C	^{1}H	Number[b]	Group	^{13}C	^{1}H
11	CH	133.8	5.89	9	CH_2	70.8	3.50
12	CH_2	118.3	5.19				3.43
1	CH	99.3	4.71	8	CH	70.4	3.80
3	CH	73.1	3.57	4	CH	69.6	3.31
10	CH_2	72.0	3.98	7	CH_2	62.7	3.53
5	CH	71.6	3.55				3.48
2	CH	71.2	3.47	$6_{AHP\text{-}MG}$	CH_2	62.5	3.56
				6_{MG}	CH_2	60.5	3.78

[a] *MG, methyl α-D-glucopyranoside.*
[b] *Numbering of atoms as shown in Figure 6.4; 6_{MG} and 6_{AHP} indicate the C-6 atom in MG and AHP-MG, respectively.*

elucidated by the HMBC section at the C-7 proton frequencies, 3.48 and 3.53 ppm, and C-5 proton frequency, 3.55 ppm. In the cross section of proton frequency of C-5, the correlations between H-5 and the matching C-atoms are visible (Figure 6.4). The important peak is at 62.5 ppm, which appeared to be a CH_2 group according to the DEPT135. After correlation of all the spectra, we can conclude that the peak at 62.5 ppm must be assigned to the C-$6_{AHP\text{-}MG}$, the primary carbon in AHP-MG. Similarly, the cross section of proton frequencies of C-7 elucidated the correlations between H-7 and C-8 and C-6_{MG} and the H-7 and C-$6_{AHP\text{-}MG}$ correlation.

This analysis shows that the AHP substitution of MG is regioselective and was preferentially located at the O-6 position of glucose. The substitution at C-2 or C-3 cannot be completely ruled out since their respective correlations might be concealed in the broad signals of the HMBC analysis. Less steric hindrance at the O-6 position may well explain the regioselectivity in this modification reaction and type of substitution instead of, in general, O-2 or O-3 position (Richardson & Gorton, 2003). Because of the agreement between the results of the NMR analysis of AHP-starch derivatives and AHP-MG, we conclude a predominantly O-6 substitution of glucose residues in AHP starch. Other techniques such as HPLC analysis and gas chromatograph-mass spectroscopy (GC-MS) might provide more complete evidence for the position of AHP group on the glucose units. These techniques require sample preparation, such as permethylation and hydrolysis of the modified starch (Mischnick, Heinrich, Gohdes, Wilke, & Rogmann, 2000).

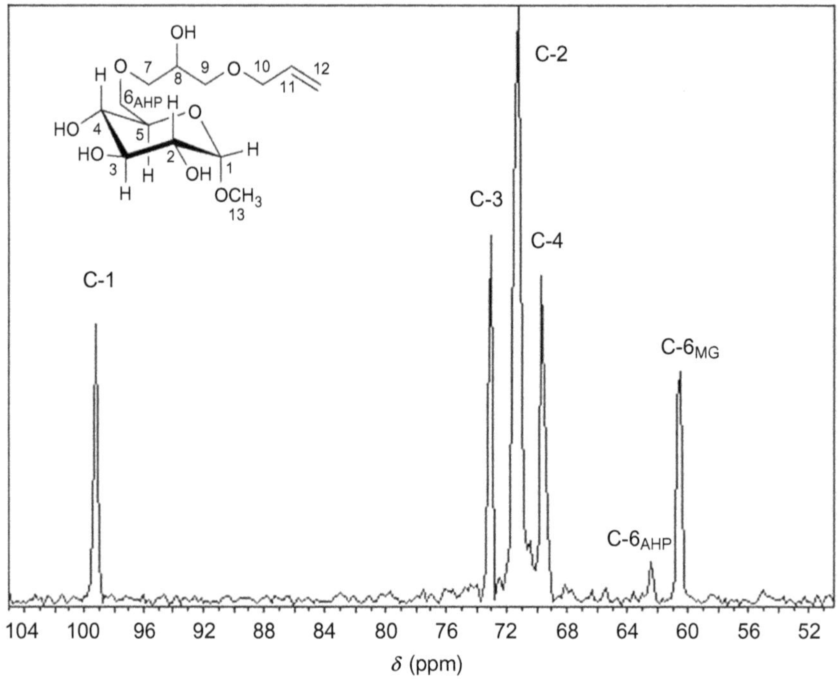

FIGURE 6.4 Cross-section of the proton frequency dimension of C-5 (3.55 ppm) of the HMBC spectrum (Huijbrechts et al., 2007).

2.4 Distribution of AGE Groups Along the Polysaccharide Chains

To determine the distribution of the AHP groups along the polysaccharide chains, the modified starches were subjected to enzyme degradation. Starch-degrading enzymes hydrolyze the glycosidic linkages between the glucose residues in the polysaccharide chain in a defined way. The introduction of the AHP groups may sterically hinder the action of the amylolytic enzymes (Chen et al., 2004). Knowledge about the enzyme degradation products of AHP starch can give significant information about the distribution of the AHP substitution along the polymer chains. For the enzymatic digestion of AHP starches, three starch-degrading enzymes with different cleavage specificities were used: pullulanase, α-amylase, and amyloglucosidase. Pullulanase is a debranching enzyme, which hydrolyzes α-(1,6) linkages of amylopectin to produce linear chains (Hizukuri, 1996). α-Amylase is an endoenzyme that hydrolyzes polysaccharides randomly at α-(1,4) D-glucosidic linkages to produce glucose and oligosaccharides containing two to seven glucose residues, and amyloglucosidase is capable of completely hydrolyzing both α-(1,4) and α-(1,6)

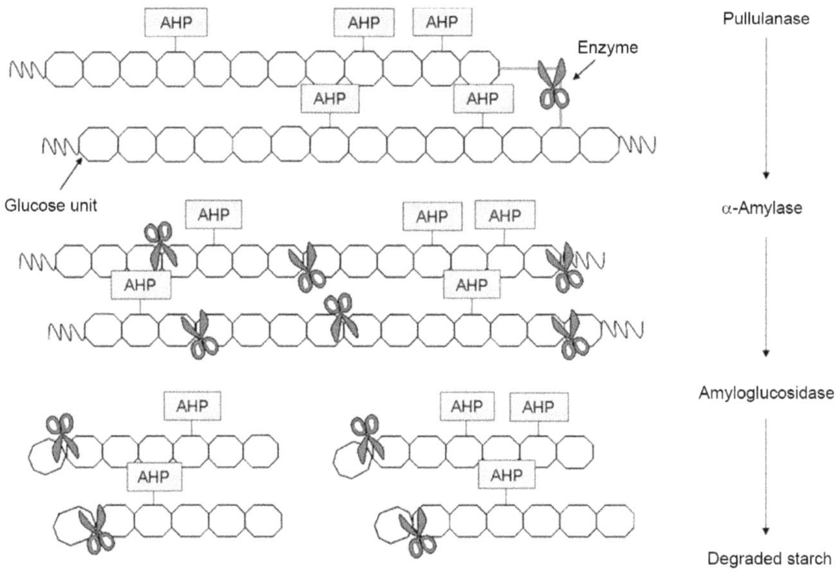

FIGURE 6.5 Diagram of the enzymatic digestion of starch with three enzymes: pullulanase, α-amylase, and amyloglucosidase.

linkages in polysaccharides, through an exomechanism from the nonreducing terminal residues, to produce β-D-glucose (Hizukuri, 1996). The action of the selected enzymes on an amylopectin chain is illustrated in Figure 6.5.

AEMS containing 70% amylose and 30% amylopectin, WMS (mainly amylopectin), and the corresponding AHP derivatives were subjected to sequential enzymatic hydrolysis using pullulanase, α-amylase, and amyloglucosidase. HPSEC elution profiles of the digested samples showed a complete conversion of AEMS and WMS into glucose and small oligomeric fragments (not shown). Higher oligomers resistant to further enzymatic digestion were present in AHP-AEMS and AHP-WMS, as clearly seen from HPAEC elution profiles (Figure 6.6). This indicates that α-amylase and amyloglucosidase digestions were hindered by AHP groups. The peaks could not be assigned because the enzymatically degraded fragments of the derivatives may have different charges, leading to different retention times.

The mass distribution of the enzyme-digested hydrolysates of AHP-AEMS and AHP-WMS was determined by MALDI-TOF MS (Huijbrechts et al., 2007). A significant difference was observed between the substitution pattern of AEMS and WMS derivatives. The MALDI-TOF mass spectrum of "AHP-AEMS digests" shows a mixture of enzyme-resistant AHP-substituted oligosaccharides with degrees of polymerization (DP) ranging

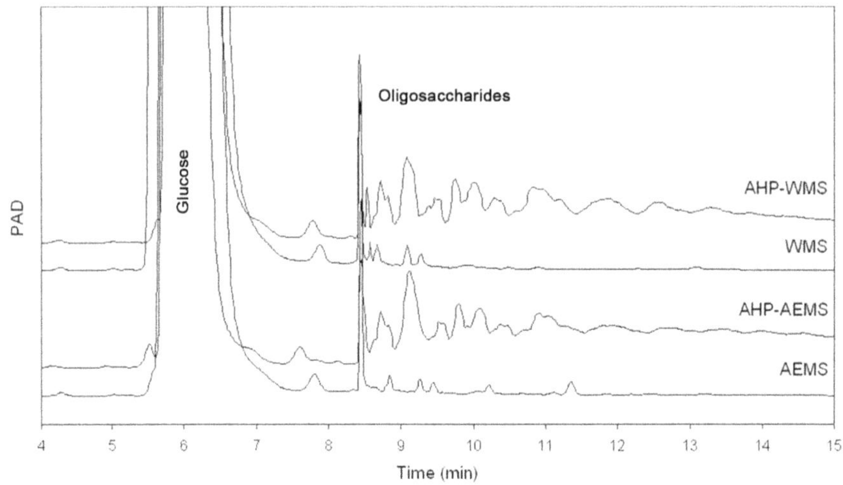

FIGURE 6.6 HPAEC elution profiles of the pullulanase, α-amylase, and amyloglucosidase hydrolysates of AEMS and WMS and their derivatives, AHP-AEMS and AHP-WMS (Huijbrechts et al., 2007).

from three to ten (Figure 6.7a). The following substituted oligosaccharides were identified: maltotriose (DP3) with one and two AHP groups; DP4 and DP5 with one, two, and three AHP groups; DP6 with two and three AHP groups; DP7 and DP8 containing two, three, and four AHP groups; DP9 containing three and four AHP groups; and DP10 containing four AHP groups. The DS of the modified oligosaccharides derived from AHP-AEMS varied with the DP of the carbohydrate oligomer. The highest substitution for the oligomers (theoretical DS values of 0.60, 0.66, and 0.75) was observed for tri-, tetra-, and pentaoligosaccharides, respectively. Larger fragments are indicative for a more clustered AHP substitution, with lower theoretical DS values ranging between 0.29 and 0.57. The MALDI-TOF mass spectrum for AHP-WMS hydrolysates shows smaller oligomers (DP ranging from three to seven) with low DS (Figure 6.7b). The following oligomers were identified: DP3, containing one AHP group (theoretical DS value of 0.33); DP4 and DP5, containing one and two AHP groups (theoretical DS between 0.20 and 0.50); and DP6 and DP7, containing two and three AHP groups (theoretical DS between 0.28 and 0.50). Both derivatives have an average DS of 0.20, corresponding to one AHP group per five glucose residues. The minimum indicative DS substitution required for enzyme hindrance is 0.20, although we also indicated that a hindrance for DS 0.28 and 0.43 (two and three AHP groups per seven glucose residues) appeared. Therefore, these results show that the enzyme-resistant carbohydrate residues contain at least one AHP group and confirm that substitution of the polysaccharide chain prevents enzyme-induced degradation due to sterical hindrance (Huijbrechts et al., 2007).

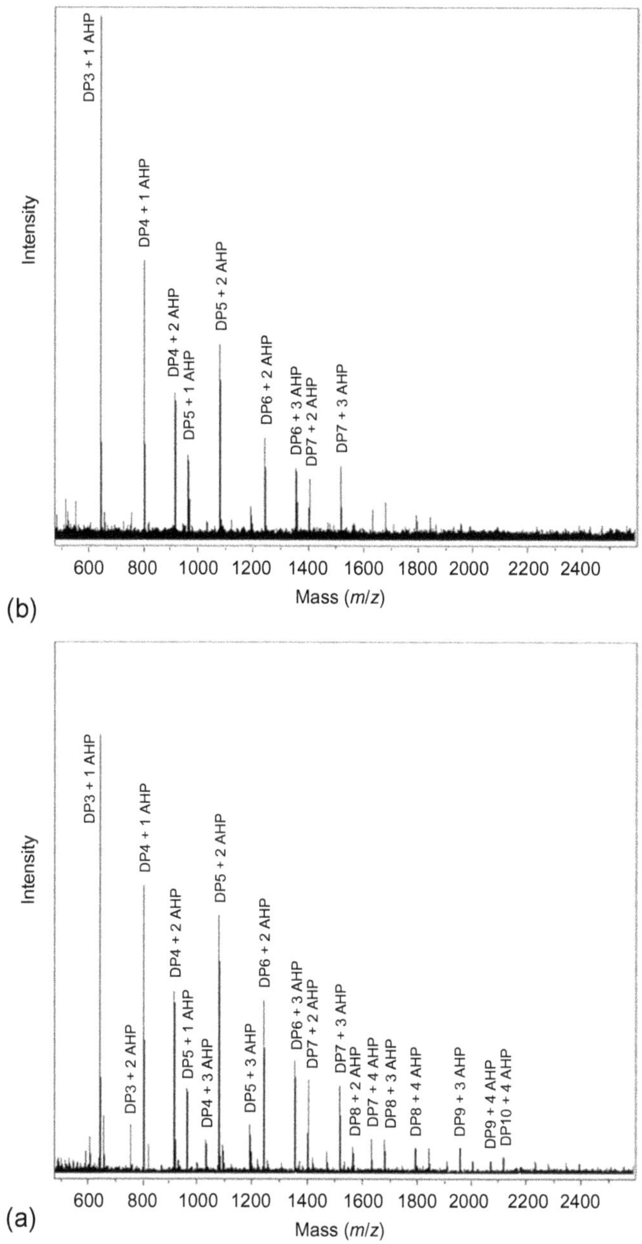

FIGURE 6.7 MALDI-TOF mass spectrum of the enzyme (pullulanase, α-amylase, and amyloglucosidase) hydrolysates of (a) AHP-AEMS and (b) AHP-WMS (AHP: 1-allyloxy-2-hydroxy-propyl-group). DP, degree of polymerization (Huijbrechts et al., 2007).

The size and the substitution pattern of the enzyme-resistant oligomers show significant differences between the susceptibilities for AHP substitution of amylose and amylopectin. Amylopectin, the main component of WMS, is more uniformly substituted and the AHP groups are homogeneously distributed throughout the polymer, essentially nonclustered. On the contrary, the AEMS starch, containing 70% amylose and 30% amylopectin, shows a more heterogeneous distribution of AHP groups, as illustrated by the heterogeneity of the oligomers formed upon enzyme degradation. In the modified oligosaccharides derived from AHP-AEMS, we have identified oligomers with low DS values, similar to those generated from AHP-WMS, and highly substituted larger oligomers with clustered AHP groups. We assume that the larger fragments with high and clustered AHP substitution are degradation products of AHP amyloses. On the basis of these results, we further assume that, in general, the amylose population is more clustered substituted than the amylopectin population, and the distribution of the AHP groups throughout the amylose chain is randomly clustered.

The ^{1}H NMR results indicate that AHP-AEMS and AHP-WMS are equally substituted by AGE, although their substitution patterns are different. According to previous substitution-distribution studies, substitutions take place preferentially in the amorphous regions of the starch granule, containing amylose and the branched part of amylopectin. The crystalline regions, containing amylopectin, are only partially accessible for substitution (Chen et al., 2004; Hood & Mercier, 1978; Richardson & Gorton, 2003; Van Der Burgt et al., 2000). Assuming that the AHP substitution of amylose taking place to a larger extent than that of amylopectin, this suggests an easier penetration of the AGE reagent throughout the amorphous regions of the starch granules and only partial penetration in the crystalline regions.

2.5 Morphological Properties

The effect of substitution on starch crystallinity and type of packing was investigated using X-ray diffraction (XRD). Because XRD traces of AEMS and WMS and their AHP derivatives showed similar pattern, we show only the XRD spectra of AEMS and AHP-AEMS as representatives (Figure 6.8). An inspection of XRD spectra showed hardly any differences in the type of packing between native starches and their derivatives after the etherification. Furthermore, the crystalline and amorphous parts seem to be similar in structure for native starches and their corresponding AHP derivatives (Huijbrechts et al., 2008).

With SEM (Figure 6.9), the particle appearance of the dry starch (ds) samples was investigated. The appearance of WMS was portrayed with distinct granules, having mixed shapes, with some almost spherical, some cubical, and sometimes "honeycomb" shapes with irregular but smooth

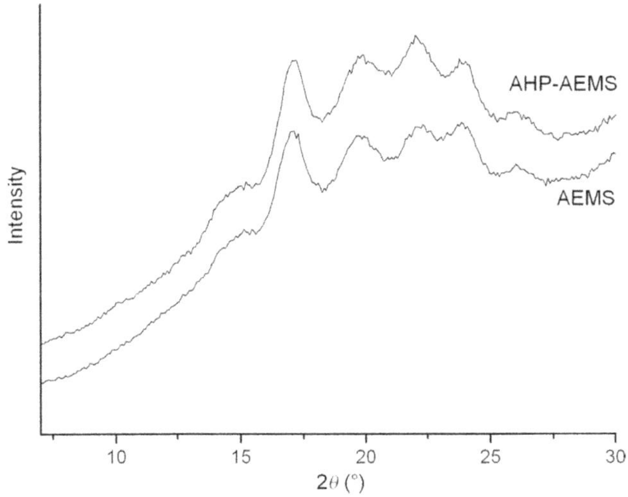

FIGURE 6.8 XRD patterns of AEMS and AHP-AEMS.

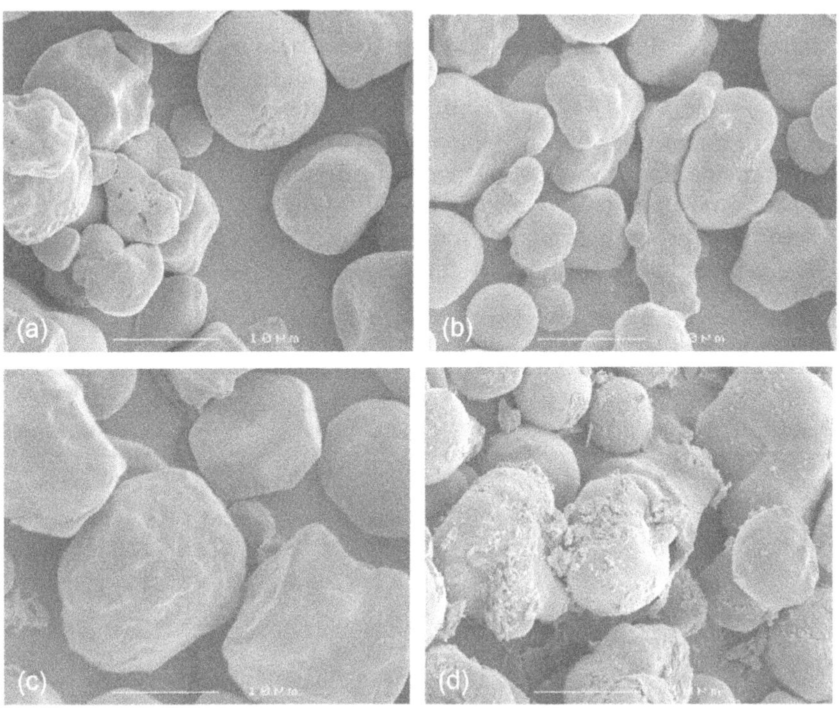

FIGURE 6.9 SEM images of (a) WMS, (b) AEMS, (c) AHP-WMS, and (d) AHP-AEMS. The micrographs were taken at a magnification of 1500×. Scale bar is 10 μm.

surfaces (Figure 6.9a), as reported before (Elfstrand et al., 2004). Likewise, the appearance of AEMS was described as being smooth and uniform (Figure 6.9b) (Li, Vasanthan, Hoover, & Rossnagel, 2004). The size and appearance of the granules from AHP-AEMS and AHP-WMS are very similar to those of their native starches (Figure 6.9c and d). However, the SEM micrographs show that the surface of the granule from AHP-AEMS and AHP-WMS are rougher and more porous than that of AEMS and WMS. It suggests that etherification of AEMS and WMS starch generates only minor surface modification upon reaction.

2.6 Physicochemical Properties of Etherified MSs

In dry granular starch, no appreciable changes in granular form for the three different MSs were obtained. High DS with AHP groups causes high decrease in starch crystallinity for AHP-WMS and AHP-MS. AHP-AEMS seems to have a slightly better degree of crystallinity (Table 6.8), which was confirmed by a higher ΔH_{gel} (Table 6.9). The incorporation of AHP substituents in the starch molecules significantly affects their physiochemical properties related to starch-water interactions. All AHP-substituted starches showed decreased gelatinization temperatures, while AHP-WMS and AHP-MS showed decreased gelatinization enthalpy (Table 6.9) (Huijbrechts et al., 2008). These differences may be ascribed to the influence of AHP groups on the interaction between the starch chains through steric hindrance by the AHP side chains, change of the hydrophobicity of the starch, or interactions between the hydroxyl functions of AHP group with starch chains. Decreases in the thermal parameters are consistent with fewer crystals being present after etherification and with a cooperative melting process enhanced by additional swelling (Jenkins & Donald, 1998; Kaur, Singh, & Singh, 2004; Liu, Corke, & Ramsden, 1999).

TABLE 6.8 X-Ray Diffraction Data of Native Starches and AHP-Starches (Huijbrechts et al., 2008)

		2θ Values (° angle)						
Sample	**DS**	**23°**	**22°**	**20°**	**18°**	**17°**	**15°**	**Degree of Crystallinity (%)**
WMS		22.84		20.06	18.18	17.07	15.16	41.9
AHP-WMS	0.23	22.84		20.04	17.92	17.12	15.15	29.5
MS		23.06		20.00	18.00	17.18	15.15	38.9
AHP-MS	0.11	23.02		19.95	18.00	17.23	15.12	22.1
AEMS		23.11	20.91	19.09		16.25		17.5
AHP-AEMS	0.13	23.64	21.95	19.49		16.92		21.5

TABLE 6.9 Thermal Properties of Native and AHP Starches During Heating[a] (Huijbrechts et al., 2008)

		Gelatinization Temperature (°C)			
Sample	**DS**	T_o[c]	T_p[d]	T_c[e]	ΔH_{gel} (J/g)[b]
WMS		57.0	69.5	75.9	13.1
AHP-WMS	0.23	50.8	57.3	65.9	4.9
MS		58.5	63.5	69.3	8.9
AHP-MS	0.11	51.4	58.1	67.5	7.9
AEMS		72.7	88.2	115.0	4.5
AHP-AEMS	0.13	59.9	83.5	110.8	7.8

[a] *Measured in diffraction scanning calorimeter.*
[b] ΔH_{gel} = *enthalpy of gelatinization.*
[c] T_o = *onset temperature.*
[d] T_p = *peak temperature.*
[e] T_c = *completion temperature.*

TABLE 6.10 Pasting Properties of Native and AHP Starches[a] (Huijbrechts et al., 2008)

				Viscosity (cP)				
Sample	**DS**	P_{onset}[b] (°C)	P_{temp}[c] (°C)	**Peak**	**Hot Paste**	**Final Paste**	**Breakdown**	**Setback**
WMS	—	68.5	78.3	3934	1637	1826	2297	189
AMP-WMS	0.23	55.5	68.0	10,841	5402	9647	5439	4245
MS	—	74.9	94.7	3648	2188	3950	1460	1762
AMP-MS	0.11	63.7	80.8	5885	1891	7097	3994	5206
AEMS	—	—	—	—	—	36	—	36
AMP-AEMS	0.13	94.1	95.0	74	—	64	10	64

[a] *Measured in Rapid Visco Analyser.*
[b] P_{onset} = *pasting onset.*
[c] P_{temp} = *pasting temperature.*

Pasting viscosity profiles of native and modified starches analyzed using Rapid Visco Analyser (RVA) are summarized in Table 6.10 (Huijbrechts et al., 2008). The RVA shows that the onset pasting temperatures ($T_{pasting}$), i.e., the temperature at which a perceptible increase in viscosity occurs, of modified and unmodified starches are higher than the onset gelatinization temperatures (T_o) determined by DSC. However, the difference between T_o and $T_{pasting}$ decreases after substitution, suggesting that the gelatinization and the increase in viscosity occur at a shorter temperature range for AHP-substituted starches than for their nonmodified starches. Furthermore, the shapes of the pasting curves differed markedly after etherification (not shown). All AHP starches show an increased setback, indicating that retrogradation takes place, resulting in a more ordered structure. The alternation in pasting properties gives shear thinning during heating and better gel network after cooling for AHP-WMS and AHP-MS (Table 6.10). The swelling power of all starches increases as DS increases (Figure 6.10a). Similarly, the solubility of WMS and MS increases as DS increases (Figure 6.10b). Microscopic

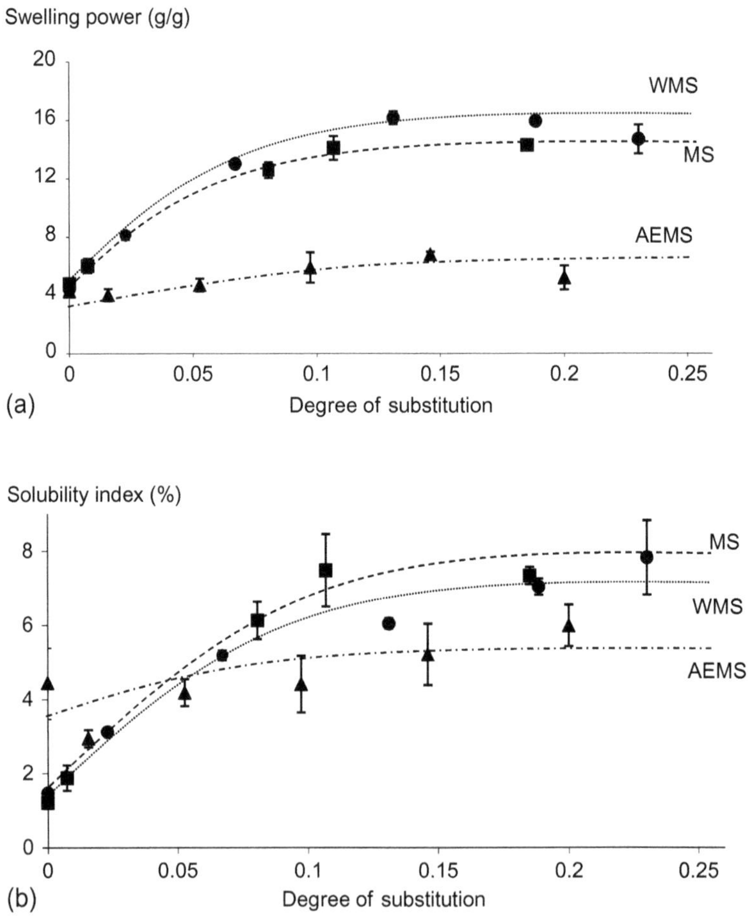

FIGURE 6.10 Swelling power (a) and solubility index (b) of native and AHP maize starches at 60 °C. The lines are given to guide the eye; dotted, dashed, and dashed-dotted lines correspond to modified • WMS, ▪ MS, and ▲ AEMS, respectively (Huijbrechts et al., 2008).

measurements confirm the swelling capacities of the three starches in order of AHP-WMS > AHP-MS > AHP-AEMS at 60 °C. Increased swollen granules induce change in the flow behavior under shear, which is interesting for biomaterials (Seidel et al., 2001).

3 SYNTHESIS AND CHARACTERIZATION OF EPOXY-STARCH DERIVATIVES

3.1 Synthesis

We have also obtained some preliminary results obtained from the epoxidation of double bonds in the allyl-WMS derivatives described before (Huijbrechts et al., 2010). The double bonds of granular AHP-WMS were epoxidized using 10 equivalents of H_2O_2 and CH_3CN in slightly alkaline suspension at 30 °C (Scheme 6.2). The combination of hydrogen peroxide and acetonitrile is a well-known reagent for the epoxidation of carbon double bonds (Chen & Reymond, 1995). A two-step colorimetric assay was developed to determine the amount of epoxy groups in the product. Furthermore, epoxy-WMS was subjected to enzymatic digestion to characterize the structure of the starch derivative.

3.2 Structural Characterization of Epoxy-WMS

The amount of epoxy groups was determined using a quantitative spectrophotometric assay (Cedrone, Bhatnagar, & Baratti, 2005). In this method, *p*-nitrobenzyl pyridine (pNBP) was used to assay epoxides through the formation of a blue chromophore. This assay was tested with epoxy-WMS and pNBP, but the colorimetric analysis of the starch suspension was impossible due to the insoluble nature of the starch. Therefore, a new two-step spectrophotometric assay was developed (Scheme 6.3). In the first step, an excess of pNBP was used to quantitatively convert all epoxide groups on the starch (Scheme 6.3, R1). A dark blue/green starch derivative was obtained. Subsequently, a small part of the supernatant containing the remaining pNBP was transferred to a tube with a high concentration of AGE. In this tube, the remaining pNBP was converted in the pNBP product **1** (Scheme 6.3, R2). The pNBP product is deprotonated

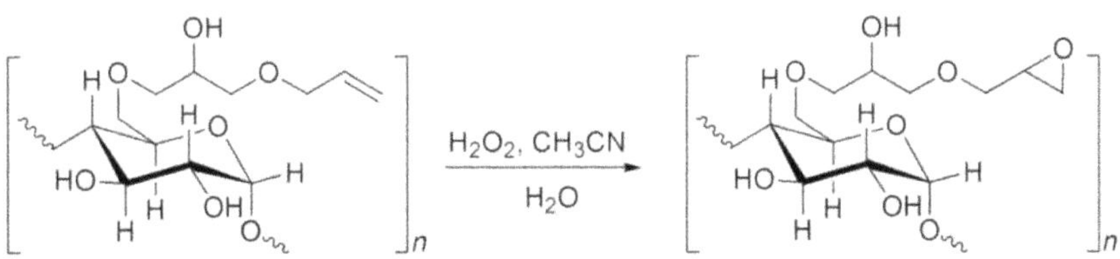

SCHEME 6.2 The synthesis of epoxy starch.

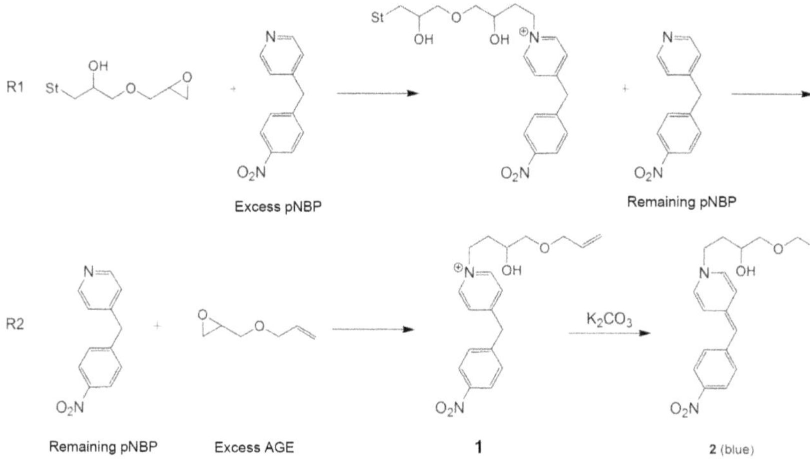

SCHEME 6.3 Two-step spectrophotometric assay to quantitative epoxy groups in modified starch, pNBP product (**1**), and blue chromophore (**2**) (Huijbrechts et al., 2010).

TABLE 6.11 Analysis of Epoxy-WMS and HD-Treated Epoxy-WMS (Huijbrechts et al., 2008)

Sample	Epoxy Groups ($mmol\ g^{-1}$)	DS	DS/DS_{allyl} (%)[a]	C (%)[b]	N (%)[b]
Epoxy-WMS	0.13±0.3	0.025[c]	11[c]	43.08	0.24[d]
HD-WMS	n.d.[e]	0.026[b]	11	39.09	0.53

[a] $DS_{allyl}=0.23$.
[b] *Determined using elemental analysis.*
[c] *Determined using pNBP titration.*
[d] *Derived from proteins that are present in WMS.*
[e] *Not determined.*

in a basic medium using K_2CO_3, giving the blue chromophore **2**. The concentration of chromophore **2** was determined by its absorption at 600 nm. The amount of pNBP initially added (R1) minus the amount of **2** formed (R2) equals the amount of epoxy groups on the starch. With this assay, DS_{epoxy} of 0.025 was determined corresponding to 0.13±0.03 mmol epoxy groups per gram of ds. (Table 6.11). This suggests that 11% of the allyl groups were converted into epoxy groups. The controls showed hardly any differences in A_{600}.

Epoxy groups on a surface are able to react with nucleophiles such as NH_2 and OH groups. Using 1,6-diaminohexane (HD), the binding of NH_2 groups to epoxy groups was investigated. In this reaction, we assume that one amino group of HD reacts with an epoxy group since there is a 10-fold excess of amino groups. The amount of NH_2 groups in HD-WMS was determined using an elemental analyzer (Table 6.11). A DS_{HD} of 0.026 was obtained. This result corresponds to DS_{epoxy} obtained by the colorimetric assay, which means that every epoxy group in epoxy starch has reacted with HD.

The results of the spectrophotometric assay of epoxy groups, and the elemental analysis of the HD reaction, imply that only a small amount of epoxy groups is available for binding to nucleophilic groups. Furthermore, reduced swelling and solubility of epoxy-WMS suggest that subsequent reactions may have taken place, such as the formation of internal cross-links in the starch granule. To study the structure of epoxy-WMS in more details, the modified starch was enzymatically hydrolyzed and the enzymatically degraded products were analyzed using MALDI-TOF MS to get information about the differences in structure between AHP-WMS and epoxy-WMS.

The extent of β-amylase hydrolysis of native starches and its derivatives was studied from HPAEC and HPSEC elution profiles. These profiles showed that the hydrolysis of the AHP-WMS and epoxy-WMS liberated less maltose than the native starch (results not shown). Furthermore, the relative amount of liberated maltose was determined at 89% for AHP-WMS and at 73% epoxy-WMS. This suggests that enzymatic degradation of epoxy-WMS was more sterically hindered due to the formation of intra- or intermolecular ether cross-linkings during reaction. According to other studies, cross-linked starch derivatives are less accessible for enzymatic hydrolysis than their etherified starches (Liu et al., 1999; Wang & Wang, 2000).

3.3 Enzymatic Digestion of Substituted WMS with Pullulanase, α-Amylase, and Amyloglucosidase

To study the structure of epoxy-WMS in more details, epoxy-WMS, AHP-WMS, and its native starch were also subjected to simultaneous enzymatic digestion using pullulanase, α-amylase, and amyloglucosidase as described before (Huijbrechts et al., 2007). The HPSEC elution profiles of both WMS derivatives showed high-molecular-weight fragments compared to the native starch (results not shown), i.e., the enzymes were sterically hindered by both WMS derivatives. A different oligomeric distribution was obtained for epoxy-WMS than for AHP-WMS.

Likewise, both HPAEC elution profiles of AHP-WMS and epoxy-WMS (Figure 6.11) show high-molecular-weight oligomers that are resistant to further enzymatic digestion. All fragments were eluted within 20 min. Again, a different molecular weight distribution of the enzymatically generated oligomeric fragments was obtained for AHP-WMS and epoxy-WMS. The oligomeric fragments of AHP-WMS eluting at 9.3, within 10.3-10.4, and at 12.2 min disappeared in the profile of digested

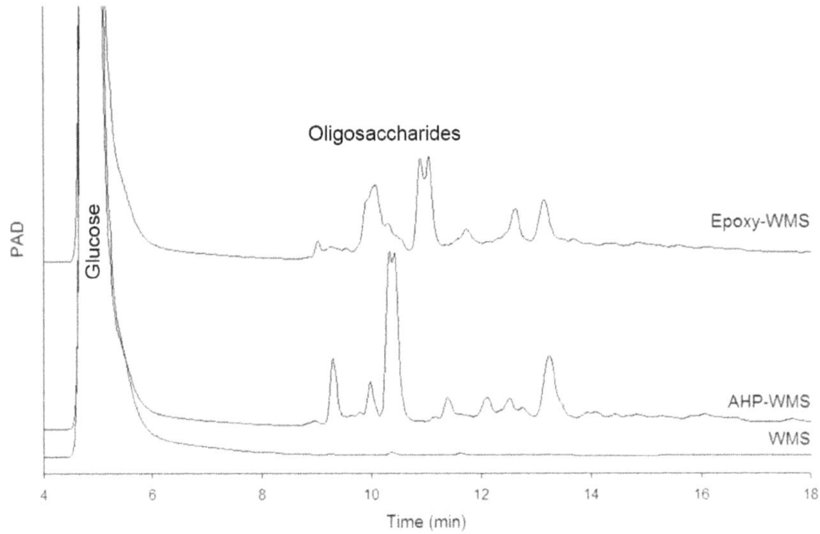

FIGURE 6.11 HPAEC elution profiles of oligomers mixture of WMS, AHP-WMS, and epoxy-WMS obtained after enzymatic degradation by pullulanase, α-amylase, and amyloglucosidase (Huijbrechts et al., 2010).

epoxy-WMS, whereas other enzyme-mediated degradation products appeared at 10.1, 10.8-11.4, and 12.7 min.

The different mass distribution of the oligomeric fragments of epoxy-WMS and AHP-WMS was determined in more details by MALDI-TOF MS (Figure 6.12). In the MALDI-TOF mass spectrum of AHP-WMS, enzyme-resistant oligosaccharides with various DP were identified, ranging from maltose (DP2) to DP9 with one to five allyl groups (AHP). The MALDI-TOF mass spectrum of epoxy-WMS showed a regular pattern with more and diverse oligomers (DP ranging from two to seven). The mass distribution of epoxy-WMS included several different enzyme-resistant oligomers having epoxy groups (Ox) and unreacted AHP groups. In addition, oligomeric fragments containing diol groups (DL) were found. Furthermore, there were fragments possessing internal cross-links (Cr_{intra}) or between two different fragments (Cr_{inter}). These oligomers with diols and cross-links were a result of subsequent reactions of epoxy groups, such as the hydration or the formation of intra- and intermolecular bridges with free hydroxyl groups in starch.

Some possible different enzyme-resistant oligomeric DP3 fragments obtained after enzymatic hydrolysis of epoxy-WMS are illustrated in Figure 6.13. Fragments containing epoxy groups (Ox) and cross-links within the oligomer (Cr_{intra}) have the same mass over charge ratios, as is illustrated in Figure 6.13. Thus, they are indistinguishable in the MS analysis. Similarly, the oligomers having diols (DL) and cross-links

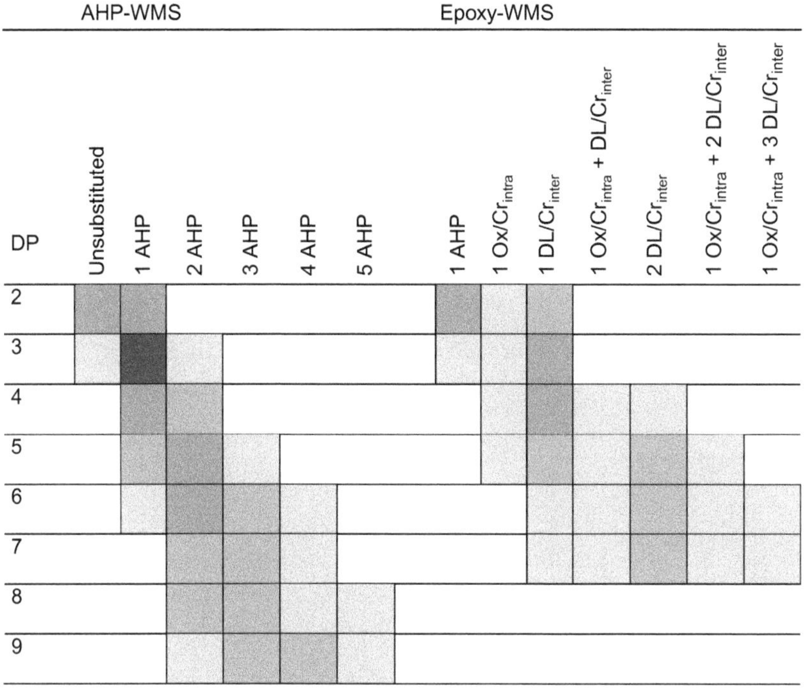

FIGURE 6.12 Mass distribution of the oligomeric fragments of AHP-WMS and epoxy-WMS. Enzyme-resistant oligomeric fragments of AHP-WMS contain allyl groups (AHP), whereas fragments of epoxy-WMS contain AHP or epoxy groups (Ox) and diol groups (DL) and cross-links within the oligomer (Cr_{intra}) or between two fragments (Cr_{inter}). DP, degree of polymerization. The total signal intensities of the oligomers in a certain DP are normalized to 100%. Relative intensities 0 ▭ 100% (Huijbrechts et al., 2010).

between two fragments (Cr_{inter}) give at the same mass over charge ratios in the MS spectrum of epoxy-WMS. The diversity of enzymatically degraded fragments becomes larger for oligomers containing more than three glucose units. The fragments at higher mass over charge ratios (Figure 6.12) could not be unambiguously assigned due to their large diversity, low relative intensities, and overlapping of signals.

In the enlargements of the MALDI-TOF mass spectra (Figure 6.14), differences between the oligomers obtained after enzymatic degradation of AHP-WMS and epoxy-WMS, respectively, are clearly shown. The fragment (DP3) containing one AHP group is still present after epoxidation. Next to this oligomer, its epoxidized DP3 or DP3 with Cr_{intra} is identified at the signal of 657 *m/z*, followed by DP3 with Cr_{inter} or DL (*m/z* 675). These differences in oligomeric fragments are also obtained for fragments containing four or five glucose units. Moreover, DP4 and DP5 with two Cr_{inter} or DL and with one epoxy group or $Cr_{intra,}$ and one cross-link between two

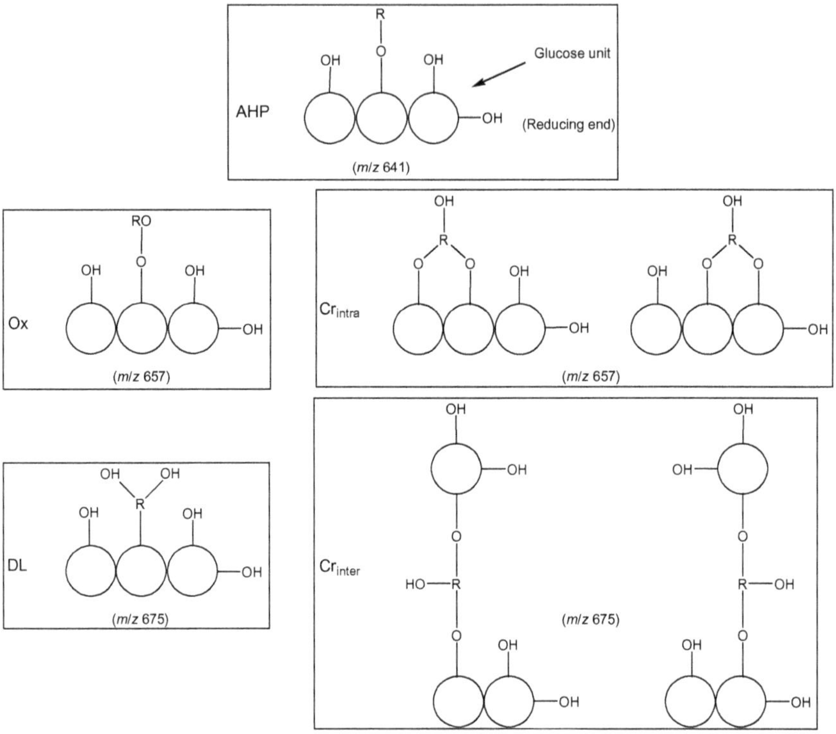

FIGURE 6.13 Simplified scheme of some possible different enzyme-resistant DP3 fragments of the pullulanase, α-amylase, and amyloglucosidase hydrolysates of epoxy-WMS. Substituted DP3 fragments may contain one allyl group (AHP, *m/z* 641), an epoxy group (Ox, *m/z* 657), a diol group (DL, *m/z* 675), a cross-link within the oligomer (Cr_{intra}, *m/z* 657), or a cross-link between two fragments (Cr_{inter}, *m/z* 675). For reasons of clarity, only fragments with a substituent at the central glucose unit are drawn. DP, degree of polymerization; R, AHP group; RO, epoxidized AHP group (Huijbrechts et al., 2010).

fragments or a diol group are found for epoxy-WMS. Furthermore, DP5 with one Ox group or Cr_{intra} and two Cr_{inter} or DL was identified in the MS analysis.

These results show significant differences in the oligomeric fragment patterns of AHP-WMS and epoxy-WMS. Small to large fragments with several allyl groups were obtained for AHP-WMS, whereas a regular pattern of more and diverse oligomers was found for epoxy-WMS. A small number of oligomers carrying unmodified AHP groups and epoxy groups were found in the MALDI-TOF MS of epoxy-WMS. Most fragments were identified as fragments decorated with diol groups and/or with ether cross-links generated, respectively, by hydration of the epoxy moieties or by the formation of covalent bonds within the oligomer or between two different fragments. These subsequent fragments may have been formed

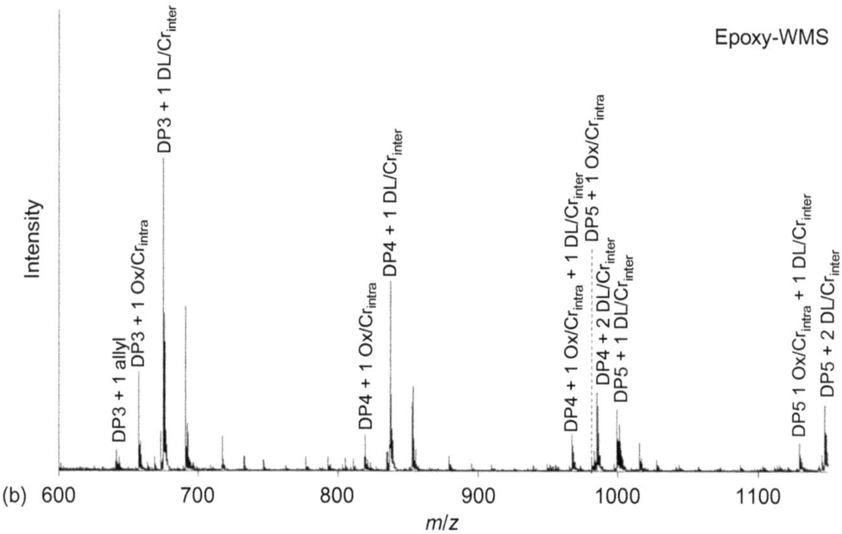

FIGURE 6.14 Enlargement of MALDI-TOF mass spectra of the pullulanase, α-amylase, and amyloglucosidase hydrolysates of AHP-WMS (a) and epoxy-WMS (b). Enzyme-resistant oligomeric fragments of AHP-WMS contain allyl groups (AHP), whereas fragments of epoxy-WMS contain AHP or epoxy groups (Ox) and cross-links within the oligomer (Cr_{intra}) and between two fragments (Cr_{inter}) or diol groups (DL). DP, degree of polymerization.

during the epoxidation or storage of the compound, but it might also be possible that these fragments were generated during the enzymatic digestion. This suggests that a larger amount of allyl groups in AHP-WMS was converted into epoxy groups than determined with pNBP test, but a significant amount of the epoxy groups reacted further with nucleophilic OH groups in glucose units or water to generate cross-links or diols, respectively.

3.4 Physicochemical Properties

The structural changes upon epoxidation of AHP starch may be elucidated by studying its physicochemical properties such as gelatinization, swelling, and solubility of the granules. Hardly any differences in the gelatinization temperatures between AHP-WMS and epoxy-WMS could be obtained (Table 6.12). However, the swelling power and solubility index both decreased after epoxidation of the allyl groups. This suggests that cross-links may have been formed in the starch, although no significant increase of gelatinization temperatures was obtained. In previous research, cross-linked starches were shown to exhibit not only reduced swelling and solubility of the granules but also increased gelatinization temperatures (Liu et al., 1999; Mischnick et al., 2000; Richardson, Nilsson, Bergquist, Gorton, & Mischnick, 2000; Wang & Wang, 2000).

TABLE 6.12 Thermal Properties[a], Swelling Power, and Solubility Index of Native Starch and Its Derivatives

Sample	T_o[a]	T_p[a]	T_c[a]	ΔH_{gel}[a]	Swelling Power (g/g)	Solubility Index (%)
WMS	57.3	69.4	75.7	13.8	4.5±0.2	0.2±0.0
AHP-WMS	49.1	56.6	65.5	5.5	14.7±1.0	7.8±1.0
Epoxy-WMS	50.3	56.5	64.7	4.9	12.9±0.2	5.4±0.2

[a] *Measured by differential scanning calorimetry;* T_o, T_p, *and* T_c *are the onset, peak, and completion temperature, respectively, and* ΔH_{gel} *is the enthalpy of gelatinization.*

4 FUTURE PERSPECTIVES

In the course of this study, several investigations were set up to synthesize and characterize epoxy-starch derivatives. Via a mild method, MSs were modified with AGE followed by epoxidation of the double bond. Most of the studies were focused on the reaction of MS and AGE to gain knowledge about the mechanism of the starch modification and to control process conditions. In principle, a range of epoxy-allyl alkenyl ethers with variable chain length can also be used for the synthesis of allyl-starch derivatives. In the optimization study, different optimal conditions were defined for the reaction of WMS and MSs with AGE, which indicated a different accessibility of the MS. The use of other type of starch or reagents in the synthesis will require new optimizations, leading to different extents of chemical modification, but our results show the usefulness of the experimental design approach for these kinds of reactions.

In the study towards the characteristics of the modified starch, the extent and position of the introduced groups were studied by NMR. Such first insight information can be supplemented by hydrolysis and HPLC analysis. Additionally, capillary GC-MS can be applied to determine the complete monomer composition of hydrolyzed and/or pretreated starch fragments. Moreover, this technique can provide additional structural information. At the polymer level, the substituent distribution of AGE in amylose-rich and amylopectin-rich MS was investigated. The observation was made that substitutions have taken place preferentially in the amorphous regions of the starch granule. Further investigation can be explored at granular level to gain detailed information about the accessibility of AGE for the different MS granules and its distribution over amylose or amylopectin. Knowledge about these structural changes in the starch granule can be of importance for understanding the altered properties and for the development of specific properties. The incorporation of a small amount of AGE led to increased swelling and solubility of the MS granules, decreased gelatinization, and altered pasting properties and flow behavior. A saturation limit of the swelling capacity for AHP starch with increasing DS was obtained. This tendency may be investigated through studying other properties such as the temperature of gelatinization and viscosity of starch with different DS.

Preliminary results were presented for the synthesis and characterization of epoxy-starch derivatives; however, more investigations are recommended to gain knowledge about the mechanism of the modification and to control the parameters in the reaction. Aside from indirect incorporation of epoxy groups, we have alternatively studied the direct introduction of epoxides into starch using bisepoxides. The reaction was performed with methyl-α-D-glucopyranoside (MG) as a model compound and 1,3-butadiene diepoxide, 1,2,5,6-diepoxyhexane, or 1,2,7,8-diepoxyoctane using the reaction conditions established earlier. The recovered products hardly contained any epoxy groups, despite using different amounts of bisepoxides. Moreover, HPLC analyses showed that a broad range of compounds were formed upon the reactions. Most of the obtained products appeared to be ring-opened and cyclized bisepoxides (Figure 6.15, for the case of 1,2,5,6-diepoxyhexane) (Wiggins & Wood, 1950). Additionally, a lot of unreacted MG was recovered. This indicated that bisepoxides rather reacted with itself than with another compound under the conditions of the experiments. Therefore, the substitution of starch with diepoxides is not an option for the synthesis of epoxy-starch derivatives with high yield. These results show that the two-step synthetic route explored by us is the only viable route for the synthesis of epoxy derivatives of starch and other carbohydrates.

When the inherent reactivity of epoxide groups leads to unavoidable cross-linking during epoxidation reaction, it may be necessary to look for other methods to utilize the modified starches for, e.g., encapsulation. We and other investigators have shown that allyl-starch derivatives can be synthesized in a mild and controllable way. It would be worth to investigate if these derivatives can be cross-linked themselves, after entrapment of, e.g., bioactive molecules, by reactions in which the double bonds are utilized, like ruthenium-catalyzed cross-alkene metathesis (Binder, Blank, & Raines, 2007; Clavier, Grela, Kirschning, Mauduit, & Nalon, 2007) or photopolymerization using dithiols (Miquelard-Garnier, Demoures, Creton, & Hourdet, 2006). Copolymerization of AHP starches

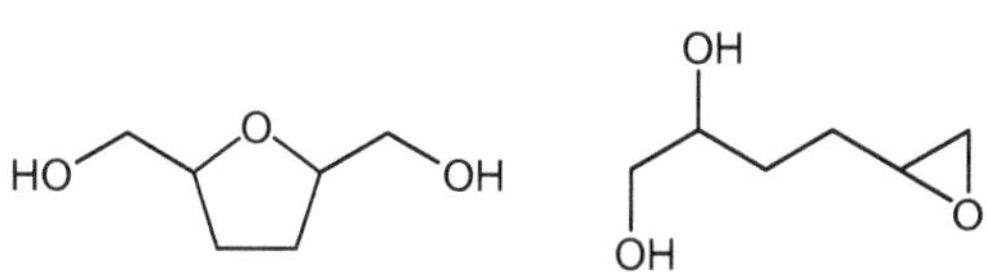

3 4

FIGURE 6.15 Main products of 1,2,5,6-diepoxyhexane, 2,5-bishydroxymethyl-tetrahydrofuran (**3**), and 1,2-dihydroxy-5,6-epoxyhexane (**4**) after the reactions.

with acrylic monomers to create hydrogels is also worth investigating. However, if the reaction conditions applied during the epoxidation are the cause of the premature cross-linking, an alternative method for epoxide formation should be sought after. For instance, the epoxidation of double bonds may be accomplished with a chemoenzymatic approach. In this alternative route, *Candida antarctica lipase* B catalyzes the generation of peracid from a carboxylic acid, which is followed by a Prilezhaev epoxidation of the corresponding allyl ether under mild conditions (Tufvesson, Adlercreutz, Lundmark, Manea, & Hatti-Kaul, 2008).

Since epoxy groups are very reactive to nucleophiles such as amino and hydroxyl groups, epoxy-starch derivatives are of interest for further functionalization with compounds having a coupling, complexing, or cross-linking function or compounds having hydrophobic, anionic, or cationic groups (Figure 6.16). The production of these multifunctional starch derivatives may be used for various purposes such as delivery systems or controlled release systems. Epoxy-starch derivatives could be tested as carrier matrix for bioactive agents such as enzymes. We performed some initial experiments on the entrapment of the thermostable β-glucosidase from *Pyrococcus furiosus* in epoxy starch, synthesized as described earlier.

In these experiments, β-glucosidase was immobilized by mixing the thermostable enzyme with the pregelatinized epoxy-WMS under mild conditions (Table 6.13). The preliminary results clearly show that the irreversible binding of the enzyme to the gelatinized epoxy starch has been accomplished since 13% enzymatic activity recovery is found after fourfold extrusion of the immobilized β-glucosidase from the gelatinized epoxy starch. Nevertheless, a much higher binding efficiency may be achieved by optimization of the procedure. For this, it may be necessary to gain more detailed knowledge about the stability of β-glucosidase under the applied reaction conditions. Other improvements in the procedure might be found in variation of the ratio of amount of enzyme and carrier and enhancing the interaction time of the enzyme with the pregelatinized starch.

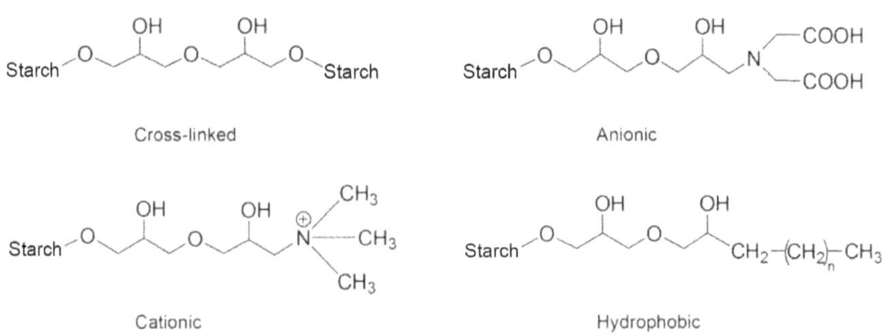

FIGURE 6.16 Functionalized epoxy-starch derivatives.

TABLE 6.13 Enzymatic Activity Recovery of β-Glucosidase by Immobilization in Epoxy-WMS[a]

β-Glucosidase	Enzymatic Activity (U)[b]	Recovery (%)
Free	25±0.5	107
Free, incubated under immobilization conditions[c]	23.0±0.4	100
Immobilized[c]	2.9±0.5	13
Extruded from the starch gel[d] first step	15.4±0.6	67
Extruded from the starch gel[d] fourth step	0.16±0.01	0.7

[a] *Sample preparation: gelatinization of epoxy starch (50 mg) in 1.0 mL 50 mM citrate buffer, pH 5.0 at 80 °C for 60 min.*

[b] *Measured in an enzymatic activity assay for β-glucosidase as described elsewhere (Hansson, Kaper, Oost, Vos, & Adlercreutz, 2001). One unit of enzyme activity (U) was defined as the amount of enzyme catalyzing the liberation of 1.0 μmol of* p-nitrophenol *per min at 50 °C under the applied conditions, using* $\varepsilon_{405} = 0.290 \times 10^3\ M^{-1}\ cm^{-1}$ *at 405 nm for* p-*nitrophenol.*

[c] *Immobilization conditions: incubation of β-glucosidase (0.05 mg, 0.9 mg* mL^{-1}*) with or without epoxy-WMS (50 mg) in 1.0 mL citrate buffer at 80 °C for 30 min and cooled to room temperature.*

[d] *Extrusion, 1-4 steps: the starch gel pellet was suspended in 0.5 mL citrate buffer followed by centrifugation at 12,000 rpm, 4 °C for 10 min. In total, 83% of enzyme activity was recovered by the extrusions.*

Acknowledgments

We thank Dr. Annemarie Huijbrechts for her contribution to this research.

References

Ameye, D., Voorspoels, J., Foreman, P., Tsai, J., Richardson, P., Geresh, S., et al. (2002). Ex vivo bioadhesion and in vivo testosterone bioavailability study of different bioadhesive formulations based on starch-g-poly(acrylic acid) copolymers and starch/poly(acrylic acid) mixtures. *Journal of Controlled Release, 79*, 173–182.

Anwar, E., Khotimah, H., & Yamar, H. (2006). An approach on pregelatinized cassava starch phosphate esters as hydrophilic excipient for controlled release tablet. *Journal of Medical Sciences, 6*, 923–929.

Athawale, V. D., & Rathi, S. C. (1997). Syntheses and characterization of starch poly (methacrylic acid) graft copolymers. *Journal of Applied Polymer Science, 66*, 1399–1403.

Bao, J., & Bergman, C. J. (2004). The functionality of rice starch. In A. C. Eliasson (Ed.), *Starch in food: Structure, function and application* (pp. 258–294). Cambridge, UK: Woodhead Publishing.

Bemiller, J. N. (1997). Starch modification: Challenges and prospects. *Starch-Starke, 49*, 127–131.

Bertoft, E. (2004). Analysing starch structure. In A. C. Elliasson (Ed.), *Starch in food: Structure, function and application* (pp. 57–96). Cambridge, UK: Woodhead Publishing.

Besemer, A. C., & Van Bekkum, H. (1994). Dicarboxy-starch by sodium-hypochlorite bromide oxidation and its calcium-binding properties. *Starch-Starke, 46*, 95–101.

Bhuniya, S. P., Rahman, S., Satyanand, A. J., Gharia, M. M., & Dave, A. M. (2003). Novel route to synthesis of allyl starch and biodegradable hydrogel by copolymerizing allyl-modified starch with methacrylic acid and acrylamide. *Journal of Polymer Science Part A: Polymer Chemistry, 41*, 1650–1658.

Bien, F., Wiege, B., & Warwel, S. (2001). Hydrophobic modification of starch by alkali-catalyzed addition of 1,2-epoxyalkanes. *Starch-Starke, 53*, 555–559.

Binder, J. B., Blank, J. J., & Raines, R. T. (2007). Olefin metathesis in homogenous aqueous media catalyzed by conventional ruthenium catalysts. *Organic Letters, 9*, 4885–4888.

Burton, S. C., & Harding, D. R. K. (1997). High-density ligand attachment to brominated allyl matrices and application to mixed mode chromatography of chymosin. *Journal of Chromatography A, 775*, 39–50.

Campechano-Carrera, E., Corona-Cruz, A., Chel-Guerrero, L., & Betancur-Ancona, D. (2007). Effect of pyrodextrinization on available starch content of Lima bean (*Phaseolus lunatus*) and Cowpea (*Vigna unguiculata*) starches. *Food Hydrocolloids, 21*, 472–479.

Cedrone, F., Bhatnagar, T., & Baratti, J. C. (2005). Colorimetric assays for quantitative analysis and screening of epoxide hydrolase activity. *Biotechnology Letters, 27*, 1921–1927.

Chen, Y., & Reymond, J. -L. (1995). Epoxidation of olefins with formamide—Hydrogen peroxide. *Tetrahedron Letters, 36*, 4015–4018.

Chen, Z. H., Schols, H. A., & Voragen, A. G. J. (2004). Differently sized granules from acetylated potato and sweet potato starches differ in the acetyl substitution pattern of their amylose populations. *Carbohydrate Polymers, 56*, 219–226.

Clavier, H., Grela, K., Kirschning, A., Mauduit, M., & Nalon, S. P. (2007). Sustainable concepts in olefin metathesis. *Angewandte Chemie International Edition, 46*, 6786–6801.

Deetae, P., Shobsngob, S., Varanyanond, W., Chinachoti, P., Naivikul, O., & Varavinit, S. (2008). Preparation, pasting properties and freeze-thaw stability of dual modified crosslink-phosphorylated rice starch. *Carbohydrate Polymers, 73*, 351–358.

Duanmu, J., Gamstedt, E. K., & Rosling, A. (2007). Synthesis and preparation of crosslinked allylglycidyl ether-modified starch-wood fibre composites. *Starch-Starke, 59*, 523–532.

Elfstrand, L., Frigard, T., Andersson, R., Eliasson, A. C., Jonsson, M., Reslow, M., et al. (2004). Recrystallisation behaviour of native and processed waxy maize starch in relation to the molecular characteristics. *Carbohydrate Polymers, 57*, 389–400.

Elomaa, M., Asplund, T., Soininen, P., Laatikainen, R., Peltonen, S., Hyvarinen, S., et al. (2004). Determination of the degree of substitution of acetylated starch by hydrolysis, H-1 NMR and TGA/IR. *Carbohydrate Polymers, 57*, 261–267.

Fang, J. M., Fowler, P. A., Sayers, C., & Williams, P. A. (2004). The chemical modification of a range of starches under aqueous reaction conditions. *Carbohydrate Polymers, 55*, 283–289.

Fanta, G. F., & Doane, W. M. (1986). Grafted starches. In O. B. Wurzburg (Ed.), *Modified starches: Properties and uses* (pp. 150–153). Boca Raton, Florida: CRC Press.

Funke, U., & Lindhauer, M. G. (2001). Effect of reaction conditions and alkyl chain lengths on the properties of hydroxyalkyl starch ethers. *Starch-Starke, 53*, 547–554.

Goff, H. D. (2004). Modified starches and the stability of frozen foods. In A. C. Eliasson (Ed.), *Starch in food: Structure, function and application* (pp. 425–440). Cambridge, UK: Woodhead Publishing.

Gonera, A., Goclik, V., Baum, M., & Mischnick, P. (2002). Preparation and structural characterisation of O-aminopropyl starch and amylose. *Carbohydrate Research, 337*, 2263–2272.

Gotlieb, K. F., & Cappelle, A. (2005). *Starch derivatization: Fascinating and unique industrial opportunities.* Wageningen, NL: Wageningen Academic Publisher.

Hamdi, G., & Ponchel, G. (1999). Enzymatic degradation of epichlorohydrin crosslinked starch microspheres by alpha-amylase. *Pharmaceutical Research, 16*, 867–875.

Hansson, T., Kaper, T., Oost, J. V. D., Vos, W. M. D., & Adlercreutz, P. (2001). Improved oligosaccharide synthesis by protein engineering of beta-glucosidase CelB from hyperthermophilic *Pyrococcus furiosus*. *Biotechnology and Bioengineering, 73*, 203–210.

Heinze, T., & Koschella, A. (2005). Carboxymethyl ethers of cellulose and starch—A review. *Macromolecular Symposia, 223*, 13–39.

Hizukuri, S. (1996). Starch: Analytical aspects. In A. C. Eliasson (Ed.), *Carbohydrates in food* (pp. 379–383). New York, New York: Marcel Dekker.

Hjermstad, E. T. & Kesler, C. C. (1962). *Starch ether derivatives and process for preparing same.* U.S. Patent No. 3,062,810.

Hood, L. F., & Mercier, C. (1978). Molecular-structure of unmodified and chemically modified manionic starches. *Carbohydrate Research, 61*, 53–66.

Huijbrechts, A. M. L., Desse, M., Budtova, T., Franssen, M. C. R., Visser, G. M., Boeriu, C. G., et al. (2008). Physicochemical properties of etherified maize starches. *Carbohydrate Polymers, 74*, 170–184.

Huijbrechts, A. A. M. L., Huang, J. R., Schols, H. A., Van Lagen, B., Visser, G. M., Boeriu, C. G., et al. (2007). 1-Allyloxy-2-hydroxy-propyl-starch: Synthesis and characterization. *Journal of Polymer Science Part A: Polymer Chemistry, 45*, 2734–2744.

Huijbrechts, A. M. L., Ter Haar, R., Schols, H. A., Franssen, M. C. R., Boeriu, C. G., & Sudholter, E. J. R. (2010). Synthesis and application of epoxy starch derivatives. *Carbohydrate Polymers, 79*, 858–866.

Huijbrechts, A. M. L., Vermonden, T., Bogaert, P., Franssen, M. C. R., Visser, G. M., Boeriu, C. G., et al. (2009). Optimization of the synthesis of 1-allyloxy-2-hydroxy-propyl-starch through statistical experimental design. *Carbohydrate Polymers, 77*, 25–31.

Jarowenko, W. (1986). Acetylated starches and miscellaneous organic acids. In W. O. Wurzburg (Ed.), *Modified starches: Properties and uses* (pp. 55–78). Boca Raton, Florida: CRC Press.

Jenkins, P. J., & Donald, A. M. (1998). Gelatinisation of starch: A combined SAXS/WAXS/DSC and SANS study. *Carbohydrate Research, 308*, 133–147.

Kaur, L., Singh, N., & Singh, J. (2004). Factors influencing the properties of hydroxypropylated potato starches. *Carbohydrate Polymers, 55*, 211–223.

Kavitha, R., & Bemiller, J. N. (1998). Characterization of hydroxypropylated potato starch. *Carbohydrate Polymers, 37*, 115–121.

Kesler, C. C. & Hjermstad, E. T. (1950a). *Preparation of starch ethers in original granule form.* U.S. Patent No. 2,516,633 (A).

Kesler, C. C. & Hjermstad, E. T. (1950b). *Starch ethers in original granule form.* U.S. Patent No. 2,516,632.

Khalil, M. I., & Farag, S. (1998). Preparation of some cationic starches using the dry process. *Starch-Starke, 50*, 267–271.

Kovats, P. L. (1971). *Dextrin and starch ethers.* U.S.A. patent application. U.S. Patent No. 3,746,699.

Laakso, T., & Sjoholm, I. (1987). Biodegradable microspheres. 10. Some properties of polyacryl starch microparticles prepared from acrylic acid-esterified starch. *Journal of Pharmaceutical Sciences, 76*, 935–939.

Lawal, O. S. (2004). Composition, physicochemical properties and retrogradation characteristics of native, oxidised, acetylated and acid-thinned new cocoyam (*Xanthosoma sagittifolium*) starch. *Food Chemistry, 87*, 205–218.

Lawal, O. S., Adebowale, K. O., Ogunsanwo, B. M., Barba, L. L., & Ilo, N. S. (2005). Oxidized and acid thinned starch derivatives of hybrid maize: Functional characteristics, wide-angle X-ray diffractometry and thermal properties. *International Journal of Biological Macromolecules, 35*, 71–79.

Le Bail, P., Morin, F. G., & Marchessault, R. H. (1999). Characterization of a crosslinked high amylose starch excipient. *International Journal of Biological Macromolecules, 26*, 193–200.

Li, J. H., Vasanthan, T., Hoover, R., & Rossnagel, B. G. (2004). Starch from hull-less barley: IV. Morphological and structural changes in waxy, normal and high-amylose starch granules during heating. *Food Research International, 37*, 417–428.

Lin, J. H., Lee, S. Y., & Chang, Y. H. (2003). Effect of acid-alcohol treatment on the molecular structure and physicochemical properties of maize and potato starches. *Carbohydrate Polymers, 53*, 475–482.

Liu, H. J., Corke, H., & Ramsden, L. (1999). Functional properties and enzymatic digestibility of cationic and cross-linked cationic ae, wx, and normal maize starch. *Journal of Agricultural and Food Chemistry, 47*, 2523–2528.

Manelius, R., Buleon, A., Nurmi, K., & Bertoft, E. (2000). The substitution pattern in cationised and oxidised potato starch granules. *Carbohydrate Research, 329*, 621–633.

Martin, R. W. (1949). Rapid estimation of ethylenes. *Analytical Chemistry, 21*, 921–922.

Massicotte, L. P., Baille, W. E., & Mateescu, M. A. (2008). Carboxylated high amylose starch as pharmaceutical excipients—Structural insights and formulation of pancreatic enzymes. *International Journal of Pharmaceutics, 356*, 212–223.

Miquelard-Garnier, G., Demoures, S., Creton, C., & Hourdet, D. (2006). Synthesis and rheological behavior of new hydrophobically modified hydrogels with tunable properties. *Macromolecules, 39*, 8128–8139.

Mischnick, P., Heinrich, J., Gohdes, M., Wilke, O., & Rogmann, N. (2000). Structure analysis of 1,4-glucan derivatives. *Macromolecular Chemistry and Physics, 201*, 1985–1995.

Moorty, S. N. (2004). Tropical sources of starch. In A. C. Eliasson (Ed.), *Starch in food structure, function and applications* (pp. 321–325). Cambridge, UK: Woodhead Publishing.

Moser, B. K. (1986). Hydroxyethylated starches. In O. B. Wurzburg (Ed.), *Modified starches: Properties and uses* (pp. 79–88). Boca Raton, Florida: CRC Press.

Nud'ga, L. A., Petrova, V. A., & Lebedeva, M. F. (2003). Effect of allyl substitution in chitosan on the structure of graft copolymers. *Russian Journal of Applied Chemistry, 76*, 1978–1982.

Radosta, S., Vorwerg, W., Ebert, A., Begli, A. H., Grulc, D., & Wastyn, M. (2004). Properties of low-substituted cationic starch derivatives prepared by different derivatisation processes. *Starch-Starke, 56*, 277–287.

Richardson, S., & Gorton, L. (2003). Characterisation of the substituent distribution in starch and cellulose derivatives. *Analytica Chimica Acta, 497*, 27–65.

Richardson, S., Nilsson, G. S., Bergquist, K. E., Gorton, L., & Mischnick, P. (2000). Characterisation of the substituent distribution in hydroxypropylated potato amylopectin starch. *Carbohydrate Research, 328*, 365–373.

Roper, H. (2002). Renewable raw materials in Europe—Industrial utilisation of starch and sugar. *Starch-Starke, 54*, 89–99.

Rutenberg, M. W., & Solarek, D. (1984). Hydroxyalkylated starches. In R. L. Whistler, J. N. Bemiller, & E. F. Paschall (Eds.), *Starch: Chemistry and technology* (pp. 311–366) (2nd ed.). Orlando, Florida: Florida Academic Press.

Seidel, C., Kulicke, W. M., Hess, C., Hartmann, B., Lechner, M. D., & Lazik, W. (2001). Influence of the cross-linking agent on the gel structure of starch derivatives. *Starch-Starke, 53*, 305–310.

Singh, N., Chawla, D., & Singh, J. (2004). Influence of acetic anhydride on physicochemical, morphological and thermal properties of corn and potato starch. *Food Chemistry, 86*, 601–608.

Singh, J., Kaur, L., & McCarthy, O. J. (2007). Factors influencing the physico-chemical, morphological, thermal and rheological properties of some chemically modified starches for food applications—A review. *Food Hydrocolloids, 21*, 1–22.

Solarek, D. (1986). Phosphorylated starches and miscellaneous inorganic esters. In O. B. Wurzburg (Ed.), *Modified starches: Properties and uses* (pp. 97–112). Boca Raton, Florida: CRC Press.

Swinkels, J. J. M. (1990). *Industrial starch chemistry: Properties, modifications and applications of starches*. Foxhol, NL: Avebe.

Taggart, P. (2004). Starch as an ingredient: Manufacture and applications. In A. C. Eliasson (Ed.), *Starch in food: Structure, function and application* (pp. 363–392). Cambridge, UK: Woodhead Publishing.

Taylor, N. W. (1979). Rheology and water swelling of carboxymethyl starch gels. *Journal of Applied Polymer Science, 24*, 2031–2040.

Thomas, D. J., & Atwell, W. A. (1999). *Starches.* St. Paul, Minnesota: Eagen Press, pp. 31–47.

Tijsen, C. J., Voncken, R. M., & Beenackers, A. (2001). Design of a continuous process for the production of highly substituted granular carboxymethyl starch. *Chemical Engineering Science, 56*, 411–418.

Tomasik, P., & Schilling, C. H. (2004). Chemical modification of starch. *Advances in Carbohydrate Chemistry and Biochemistry, 59*, 175–403.

Trubiano, P. C. (1986). Succinate and substituted succinate derivatives of starch. In O. B. Wurzburg (Ed.), *Modified starches: Properties and uses* (pp. 131–148). Boca Raton, Florida: CRC Press.

Tsai, J. J. & Meier, E. A. (1990). *Polysaccharide graft copolymers containing reactive aminoethyl halide group.* U.S. Patent No. 4,973,641.

Tufvesson, P., Adlercreutz, D., Lundmark, S., Manea, M., & Hatti-Kaul, R. (2008). Production of glycidyl ethers by chemo-enzymatic epoxidation of allyl ethers. *Journal of Molecular Catalysis B: Enzymatic, 54*, 1–6.

Tuschhoff, J. V. (1986). Hydroxypropylated starch. In O. B. Wurzburg (Ed.), *Modified starches: Properties and uses* (pp. 89–112). Boca Raton, Florida: CRC Press.

Tuting, W., Wagemann, K., & Mischnick, P. (2004). Enzymatic degradation and electrospray tandem mass spectrometry as tools for determining the structure of cationic starches prepared by wet and dry methods. *Carbohydrate Research, 339*, 637–648.

Van Der Burgt, Y. E. M., Bergsma, J., Bleeker, I. P., Mijland, P., Van Der Kerk-Van Hoof, A., Kamerling, J. P., et al. (2000). Distribution of methyl substituents in amylose and amylopectin from methylated potato starches. *Carbohydrate Research, 325*, 183–191.

Van Soest, J. J. G., Hulleman, S. H. D., Dewit, D., & Vliegenthart, J. F. G. (1996). Crystallinity in starch bioplastics. *Industrial Crops and Products, 5*, 11–22.

Vanwarners, A., Stamhuis, E. J., & Beenackers, A. (1989). Kinetics of the hydroxyethylation of starch in aqueous slurries. *Starch-Starke, 41*, 21–24.

Wach, R. A., Mitomo, H., Nagasawa, N., & Yoshii, F. (2003). Radiation crosslinking of carboxymethylcellulose of various degree of substitution at high concentration in aqueous solutions of natural pH. *Radiation Physics and Chemistry, 68*, 771–779.

Wang, Y. J., Truong, V. D., & Wang, L. F. (2003). Structures and rheological properties of corn starch as affected by acid hydrolysis. *Carbohydrate Polymers, 52*, 327–333.

Wang, Y. J., & Wang, L. F. (2000). Effects of modification sequence on structures and properties of hydroxypropylated and crosslinked waxy maize starch. *Starch-Starke, 52*, 406–412.

Wang, Y. J., & Wang, L. F. (2002). Characterization of acetylated waxy maize starches prepared under catalysis by different alkali and alkaline-earth hydroxides. *Starch-Starke, 54*, 25–30.

Whistler, R. L., & Bemiller, J. N. (1997). *Carbohydrate chemistry for food scientists.* St. Paul, Minnesota: Eagen Press, pp. 117–151.

White, P. J., & Tziotis, A. (2004). New corn starches. In A. C. Eliasson (Ed.), *Starch in food: Structure, function and application* (pp. 295–314). Cambridge, UK: Woodhead Publishing.

Wiggins, L. F., & Wood, D. J. C. (1950). Anhydrides of polyhydric alcohols. 14. Observations on the ring scission of 1,2,5,6-diepoxyhexane and 3,4-isopropylidene 1,2,5,6-dianhydromannitol. *Journal of the Chemical Society*, 1566–1575.

Wilham, C. A., McGuire, T. A., Rudolphi, A. S., & Mehltretter, C. L. (1963). Polymerization studies with allyl starch. *Journal of Applied Polymer Science, 7*, 1403–1410.

Wu, Z. H., Pittman, C. U., & Gardner, S. D. (1995). Bonding epoxide functions to nitric acid-oxidized carbon-fibers using epichlorohydrin—A reliable analytical determination of epoxide functions on fiber surfaces. *Carbon, 33*, 607–616.

Wu, Y. S., & Seib, P. A. (1990). Acetylated and hydroxypropylated distarch phosphates from waxy barley—Paste properties and freeze-thaw stability. *Cereal Chemistry, 67*, 202–208.

Wurzburg, O. B. (1986a). Converted starches. In O. B. Wurzburg (Ed.), *Modified starches: Properties and uses* (pp. 17–40). Boca Raton, Florida: CRC Press.
Wurzburg, O. B. (1986b). Crosslinked starches. In O. B. Wurzburg (Ed.), *Modified starches: Properties and uses* (pp. 41–53). Boca Raton, Florida: CRC Press.
Wurzburg, O. B. (1986c). Introduction. In O. B. Wurzburg (Ed.), *Modified starches: Properties and uses* (pp. 3–16). Boca Raton, Florida: CRC Press.
Xu, A., & Seib, P. A. (1997). Determination of the level and position of substitution in hydroxypropylated starch by high-resolution H-1-NMR spectroscopy of alpha-limit dextrins. *Journal of Cereal Science, 25*, 17–26.
Zhang, D. F., Ju, B. Z., Zhang, S. F., & Yang, J. Z. (2008). The study on the synthesis and action mechanism of starch succinate half ester as water-reducing agent with super retarding performance. *Carbohydrate Polymers, 71*, 80–84.
Zhang, L. M., Yang, C., & Yan, L. (2005). Perspectives on: Strategies to fabricate starch-based hydrogels with potential biomedical applications. *Journal of Bioactive and Compatible Polymers, 20*, 297–314.
Zhu, Z. (2003). Starch mono-phosphorylation for enhancing the stability of starch/PVA blend pastes for warp sizing. *Carbohydrate Polymers, 54*, 115–118.

PART V

STRUCTURES AND PROPERTIES OF NOVEL STARCH-BASED MATERIALS

CHAPTER

7

Mechanical Properties of Starch-Based Plastics

C. Chaléat[1], Peter J. Halley[2], R.W. Truss[3,4]

[1]Australian Institute of Bioengineering and Nanotechnology, The University of Queensland, Brisbane, Queensland, Australia

[2]School of Chemical Engineering and Australian Institute of Bioengineering and Nanotechnology, The University of Queensland, Brisbane, Queensland, Australia

[3]School of Mechanical and Mining Engineering, The University of Queensland, Brisbane, Queensland, Australia

[4]School of Chemical Engineering, The University of Queensland, Brisbane, Queensland, Australia

1 INTRODUCTION

For starch-based plastics to be used for industrial applications, a thorough understanding of their mechanical properties and the factors that will influence their properties is required.

Although many studies have investigated the mechanical properties and structure of starch-based plastics, definitive data are difficult to obtain since individual studies have been made on a wide range of starches, using a wide range of plasticizers and processing conditions and using a wide range of test conditions. Unfortunately, in many of these studies, the materials have been insufficiently characterized and the test conditions poorly defined or controlled, which makes the data of dubious use. Nevertheless, various parameters and factors have been observed to show a significant influence on their behavior. These factors can be classified into two primary groups: the composition and structure of the starch-based plastic and the conditions of testing. The majority of work has been focused on the first of these groups as selection, and the processing of starch was critical in the development of commercial products. This has also lead to blending of starch with other synthetic and natural polymers in an attempt to alleviate shortcomings in the properties of the starch. Few systematic studies have been undertaken to address the effect of test conditions, but with the growing maturity of these products, such information will be required for more advanced applications.

2 PROCESSING OF STARCH INTO THERMOPLASTIC MATERIALS

Starch consists essentially of two types of polymers of α-D-glucose, an essentially linear molecule, amylose, and a highly branched molecule, amylopectin. The ratio of amylose to amylopectin varies with the source of starch. Plants store starch in many types of storage organs such as seeds, tubers, roots, and fruits. The starch occurs as tiny white granules that are insoluble in cold water and in most organic solvents. Amylose and amylopectin are packed together inside the semicrystalline granule with an overall crystallinity of about 15-45% depending not only on its origin but also on the techniques used to determine it (Buléon, Colonna, Planchot, & Ball, 1998; Zobel, 1988). The short side chains of amylopectin (DP ranging between 15 and 18) adopt double-helical conformations and associate into clusters that form ordered regions or "crystalline lamellae," while the branching points constitute the "amorphous lamellae" as shown on Figure 7.1.

To generate thermoplastic starch polymers (TPSs), this complex semicrystalline structure needs to be destroyed to produce an amorphous material. This process called gelatinization is achieved by heating the starch in the presence of water and/or some other plasticizer. In the extrusion of TPS, heat for gelatinization is supplied by the heated barrel of the extruder. Gelatinization is also assisted by the mechanical work imparted by the extruder screw. If during processing, gelatinization is incomplete, remnants of the original semicrystalline structure will be retained in the TPS with consequences for the mechanical properties of the material. It should also be noted that the gelatinization temperature changes with the amylose/amylopectin ratio of the starch (Shogren, 1998), so processing conditions may need to be varied for each starch source.

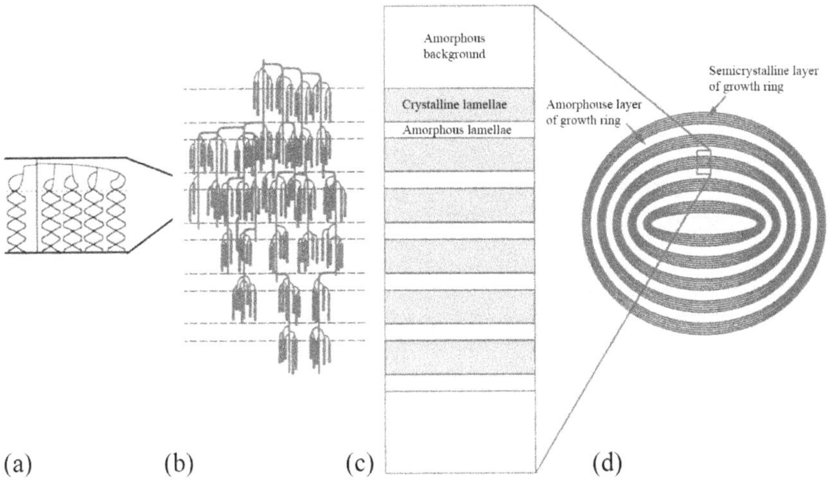

FIGURE 7.1 Schematic diagram of the starch granule organization. (a) Double-helical conformation between amylopectin side chains in a cluster; (b) chain of amylopectin giving rise to the crystalline (side-chain cluster) and amorphous lamellae (branching regions); (c) stack of alternating amorphous and crystalline lamellae forming a semicrystalline growth ring; (d) single granule comprising alternating semicrystalline and amorphous growth rings. *Adapted from Jenkins and Donald (1995) and Ball et al. (1996).*

3 FACTORS AFFECTING THE MECHANICAL PROPERTIES OF STARCH-BASED MATERIALS

3.1 Effect of Starch Source

Starch source has been shown to influence the properties of TPS, and these differences in properties have been ascribed partly to the differences in amylose and amylopectin contents and their molecular structure (Hulleman, Janssen, & Feil, 1998; Lourdin, Dellavalle, & Colonna, 1995; Van Soest, Benes, Wit, & Vliegenthart, 1996; Van Soest, Hulleman, Dewit, & Vliegenthart, 1996). The ratio of amylopectin to amylose varies significantly with botanical source of the starch. In most common types of cereal starches, the ratio of the two polysaccharides ranges between 70% and 82% of amylopectin and 18% and 30% of amylose. However, in some plant species, mutant genotypes of maize, barley, and rice are bred with an altered amylose content such as amylomaize starches containing as much as 80% amylose and waxy starches with less than 1% amylose (Buléon et al., 1998).

The effect of starch source on the stress-strain behavior is shown in Figure 7.2. These materials contained 30 wt% glycerol and 20 wt% water as plasticizers reducing the T_g of the starch to below the test temperature.

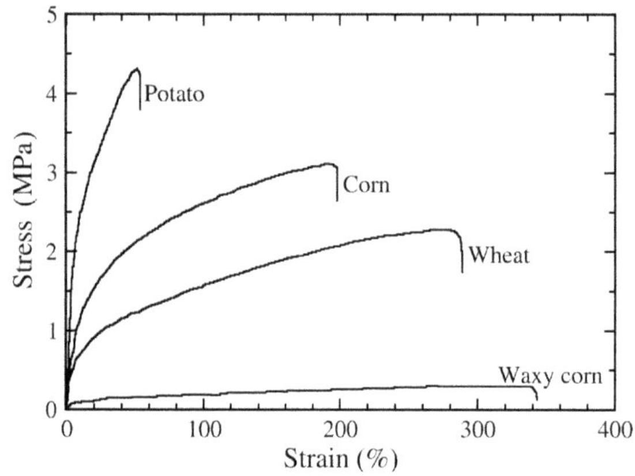

FIGURE 7.2 Stress-strain curves of various plasticized starch films (with 30 wt% of glycerol and 20 wt% water). *Adapted from Hulleman et al. (1998).*

These materials thus showed a relatively high elongation to break, low tensile strength, and low Young's modulus, characteristics of a rubbery state. However, differences between the TPSs from different sources are obvious: TPS containing around 25% amylose (e.g., wheat, corn, and potato starches) presented higher tensile strength than TPS containing 99% amylopectin (e.g., waxy corn starch). The differences between the TPSs with nominally equivalent amylose content (wheat, corn, and potato) were attributed to two factors. Firstly, there were small variations in the amount of B-type crystallinity in the samples, and the presence of crystallinity is known to increase the stiffness and stress at break in rubbery materials. Hulleman et al. (1998) also attributed the differences to the mobility of the molecules present and their ability to form association or entanglements. They varied the water content of the starch materials during compression molding of the samples and noted that the water content to achieve maximum strain at break was higher for potato starch, which has a higher degree of polymerization than corn or wheat.

High-amylose materials generally result in products with higher strength and stiffness than high-amylopectin materials (Figure 7.3). The linear amylose chains entangle effectively and form strong physical cross-links, mainly by hydrogen bonding and local crystallization, resulting in strong and coherent materials. Lourdin et al. (1995) found that in unplasticized films, tensile strength increased linearly with the amount of amylose, from 38 MPa for pure amylopectin film up to 69 MPa for pure amylose film. In plasticized material (with 20% glycerol), the efficiency of plasticization with the same amount of plasticizers was dependent on the amylose/amylopectin ratio. Amylopectin was more sensitive than amylose to the action of plasticizer, and the mechanical properties of films containing glycerol remained constant between 40% and 100% amylose.

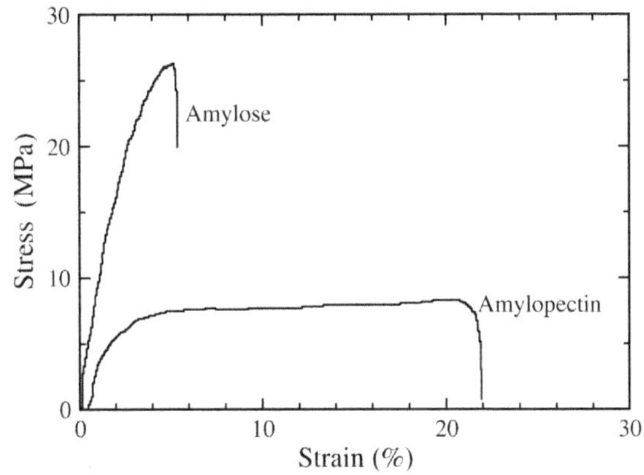

FIGURE 7.3 Typical stress-strain behavior of pure plasticized amylopectin and amylose films. *Adapted from Lourdin et al. (1995).*

Myllarinen, Partanen, Seppala, and Forssell (2002) investigated the plasticization of amylose and amylopectin films with water and glycerol. They observed, under low glycerol contents, an increase in the brittleness of the films (i.e., low tensile strength and elongation). They also noticed an increase in elongation when more than 20% glycerol was added to amylose, which was suggested to be the result of a phase-separated system (Lourdin, Ring, & Colonna, 1998). It was also found that amylose could form films with up to 70% glycerol, but for amylopectin, only a fluid-like material was produced above 30% glycerol. This difference indicated that the polymer network primarily defined the material properties of the highly plasticized starch films.

Using light microscopy on iodine-stained amylose/amylopectin films, Rindlav-Westling, Stading, and Gatenholm (2002) demonstrated that phase separation occurred between amylose and amylopectin at low amylose content. At higher amylose proportions (i.e., 25% amylose), however, a more homogeneous system was obtained and the phase separation was apparently prevented by amylose gelation and the formation of a continuous amylose network.

It should also be noted that variations in structure of amylose and amylopectin (such as molecular weights, branching degree of amylopectin, and occurrence of cocrystallization between amylose and amylopectin) also influence significantly properties of TPS (Lourdin et al., 1995; Van Soest & Borger, 1997).

3.2 Effect of Plasticizers

Plasticizers have two important roles in TPS materials. First, most starch polymers require plasticizers to aid in gelation and to decrease the processing temperatures during extrusion to avoid degradation of the starch. In addition, unplasticized starch polymers are usually brittle at room temperature due to their relatively high T_g and the lack of main chain sub-T_g relaxation (Donovan, 1979) due to the glucose ring. This parallels the behavior of synthetic polymers where only those amorphous glassy polymers, such as polycarbonate and PVC, with a strong β relaxation below room temperature show any significant toughness at room temperature. Several authors (Bradley & Carr, 1976; Nishinari, 1980) have reported a β transition in amylose that has been assigned to rotation of methylol-water complexes. Indeed, in amylose films containing approximately 6% of water, Bradley and Carr (1976) found a β relaxation at around −55 °C, which strengthened with increasing water content. They also noticed a γ transition assigned to rotation of methylol groups.

The addition of plasticizers, which are often low-molecular-weight molecules, increases the interchain spacing (increase of free volume) and reduces the interactions between the polymer chains. Thus, polymer chains have increased chain mobility with the polymer T_g shifting to lower temperature, resulting in a less brittle material. Plasticization of starch has been extensively investigated. A broad range of plasticizers have been used with starch such as water (Hulleman et al., 1998; Kalichevsky, Jaroszkiewicz, Ablett, Blanshard, & Lillford, 1992; Shogren, 1992; Zeleznak & Hoseney, 1987), polyols such as glycerol (De Graaf, Karman, & Janssen, 2003; Forssell, Mikkila, Suortti, Seppala, & Poutanen, 1996; Lourdin, Bizot, & Colonna, 1997a; Lourdin, Coignard, Bizot, & Colonna, 1997; Lourdin et al., 1998; Mathew & Dufresne, 2002; Shogren, Swanson, & Thompson, 1992; Van Soest, 1996) and sorbitol (Gaudin, Lourdin, Le Botlan, Ilari, & Colonna, 1999; Lourdin, Coignard, et al., 1997; Mathew & Dufresne, 2002), low-molecular-weight sugars (Kalichevsky, Jaroszkiewicz, & Blanshard, 1993), and urea (Lourdin, Coignard, et al., 1997; Shogren et al., 1992).

The glass transition of thermoplastic starch depends on the type and concentration of plasticizer (Figure 7.4). The behavior is broadly similar for all additives except glycerol diacetate, which displayed a phase separation. Figure 7.4a demonstrates the efficiency of water as a plasticizer for amorphous amylopectin with a drop in the T_g of 80 °C over 10-20% water content range. This is an important consideration when testing the mechanical properties of thermoplastic starch since the moisture content (MC) of a starch sample can vary significantly with changes in storage relative humidity (RH). This is demonstrated in Figure 7.5 that shows the measured MC for a high-amylose starch plasticized solely with water and stored for 14 days at room temperature under different RHs. Moreover, the MC clearly has a strong influence on the mechanical properties as shown by the decrease in yield stress from >40 to 1 MPa when the MC varies from 10% (at 23% RH) to 29% (at 98% RH) (Figure 7.5) (Chaléat, Halley, & Truss, 2008). Bader and Dietmar Goritz (1994) reported that

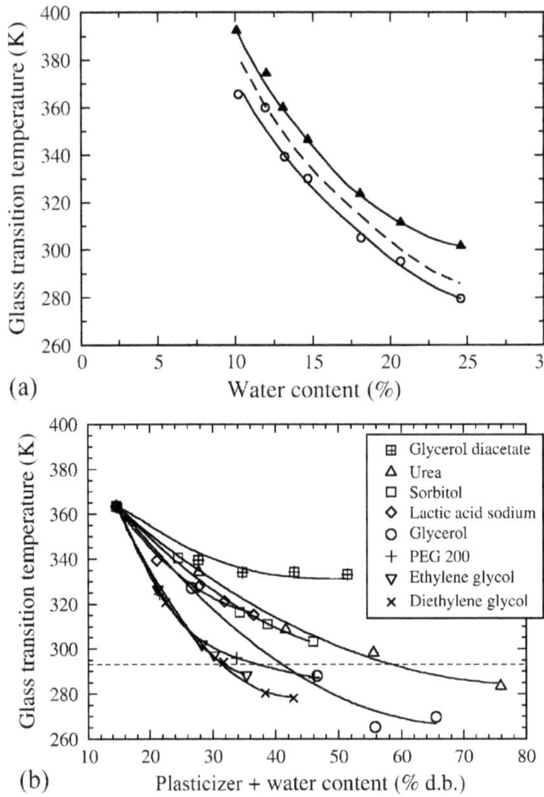

FIGURE 7.4 (a) Glass transition temperature of amorphous amylopectin (gelatinized with water) measured from DMTA tan δ peak (▲), change in DMTA E' slope (○), and DSC scans (– – –) as a function of water content. (b) Glass transition temperature versus total water and plasticizer content for cast potato starch films plasticized with various plasticizers and equilibrated at 57% RH. *Panel a: Adapted from Kalichevsky et al. (1992). Panel b: Adapted from Lourdin, Coignard, et al. (1997).*

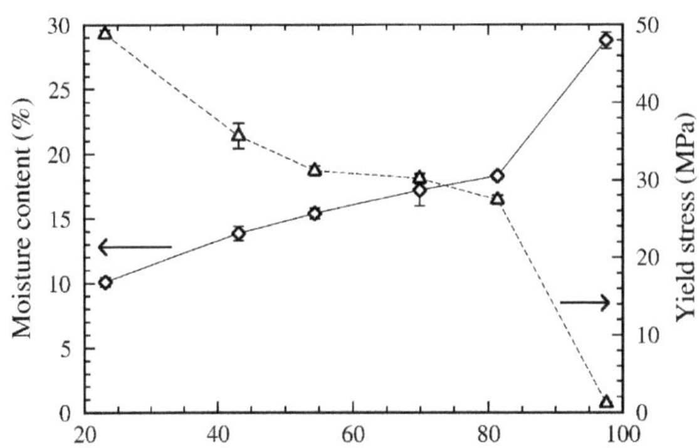

FIGURE 7.5 Moisture content and yield stress for a TPS-based material (processed with water as only plasticizer) as a function of RH conditions after storage for 14 days. *Adapted from Chaléat, Halley, et al. (2008).*

the modulus of a high-amylose corn starch film decreased from a plateau value of ~3.5 GPa at water contents below 5% water to <1 GPa at water content of around 20%. This was accompanied by a change in the tensile failure mode from brittle at low water content to ductile at higher water content. It is thus extremely important to condition samples for adequate times before testing and to ensure that the water content does not change during testing.

Avérous (2004) had shown that the mechanical properties of TPS are intimately linked with the addition of plasticizers and the resulting T_g; Table 7.1 lists tensile properties of starch plasticized with different amounts of water and glycerol. It can be observed that a wide range of mechanical properties can be achieved by varying the plasticizer content. The T_g, tensile strength, and modulus decrease with increasing total plasticizer content, while the elongation to break increases and then begins to decrease with increasing plasticizer content.

The effect of plasticizers can be quite complex since many of the substances used to plasticize starch are themselves hydrophilic and can affect the availability and interaction of water with starch. Lourdin, Coignard, et al. (1997) and Mathew and Dufresne (2002) determined the sorption behavior and the calorimetric glass transition of starch plasticized with varying plasticizer content. They observed that the water content of the materials was far more dependent on RH when a plasticizer was added. For low RH, water content decreased with increasing plasticizer content. For higher water activities (>43% RH), water content increased markedly with the plasticizer content owing to the creation of hydrogen bonds between plasticizer and water. Above a certain amount of plasticizer (e.g., more than 18% for glycerol), phase separation between starch and

TABLE 7.1 Example of Tensile Properties of PlS from Avérous, Fauconnier, Moro, and Fringant (2000) and Avérous, Moro, Dole, and Fringant (2000)

	S74G10W16	S70G18W12	S67G24W9	S65G35W0
Glycerol content (%)	14	25	35	30
Water content (%)	9	9	12	13
T_g (DSC) (°C)	43	8	−7	−20
Young's modulus (MPa)	1144 (42)	116 (11)	45 (5)	11 (1)
Tensile strength (MPa)	21.4 (5.2)	4.0 (1.7)	3.3 (0.1)	1.4 (0.1)
Elongation at break (%)	3 (0.4)	104 (4.7)	98 (5.2)	60 (5.2)

Contents and properties are given after equilibrium at 23 °C and 50% RH during 6 weeks. Initial formulation "S*x*G*y*W*z*" is expressed as starch *x* (%), glycerol *y* (%), and water *z* (%) (standard deviations are shown in brackets).

plasticizer occurred with formation of a plasticizer-rich phase (Lourdin, Bizot, & Colonna, 1997b). Mathew and Dufresne (2002) also found that the T_g of TPS decreased with decreasing molecular weight of the plasticizer.

Lourdin, Bizot, et al. (1997a) had shown also that, at low plasticizer content, an "antiplasticization" effect occurred. Mechanical properties exhibited a minimum in elongation at break for glycerol content of 12%. Even though this phenomenon of antiplasticization has long been observed in synthetic polymers such as polycarbonate (Jackson & Caldwell, 1967a, 1967b), the mechanisms involved are not completely known. In glycerol-containing potato starch films, Lourdin, Bizot, et al. (1997a) considered that the presence of strong interactions between starch and plasticizer impeded movements of the polymer chains as revealed by a low amplitude of the tan δ β-relaxation peak on the DMTA thermogram and, therefore, leading to a higher modulus in the material. The preponderant role of the local molecular mobilities (β relaxation) involved in this mechanism has also been discussed by Gaudin et al. (1999) and Gaudin, Lourdin, Forssell, and Colonna (2000) in starch-sorbitol-water systems. The authors observed that sorbitol, up to 27% in concentration, induced an antiplasticization effect. Beyond this value, sorbitol acted as a plasticizer. Chang, Cheah, and Seow (2000) observed a similar antiplasticization effect on tapioca starch-water films, prevalent over the range of MC (4-8%) where such films existed in the glassy state. The conventional plasticizing effects of water could only be observed with further hydration beyond the limit imposed by the antiplasticization range.

3.3 Crystallinity and Postprocessing Aging

The degree of crystallinity will affect the mechanical properties of any semicrystalline polymer. The modulus and strength will increase with increasing crystallinity, and there is a tendency to brittle in highly crystalline materials. The type and amount of crystalline structures found in TPS are influenced by the composition, processing methods and conditions and postprocessing recrystallization, and aging.

According to Van Soest, Hulleman, De Wit, and Vliegenthart (1996), two types of crystallinity can be distinguished in TPS directly after thermal processing by extrusion, compression molding, injection molding, or kneading. The first is residual crystallinity from native starch (A, B, or C type) caused by incomplete melting of starch during processing. Van Soest, Hulleman, De Wit, et al. (1996) showed that the amount of residual crystallinity was affected by the processing conditions (i.e., temperature and shear stress) and indirectly by the composition. Lower amounts of plasticizers (glycerol in this case) caused a reduction in residual crystallinity due to the increase in melt viscosity increasing the shear stress on the melt, which aided in breaking down the crystallites.

The second type of crystallinity was processing-induced crystallinity. This was caused by fast recrystallization of amylose into single-helical structures such as V_H and V_A types, and also the E_H-type, stable at low MC and shown to transform in the V_H type with increasing water content (Mercier, Charbonniere, Grebaut, & Gueriviere, 1980). Van Soest, Hulleman, De Wit, et al. (1996) found that increasing the screw speed during extrusion or increasing residence time during kneading caused an increase in single-helical crystallinity. In waxy starches containing no amylose, no V- or E-type crystallinity is formed. The authors also noticed that the amount of E and V type increased with decreasing water content (Van Soest & Knooren, 1997).

In amylose films obtained by casting, B-type crystallinity developed upon film formation (Rindlav, Hulleman, & Gatenholm, 1997). The difference in crystalline morphology can be explained by the difference in molecular mobility present in solvent casting processes.

Different authors (Avérous, Fauconnier, et al., 2000; Avérous, Moro, et al., 2000; Van Soest & Borger, 1997) have shown that, after processing, TPS displays a strong aging evolution with changes in mechanical behavior such as an increase in the Young's modulus over time. TPS presents two kinds of aging behavior depending on whether TPS is in a glassy or rubbery state. Below T_g, TPS undergoes a physical aging. Volume relaxation occurs and the mechanical behavior of the material shows a progressive embrittlement (Lourdin, Colonna, Brownsey, Noel, & Ring, 2002). At temperatures above T_g, retrogradation occurs with recrystallization and rearrangements of amylose and amylopectin helices.

Physical aging below T_g occurs very slowly by molecular relaxations as glasses tend toward their equilibrium state. The effects of physical aging on the material properties of amorphous carbohydrates were examined by Lourdin et al. (2002) on a potato starch-water-sorbitol (16:3:1) system. The mechanical behavior of this blend was characterized by three-point bend tests on samples immersed in a temperature-controlled bath of silicone oil at 28 °C. This is below the T_g of this material (~50 °C). The stress at break increased with increasing aging time of 5 min to 60 days from 3.1 to 4.2 MPa, whereas the deformation at break decreased from 22% to 7.3%. The physical aging, thereby, induced a progressive embrittlement. Shogren (1992) and Shogren and Jasberg (1994) investigated physical aging in extruded corn starch and noticed less significant changes in mechanical properties.

Aging of starch materials above the glass transition occurs by retrogradation (Shogren, 1992; Van Soest, 1996; Van Soest & Borger, 1997; Van Soest, De Wit, Tournois, & Vliegenthart, 1994; Van Soest, Dewit, Tournois, & Vliegenthart, 1994; Van Soest, Hulleman, De Wit, et al.,

1996; Van Soest, Hulleman, Dewit, et al., 1996). Here, the starch molecules have much more mobility and can reassociate in more ordered structures, such as the various helical structures and ordered packing arrangements found in the crystal structures. During postprocessing aging of glycerol-containing starch plastics above T_g, both amylose and amylopectin recrystallize in the double-helical B-type crystal structure (Van Soest, Hulleman, Dewit, et al., 1996; Van Soest & Knooren, 1997). In a study of potato starch processed by extrusion in the presence of water and glycerol, the time-dependent changes in the product's mechanical properties were correlated to the changes in crystallinity (Van Soest, Hulleman, Dewit, et al., 1996). An increase in B-type crystallinity to 30% relative crystallinity compared to native potato starch was shown to lead to an increase in Young's modulus from 40 to 70 MPa and in tensile strength from 5.5 to 7 MPa and a decrease in elongation from 90% to 55% for TPS materials with a water content of 14%. For materials with the highest crystallinity (more than 30% B-type relative crystallinity), a sharp decrease in elongation was observed and the samples cracked during storage.

It has been shown that the plasticizer content affects the crystallization rate. Higher amounts of plasticizer increase the mobility of the starch chains and lower the T_g. The rate of crystallization increases with increasing water content (Van Soest, De Wit, et al., 1994; Van Soest, Dewit, et al., 1994). Conversely, Van Soest, De Wit, et al. (1994) found that glycerol reduced the crystallization rate at a constant water content due to the formation of starch-glycerol interactions and stabilization of water resulting in a decreased mobility of the starch chains. However, due to the hygroscopicity of glycerol, the water content increases with increasing amount of glycerol, thereby lowering the T_g and increasing the crystallization rate.

Analysis of extruded samples prepared from high-amylose and high-amylopectin starches revealed that, in amylose-enriched materials, crystallization occurs much faster than in amylopectin-enriched materials (Van Soest & Borger, 1997; Van Soest & Essers, 1997). These differences were attributed to high-amylose starches containing more amorphous structures capable of additional crystallization. Van Soest (1996) proposed a complex model describing, at a molecular level, a series of structural changes occurring during aging above T_g. Within the first days following processing, amylose single helices and double helices are formed. The formation of single helices is related to processing parameters, amylose content, and plasticizers (process-induced crystallinity) and double helices to storage conditions (retrogradation). Aggregation of double helices and then B-type crystallization (long-range ordering) occur (Van Soest, De Wit, et al., 1994; Van Soest, Dewit, et al., 1994) increasing the strength and stiffness of the material. The reduction in elongation can be explained by the presence of crystallites acting as permanent cross-links. Amylose crystallization is fast compared to the crystallization of double helices composed of amylopectin outer chains, which can be either intra- or intermolecular (Van Soest, Hulleman, Dewit, et al., 1996). Cocrystallization of amylopectin with amylose probably also occurs. Van Soest (1996) concluded that processing and retrogradation induce a complex starch network with strong interactions by the formation of hydrogen bonds, helices, and crystals.

3.4 Properties Limited by Flaws

Many mechanical properties, particularly toughness, elongation to break, and strength in brittle polymers, are dependent on the presence of flaws in the materials. These may range from the presence of internal voids or actual imperfections of the surface to local variations in the microstructure, density, or composition. For TPSs, another range of flaw can exist resulting from the processing. If the starch is incompletely gelatinized, then remnants of the initial crystallinity can provide hard particles that can initiate fracture in the sample. In addition, ghosts of the initial starch granules can survive processing and these too provide inhomogeneities in the structure that can limit ductility.

Figure 7.6 shows an SEM image of an extruded starch sheet. The sheet displays a rather uniform surface but contains a number of hard particles that have left voids in their wake appearing to be aligned in the extrusion direction. These hard particles are ungelatinized starch granules and their presence with associated voids creates a significant number of flaws, which can lead to low ductility.

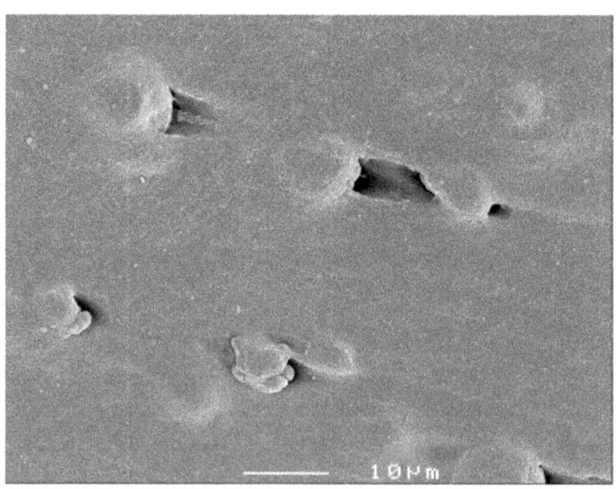

FIGURE 7.6 SEM image of a surface of extruded starch-based sheet.

4 EFFECT OF TESTING CONDITIONS

All polymer properties are time- and temperature-dependent, and an understanding of these dependencies is important in service behavior of the materials. However, there have been few systematic studies on how mechanical properties of TPS material vary with test conditions with the possible exception of the effect of RH on water plasticization of TPS.

A recent study by Chaléat, Michel-Amadry, Halley, and Truss (2008) on a commercial starch-based blend sheet systematically measured the yield stress as a function of strain rate, temperature, and RH. They showed that the equilibrium water content changed with the storage RH and that the yield stress was highly dependent on the storage RH. This work showed that the data at different MCs could be shifted to a master curve by a shift along the strain rate axis in a so-called "time-moisture" superposition similar to the well-known time-temperature superposition of the Williams-Landel-Ferry equation for polymers. The master curves at two different temperatures are shown in Figure 7.7. The strain rate and temperature dependence of this master curve data could be modeled well by the Eyring rate theory (Eyring, 1936) in which the stress, σ, is related to the strain rate, $\dot{\varepsilon}$, and absolute temperature, T, by

$$\dot{\varepsilon} = \dot{\varepsilon}_0 \exp\left(-\frac{\Delta H}{RT}\right) \sin h\left(\frac{\upsilon\sigma}{RT}\right),$$

where ΔH is the activation energy, v is a constant with units of volume, $\dot{\varepsilon}_0$ is a constant, and R is the gas constant. In many polymers, two or more

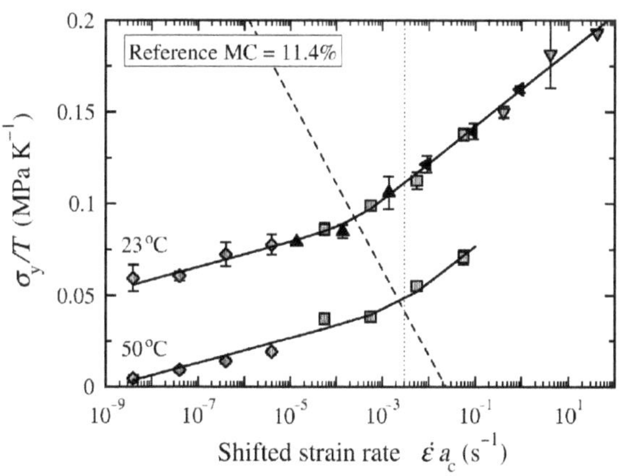

FIGURE 7.7 Master curves of the yield stress to temperature ratio versus the shifted strain rate, obtained at 23 °C and 50 °C, for a reference MC of 11.4%. The transition boundary from process 1-dominated yielding to process 2-activated yielding-based on Equation (4.5) is indicated by a dashed line. The dotted line shows the equivalent strain rate of the DMTA experiments (Chaléat, Michel-Amadry, et al., 2008).

activated rate processes are required to fit data over a large range of temperatures and strain rates (Bauwens-Crowet, Bauwens, & Homès, 1969; Ree & Eyring, 1955; Truss, Clarke, Duckett, & Ward, 1984) and that is the case with this TPS blend. The transition between the two regions corresponded with the $T\alpha$ of the dispersed phase in the blend. This work is significant in that it suggests that TPSs respond in a similar fashion to conventional polymers to changes in testing rate and temperature.

5 PROPERTIES OF STARCH-BASED BLENDS

Pure thermoplastic starch materials are extremely sensitive to water and display rather low mechanical properties when wet. The properties tend to vary over time, either permanently due to crystallization or transiently as a result of varying levels of plasticization due to water content that changes with ambient RH. Materials based solely on TPS are thus of limited practical value. Therefore, much of the commercial research on TPS has involved associating TPS with polymers (synthetic or natural), fillers, and nano-fillers and/or modified starches (Avérous, 2004; Halley, 2005).

Early work on TPS blends includes the work of Otey, Westhoff, and Doane (1980, 1987) who investigated the development of blown agricultural mulch films based on gelatinized corn starch, polyethylene (PE) and poly(ethylene acrylic acid) (PEAA) polymers, and various additives. Good film blowing performance was achieved with increased ammonia and urea content improving starch-PEAA interactions. George, Sullivan, and Park (1994) also looked at the effect of poly(ethylene-*co*-vinyl alcohol) (EVOH) concentration on the processability and physical properties of starch plasticized with glycerin and water. These studies investigating TPS blending with synthetic polymers (i.e., PE and EVOH) resulted in nonfully biodegradable but fragmentable materials. Nevertheless, it is important to note that much of this earlier work has led to the science behind modern biodegradable blends.

To maintain biodegradability, many different TPS blends have been developed using biodegradable polymers. Some of these have been with natural biopolymers such as pectins (Fishman, Coffin, Konstance, & Onwulata, 2000; Otaigbe, Goel, Babcock, & Jane, 1999), but most of the research has focused on blending TPS with biodegradable polyesters such as PCL (Avérous & Fringant, 2001; Avérous, Moro, et al., 2000; Bastioli, Cerutti, Guanella, Romano, & Tosin, 1995; Koenig & Huang, 1995; Matzinos, Tserki, Kontoyiannis, & Panayiotou, 2002; Myllymäki et al., 1998; Vikman, Hulleman, Van der Zee, Myllarinen, & Feil, 1999), PEA (Avérous, Fauconnier, et al., 2000; Avérous & Fringant, 2001), PHBV (Ramsay, Langlade, Carreau, & Ramsay, 1993; Verhoogt et al., 1995), PBSA

(Avérous & Fringant, 2001), PLA (Martin & Avérous, 2001), and PHEE (Walia, Lawton, & Shogren, 2002; Walia, Lawton, Shogren, & Felker, 2000). These TPS/polyester blends show an increased hydrophobic character, a lower water permeability, and improved mechanical properties when compared to pure TPS.

Starch and biodegradable polyester are rather immiscible. Their blends generally demonstrate phase separation and poor interfacial properties. Nevertheless, an interesting behavior results from this low compatibility. Several authors (Avérous, Moro, et al., 2000; Vikman et al., 1999; Walia et al., 2000) have observed in TPS/polyester blends the migration of the low-viscosity polymer (i.e., the polyester) toward the surface leading to the formation of a multilayer structure with a polyester-rich skin decreasing significantly the water sensitivity (Avérous & Fringant, 2001).

Various strategies have been followed to induce compatibilization between starch and polyester, such as functionalization of polyester with, e.g., maleic anhydride (Mani & Bhattacharya, 2001; Mani, Bhattacharya, & Tang, 1999), or functionalization of starch with, for instance, poly(glycidyl methacrylate) (Willett, Kotnis, O'Brien, Fanta, & Gordon, 1998). In another work, graft copolymers have been made such as starch-graft polyester or PCL-grafted starch (Mani, Tang, & Bhattacharya, 1998).

Since the 1970s, starch/PVOH blends have been studied primarily for producing films by means of solution casting or calendaring (Lenk & Merrall, 1981; Otey & Mark, 1976; Otey, Mark, Mehltretter, & Russell, 1974; Takahashi & Nakamura, 1970; Young, 1967) and starch-based packaging foams as a replacement for expanded polystyrene (Shogren, Lawton, Tiefenbacher, & Chen, 1998). Previous work has shown that the presence of PVOH improves the strength and flexibility of gelatinized starch. Young (1967) found that films cast from aqueous solutions of amylose or high-amylose starch and PVOH had higher strength and elongation to break at 23% and 50% ERH than films cast from starch alone. Otey and Mark (1976) found similar improvements in the mechanical properties for film cast from mixture solutions of normal (27% amylose) cornstarch and PVOH and also incorporated formaldehyde as a cross-linking agent.

From an economic point of view, solution processing is unlikely to be acceptable for commercial products due to the cost and low efficiency of this process. Melt processing via extrusion using conventional plastic processing equipment is required for these blends to be commercially exploited. Mao, Imam, Gordon, Cinelli, and Chiellini (2000) examined the extrusion of thermoplastic corn starch/glycerol/PVOH blends and noted significant improvements in the mechanical properties for a starch-glycerol blend containing 10% PVOH. A tensile strength of 4 MPa and an elongation 150%, respectively, were achieved, instead of 1.8 MPa and 113% obtained for a starch material plasticized with 30% glycerol. Liu, Feng, and Yi (1999) investigated melt processing (by extrusion) and plasticization of starch/PVOH blends and found that a blend plasticized with a 12% mixture of glycerol and water (50/50 wt%) showed optimum rheological and mechanical properties. The authors found that the incorporation of PVOH in a TPS matrix can enhance the mechanical properties. The improvement is, however, limited due to the poor interface adhesion between the two components.

Park, George, Muldoon, and Flammino (1994) investigated the effect of PVOH hydrolysis level and molecular weight on the properties of high-amylose starch plasticized with glycerol and water processed by twin-screw extrusion. They noted that higher PVOH hydrolysis levels and higher molecular weight led to higher tensile strength. At the same molecular weight, the fully hydrolyzed (98%) PVOH/TPS blend exhibited higher tensile strength than the partially hydrolyzed (88%) blend, for all compositions.

Relatively little information is available on the compatibility, miscibility, and morphology of blends based on starch/PVOH mixtures. Characterization of solution mixtures indicated that the polysaccharide and the synthetic polymer are largely incompatible (Okaya et al., 1992). In corn starch/PVOH composites, Nwufo and Griffin (1985) noticed the presence of starch granules distributed more uniformly within the matrix with an increase in starch content. In the case of starch/PVOH foamed particles (Shogren et al., 1998), SEM observations indicated that the two components were mostly phase-separated with starch granules, which were only gelatinized and immersed in a continuous PVOH phase. However, the improvement of mechanical properties of both cast and blown-extruded films suggested that there is at least a "mechanical compatibility" between starch and PVOH (Otey et al., 1974). Cascone et al. (2001) also reported cast starch/PVOH films behaving mainly as a two-phase system, while significant interactions occur between PVOH and the linear component of starch, amylose.

Chaléat et al. (2012) studied the miscibility and morphology of a range of high-amylose starch/PVOH blends. DMTA analysis displayed two distinct T_gs suggesting a phase-separated system. A small change in the T_g of the PVOH phase with the introduction of the starch suggested a small degree of miscibility. High-resolution solid-state NMR also indicated some interaction between the two phases, while proton spin-lattice relaxation time measurements confirmed a phase-separated microstructure with a small degree of intermixing. This work also showed that in blends of starch and PVOH, the water will partition unequally between the two phases and that the molecular transitions in each phase will respond independently to the overall water content.

In a related paper, the toughness of a high-amylose starch sheet contained <10% of a high-molecular-weight polyol was measured as a function of storage RH (or MC) (Chaléat, Halley, et al., 2008). The toughness

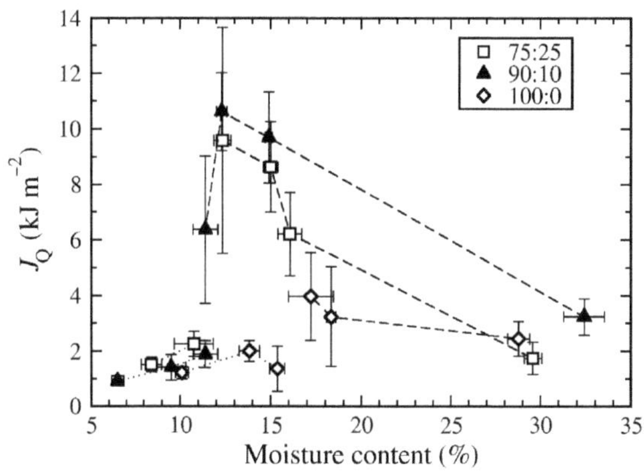

FIGURE 7.8 Variations of the strain energy release rate, J_Q for plane-stress (dashed lines) and mixed-mode conditions (dotted lines) with moisture content for 75:25, 90:10, and 100:0 TPS:PVOH blends (lines are only included to highlight trends) (Chaléat, Halley, et al., 2008).

was measured using a work of fracture technique at high MC and a compact tension geometry at low MC. The results were presented as a nominal strain energy release rate, J_Q, as a function of MC (Figure 7.8). At low MC, the material was quite brittle but with a slightly increasing toughness with increasing MC. There was a large peak in the toughness at ~11-12% moisture after which J_Q decreased as the material became increasingly soft with increasing water content. Examination of the fracture surfaces of these materials in the SEM (Figure 7.9) showed that at low MC, the fracture behavior was quite brittle. At the peak in the levels of J_Q, considerable drawing of the dispersed phase occurred, while at higher MC, the whole of the sample was deforming. The peak in the toughness at 11-13% water corresponded with the $T\alpha$ of the dispersed phase, which meant that the material consisted of a glassy matrix with a dispersed rubbery phase (Chaléat, 2008). The toughness of the blend was thus a form of rubber toughening that is well known in synthetic systems such as high-impact polystyrene, ABS, and rubbery toughened epoxies.

Extruded starch/PVOH blends could be expected to result in more homogeneous dispersions because of the intense mixing and shearing during processing, helping the destructurization of starch granules. SEM micrographs of starch/PVOH blends with 13% PVOH from Liu et al. (1999) showed the starch component forming a continuous phase and PVOH as a discrete phase especially on the surface. In the core area, long PVOH fibers of 18 μm were seen poorly embedded in the starch matrix. Surprisingly, the authors reported a more homogeneous structure for the blends containing 19% PVOH. At high content, PVOH seemed to associate strongly with itself limiting its compatibility with other polymers (Stenhouse, 1993).

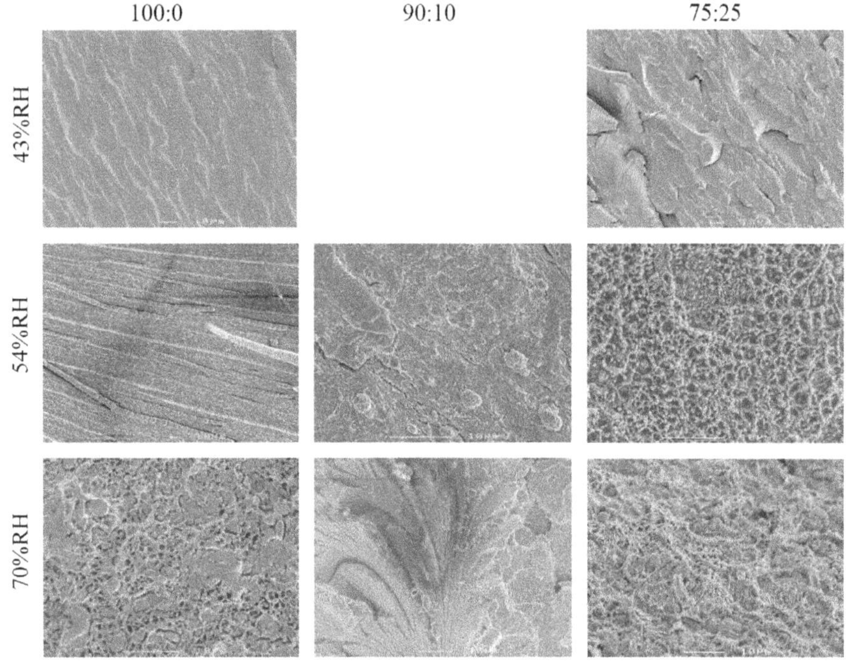

FIGURE 7.9 SEM micrographs of fractured surfaces of 100:0, 90:10, and 75:25 TPS:PVOH blends conditioned at various relative humidities.

6 SUMMARY AND PERSPECTIVES

Many of the properties of starch-based plastics follow the trends that are now well established for synthetic polymers. For example, yield strength and modulus increase with crystallinity. Plasticizers reduce the T_g and lower the modulus and strength. Toughness is increased by high-molecular-weight linear molecules. What sets starch-based plastics apart from most high-volume synthetic polymers is the nature of the natural starting material. The multilevel structure in the native starch granules must be destroyed to give a homogeneous melt during processing, and if this is incomplete, the mechanical properties will be variable and poor. Moreover, the amorphous form of the solid thermoplastic starch is not the thermodynamically stable form, and there is a natural tendency for the starch molecules to reestablish some of the ordered and crystalline structure of its native state. This behavior is no different to a semicrystalline synthetic polymer that has been rapidly cooled, which can also recrystallize if the temperature is above the T_g. For TPS materials, the plasticizers, including water absorbed from the atmosphere, can reduce the T_g of starch sufficiently for this to occur at normal storage conditions.

Water content also acts as an additional variable to temperature and strain rate in determining the solid-state mechanical properties. Just as there is a well-known time/temperature equivalence for properties of synthetic polymer, an MC/time or MC/temperature equivalence exists for starch materials.

Thus, the mechanical properties of starch-based plastics are highly depended on the source of the starch, the processing and aging conditions, and the test conditions. Consequently, any reporting of mechanical properties must specify these conditions for them to have any worth in comparing with other starch-based products.

Future direction for starch-based polymers will be directed at overcoming the major issues regarding the water sensitivity of their properties. This is already occurring in the formulation of blends and layered materials. The challenges will be to maintain a sustainable and eco-friendly product. Other approaches will be to chemically modify the starch to make it inherently more hydrophobic. This will pose challenges both in processing the modified starch and in avoiding the production of a brittle material.

Acknowledgments

We gratefully acknowledge over 40 researchers who have worked in our laboratories on this work since 1995 and to the funding from the Cooperative Research Centre for International Food Manufacture and Packaging Science, the Australian Research Council, the University of Queensland, Plantic Technologies Limited, and the Cooperative Research Centre for Polymers.

References

Avérous, L. (2004). Biodegradable multiphase systems based on plasticized starch: A review. *Journal of Macromolecular Science, Polymer Reviews, C44*, 231–274.

Avérous, L., Fauconnier, N., Moro, L., & Fringant, C. (2000). Blends of thermoplastic starch and polyesteramide: Processing and properties. *Journal of Applied Polymer Science, 76*, 1117–1128.

Avérous, L., & Fringant, C. (2001). Association between plasticized starch and polyesters: Processing and performances of injected biodegradable systems. *Polymer Engineering and Science, 41*, 727–734.

Avérous, L., Moro, L., Dole, P., & Fringant, C. (2000). Properties of thermoplastic blends: Starch-polycaprolactone. *Polymer, 41*, 4157–4167.

Bader, H. G., & Dietmar Goritz, D. (1994). Investigations on high amylose corn starch films. Part 3: Stress strain behaviour. *Starch/Stärke, 46*, 435–439.

Ball, S., Guan, H. P., James, M., Myers, A., Keeling, P., Mouille, G., et al. (1996). From glycogen to amylopectin: A model for the biogenesis of the plant starch granule. *Cell, 86*, 349–352.

Bastioli, C., Cerutti, A., Guanella, I., Romano, G. C., & Tosin, M. (1995). Physical state and biodegradation behavior of starch-polycaprolactone systems. *Journal of Environmental Polymer Degradation, 3*, 81–95.

Bauwens-Crowet, C., Bauwens, J. C., & Homès, G. (1969). Tensile yield stress behavior of glassy polymers. *Journal of Polymer Science, Part A: Polymer Chemistry, 7*, 735–742.

Bradley, S. A., & Carr, S. H. (1976). Mechanical loss processes in polysaccharides. *Journal of Polymer Science, Part B: Polymer Physics, 14*, 111–124.

Buléon, A., Colonna, P., Planchot, V., & Ball, S. (1998). Starch granules: Structure and biosynthesis. *International Journal of Biological Macromolecules, 23*, 85–112.

Cascone, M. G., Barbani, N., Cristallini, C., Giusti, P., Ciardelli, G., & Lazzeri, L. (2001). Bioartificial polymeric materials based on polysaccharides. *Journal of Biomaterials Science-Polymer Edition, 12*, 267–281.

Chaléat, C. M. (2008). Structure–property relationships of plasticised starch/poly (vinyl alcohol) blends, Ph D Thesis, The University of Queensland.

Chaléat, C. M., Halley, P. J., & Truss, R. W. (2008). Properties of a plasticised starch blend. Part 1: Influence of moisture content on fracture properties. *Carbohydrate Polymers, 71*, 535–543.

Chaléat, C. M., Halley, P. J., & Truss, R. W. (2012). Study on the phase separation of plasticised starch/poly(vinyl alcohol) blends. *Polymer Degradation and Stability, 97*, 1930–1939.

Chaléat, C. M., Michel-Amadry, G., Halley, P. J., & Truss, R. W. (2008). Properties of a plasticised starch blend—Part 2: Influence of strain rate, temperature and moisture on the tensile yield behaviour. *Carbohydrate Polymers, 74*, 366–371.

Chang, Y. P., Cheah, P. B., & Seow, C. C. (2000). Plasticizing-antiplasticizing effects of water on physical properties of tapioca starch films in the glassy state. *Journal of Food Science, 65*, 445–451.

De Graaf, R. A., Karman, A. P., & Janssen, L. (2003). Material properties and glass transition temperatures of different thermoplastic starches after extrusion processing. *Starch/Stärke, 55*, 80–86.

Donovan, J. W. (1979). Phase transitions of the starch-water system. *Biopolymers, 18*, 263–275.

Eyring, H. (1936). Viscosity, plasticity, and diffusion as examples of absolute reaction rates. *Journal of Chemical Physics, 4*, 283–291.

Fishman, M. L., Coffin, D. R., Konstance, R. P., & Onwulata, C. I. (2000). Extrusion of pectin/starch blends plasticized with glycerol. *Carbohydrate Polymers, 41*, 317–325.

Forssell, P., Mikkila, J., Suortti, T., Seppala, J., & Poutanen, K. (1996). Plasticization of barley starch with glycerol and water. *Journal of Macromolecular Science, Pure and Applied Chemistry, A33*, 703–715.

Gaudin, S., Lourdin, D., Forssell, P. M., & Colonna, P. (2000). Antiplasticisation and oxygen permeability of starch-sorbitol films. *Carbohydrate Polymers, 43*, 33–37.

Gaudin, S., Lourdin, D., Le Botlan, D., Ilari, J. L., & Colonna, P. (1999). Plasticisation and mobility in starch-sorbitol films. *Journal of Cereal Science, 29*, 273–284.

George, E. R., Sullivan, T. M., & Park, E. H. (1994). Thermoplastic starch blends with a poly(ethylene-co-vinyl alcohol). Processability and physical properties. *Polymer Engineering and Science, 34*, 17–23.

Halley, P. J. (2005). Thermoplastic starch biodegradable polymers. In Ray Smith (Ed.), *Biodegradable polymers for industrial applications* (pp. 139–162). Boca Raton, Cambridge: CRC Press, Woodhead Publishing Ltd.

Hulleman, S. H. D., Janssen, F. H. P., & Feil, H. (1998). The role of water during plasticization of native starches. *Polymer, 39*, 2043–2048.

Jackson, W. J., & Caldwell, J. R. (1967a). Antiplasticization II. Characteristics of antiplasticizers. *Journal of Applied Polymer Science, 11*, 211–226.

Jackson, W. J., & Caldwell, J. R. (1967b). Antiplasticization III. Characteristics and properties of antiplasticizable polymers. *Journal of Applied Polymer Science, 11*, 227–244.

Jenkins, P. J., & Donald, A. M. (1995). The influence of amylose on starch granule structure. *International Journal of Biological Macromolecules, 17*, 315–321.

Kalichevsky, M. T., Jaroszkiewicz, E. M., Ablett, S., Blanshard, J. M. V., & Lillford, P. J. (1992). The glass-transition of amylopectin measured by Dsc, Dmta and Nmr. *Carbohydrate Polymers, 18*, 77–88.

Kalichevsky, M. T., Jaroszkiewicz, E. M., & Blanshard, J. M. V. (1993). A study of the glass transition of amylopectin sugar mixtures. *Polymer, 34,* 346–358.
Koenig, M. F., & Huang, S. J. (1995). Biodegradable blends and composites of polycaprolactone and starch derivatives. *Polymer, 36,* 1877–1882.
Lenk, R. S., & Merrall, R. E. (1981). Starch-poly(vinyl alcohol) heat-sealable film. 2. *Polymer, 22,* 1279–1282.
Liu, Z. Q., Feng, Y., & Yi, X. S. (1999). Thermoplastic starch/PVAl compounds: Preparation, processing, and properties. *Journal of Applied Polymer Science, 74,* 2667–2673.
Lourdin, D., Bizot, H., & Colonna, P. (1997a). "'Antiplasticization" in starch-glycerol films? *Journal of Applied Polymer Science, 63,* 1047–1053.
Lourdin, D., Bizot, H., & Colonna, P. (1997b). Correlation between static mechanical properties of starch-glycerol materials and low-temperature relaxation. *Macromolecular Symposia, 114,* 179–185.
Lourdin, D., Coignard, L., Bizot, H., & Colonna, P. (1997). Influence of equilibrium relative humidity and plasticizer concentration on the water content and glass transition of starch materials. *Polymer, 38,* 5401–5406.
Lourdin, D., Colonna, P., Brownsey, G. J., Noel, T. R., & Ring, S. G. (2002). Structural relaxation and physical ageing of starchy materials. *Carbohydrate Research, 337,* 827–833.
Lourdin, D., Dellavalle, G., & Colonna, P. (1995). Influence of amylose content on starch films and foams. *Carbohydrate Polymers, 27,* 261–270.
Lourdin, D., Ring, S. G., & Colonna, P. (1998). Study of plasticizer-oligomer and plasticizer-polymer interactions by dielectric analysis: Maltose-glycerol and amylose-glycerol-water systems. *Carbohydrate Research, 306,* 551–558.
Mani, R., & Bhattacharya, M. (2001). Properties of injection moulded blends of starch and modified biodegradable polyesters. *European Polymer Journal, 37,* 515–526.
Mani, R., Bhattacharya, M., & Tang, J. (1999). Functionalization of polyesters with maleic anhydride by reactive extrusion. *Journal of Polymer Science, Part A: Polymer Chemistry, 37,* 1693–1702.
Mani, R., Tang, J., & Bhattacharya, M. (1998). Synthesis and characterization of starchgraft-polycaprolactone as compatibilizer for starch/polycaprolactone blends. *Macromolecular Rapid Communications, 19,* 283–286.
Mao, L. J., Imam, S., Gordon, S., Cinelli, P., & Chiellini, E. (2000). Extruded cornstarch-glycerol-polyvinyl alcohol blends: Mechanical properties, morphology, and biodegradability. *Journal of Polymers and the Environment, 8,* 205–211.
Martin, O., & Avérous, L. (2001). Poly(lactic acid): Plasticization and properties of biodegradable multiphase systems. *Polymer, 42,* 6209–6219.
Mathew, A. P., & Dufresne, A. (2002). Plasticized waxy maize starch: Effect of polyols and relative humidity on material properties. *Biomacromolecules, 3,* 1101–1108.
Matzinos, P., Tserki, V., Kontoyiannis, A., & Panayiotou, C. (2002). Processing and characterization of starch/polycaprolactone products. *Polymer Degradation and Stability, 77,* 17–24.
Mercier, C., Charbonniere, R., Grebaut, J., & Gueriviere, J. (1980). Formation of amylose-lipid complexes by twin-screw extrusion cooking of Manioc starch. *Cereal Chemistry, 57,* 4–9.
Myllarinen, P., Partanen, R., Seppala, J., & Forssell, P. (2002). Effect of glycerol on behaviour of amylose and amylopectin films. *Carbohydrate Polymers, 50,* 355–361.
Myllymäki, O., Myllärinen, P., Forssell, P., Suortti, T., Lähteenkorva, K., Ahvenainen, R., et al. (1998). Mechanical and permeability properties of biodegradable extruded starch/polycaprolactone films. *Packaging Technology & Science, 11,* 265–274.
Nishinari, K. (1980). Viscoelastic, dielectric, and piezoelectric behavior of solid amylose. *Journal of Polymer Science, Part B: Polymer Physics, 18,* 1609–1619.
Nwufo, B. T., & Griffin, G. J. L. (1985). Microscopic appearance of fractured surfaces and mechanical properties of starch extended poly(vinyl alcohol). *Journal of Polymer Science, Part A: Polymer Chemistry, 23,* 2023–2031.

Okaya, T., Kohno, H., Terada, K., Sato, T., Maruyama, H., & Yamauchi, J. (1992). Specific interaction of starch and polyvinyl alcohols having long alkyl-groups. *Journal of Applied Polymer Science*, *45*, 1127–1134.

Otaigbe, J. U., Goel, H., Babcock, T., & Jane, J. (1999). Processability and properties of biodegradable plastics made from agricultural biopolymers. *Journal of Elastomers and Plastics*, *31*, 56–71.

Otey, F. H. & Mark, A. M. (1976). Degradable starch-based agricultural mulch films. U.S. Patent, 3,949,145, 1976.

Otey, F. H., Mark, A. M., Mehltretter, C. L., & Russell, C. R. (1974). Starch-based film for degradable agricultural mulch. *Industrial and Engineering Chemistry Product Research and Development*, *13*, 90–92.

Otey, F. H., Westhoff, R. P., & Doane, W. M. (1980). Starch-based blown films. *Industrial and Engineering Chemistry Product Research and Development*, *19*, 592–595.

Otey, F. H., Westhoff, R. P., & Doane, W. M. (1987). Starch-based blown films. *Industrial and Engineering Chemistry Product Research and Development*, *26*, 1659–1663.

Park, E. H., George, E. R., Muldoon, M. A., & Flammino, A. (1994). Thermoplastic starch blends with poly(vinyl alcohol): Processability, physical properties, and biodegradability. *Polymer News*, *19*, 230–238.

Ramsay, B. A., Langlade, V., Carreau, P. J., & Ramsay, J. A. (1993). Biodegradability and mechanical properties of poly-(β-hydroxybutyrate-co-β-hydroxyvalerate) starch blends. *Applied and Environmental Microbiology*, *59*, 1242–1246.

Ree, T., & Eyring, H. (1955). Theory of non-Newtonian flow. I. Solid plastic system. *Journal of Applied Physics*, *26*, 793–800.

Rindlav, A., Hulleman, S. H. D., & Gatenholm, P. (1997). Formation of starch films with varying crystallinity. *Carbohydrate Polymers*, *34*, 25–30.

Rindlav-Westling, A., Stading, M., & Gatenholm, P. (2002). Crystallinity and morphology in films of starch, amylose and amylopectin blends. *Biomacromolecules*, *3*, 84–91.

Shogren, R. L. (1992). Effect of moisture content on the melting and subsequent physical aging of cornstarch. *Carbohydrate Polymers*, *19*, 83–90.

Shogren, R. (1998). Starch: Properties and materials applications. In D. Kaplan (Ed.), *Biopolymers from renewable resources*. Berlin, New York: Springer.

Shogren, R. L., Lawton, J. W., Tiefenbacher, K. F., & Chen, L. (1998). Starch poly(vinyl alcohol) foamed articles prepared by a baking process. *Journal of Applied Polymer Science*, *68*, 2129.

Shogren, R. L., & Jasberg, B. K. (1994). Aging properties of extruded high-amylose starch. *Journal of Environmental Polymer Degradation*, *2*, 99–109.

Shogren, R. L., Swanson, C. L., & Thompson, A. R. (1992). Extrudates of cornstarch with urea and glycols—Structure mechanical property relations. *Starch/Stärke*, *44*, 335–338.

Stenhouse, P. J. (1993). Starch-based blown films. In C. Ching, D. Kaplan, & E. Thomas (Eds.), *Biodegradable polymers and packaging* (pp. 151–158). Lancaster: Technomic Publishing Company.

Takahashi, R., & Nakamura, A. (1970). Properties of starch films and their improvement. *Starch/Stärke*, *22*, 309.

Truss, R. W., Clarke, P. L., Duckett, R. A., & Ward, I. M. (1984). The dependence of yield behavior on temperature, pressure, and strain rate for linear polyethylenes of different molecular-weight and morphology. *Journal of Polymer Science, Part B: Polymer Physics*, *22*, 191–209.

Van Soest, J. J. G. (1996). Starch plastics: Structure–property relationships, PhD thesis, Utrecht University.

Van Soest, J. J. G., Benes, K., Wit, D. D., & Vliegenthart, J. F. G. (1996). The influence of starch molecular mass on the properties of extruded thermoplastic starch. *Polymer*, *37*, 3543–3552.

Van Soest, J. J. G., & Borger, D. B. (1997). Structure and properties of compression-molded thermoplastic starch materials from normal and high-amylose maize starches. *Journal of Applied Polymer Science, 64*, 631–644.
Van Soest, J. J. G., De Wit, D., Tournois, H., & Vliegenthart, J. F. G. (1994). The influence of glycerol on structural changes in waxy maize starch as studied by Fourier transform infrared spectroscopy. *Polymer, 35*, 4722–4727.
Van Soest, J. J. G., Dewit, D., Tournois, H., & Vliegenthart, J. F. G. (1994). Retrogradation of potato starch as studied by Fourier-Transform Infrared-Spectroscopy. *Starch/Stärke, 46*, 453–457.
Van Soest, J. J. G., & Essers, P. (1997). Influence of amylose-amylopectin ratio on properties of extruded starch plastic sheets. *Journal of Macromolecular Science, Pure and Applied Chemistry, A34*, 1665–1689.
Van Soest, J. J. G., Hulleman, S. H. D., De Wit, D., & Vliegenthart, J. F. G. (1996). Crystallinity in starch bioplastics. *Industrial Crops and Products, 5*, 11–22.
Van Soest, J. J. G., Hulleman, S. H. D., Dewit, D., & Vliegenthart, J. F. G. (1996). Changes in the mechanical properties of thermoplastic potato starch in relation with changes in B-type crystallinity. *Carbohydrate Polymers, 29*, 225–232.
Van Soest, J. J. G., & Knooren, N. (1997). Influence of glycerol and water content on the structure and properties of extruded starch plastic sheets during aging. *Journal of Applied Polymer Science, 64*, 1411–1422.
Verhoogt, H., Stpierre, N., Truchon, F. S., Ramsay, B. A., Favis, B. D., & Ramsay, J. A. (1995). Blends containing poly(hydroxybutyrate-co-12%-hydroxyvalerate) and thermoplastic starch. *Canadian Journal of Microbiology, 41*, 323–328.
Vikman, M., Hulleman, S. H. D., Van der Zee, M., Myllarinen, P., & Feil, H. (1999). Morphology and enzymatic degradation of thermoplastic starch-polycaprolactone blends. *Journal of Applied Polymer Science, 74*, 2594–2604.
Walia, P. S., Lawton, J. W., & Shogren, R. L. (2002). Mechanical properties of thermoplastic starch/poly(hydroxy ester ether) blends: Effect of moisture during and after processing. *Journal of Applied Polymer Science, 84*, 121–131.
Walia, P. S., Lawton, J. W., Shogren, R. L., & Felker, F. C. (2000). Effect of moisture level on the morphology and melt flow behavior of thermoplastic starch/poly(hydroxy ester ether) blends. *Polymer, 41*, 8083–8093.
Willett, J. L., Kotnis, M. A., O'Brien, G. S., Fanta, G. F., & Gordon, S. H. (1998). Properties of starch-graft-poly(glycidyl methacrylate) PHBV composites. *Journal of Applied Polymer Science, 70*, 1121–1127.
Young, A. H. (1967). Polyvinyl alcohol plasticised amylose compositions, U.S. Patent, 1967.
Zeleznak, K. J., & Hoseney, R. C. (1987). The glass-transition in starch. *Cereal Chemistry, 64*, 121–124.
Zobel, H. F. (1988). Molecules to granules: A comprehensive starch review. *Starch/Stäke, 40*, 44–50.

CHAPTER

8

Nanobiocomposites Based on Plasticized Starch

Luc Avérous, Eric Pollet

BioTeam/ICPEES-ECPM, Université de Strasbourg, Strasbourg Cedex 2, France

1 PLASTICIZED STARCH

1.1 Generalities

Except for applications as fillers to produce reinforced plastics (Shah, Bandopadhyay, & Bellare, 1995), native starch is typically chemically and/or physically modified and used as destructured starch for food or nonfood applications. Starch is gelatinized with heat combined with high water content, which is the destructuring agent. Gelatinization causes the disruption of the granule organization and the starch swells, forming a viscous paste with destruction of most of inter-macromolecule hydrogen links and a reduction of the melting and the glass transition temperature (T_g). Depending on the level of destructuration (thermomechanical input associated with the water content), one can obtain different products and applications. For instance, we can achieve expanded structures at a medium water content and medium destructuring input. Such closed cells structures (foams) have been developed to obtain shock absorbable (e.g., loose fill) and isothermal packaging. At high destructuring level and low water content, we obtain plasticized starch (PLS) (Shogren, 1992). The first patents and articles on PLS were published at the end of the 1980s (Averous, 2004; Avérous & Halley, 2009).

Most starch applications require water addition and partial or complete gelatinization. By decreasing the moisture content (to less than 20 wt%), the melting temperature tends to be close to the degradation temperature. For instance, the melting temperature of pure and dry starch is 220-240 °C compared to the temperature of the beginning of starch decomposition, 220 °C

(Russell, 1987). To overcome this issue, nonvolatile (at the process temperature) plasticizers such as polyols are added to decrease the melting and process temperature (Gaudin, Lourdin, Le Botlan, Ilari, & Colonna, 1999; Mathew & Dufresne, 2002b). A mixture of different polyols can be utilized, e.g., glycerol with sorbitol (Chivrac, Pollet, Dole, & Avérous, 2010). Other compounds such as those containing nitrogen (urea, ammonium derived, and amines) can also be used (Ma, Yu, & Jin, 2004; Ma, Yu, & Ma, 2005; Ma, Yu, & Wan, 2006; Shogren, Swanson, & Thompson, 1992).

1.2 Process

The disruption of granular starch is the transformation of the semicrystalline granule into a homogeneous, mainly amorphous, material with a partial destruction of inter-macromolecular hydrogen bonds. Starch disruption can be obtained either by solvent-casting process or by a molten-state processing, where starch and the plasticizers are mixed under thermomechanical treatment.

Solvent-assisted starch granule disruption is mostly carried out with water. At ambient temperature, starch remains insoluble in water and keeps its granular structure. A water temperature increase induces an irreversible swelling named "gelatinization." During this gelatinization, the amylose is rather solubilized, the granule semicrystalline structure disappears, and the granules swell rapidly. This phenomenon occurs at a given temperature defined as "gelatinization temperature," which depends on the starch botanical origin (Cooke & Gidley, 1992; Genkina, Wikman, Bertoft, & Yuryev, 2007; Stevens & Elton, 1971). To obtain a quite full starch solubilization, hot DMSO is often used as solvent. Then, this solvent is volatilized under vacuum and heat.

Molten-state processing can be done in one or two stages. In a one-stage process, usually a twin-screw extruder is fed with native starch. Along the barrel, water and plasticizer are then successively introduced. The water in excess is often eliminated before the extruder exit by vacuum venting to avoid the formation of a foam structure at the die exit. The starch is fully destructured and melted. In a two-stage process (Figure 8.1), the first stage is a dry-blend preparation (Averous, Fauconnier, Moro, & Fringant, 2000; Averous, Moro, Dole, & Fringant, 2000). For example, in a turbo mixer, under high speed, the plasticizer is slowly added into the native starch. Then, the mixture is placed in a vented oven, allowing the diffusion of the plasticizer into the granule, where the plasticizer swells the starch. After cooling, the right amount of water is added to the mixture using a turbo mixer. This dry blend is then introduced into an extruder. During the extrusion, the starch granules are initially fragmented, and then under temperature and shearing, starch is destructured, plasticized, and melted.

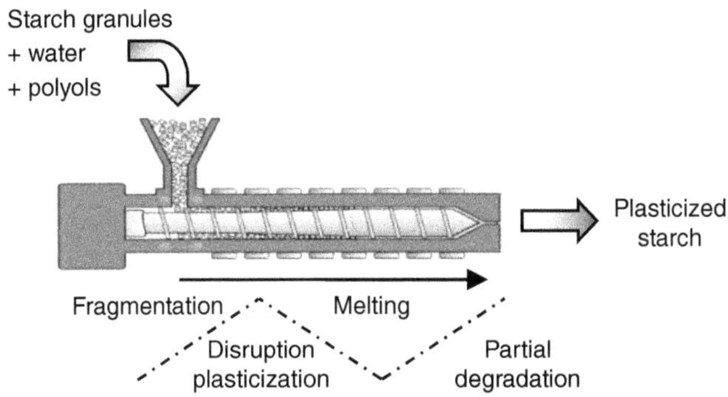

FIGURE 8.1 Schematic representation of the starch extrusion process. (For color version of this figure, the reader is referred to the online version of this chapter.)

Under the thermomechanical input, starch chains are also partially depolymerized. At the end, a homogeneous molten phase is obtained (Averous, 2004).

The starch granule disruption being dependent on the specific mechanical energy (SME) provided during the process, this material could be described as a "thermomechanical plastic" material (Martin, Averous, & Della Valle, 2003b; Ollett, Parker, Smith, Miles, & Morris, 1990). During the process, amylose and amylopectin degradation occurs and this phenomenon is obviously dependent on the thermal and mechanical energy brought to the system. Numerous studies were conducted to determine the degradation mechanism and to understand the contribution of each parameter (Baud, Colonna, Della Valle, & Roger, 1999; Davidson, Parker, Diosady, & Rubin, 1984; Davidson, Paton, Diosady, & Larocque, 1984; Della Valle, Boche, Colonna, & Vergnes, 1995; Orford, Parker, & Ring, 1993; Sagar & Merrill, 1995; Vergnes, Villemaire, Colonna, & Tayeb, 1987; Wang, Chiang, Yeh, Zhao, & Kim, 1989).

1.3 Properties

1.3.1 Postprocessing Aging

Different authors have shown that after processing (Avérous, Fringant, & Moro, 2001; Doungjai & Sanguansri, 2007; Shogren, 1992; Thirathumthavorn & Charoenrein, 2007; van Soest & Knooren, 1997), PLS exhibits aging with, for instance, a strong effect on the mechanical properties such as the increase of Young's modulus over several weeks of aging. PLS presents two kinds of aging behavior depending on T_g value. In the sub-T_g domain, PLS shows a physical aging versus time, with a material densification. At a temperature above T_g, PLS presents retrogradation phenomena with the evolution of the crystallinity and rearrangements of plasticizer molecules into the material as a function of storage time (van Soest & Knooren, 1997). The retrogradation kinetics depend on the macromolecules mobility and the plasticizer type and content (Smits, Hulleman, van Soest, Feil, & Vliegenthart, 1999).

1.3.2 Crystallinity

Compared to native starch, PLS shows a reduced crystallinity with a strong modification of the types and the organization of the crystallinity. According to van Soest, Hulleman, Dewit, and Vliegenthart (1996a, 1996b), two kinds of crystallinity are obtained after PLS processing, (i) residual crystallinity from native starch (A, B, and C types) and (ii) processing-induced crystallinity (V and E types). Crystallinity induced by processing is influenced by parameters such as the extrusion residence time, the screw speed, and the process temperature. It is mainly caused by the fast recrystallization of amylose into single-helical structures. After postprocessing aging, van Soest et al. proposed a complex model for crystallization of PLS with amorphous and crystalline amylose and amylopectin and probable cocrystallization between amylose and amylopectin (van Soest & Knooren, 1997).

1.3.3 Plasticizers Interactions

The evolution of the moisture content impacts the properties and the transition temperatures of PLS, such as the T_g, because water acts also as a plasticizer (Hulleman, Janssen, & Feil, 1998). Water is a volatile plasticizer that is equilibrated by sorption-desorption with the environment. Lourdin et al. and Mathew and Dufresne have determined the moisture content after equilibrium on starch as a function of the relative humidity and the nature and content of the plasticizer (Lourdin, Coignard, Bizot, & Colonna, 1997; Mathew & Dufresne, 2002b). Godbillot et al. have proposed a phase diagram for a water/glycerol/starch system (Godbillot, Dole, Joly, Roge, & Mathlouthi, 2006), as a function of the plasticizer content. These authors highlighted the different interactions taking place in these multiphase systems (Figure 8.2). Depending on the glycerol content and the relative humidity, the plasticizer can be more or less linked with the polysaccharide chains. The plasticizer could occupy specific sorption sites when the water and plasticizer content are low or become free when the sorption sites are fulfilled, at high relative humidity and glycerol content. Between these two limiting cases, complex interactions between the different components could be established. Thus, in certain conditions, a phase separation occurs, generating multiphase structures with rich and poor plasticizer domains (Lourdin et al., 1997; Lourdin, Ring, & Colonna, 1998).

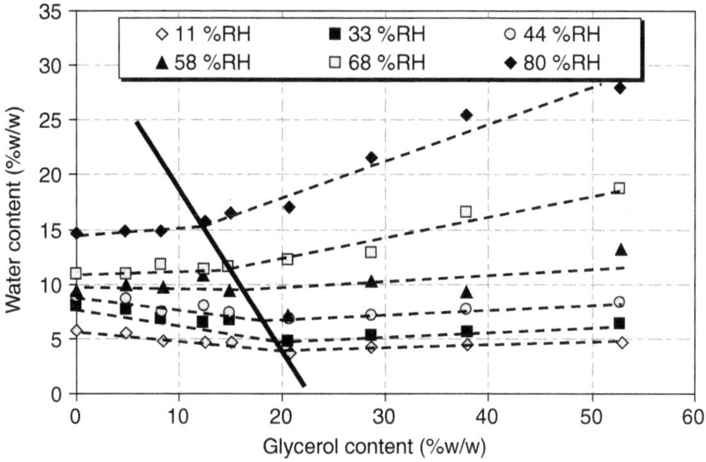

FIGURE 8.2 Water content at equilibrium as a function of glycerol content and relative humidity (%RH). *Source: Godbillot et al. (2006).*

1.3.4 Physical Properties

Compared to native starch, PLS shows some particular properties. Depending on the plasticizer/dry starch ratio, PLS presents a wide range of properties. Table 8.1 shows that increasing "glycerol/dry starch" ratios modify different properties, including physical and mechanical behaviors.

In some cases, the Cp variations at glass transitions are very low and difficult to be determined via DSC, thus DMTA determinations are often more desirable to obtain PLS glass transitions. By DMTA, the broad and main relaxation (α transition) can be linked to the PLS glass transition (Averous, Fauconnier, Moro, et al., 2000; Averous, Moro, et al., 2000). The secondary relaxation (between −50 and −60 °C) could be connected to the glass transition of glycerol (Averous, Fauconnier, et al., 2000; Averous, Moro, et al., 2000). According to Lourdin et al. (1997), this latter relaxation could be an indicator of the level of interactions between the plasticizer and the polysaccharides. The relaxation temperature decreases when glycerol content increases along with the phase segregation.

Table 8.1 shows that we can achieve a large range of mechanical properties and permeability (moisture and oxygen) by variation of the plasticizer content (Gaudin et al., 1999a, 1999b; Lourdin et al., 1997). Although in PLS the water permeability is high due to its polar character, note the oxygen permeability is found lower than most polyesters (van Tuil, Fowler, Lawther, & Weber, 2000). Thus, there appears to be potential for the use of PLS as an oxygen barrier material in multilayer structures (Dole, Avérous, Joly, Della Valle, & Bliard, 2005), to try to replace, e.g., EVOH. With increasing plasticizer content, the permeability increases drastically at the glass transition and continues to rise through the rubber plateau.

TABLE 8.1 Plasticized Wheat Starch—Formulations and Physical Properties Determined After Equilibrium at 23 °C and 50% RH, 6 weeks

	Glycerol/Starch Ratio [a](w/w)	**Water (wt%)**	**Density**	**T_g DSC (°C)**	**α Transition DMTA (C°)**	**Modulus (MPa)**	**Max. Tensile Strength (MPa)**	**Elongation at Break (%)**
S74G10W10[a]	0.14	9	1.39	43	63	1144 ± 42	21.4 ± 5.2	3 ± 0
S75G18W12[a]	0.25	9	1.37	8	31	116 ± 11	4.0 ± 1.7	104 ± 5
S67G24W9[a]	0.35	12	1.35	−7	17	45 ± 5	3.3 ± 0.1	98 ± 5
S65G35W0[a]	0.50	13	1.34	−20	1	11 ± 1	1.4 ± 0.1	60 ± 5

[a] *Initial PLS formulation (SxGyWz): S, starch (x %wt); G, glycerol (y %wt); W, water(z %wt).*
Source: Avérous and Halley (2009).

To understand the behavior of PLS during processing, the rheological behavior of this material has been studied in the molten state. A previous publication has shown the high dependency of the SME on the PLS viscosity (Martin, Averous, & Della Valle, 2003a).

1.4 Plasticized Starch: Issues and Strategies

As a renewable material, PLS shows strong end-use attributes like total compostability without toxic residues. Besides, compared to synthetic thermoplastics, it is a rather cheap material. Compared to fossil resources, the price of starch resources remains currently rather stable. In addition, PLS can be easily processed with plastic processing machines and does not need the development of special equipment. Also, depending on the plasticizer level and the starch botanical source, a wide properties' range may be obtained. However, unfortunately, PLS shows high moisture sensitivity and rather weak mechanical properties compared to conventional synthetic polymer, and these latter are important properties for more widespread polymer applications.

To overcome these weaknesses, different strategies have been examined such as chemical modification of starch. This approach has been carried out since the first half of the 1970s focused on modified starches and cellulose. For example, starch esterification (e.g., by acetylation) improves the water resistance of this material (Fringant, Desbrières, & Rinaudo, 1996; Fringant, Rinaudo, Foray, & Bardet, 1998). In addition, by adjusting the degree of substitution (between 0 and 3), a controlled hydrophobic character can be obtained. Additionally, these modified starches can be plasticized (e.g., with ester citrate). However, the strategy of chemical modification is strongly limited as far as potential toxicity and diversity of by-products obtained during the chemical reactions are concerned. Another limitation lies in the cost of the additional process stages for modification and product purification (to eliminate the by-products). Moreover, the chemical reactions often lead to the decrease in polysaccharide molecular weight due to chain scission. Consequently, the final properties are altered. However, a more promising strategy has been developed that consists in the association of PLS with other compounds to obtain compostable multiphase materials. Additionally, depending on the processing conditions, we can obtain different structures for compostable multiphase materials, such as blends with, e.g., biodegradable polyesters (PCL, PLA, and PHA) or biocomposites (biodegradable composites) (Averous, 2004; Averous & Boquillon, 2004; Averous, Fauconnier, et al., 2000; Avérous & Fringant, 2001; Avérous et al., 2001; Avérous & Le Digabel, 2006; Averous, Moro, et al., 2000). This approach, mainly based on materials formulation, induces some issues linked with the final nano-/microstructure, the quality of the interfaces/interphases, the continuity, and the compatibility between the phases, which need to be controlled.

2 STARCH-BASED NANOBIOCOMPOSITES

2.1 From Macro- to Nanobiocomposites

Biocomposites are a special class of composite materials. They are obtained by blending biodegradable polymers with fillers (e.g., lignocellulose fibers). Tailoring new composites within a perspective of ecodesign or sustainable development is a philosophy that is applied to more and more materials. Depending on the filler size, we can obtain macrobiocomposite or nanobiocomposites. For this latter form, the filler is submicron in size and controls the matrix nanostructure.

To elaborated macro-/microcomposites, various types of (ligno-) cellulose fibers or microfibrils were tested in association with PLS such as microfibrils from potato pulp (Dufresne, Dupeyre, & Vignon, 2000), bleached leafwood fibers (Averous, 2007; Averous & Boquillon, 2004; Avérous et al., 2001; Funke, Bergthaller, & Lindhauer, 1998), and fibers from bleached eucalyptus pulp (Curvelo, De Carvalho, & Agnelli, 2001). These different authors have shown high compatibility between starch and these fibers. Different types of fractioned or modified lignins (e.g., Kraft lignins) have also been tested in association with PLS. Mineral microfillers were tested into a PLS matrix (De Carvalho, Curvelo, & Agnelli, 2001). Kaolin particles, with micron size, were incorporated by extrusion. Due to a significant compatibility between matrix and filler, subsequent behavior such as a glass transition increase, a reduction of water uptake, and an increase of the stiffness can be observed.

Nanobiocomposites are organic/inorganic hybrid biomaterials composed of nanosized fillers (nanofillers) incorporated into a biopolymer matrix (Bordes, Pollet, & Avérous, 2009; Chivrac, Pollet, & Averous, 2009). Depending on the chosen nanofiller, the nanocomposite materials could exhibit drastic modifications in their properties, like improved mechanical properties, higher barrier properties, or change in their thermal and electrical conductivity (Alexandre & Dubois, 2000). Such properties' enhancements rely on the nanofiller geometry, the nanofiller surface area (e.g., 700 m^2/g for the montmorillonite (MMT) when the nanofiller is fully exfoliated), and nanofiller surface chemistry. According to the literature, three main types of nanoparticles have been incorporated into PLS nanocomposites, (i) whiskers obtained from cellulose, (ii) nanocrystals from starch, and (iii) natural or organomodified nanoclays (Chivrac et al., 2009).

2.2 Nanobiocomposites with Polysaccharides Nanofillers

2.2.1 *Whisker-Based Nanobiocomposites*

Through a long pretreatment, whiskers can be isolated from their original biomass through acid hydrolysis with concentrated mineral acids under strictly controlled conditions of time and temperature

(Azizi Samir, Alloin, & Dufresne, 2005; Azizi Samir, Alloin, Sanchez, El Kissi, & Dufresne, 2004). Acid action results in a decrease of the amorphous parts by removing polysaccharide material closely bonded to the crystallite surface and breaks down portions of glucose chains in most accessible, noncrystalline regions and by acid hydrolysis of cellulosic materials. Although chitin can be used, whiskers are typically cellulose monocrystals. Some authors used tunicin (seafood cellulose) whiskers (Angles & Dufresne, 2000, 2001; Mathew & Dufresne, 2002a; Mathew, Thielemans, & Dufresne, 2008), which are slender parallelepiped rods of 500 nm to 1-2 μm length and 10 nm width, into PLS matrixes. The whisker-matrix interactions are important and the high aspect ratio of the nanoparticles (50-200) and the high specific area ($\approx 170\ m^2/g$) increase the interfacial phenomena. Compared to the common PLS-cellulose macrocomposites, the global behavior of nanowhisker-based material is primarily driven by the matrix/nanofiller interface, which in turn controls the subsequent performance properties (mechanical properties and permeability). For instance, tunicin whiskers favor starch crystallization due to the nucleating effect of the nanofiller (Mathew & Dufresne, 2002a).

2.2.2 Starch Nanocrystals-Based Nanobiocomposites

Waxy maize starch nanocrystals were obtained by acid hydrolysis of native granules by strictly controlling the temperature of the process, the acid and starch concentrations, the hydrolysis duration, and the stirring speed. Waxy maize starch nanocrystals consist of 5-7 nm thick platelet-like particles with a length ranging from 20 to 40 nm and a width in the range 15-30 nm. They were used as a reinforcing agent in a waxy maize starch matrix plasticized with glycerol, where Angellier et al. have shown that the reinforcing effect of starch nanocrystals can be attributed to strong filler/filler and filler/matrix interactions due to the establishment of increased hydrogen bonding. The presence of starch nanocrystals leads to a slowing down of the recrystallization of the matrix during aging in atmospheres with high humidity (Angellier, Molina-Boisseau, Dole, & Dufresne, 2006).

2.3 Clay-Based Nanobiocomposites

The most intensive research on nanobiocomposites is focused on nanoclays (Chivrac et al., 2009; Raquez, Nabar, Narayan, & Dubois, 2007) and especially on MMT or sepiolite (SEP) due to their availability, versatility, and low environment and health concerns.

2.3.1 Nanoclays

Phyllosilicates are a wide family in which clays with different structure, texture, or morphology can be found. Clays are commonly referred to as 1:1 or 2:1. Clays are fundamentally built of tetrahedral sheets (T) and octahedral sheets (O). A 1:1 clay would consist of one tetrahedral sheet and one octahedral sheet, and examples would be kaolinite and serpentine. A 2:1 clay consists of an octahedral sheet sandwiched between two tetrahedral sheets, such as the illite, smectite, and chlorite (although chlorite has an external octahedral sheet often referred to as "brucite"). Table 8.2 presents a conventional phyllosilicates classification.

The phyllosilicate crystal structure can be based on the pyrophyllite structure $Si_4Al_2O_{10}(OH)_2$ and can be described as a crystalline 2:1 layered clay mineral with a central alumina octahedral sheet sandwiched between two silica tetrahedral sheets corresponding to seven atomic layers superposed (Figure 8.3) (Hendricks, 1942). This structure becomes $(Si_8)(Al_{4-y}Mg_y)O_{20}(OH)_4$, M_y^+ for the MMT or $(Si_8)(Al_{6-y}Li_y)O_{20}(OH)_4$ and M_y^+ for the hectorite. These differentiations are mainly due to the isomorphic substitutions that take place inside the aluminum oxide layer. These substitutions induce a negative charge inside the clay platelet, which is naturally counterbalanced by inorganic cations (Li^+, Na^+, Ca^{2+}, K^+, and Mg^{2+}) located into the interlayer spacing. The global charge varies depending on the phyllosilicates. The charge amount is characterized by the cationic exchange capacity (CEC) and corresponds to the amount of monovalent cations necessary to compensate the platelets' negative charge, which is usually given in milliequivalent per 100 g (meq/100 g) (Sinha Ray and Okamoto, 2003). For instance, the MMT CEC varies from 70 to 120 meq/100 g depending on their extraction site (Thomas et al., 1999).

The distance observed between two platelets of the primary particle, named interlayer spacing or *d*-spacing (d_{001}), depends on the silicate type. This value does entirely depend not only on the layer crystal structure but also on the type of the countercation and on the hydration state of the silicate. For instance, $d_{001} = 0.96$ nm for anhydrous MMT with sodium as counter ion, but $d_{001} = 1.2$ nm in usual conditions. This increase is linked to the adsorption of one layer of water molecules between the clay platelets (Alexandre & Dubois, 2000).

Recently, some studies have been focused on the use of SEP, a nanofiller with a needlelike structure, to produce nanocomposites (Bilotti, Fischer, & Peijs, 2008; Chen, Zheng, Sun, & Jia, 2007; Darder et al., 2006; Duquesne, Moins, Alexandre, & Dubois, 2007; Ma, Bilotti, Peijs, & Darr, 2007; Xie et al., 2007; Zheng, 2006). This nanofiller is a microcrystalline-hydrated magnesium silicate of theoretical unit cell formula $Mg_8Si_{12}O_{30}(OH)_4g(H_20)_4 \cdot 8H_20$. It shows an alternation of blocks and channels that grow up in the fiber direction. The blocks are basically constituted by two layers of tetrahedral silica sandwiching a central octahedral magnesium oxide-hydroxide layer (Figure 8.4). Some isomorphic substitutions occur inside these central layers, creating a negative charge naturally counterbalanced by the presence of inorganic cations (Na^+, Ca^{2+}). The SEP channels are filled both with the coordinated water molecules, which are bonded to the Mg^{2+} ions located at the edges of octahedral sheets, and with the

TABLE 8.2 Conventional Phyllosilicate Classification

Type	System	Species	Di- or Tri Octahedral Character	Interlayer Cations	Octahedron	Tetrahedron
2:1 $Si_40_{10}(0H)_2$	Pyrophyllite-talc ($X=0$)	Pyrophyllite	Di		Al_2	Si_4
		Talc	Tri		Mg_3	Si_4
	Smectite ($0.25<X<0.6$)	Montmorillonite	Di	Na, Ca, H_2O	$(Al_2, Mg)_2$	Si_4
		Hectorite	Di	Na, Ca, H_2O	$(Mg_2, Li)_2$	Si_4
		Beidellite	Di	Na, Ca, H_2O	Al_2	$(Si, Al)_4$
		Saponite	Tri	Na, Ca, H_2O	Mg_3	$(Si, Al)_4$
	Vermiculite ($0.25<X<0.9$)	Vermiculite	Di	K, Al, H_2O	$(Al, Mg)_2$	$(Si, Al)_4$
		Vermiculite	Tri	K, Mg, H_2O	$(Mg, Al)_3$	$(Si, Al)_4$
	Mica ($X\sim1$)	Muscovite	Di	K	Al_2	Si_3, Al
		Paragonite		Na	Al_2	Si_3, Al
	Brittle mica ($X\sim2$)	Phlogopite	Tri	K	$(Mg, Fe^{2+})_3$	Si_3, Al
		Biotite		K	$(Mg, Fe^{2+})_3$	Si_3, Al
2:1:1 $Si_40_{10}(0H)_8$	Chlorite (large variation of X)	Donbassite	Di	$Al_2(OH)_6$	Al_2	$(Si, Al)_4$
		Sudoite	Di, tri	$(Mg, Al)_3(OH)_6$	$(Al, Mg)_2$	$(Si, Al)_4$
		Clinochlore	Tri	$(Mg, Al)_3(OH)_6$	$(Al, Mg)_3$	$(Si, Al)_4$

Continued

TABLE 8.2 Conventional Phyllosilicate Classification—cont'd

Type	System	Species	Di- or Tri Octahedral Character	Interlayer Cations	Octahedron	Tetrahedron
		Chamosite		(Fe, Al)$_3$(OH)$_6$	(Al, Fe)$_2$	(Si, Al)$_4$
1:1 $Si_20_5(0H)_4$	Kaolin mineral serpentine ($X \sim 0$)	Kaolinite	Di		Al_2	Si_2
		Halloysite		H_2O	Al_2	Si_2
		Chrysotile	Tri		Mg_3	Si_2
Needle	Sepiolite-palygorskite ($X \sim 0$)	Sepiolite	Tri	$(H_2O)_4{\cdot}H_2O$	Mg_8	Si_{12}
		Palygorskite		$(H_2O)_4{\cdot}H_2O$	Mg_8	Si_8
Amorphous or low crystalline		Imogolite			$Al(OH)_3$	SiO_3OH

X, degree of isomorphous substitution.

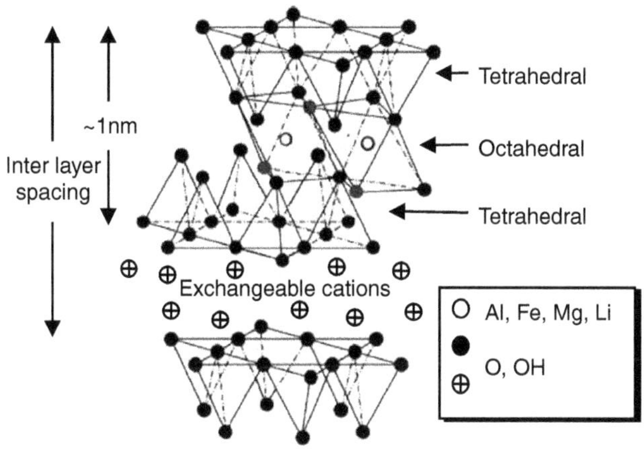

FIGURE 8.3 Structure of 2:1 phyllosilicates (e.g., montmorillonite).

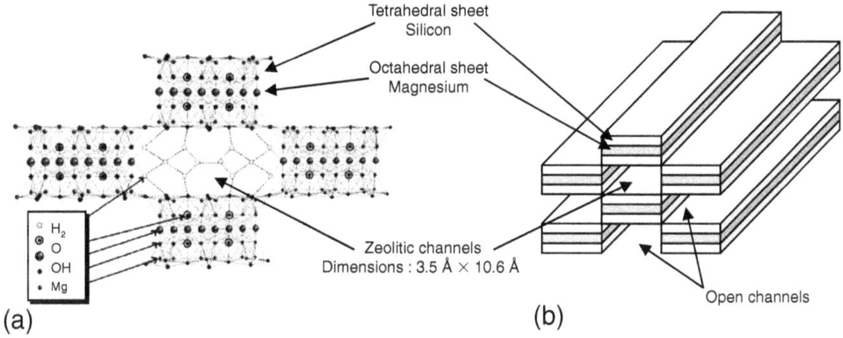

FIGURE 8.4 Sepiolite: (a) projection onto the (001) plane, (b) fibrous structure.

zeolitic water, which is associated to the structure by hydrogen bonding. The discontinuity of the silica sheets gives rise to the presence of silanol groups (Si-OH) at the edges of the external surface of the SEP nanoparticles.

2.3.2 *Organomodification of Phyllosilicates*

The phyllosilicate multiscale structure has different porosity levels, which drive its swelling properties. The water absorption occurs, thanks to the intercalated cation hydration, which lowers the attractive forces between the clay layers (Sposito & Grasso, 1999), and also thanks to the water capillarity phenomena, which take place into the interparticle and interaggregate porosities (Luckham & Rossi, 1999; Méring, 1946). For a given pressure, this swelling is characterized by a d_{001} increase until an equilibrium distance (Cases et al., 1992). In general, the smaller is the cation and the lower is its charge, the higher the clay swelling is. For MMT, the swelling decreases depending on the cation chemical type according to the following trend: $Li^+ > Na^+ > Ca^{2+} > Fe^{2+} > K^+$ (Powell, Fischer, & Skipper, 1998; Powell, Tongkhao, Kennedy, & Slade, 1998; Tettenhorst, 1962). The potassium cation is a specific case because its size is equal to the dimension of the platelet surface cavity. Thus, the potassium is trapped into these cavities, leading to a lowering of its hydration ability.

To enhance the intercalation-exfoliation process into a polymer matrix, a chemical modification of the clay surface, with the aim to match the polymer polarity, is often carried out (Alexandre & Dubois, 2000; Sinha Ray & Okamoto, 2003). The cationic exchange is the most common technique, but other original techniques as the organosilane grafting (Dai & Huang, 1999; Ke, Lü, Yi, Zhao, & Qi, 2000), the use of ionomers (Lagaly, 1999; Shen, Simon, & Cheng, 2002), and block copolymer adsorption (Fischer, Gielgens, & Koster, 1999) are also used. The cationic exchange consists in the inorganic cation substitution by organic ones. These cations are often alkylammonium surfactants having at least one long alkyl chain (Table 8.3). Phosphonium salts are also interesting clay modifiers, thanks to their higher thermal stability, but they are not often used (Wilkie, Zhu, & Uhl, 2001). The ionic substitution is performed into water because of the clay swelling, which facilitates the organic cation insertion between the platelets. Then, the solution is filtered, washed with distilled water (to remove the salt formed during the surfactant adsorption and the surfactant excess), and lyophilized to obtain the organomodified clay. In addition to the modification of the clay surface polarity, organomodification increases the d_{001}, which will also further facilitate the polymer chain intercalation (Lagaly, 1986). Various commercially available organomodified montmorillonites (OMMT), which mainly differ from the nature of their countercation (Table 8.3) and their CEC, are produced with this technique (e.g., Cloisite® 15A, 20A, and 30B).

2.3.3 Starch-Based Nanobiocomposites Based on MMTs

The nanofiller incorporation into the polymer matrix can be carried out with three main techniques (Alexandre & Dubois, 2000):

i. *In situ polymerization process*: In this method, layered silicates are swollen into a monomer solution. Then, the monomer polymerization is initiated and propagated. The macromolecules' molecular weight increases, leading to a d_{001} increase and sometimes to an almost fully exfoliated morphology for some studied systems (Sinha Ray & Okamoto, 2003). However, since polysaccharide chains are synthesized during the plant growth and then extracted from the vegetal, this technique cannot be used to prepare polysaccharides nanobiocomposites.

TABLE 8.3 Some Organomodified MMT Types and the Counterion Chemical Structures

Code	Name	Countercation
MMT-Na	Natural sodium montmorillonite	Na^+
OMMT-Alk1	Cloisite® 15A—Southern Clay	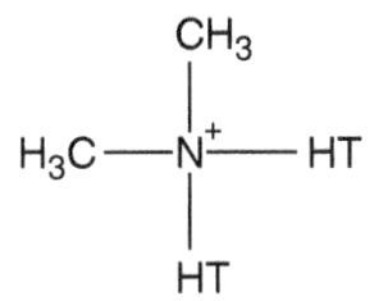 Dimethyl dihydrogenated tallow ammonium
OMMT-Alk2	Cloisite® 6A—Southern Clay	
OMMT-Bz	Cloisite® 10A—Southern Clay	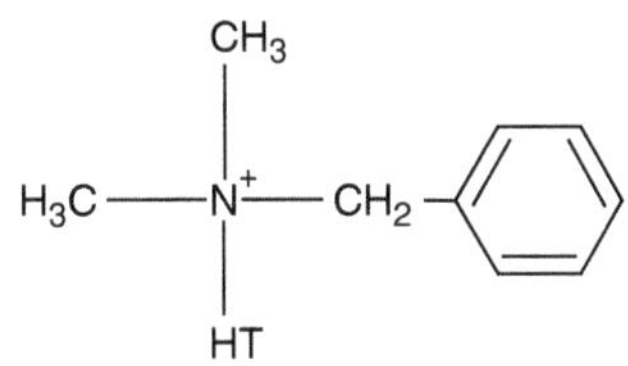 Dimethyl benzyl hydrogenated tallow ammonium
OMMT-OH	Cloisite® 30B—Southern Clay	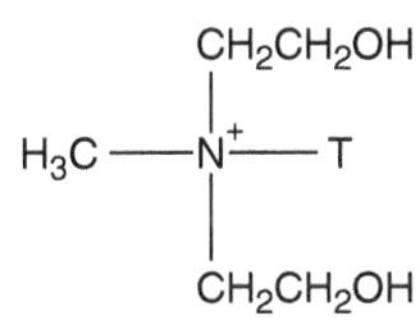 Methyl tallow bis-2-hydroxyethyl ammonium
OMMT-CS	–	Cationic starch

T, tallow (~65% C18, ~30% C16, and ~5% C14); HT, hydrogenated tallow.

ii. *Solvent intercalation process*: This elaboration process is based on a solvent system in which the polymer is soluble and the silicate layers are swellable. The polymer is first dissolved in an appropriate solvent. In parallel, the clay (organomodified or not) is swollen and dispersed into the same solvent or another one to obtain a miscible solution. Both systems are pooled together leading to a polymer chain intercalation. Then, the solvent is evaporated to obtain nanocomposite materials. Nevertheless, for non-water-soluble polymers, this process involves the use of large amount of organic solvents, which is environmentally unfriendly and cost-prohibitive. Moreover, a small amount of solvent may remain in the final product at the polymer/clay interface creating lower interfacial interaction between the polymer and the clay surfaces (Jin, Park, Im, Kwak, & Kwak, 2002). Thus, this technique is mainly used in academic studies. Since some polysaccharides, such as chitosan or pectin, cannot be melt processed due to high thermal or thermomechanical degradations, the solvent process has been extensively used to produce polysaccharide/clay hybrid materials.

iii. *Melt intercalation process*: Both the polymer and the clay are introduced simultaneously into a melt mixing device (extruder and internal mixer). According to Dennis et al. (Dennis et al., 2001), in addition to the polymer/nanofiller affinity, two main process parameters favor the nanodispersion of the nanoclay. These parameters, which are the driving force of the intercalation-exfoliation process into the matrix, are (i) the residence time and (ii) the shearing. The shearing is necessary to induce the platelets' delamination from the clay tactoids. The extended residence time is needed to allow the polymer chain diffusion into the interlayer gallery and then to obtain an exfoliated morphology.

This latter simple process has extensively been used to prepare polysaccharide nanobiocomposite materials. Nevertheless, the thermal or thermomechanical inputs lead to partial chains' degradation. Moreover, the high residence times needed to enhance the clay exfoliation process favor the matrix degradation. Therefore, it is necessary to balance the process parameters to minimize the chains' degradation and to obtain a well-exfoliated morphology. Depending on the process conditions and on the polymer/nanofiller affinity, different morphologies can be obtained. These morphologies can be divided in three distinct main categories: (i) microcomposites, (ii) intercalated nanocomposites, or (iii) exfoliated nanocomposites (Alexandre & Dubois, 2000; Sinha Ray & Okamoto, 2003; Vaia & Giannelis, 1997). For microcomposites, the polymer chains have not penetrated into the interlayer spacing and the clay particles are aggregated. In this case, the designation as nanocomposite is abusive. In the intercalated structures, the polymer chains have diffused between the platelets leading to a d_{001} increase. In the exfoliated state, the clay layers are individually delaminated and homogeneously dispersed into the polymer matrix. Intermediate dispersion states are often observed, such as intercalated-exfoliated structures. This classification does not take into account the dispersion multiscale structure, such as percolation phenomenon and preferential orientation of the clay layers (Sinha Ray & Okamoto, 2003).

To reach exfoliation, different nanofillers and dispersion protocols have been tested to produce nanobiocomposites based on PLS and MMT (Chivrac et al., 2009). Rather, hydrophobic nanofillers were incorporated into PLS based on wheat (Chiou, Yee, Glenn, & Orts, 2005), potato (Chen & Evans, 2005; Park, Lee, Park, Cho, & Ha, 2003; Park et al., 2002), or corn (Zhang, Yu, Xie, Naito, & Kagawa, 2007), with MMT content varying from 0 to 9 wt%. It was clearly demonstrated that the incorporation of OMMT-Alk1, OMMT-Alk2, or OMMT-Bz (see Table 8.3 for the OMMT designations) led to the formation of microbiocomposites (Chen & Evans, 2005; Chiou et al., 2005; Park et al., 2002; Zhang et al., 2007), evidenced by the unchanged values of the d_{001}. Higher spacing results were obtained with OMMT-OH, which presents a more hydrophilic character, with a slight d_{001} shift and a strong decrease in the diffraction peak intensity (Chiou et al., 2005, 2006; Park et al., 2003), corresponding to a higher dispersion. This state was likely achieved due to the hydrogen bonds established between the clay surfactant and the starch chains (Park et al., 2003).

Nanobiocomposites were also elaborated with natural sodium MMT (MMT-Na) by trying to take advantage of the hydrophilic character of both the starch matrix and the MMT-Na nanofiller (Avella et al., 2005; Chen, Chen, & Evans, 2005; Chen & Evans, 2005; Chiou et al., 2007, 2005, 2006; Cyras, Manfredi, Ton-That, & Vazquez, 2008; Huang, Yu, & Ma, 2004; Pandey & Singh, 2005; Park et al., 2003, 2002; Zhang et al., 2007). These materials were prepared with corn starch (Huang et al., 2004; Pandey & Singh, 2005; Zhang et al., 2007), wheat (Chiou et al., 2005, 2007), or potato (Avella et al., 2005; Chen et al., 2005; Chen & Evans, 2005; Cyras et al., 2008; Park et al., 2003, 2002). It was highlighted that for glycerol content higher than 10 wt%, such systems led to the formation of an intercalated structure with a d_{001} increasing to 18 Å, a value that is commonly reported in the literature as generally attributed to glycerol intercalation (Chiou et al., 2006; Wilhelm, Sierakowski, Souza, & Wypych, 2003). The influence of the plasticizer on the MMT dispersion and on the exfoliation state was also highlighted by Dean, Yu, and Wu (2007). Corresponding results showed a homogeneous dispersion with an exfoliated structure in agreement with results of Cyras et al. (2008), which have demonstrated that for glycerol content lower than 10 wt%, exfoliation was achieved. In certain conditions, MMT-Na seemed suitable to achieve exfoliation. These results were in agreement with some studies that have highlighted the formation of hydrogen bonds and strong interactions between glycerol and MMT platelets (Huang et al., 2004; Pandey & Singh, 2005; Wilhelm et al., 2003). Huang et al. (Huang, Yu, & Ma, 2005a, 2005b, 2006; Huang, Yu, Ma, & Jin, 2005) have demonstrated that by changing the plasticizer nature, with, e.g., urea or urea/formamide, exfoliation can also be reached. Nevertheless, these compounds generate ecotoxic residues after biodegradation or composting and cannot be used for safe biodegradable materials. Kampeerapappun, Aht-Ong, Pentrakoon, and Srikulkit (2007) had focused their attention on the use of a new eco-friendly compatibilizer, chitosan, to promote the MMT platelets' exfoliation. But, only a small increase in the d_{001} was achieved. Nevertheless, a similar approach was successfully applied (Chivrac, Pollet, & Averous, 2008), with cationic starch (CS) as MMT organomodifier with a multistep protocol (Figure 8.5). According to these authors, no diffraction peak was observed by X-ray diffraction, suggesting an exfoliated morphology, which was confirmed by TEM analysis (Figure 8.6).

As usual, nanocomposites displayed substantial improvement in mechanical properties like Young's modulus, which is correlated with the clay loading for MMT-Na (Chivrac, Pollet, & Averous, 2008) (with corn and wheat starch) or OMMT-CS (with wheat starch). The mechanical improvement depends on the nanobiocomposites structure. The modulus increase is higher in the case of exfoliation with CS-modified MMT compared to MMT-Na. For instance, Figure 8.7 shows the different modulus evolutions determined by uniaxial tensile test between an exfoliated structure based on OMMT-CS and a nonexfoliated one, based on MMT-Na.

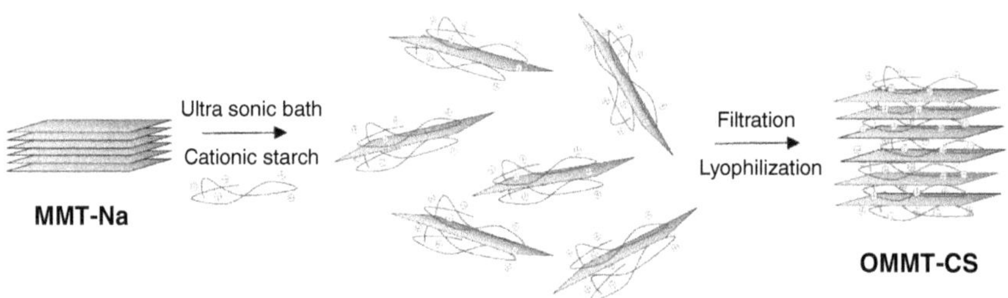

FIGURE 8.5 Schematic representation of the MMT-Na organomodification by exfoliation/adsorption technique. (For color version of this figure, the reader is referred to the online version of this chapter.) *Source: Chivrac, Pollet, and Averous (2008a).*

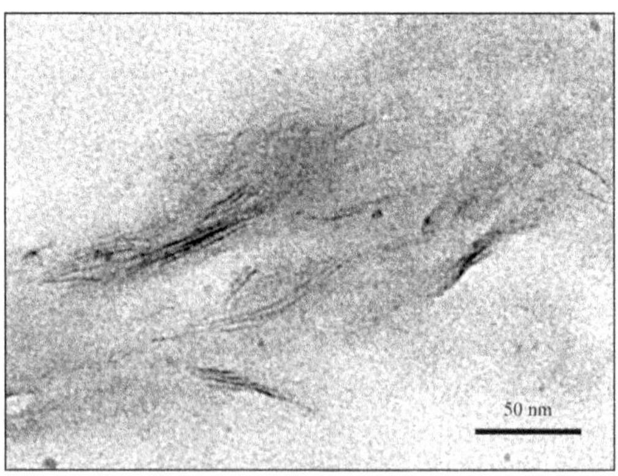

FIGURE 8.6 TEM micrograph of nanobiocomposites based on plasticized starch/OMMT-CS 3 wt% (black scale = 50 nm). *Source: Chivrac, Pollet, and Averous (2008a).*

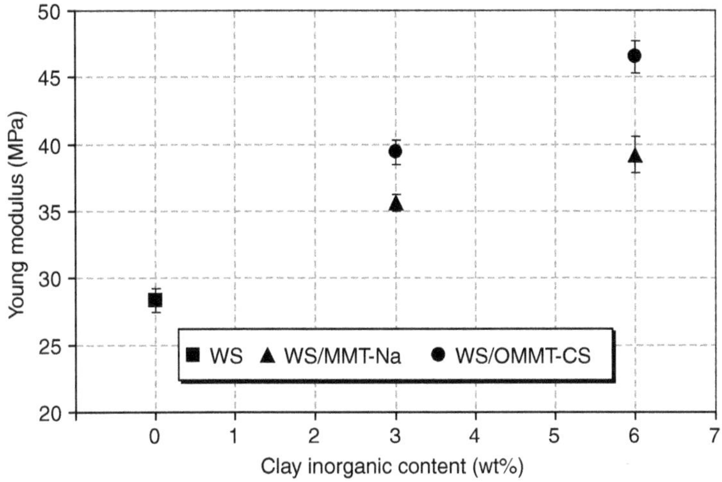

FIGURE 8.7 Variations of Young's modulus vs. clay content, for plasticized wheat starch with MMT-Na (microbiocomposite) and OMMT-CS (exfoliated nanobiocomposite), stabilized at 57% RH at room temperature. *Source: Chivrac, Pollet, and Averous (2008a).*

Similar behavior is obtained on the elongation at break, which slightly increases with the clay content in the case of exfoliation and decrease in the case of a microcomposites structure (Table 8.4). In the same way, the energy at break increases in the case of exfoliation and decreases in the case of a microcomposite structure.

From thermomechanical measurements based on DMTA characterization, we can determine the influence of the nanofillers on the local mobility of the chains and thus on their relaxation temperatures that could in turn be associated with the glass transitions. In the case of PLS/MMT-Na nanobiocomposites, the temperatures of two main relaxation peaks shifted toward higher temperatures, indicating that the layered clays strongly restricted the starch chain mobility. This tendency (Park et al., 2002) was attributed to the MMT-Na higher affinity with the starch chains. The same trends were observed by DSC (Huang et al., 2006), meaning that starch/clay hybrids were strongly affected by the clay surface polarity and the clay/matrix interactions.

Some authors studied in detail the thermal stability of nanobiocomposites by TGA. Park et al. (2003) showed that the potato starch/MMT-Na and OMMT-OH hybrids have a higher degradation temperature than the neat matrix. The MMT-Na thermal stability was higher than the OMMT-OH nanobiocomposites one. Such a result highlighted a relationship between the clay dispersion state and the thermal stability. Such behavior is observed in most nanocomposite systems and is linked to the clay aspect ratio and the dispersion state. The exfoliation of the

TABLE 8.4 Mechanical Properties of Wheat Starch Nanobiocomposites with MMT and SEP

	Young's Modulus (MPa)	Strain at Break (%)	Stress at Break (MPa)
WS	28.3 ± 0.9	31.7 ± 1.5	2.24 ± 0.04
WS/CS 3 wt%	27.9 ± 0.5	32.7 ± 2.0	2.26 ± 0.12
WS/CS 6 wt%	28.1 ± 1.8	32.5 ± 3.1	2.24 ± 0.11
WS/SEP-Na 3 wt%	45.3 ± 0.7	36.5 ± 2.1	2.91 ± 0.06
WS/SEP-Na 6 wt%	67.3 ± 2.3	31.0 ± 1.0	2.99 ± 0.04
WS/OSEP-1CS 6 wt%	71.9 ± 8.9	27.4 ± 1.4	3.09 ± 0.18
WS/OSEP-2CS 6 wt%	74.8 ± 4.5	32.6 ± 0.7	3.17 ± 0.09
WS/OSEP-4CS 3 wt%	49.1 ± 1.5	36.1 ± 2.1	2.85 ± 0.12
WS/OSEP-4CS 6 wt%	76.0 ± 1.9	35.2 ± 1.5	3.21 ± 0.13
WS/OSEP-6CS 6 wt%	74.8 ± 2.6	34.6 ± 0.7	3.19 ± 0.09
WS/MMT-Na 3 wt%	35.6 ± 0.6	27.3 ± 0.6	2.32 ± 0.08
WS/MMT-Na 6 wt%	39.2 ± 1.4	21.0 ± 0.8	1.90 ± 0.06
WS/OMMT-CS 3 wt%	39.4 ± 0.9	32.5 ± 0.8	2.43 ± 0.06
WS/OMMT-CS 6 wt%	46.5 ± 1.2	33.3 ± 2.0	2.63 ± 0.16

Source: Chivrac, Pollet, Schmutz, and Avérous (2010c) and Chivrac, Pollet, and Averous (2008a).

MMT nanoplatelets into the starch matrix increases the tortuosity of the combustion gas diffusion pathway and the formation of a char at the surface.

Nanoclays also impact the water vapor permeability of the corresponding nanocomposite materials. Park et al. (2002) examined the potato starch nanobiocomposite water vapor permeabilities with different type of clays. According to these authors, all the clay-based films showed lower water vapor permeability compared to the neat matrix. Best results are obtained with MMT-Na nanobiocomposites, which present the higher dispersion state according to these authors. The same trends are observed in different papers (Huang & Yu, 2006; Park et al., 2003). Unfortunately, a recent paper shows that with OMMT-CS, results obtained with high glycerol content are rather poor (Chivrac, Angellier-Coussy, Guillard, Pollet, & Avérous, 2010). The relatively high plasticizer content (23 wt% glycerol) induces a phase separation, with plasticizer-rich and carbohydrate-rich phases, resulting in the nanoclay being preferentially located in the carbohydrate-rich domains. As a consequence, a preferential way for water transfer was more likely created in the very hydrophilic glycerol-rich domains where the nanoclay platelets were almost totally absent. Thus, even if exfoliated morphology is achieved, the heterogeneous clay distribution and phase separation phenomena explain the lack of improvement and even the decline in the moisture barrier properties for these glycerol-PLS nanobiocomposites. For the microbiocomposites based on OMMT-Bz, OMMT-Alk2, or OMMT-OH, the barrier properties' enhancement was linked to a decrease in the water solubility due to the surfactant hydrophobic nature. This behavior is induced by these two distinct phenomena, i.e., (i) the silicate layers dispersion and (ii) the solubility of the gas into the material (Alexandre & Dubois, 2000).

2.3.4 *Starch-Based Nanobiocomposites Based on SEP*

Recently, new nanocomposites have been elaborated from needle-shaped nanofillers (e.g., SEP and palygorskite) (Bilotti et al., 2008; Bokobza, Burr, Garnaud, Perrin, & Pagnotta, 2004; Duquesne et al., 2007). Compared to more conventional layered silicates, they present substantial differences, namely, their aspect ratio and their surface properties (silanol groups located at the nanofiller surface), and thus could permit the elaboration of innovative polysaccharide-based nanohybrid materials with extended properties.

A recent study (Chivrac, Pollet, Schmutz, & Avérous, 2010) has shown the behavior of natural sepiolite (SEP-Na) and SEP organomodified with cationic starch (OSEP-CS), prepared according to the multistep protocol shown in Figure 8.8, to produce a new class of starch-based

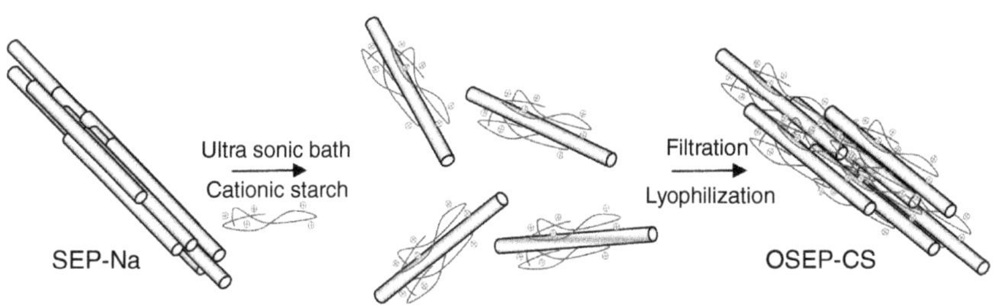

FIGURE 8.8 Schematic representation of the SEP-Na organomodification by dispersion/adsorption technique (Chivrac, Pollet, Schmutz, & Avérous, 2010c). (For color version of this figure, the reader is referred to the online version of this chapter.)

nanobiocomposites. Morphological analyses have highlighted a homogeneous nanoscale dispersion of SEP-Na. Surprisingly, the dispersion of OSEP-CS has led to different results depending on the cationic starch content adsorbed on the clay surface. The incorporation of OSEP-1CS into PLS-based materials leads to the formation of big clay aggregates. Then, at higher cationic starch content, this surfactant fully acts as a compatibilizer and greatly favors the clay nanodispersion within the PLS matrix. The TEM micrographs have also shown a phase separation between domains with and without SEP for the nanobiocomposites materials based on SEP-Na or OSEP-CS. This inhomogeneity is induced by the high glycerol content of the formulation, which leads to phase separation between domains rich in glycerol and domains with low glycerol content, the SEP being dispersed into the domain rich in glycerol.

Figure 8.9 shows X-ray diffraction measurements. Transcrystallization phenomenon occurs at the SEP/matrix interface, thanks to the hydrogen bonds established between the nanofiller and the starch chains. The diffractogram obtained for the neat matrix displays E_H-type crystallization peak at $2\theta = 17.2°$, corresponding to the amylopectin recrystallization. V_H-type crystallization peaks are also observed at $2\theta = 19.9°$ and 22.5° and correspond to the process-induced amylose crystallization into single-helical structure (van Soest et al., 1996a, 1996b). No significant evolution of these E_H-type and V_H-type crystallization peaks are observed for the starch/SEP nanobiocomposites samples. However, their diffractograms display a new diffraction peak located at 26.4°, corresponding to a crystal lattice structure of 3.4 Å (calculated from Bragg's law). Such a new diffraction peak has already been reported into starch/tunicin whisker composites. It is attributed to amylopectin crystallization at the filler

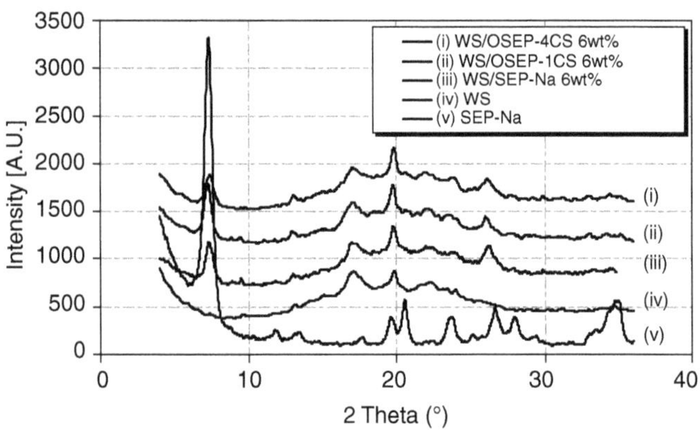

FIGURE 8.9 XRD patterns for SEP-Na, WS (wheat starch), WS/SEP-Na, and WS/OSEP-CS nanobiocomposites with various inorganic contents (3 and 6 wt%). 1CS and 4CS means CS contents of 1 and 4 times the CEC, respectively (Chivrac, Pollet, Schmutz, & Avérous, 2010c).

interface and is favored by hydrogen bonds established between the filler and the macromolecules (Neus Angles & Dufresne, 2000). Thus, these results suggest the occurrence of a new crystal structure likely induced by the interactions established between the numerous silanol groups located at the edge of the SEP needles and the hydroxyl groups of the polysaccharide chains. Such behavior of SEP inducing or promoting a specific type of crystallinity at the polymer/clay interface has already been reported in the case of PP/SEP nanocomposites (Bilotti et al., 2008; Ma et al., 2007) and for PA6/SEP nanocomposites (Bilotti et al., 2008). Interestingly, the XRD analyses performed on starch/MMT nanobiocomposites (Chivrac, Pollet, Schmutz, & Avérous, 2008) did not show such filler induced new type of crystallization.

Thermal analyses have demonstrated that the SEP-Na enhances the thermal stability of the starch matrix, thanks to dehydration mechanism that locally reduces the sample temperature. On the contrary, it has been pointed out that cationic starch thermal degradation by-products may favor the PLS degradation mechanism. Uniaxial tensile tests have been performed and have demonstrated that the incorporation of well-dispersed SEP nanofiller can increase the material rigidity without affecting its strain at break properties as for previous results on MMT-based nanobiocomposites (Chivrac, Pollet, et al. 2010) (Table 8.4).

3 CONCLUSIONS

In conclusion, the macro- or nanostructure and subsequent processing, thermomechanical properties, and performance properties of these materials have been examined. This examination included a detailed review of the complexity of starch polymers, recent advances in novel starch modifications and compounds, and a detailed examination of the effects of plasticized starch microbiocomposites and nanobiocomposites. Specific structures and subsequent properties are controlled by many specific factors such as filler shape, size, and surface chemistry; processing conditions; and environmental aging. In the case of nanobiocomposites, it is evident that nanofiller-polymer matrix interfacial interactions are extremely important to the final nanostructures and performance of these materials.

Large improvements in the material performance were found with polysaccharide nanofillers and nanoclay-based nanobiocomposites, some of them are linked to traditional matrix reinforcement, and some others are brought by the high interface area between the nanofillers and the matrix and by the corresponding dispersion state.

The different structures, which are obtained, can fulfill the requirements of different applications, such as packaging or other short-lived applications (catering, agriculture, leisure, and hygiene) where long-lasting polymers are not entirely adequate. In some special cases, these materials can also find biomedical applications linked to with their intrinsic properties.

References

Alexandre, M., & Dubois, P. (2000). Polymer-layered silicate nanocomposites: Preparation, properties and uses of a new class of materials. *Materials Science and Engineering R Reports, 28*, 1–63.

Angellier, H., Molina-Boisseau, S., Dole, P., & Dufresne, A. (2006). Thermoplastic starch-waxy maize starch nanocrystals nanocomposites. *Biomacromolecules, 7*, 531–539.

Angles, M. N., & Dufresne, A. (2000). Plasticized starch/tunicin whiskers nanocomposites. 1. Structural analysis. *Macromolecules, 33*, 8344–8353.

Angles, M. N., & Dufresne, A. (2001). Plasticized starch/tunicin whiskers nanocomposite materials. 2. Mechanical behavior. *Macromolecules, 34*, 2921–2931.

Avella, M., De Vlieger, J. J., Errico, M. E., Fischer, S., Vacca, P., & Volpe, M. G. (2005). Biodegradable starch/clay nanocomposite films for food packaging applications. *Food Chemistry, 93*, 467–474.

Averous, L. (2004). Biodegradable multiphase systems based on plasticized starch: A review. *Materials Science and Engineering R Reports, 44*, 231–274.

Averous, L. (2007). Cellulose-based biocomposites: Comparison of different multiphasic systems. *Composite Interfaces, 14*, 787–805.

Averous, L., & Boquillon, N. (2004). Biocomposites based on plasticized starch: Thermal and mechanical behaviours. *Carbohydrate Polymers, 56*, 111–122.

Averous, L., Fauconnier, N., Moro, L., & Fringant, C. (2000). Blends of thermoplastic starch and polyesteramide: Processing and properties. *Journal of Applied Polymer Science, 76*, 1117–1128.

Avérous, L., & Fringant, C. (2001). Association between plasticized starch and polyesters: Processing and performances of injected biodegradable systems. *Polymer Engineering and Science, 41*, 727–734.

Avérous, L., Fringant, C., & Moro, L. (2001). Plasticized starch-cellulose interactions in polysaccharide composites. *Polymer, 42*, 6565–6572.

Avérous, L., & Halley, P. J. (2009). Biocomposites based on plasticized starch. *Biofuels, Bioproducts and Biorefining, 3*, 329–343.

Avérous, L., & Le Digabel, F. (2006). Properties of biocomposites based on lignocellulosic fillers. *Carbohydrate Polymers, 66*, 480–493.

Averous, L., Moro, L., Dole, P., & Fringant, C. (2000). Properties of thermoplastic blends: Starch-polycaprolactone. *Polymer, 41*, 4157–4167.

Azizi Samir, M. A. S., Alloin, F., & Dufresne, A. (2005). Review of recent research into cellulosic whiskers, their properties and their application in nanocomposite field. *Biomacromolecules, 6*, 612–626.

Azizi Samir, M. A. S., Alloin, F., Sanchez, J. -Y., El Kissi, N., & Dufresne, A. (2004). Preparation of cellulose whiskers reinforced nanocomposites from an organic medium suspension. *Macromolecules, 37*, 1386–1393.

Baud, B., Colonna, P., Della Valle, G., & Roger, P. (1999). Macromolecular degradation of extruded starches measured by HPSEC-MALLS. In P. Colonna & S. Guilbert (Eds.), *Biopolymer science: Food and non food applications, 91*, 217–221. Paris: Les Colloques de l'INRA.

Bilotti, E., Fischer, H. R., & Peijs, T. (2008). Polymer nanocomposites based on needle-like sepiolite clays: Effect of functionalized polymers on the dispersion of nanofiller, crystallinity, and mechanical properties. *Journal of Applied Polymer Science, 107*, 1116–1123.

Bokobza, L., Burr, A., Garnaud, G., Perrin, M. Y., & Pagnotta, S. (2004). Fibre reinforcement of elastomers: Nanocomposites based on sepiolite and poly(hydroxyethyl acrylate). *Polymer International, 53*, 1060–1065.

Bordes, P., Pollet, E., & Avérous, L. (2009). Nano-biocomposites: Biodegradable polyester/nanoclay systems. *Progress in Polymer Science (Oxford), 34*, 125–155.

Cases, J. M., Brend, I., Besson, G., François, M., Uriot, J. P., Thomas, F., et al. (1992). Mechanism of adsorption and desorption of water vapor by homoionic montmorillonite. 1. The sodium-exchanged form. *Langmuir, 8*, 2730–2739.

Chen, M., Chen, B., & Evans, J. R. G. (2005). Novel thermoplastic starch-clay nanocomposite foams. *Nanotechnology, 16*, 2334–2337.

Chen, B., & Evans, J. R. G. (2005). Thermoplastic starch-clay nanocomposites and their characteristics. *Carbohydrate Polymers, 61*, 455–463.

Chen, H., Zheng, M., Sun, H., & Jia, Q. (2007). Characterization and properties of sepiolite/polyurethane nanocomposites. *Materials Sciences and Engineering A Structural Materials: Properties, Microstructure and Processing, 445–446*, 725–730.

Chiou, B. S., Wood, D., Yee, E., Imam, S. H., Glenn, G. M., & Orts, W. J. (2007). Extruded starch-nanoclay nanocomposites: Effects of glycerol and nanoclay concentration. *Polymer Engineering and Science, 47*, 1898–1904.

Chiou, B. -S., Yee, E., Glenn, G. M., & Orts, W. J. (2005). Rheology of starch-clay nanocomposites. *Carbohydrate Polymers, 59*, 467–475.

Chiou, B. -S., Yee, E., Wood, D., Shey, J., Glenn, G., & Orts, W. (2006). Effects of processing conditions on nanoclay dispersion in starch-clay nanocomposites. *Cereal Chemistry, 83*, 300–305.

Chivrac, F., Angellier-Coussy, H., Guillard, V., Pollet, E., & Avérous, L. (2010). How does water diffuse in starch/montmorillonite nano-biocomposite materials? *Carbohydrate Polymers, 82*, 128–135.

Chivrac, F., Pollet, E., & Averous, L. (2008). New approach to elaborate exfoliated starch-based nanobiocomposites. *Biomacromolecules, 9*, 896–900.

Chivrac, F., Pollet, E., & Averous, L. (2009). Progress in nano-biocomposites based on polysaccharides and nanoclays. *Materials Science and Engineering R Reports, 67*, 1–17.

Chivrac, F., Pollet, E., Dole, P., & Avérous, L. (2010). Starch-based nano-biocomposites: Plasticizer impact on the montmorillonite exfoliation process. *Carbohydrate Polymers, 79*, 941–947.

Chivrac, F., Pollet, E., Schmutz, M., & Avérous, L. (2010). Starch nano-biocomposites based on needle-like sepiolite clays. *Carbohydrate Polymers, 80*, 145–153.

Cooke, D., & Gidley, M. J. (1992). Loss of crystalline and molecular order during starch gelatinisation: Origin of the enthalpic transition. *Carbohydrate Research, 227*, 103–112.

Curvelo, A. A. S., De Carvalho, A. J. F., & Agnelli, J. A. M. (2001). Thermoplastic starch-cellulosic fibers composites: Preliminary results. *Carbohydrate Polymers, 45*, 183–188.

Cyras, V. P., Manfredi, L. B., Ton-That, M. T., & Vazquez, A. (2008). Physical and mechanical properties of thermoplastic starch/montmorillonite nanocomposite films. *Carbohydrate Polymers, 73*, 55–63.

Dai, J. C., & Huang, J. T. (1999). Surface modification of clays and clay-rubber composite. *Applied Clay Science, 15*, 51–65.

Darder, M., Lopez-Blanco, M., Aranda, P., Aznar, A. J., Bravo, J., & Ruiz-Hitzky, E. (2006). Microfibrous chitosan—Sepiolite nanocomposites. *Chemistry of Materials, 18*, 1602–1610.

Davidson, V. J., Parker, R., Diosady, L. L., & Rubin, L. T. (1984). A model for mechanical degradation of wheat starch in a single-screw extruder. *Journal of Food Science, 49*, 1154–1169.

Davidson, V. J., Paton, D., Diosady, L. L., & Larocque, G. (1984). Degradation of wheat starch in a single screw extruder: Characteristics of extruded starch polymers. *Journal of Food Science, 49*, 453–458.

De Carvalho, A. J. F., Curvelo, A. A. S., & Agnelli, J. A. M. (2001). A first insight on composites of thermoplastic starch and kaolin. *Carbohydrate Polymers, 45*, 189–194.

Dean, K., Yu, L., & Wu, D. Y. (2007). Preparation and characterization of melt-extruded thermoplastic starch/clay nanocomposites. *Composites Science and Technology, 67*, 413–421.

Della Valle, G., Boche, Y., Colonna, P., & Vergnes, B. (1995). The extrusion behaviour of potato starch. *Carbohydrate Polymers, 28*, 255–264.

Dennis, H. R., Hunter, D. L., Chang, D., Kim, S., White, J. L., Cho, J. W., et al. (2001). Effect of melt processing conditions on the extent of exfoliation in organoclay-based nanocomposites. *Polymer, 42*, 9513–9522.

Dole, P., Avérous, L., Joly, C., Della Valle, G., & Bliard, C. (2005). Evaluation of starch-PE multilayers: Processing and properties. *Polymer Engineering and Science, 45*, 217–224.

Doungjai, T., & Sanguansri, C. (2007). Aging effects on sorbitol- and non-crystallizing sorbitol-plasticized tapioca starch films. *Starch/Staerke, 59*, 493–497.

Dufresne, A., Dupeyre, D., & Vignon, M. R. (2000). Cellulose microfibrils from potato tuber cells: Processing and characterization of starch-cellulose microfibril composites. *Journal of Applied Polymer Science, 76*, 2080–2092.

Duquesne, E., Moins, S., Alexandre, M., & Dubois, P. (2007). How can nanohybrids enhance polyester/sepiolite nanocomposite properties? *Macromolecular Chemistry and Physics, 208*, 2542–2550.

Fischer, H. R., Gielgens, L. H., & Koster, T. P. M. (1999). Nanocomposites from polymers and layered minerals. *Acta Polymerica, 50*, 122–126.

Fringant, C., Desbrières, J., & Rinaudo, M. (1996). Physical properties of acetylated starch-based materials: Relation with their molecular characteristics. *Polymer, 37*, 2663–2673.

Fringant, C., Rinaudo, M., Foray, M. F., & Bardet, M. (1998). Preparation of mixed esters of starch or use of an external plasticizer: Two different ways to change the properties of starch acetate films. *Carbohydrate Polymers, 35*, 97–106.

Funke, U., Bergthaller, W., & Lindhauer, M. G. (1998). Processing and characterization of biodegradable products based on starch. *Polymer Degradation and Stability, 59*, 293–296.

Gaudin, S., Lourdin, D., Le Botlan, D., Forssell, P., Ilari, J. L., & Colonna, P. (1999a). Effect of polymer-plasticizer interactions on the oxygen permeability of starch-sorbitol-water films. *Macromolecular Symposia, 138*, 245–248.

Gaudin, S., Lourdin, D., Le Botlan, D., Forssell, P., Ilari, J. L., & Colonna, P. (1999b). Relationships between oxygen permeability and polymer-plasticizer interactions on starch-sorbitol-water films. In P. Colonna & S. Guilbert (Eds.), *Biopolymer science: Food and non food applications, 91*, 173–178. Paris: Les Colloques de l'INRA.

Gaudin, S., Lourdin, D., Le Botlan, D., Ilari, J. L., & Colonna, P. (1999). Plasticisation and mobility in starch-sorbitol films. *Journal of Cereal Science, 29*, 273–284.

Genkina, N. K., Wikman, J., Bertoft, E., & Yuryev, V. P. (2007). Effects of structural imperfection on gelatinization characteristics of amylopectin starches with A- and B-type crystallinity. *Biomacromolecules, 8*, 2329–2335.

Godbillot, L., Dole, P., Joly, C., Roge, B., & Mathlouthi, M. (2006). Analysis of water binding in starch plasticized films. *Food Chemistry, 96*, 380–386.

Hendricks, S. B. (1942). Lattice structure of clay minerals and some properties of clay. *Journal of Geology, 50*, 276–293.

Huang, M., & Yu, J. (2006). Structure and properties of thermoplastic corn starch/montmorillonite biodegradable composites. *Journal of Applied Polymer Science, 99*, 170–176.

Huang, M. F., Yu, J. G., & Ma, X. F. (2004). Studies on the properties of montmorillonite-reinforced thermoplastic starch composites. *Polymer, 45*, 7017–7023.

Huang, M., Yu, J., & Ma, X. (2005a). Studies on properties of the thermoplastic starch/montmorillonite composites. *Acta Polymerica Sinica, 1*, 862–867.

Huang, M. F., Yu, J. G., & Ma, X. F. (2005b). Preparation of the thermoplastic starch/montmorillonite nanocomposites by melt-intercalation. *Chinese Chemical Letters, 16*, 561–564.

Huang, M., Yu, J., & Ma, X. (2006). High mechanical performance MMT-urea and formamide-plasticized thermoplastic cornstarch biodegradable nanocomposites. *Carbohydrate Polymers, 63*, 393–399.

Huang, M. F., Yu, J. G., Ma, X. F., & Jin, P. (2005). High performance biodegradable thermoplastic starch—EMMT nanoplastics. *Polymer, 46*, 3157–3162.

Hulleman, S. H. D., Janssen, F. H. P., & Feil, H. (1998). The role of water during plasticization of native starches. *Polymer, 39*, 2043–2048.

Jin, Y. H., Park, H. J., Im, S. S., Kwak, S. Y., & Kwak, S. (2002). Polyethylene/clay nanocomposite by in-situ exfoliation of montmorillonite during Ziegler-Natta polymerization of ethylene. *Macromolecular Rapid Communications, 23*, 135–140.

Kampeerapappun, P., Aht-Ong, D., Pentrakoon, D., & Srikulkit, K. (2007). Preparation of cassava starch/montmorillonite composite film. *Carbohydrate Polymers, 67*, 155–163.

Ke, Y., Lü, J., Yi, X., Zhao, J., & Qi, Z. (2000). The effects of promoter and curing process on exfoliation behavior of epoxy/clay nanocomposites. *Journal of Applied Polymer Science, 78*, 808–815.

Lagaly, G. (1986). Interaction of alkylamines with different types of layered compounds. *Solid State Ionics, 22*, 43–51.

Lagaly, G. (1999). Introduction: From clay mineral-polymer interactions to clay mineral-polymer nanocomposites. *Applied Clay Science, 15*, 1–9.

Lourdin, D., Coignard, L., Bizot, H., & Colonna, P. (1997). Influence of equilibrium relative humidity and plasticizer concentration on the water content and glass transition of starch materials. *Polymer, 38*, 5401–5406.

Lourdin, D., Ring, S. G., & Colonna, P. (1998). Study of plasticizer-oligomer and plasticizer-polymer interactions by dielectric analysis: Maltose-glycerol and amylose-glycerol-water systems. *Carbohydrate Research, 306*, 551–558.

Luckham, P. F., & Rossi, S. (1999). The colloid and rheological properties of bentonite suspensions. *Advances in Colloid and Interface Science, 82*, 43–92.

Ma, J., Bilotti, E., Peijs, T., & Darr, J. A. (2007). Preparation of polypropylene/sepiolite nanocomposites using supercritical CO2 assisted mixing. *European Polymer Journal, 43*, 4931–4939.

Ma, X. F., Yu, J., & Jin, F. (2004). Urea and formamide as a mixed plasticizer for thermoplastic starch. *Polymer International, 53*, 1780–1785.

Ma, X. F., Yu, J. G., & Ma, Y. B. (2005). Urea and formamide as a mixed plasticizer for thermoplastic wheat flour. *Carbohydrate Polymers, 60*, 111–116.

Ma, X. F., Yu, J. G., & Wan, J. J. (2006). Urea and ethanolamine as a mixed plasticizer for thermoplastic starch. *Carbohydrate Polymers, 64*, 267–273.

Martin, O., Averous, L., & Della Valle, G. (2003a). In-line determination of plasticized wheat starch viscoelastic behavior: Impact of processing. *Carbohydrate Polymers, 53*, 169–182.

Mathew, A. P., & Dufresne, A. (2002a). Morphological investigation of nanocomposites from sorbitol plasticized starch and tunicin whiskers. *Biomacromolecules, 3*, 609–617.

Mathew, A. P., & Dufresne, A. (2002b). Plasticized waxy maize starch: Effect of polyols and relative humidity on material properties. *Biomacromolecules, 3*, 1101–1108.

Mathew, A. P., Thielemans, W., & Dufresne, A. (2008). Mechanical properties of nanocomposites from sorbitol plasticized starch and tunicin whiskers. *Journal of Applied Polymer Science, 109*, 4065–4074.

Méring, J. (1946). The hydration of montmorillonite. *Transactions of the Faraday Society, 42B*, 205–219.

Neus Angles, M., & Dufresne, A. (2000). Plasticized starch/tunicin whiskers nanocomposites. 1. Structural analysis. *Macromolecules, 33*, 8344–8353.

Ollett, A. L., Parker, R., Smith, A. C., Miles, M. J., & Morris, V. J. (1990). Microstructural changes during the twin-screw extrusion cooking of maize grits. *Carbohydrate Polymers, 13*, 69–84.

Orford, P. D., Parker, R., & Ring, S. G. (1993). The functional properties of extrusion-cooked waxy-maize starch. *Journal of Cereal Science, 18*, 277–286.

Pandey, J. K., & Singh, R. P. (2005). Green nanocomposites from renewable resources: Effect of plasticizer on the structure and material properties of clay-filled starch. *Starch-Starke, 57*, 8–15.

Park, H. M., Lee, W. K., Park, C. Y., Cho, W. J., & Ha, C. S. (2003). Environmentally friendly polymer hybrids Part I mechanical, thermal, and barrier properties of thermoplastic starch/clay nanocomposites. *Journal of Materials Science, 38*, 909–915.

Park, H. M., Li, X., Jin, C. Z., Park, C. Y., Cho, W. J., & Ha, C. S. (2002). Preparation and properties of biodegradable thermoplastic starch/clay hybrids. *Macromolecular Materials and Engineering, 287*, 553–558.

Powell, D. H., Fischer, H. E., & Skipper, N. T. (1998). The structure of interlayer water in Li-montmorillonite studied by neutron diffraction with isotopic substitution. *Journal of Physical Chemistry B, 102*, 10899–10905.

Powell, D. H., Tongkhao, K., Kennedy, S. J., & Slade, P. G. (1998). Interlayer water structure in Na- and Li-montmorillonite clays. *Physica B, 241–243*, 387–389.

Raquez, J. M., Nabar, Y., Narayan, R., & Dubois, P. (2007). New developments in biodegradable starch-based nanocomposites. *International Polymer Processing, 22*, 463–470.

Russell, P. L. (1987). Gelatinization of starches of different amylose amylopectin content—A study by differential scanning calorimetry. *Journal of Cereal Science, 6*, 133–145.

Sagar, A. D., & Merrill, E. W. (1995). Starch fragmentation during extrusion processing. *Polymer, 36*, 1883–1886.

Shah, P. B., Bandopadhyay, S., & Bellare, J. R. (1995). Environmentally degradable starch filled low-density polyethylene. *Polymer Degradation and Stability, 47*, 165–173.

Shen, Z., Simon, G. P., & Cheng, Y. B. (2002). Comparison of solution intercalation and melt intercalation of polymer-clay nanocomposites. *Polymer, 43*, 4251–4260.

Shogren, R. L. (1992). Effect of moisture-content on the melting and subsequent physical aging of cornstarch. *Carbohydrate Polymers, 19*, 83–90.

Shogren, R. L., Swanson, C. L., & Thompson, A. R. (1992). Extrudates of cornstarch with urea and glycols—Structure mechanical property relations. *Starch-Starke, 44*, 335–338.

Sinha Ray, S., & Okamoto, M. (2003). Polymer/layered silicate nanocomposites: A review from preparation to processing. *Progress in Polymer Science, 28*, 1539–1641.

Smits, A. L. M., Hulleman, S. H. D., van Soest, J. J. G., Feil, H., & Vliegenthart, F. G. (1999). The influence of polyols on the molecular organization in starch-based plastics. *Polymers for Advanced Technologies, 10*, 570–573.

Sposito, G., & Grasso, D. (1999). Electrical double layer structure, forces, and fields at the clay water interface. *Surfactant Science Series, 85*, 107–249.

Stevens, D. J., & Elton, G. A. H. (1971). Thermal properties of the starch/water system Part I. Measurement of heat of gelatinisation by differential scanning calorimetry. *Starch-Starke, 23*, 8–11.

Tettenhorst, R. (1962). Cation migration in montmorillonites. *American Mineralogist, 47*, 769–773.

Thirathumthavorn, D., & Charoenrein, S. (2007). Aging effects on sorbitol- and non-crystallizing sorbitol-plasticized tapioca starch films. *Starch-Starke, 59*, 493–497.

Thomas, F., Michot, L. J., Vantelon, D., Montarges, E., Prelot, B., Cruchaudet, M., et al. (1999). Layer charge and electrophoretic mobility of smectites. *Colloids and Surfaces A: Physicochemical and Engineering Aspects, 159*, 351–358.

Vaia, R. A., & Giannelis, E. P. (1997). Polymer melt intercalation in organically-modified layered silicates: Model predictions and experiment. *Macromolecules, 30*, 8000–8009.

van Soest, J. J. G., Hulleman, S. H. D., De Wit, D., & Vliegenthart, J. F. G. (1996). Changes in the mechanical properties of thermoplastic potato starch in relation with changes in B-type crystallinity. *Carbohydrate Polymers, 29*, 225–232.

van Soest, J. J. G., Hulleman, S. H. D., Dewit, D., & Vliegenthart, J. F. G. (1996). Crystallinity in starch bioplastics. *Industrial Crops and Products, 5*, 11–22.

van Soest, J. J. G., & Knooren, N. (1997). Influence of glycerol and water content on the structure and properties of extruded starch plastic sheets during aging. *Journal of Applied Polymer Science, 64*, 1411–1422.

van Tuil, R., Fowler, P., Lawther, M., & Weber, C. J. (2000). Properties of biobased packaging materials. In *Biobased packaging materials for the food industry—Status and perspectives.* Frederiksberg, Denmark: KVL.

Vergnes, B., Villemaire, J. P., Colonna, P., & Tayeb, J. (1987). Interrelationships between thermomechanical treatment and macromolecular degradation of maize starch in a novel rheometer with preshearing. *Journal of Cereal Science, 5*, 189.

Wang, S. S., Chiang, W. C., Yeh, A. I., Zhao, B., & Kim, I. H. (1989). Kinetics of phase transition of waxy corn starch at extrusion temperatures and moisture contents. *Journal of Food Science, 54*, 1298–1301.

Wilhelm, H. -M., Sierakowski, M. -R., Souza, G. P., & Wypych, F. (2003). Starch films reinforced with mineral clay. *Carbohydrate Polymers, 52*, 101–110.

Wilkie, C. A., Zhu, J., & Uhl, F. (2001). How do nanocomposites enhance the thermal stability of polymers? *Polymer Preprints, 42*, 392.

Xie, S., Zhang, S., Wang, F., Yang, M., Seguela, R., & Lefebvre, J. M. (2007). Preparation, structure and thermomechanical properties of nylon-6 nanocomposites with lamella-type and fiber-type sepiolite. *Composites Science and Technology, 67*, 2334–2341.

Zhang, Q. X., Yu, Z. Z., Xie, X. L., Naito, K., & Kagawa, Y. (2007). Preparation and crystalline morphology of biodegradable starch/clay nanocomposites. *Polymer, 48*, 7193–7200.

Zheng, Y. (2006). Study on sepiolite-reinforced polymeric nanocomposites. *Journal of Applied Polymer Science, 99*, 2163–2166.

PART VI

NOVEL CHARACTERISATION METHODS

CHAPTER

9

Starch NMR

Michael J. Gidley

Centre for Nutrition and Food Sciences, ARC Centre of Excellence in Plant Cell Walls, Queensland Alliance for Agriculture and Food Innovation, The University of Queensland, Brisbane, Queensland, Australia

1 GENERAL PRINCIPLES

Nuclear magnetic resonance (NMR) phenomena have their origins within the nucleus of certain individual atom types, e.g., ^{1}H and ^{13}C, which possess a net "nuclear" spin. The effects are only observed in a magnetic field (magnetic) and involve the exchange of energy between at least two levels (resonance). The simplest form of NMR event involves two energy levels—this is the case for ^{1}H, ^{13}C, and ^{31}P, the relevant NMR nuclei for studying starch, as illustrated in Figure 9.1.

As for other forms of spectroscopy, the energy level gap determines both the frequency of observation and the sensitivity of the experiment. In the case of NMR, the energy level gap depends on the size of the magnetic field applied (Figure 9.1). To compare spectral positions (conventionally termed chemical shifts) for data collected on spectrometers of different magnetic field, chemical shifts are expressed in parts per million (ppm) of the applied field. One consequence of the dependence on magnetic field strength is that if the latter is varied, then the NMR signal varies in frequency accordingly. This is the original basis for magnetic resonance imaging (MRI), where a designed magnetic field gradient is applied across the sample to be imaged, leading to a unique magnetic field strength for each region within the sample. The location of ^{1}H nuclei within the sample can then be deduced from the measured spectral frequencies.

For the NMR property of nuclear spin, the energy level gap is very small. This means that the population of nuclear spins is only slightly less in the higher energy level than in the lower, resulting in facile exchange of

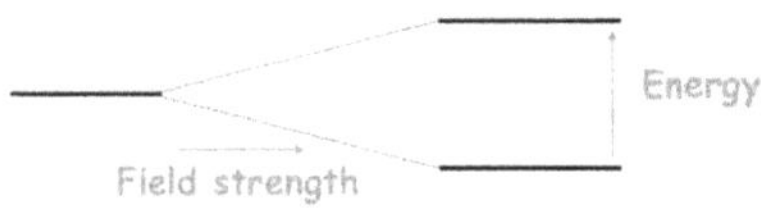

FIGURE 9.1 Schematic representation of nuclear magnetic resonance for a single nucleus. (For color version of this figure, the reader is referred to the online version of this chapter.)

nuclear spin between the energy levels (resonance). Another consequence of this small difference in populations between energy levels is that NMR is markedly less sensitive than, e.g., ultraviolet or infrared spectroscopy, which has larger energy level gaps and hence a greater difference in population levels that are perturbed by the applied energy. For dilute nuclei such as ^{13}C, the low natural abundance (~1% compared to ^{12}C) reduces sensitivity even further.

The practical operation of NMR spectroscopy involves the energetic excitation of all nuclei within the sample followed by collection of resonance and decay information as a function of time. In order to achieve sufficient signal-to-noise ratios, a large number of sequential energy inputs (known as pulses) and time domain signal acquisition steps are accumulated. The signals are then Fourier-transformed to convert from time domain to frequency domain spectral information. One reason for operating at high magnetic fields is therefore to maximize sensitivity, as typically fewer pulses and therefore shorter overall acquisition times are needed to generate adequate signal-to-noise ratios.

2 CHEMICAL AND PHYSICAL INFORMATION

There are two broad types of information that can be obtained from NMR spectroscopy. One is based on the effects of molecular structure and interactions (chemical effects) on the energy level gap, usually manifested by the chemical shift in high-resolution spectroscopy. Although chemical shifts cover only a small fraction of the applied magnetic field (e.g., 10 ppm for ^{1}H and 200 ppm for ^{13}C), effects are highly reproducible, being reliably measured to 0.001 ppm for ^{1}H and 0.01 ppm for ^{13}C by reference to chemical shift standards. Chemical shift values are primarily determined by the electronic environment of a nucleus with greater electronegativity in the vicinity, causing a downfield shift (higher ppm). For example, carbons and hydrogens in C—CH_2—C groups typically appear in the region of 1-2 ppm in ^{1}H spectra and 30-35 ppm in ^{13}C spectra; C—CH(OH)—C groups are in the region of 3-4 ppm and 60-80 ppm, respectively; and C—CH(OH)—O groups are at 4-6 ppm and 90-110 ppm, respectively. The high discriminating power of chemical shift effects and their reproducibility makes NMR a useful fingerprinting technique for complex molecular structures and compositions. For the relatively simple molecular structures within starches, both ^{1}H and ^{13}C high-resolution spectroscopy can be used to identify and quantify branch points and end groups (Gidley, 1985). Although reducing ends of starch molecules can be determined chemically and nonreducing ends can be determined by the difference in reducing power before and after debranching, ^{1}H NMR is the method of choice for quantifying the degree of branching based on the ratio of intensities observed for anomeric proton signals for (1-6)- and (1-4)-linked residues (Ao et al., 2007; Gidley, 1985). Recent advances in solubilization methods for nonaqueous solvents coupled with sharpening and shifting of relevant peaks with dilute acid have increased the reliability of the technique (Tizzotti, Sweedman, Tang, Schaefer, & Gilbert, 2011). Additional applications of NMR in chemical analysis of starches include quantification of chemical modification with organic substituents and characterization of the relative amounts of phosphate substituents using ^{31}P NMR (Kasemsuwan & Jane, 1996; Mette Bay-Smidt, Wischmann, Olsen, & Nielsen, 1994). In contrast to phosphate determination methods that measure total phosphorous, the NMR method is able to separately quantify the amounts of 6-linked and 3-linked phosphate, e.g., in transgenic lines showing a pleiotropic increase in phosphorylation (Safford et al., 1998).

The second general type of information available from NMR spectroscopy is based on relaxation properties and gives information on physical effects operating within the system. Following the application of energy to promote nuclear spins to the higher level (Figure 9.1), energy is lost as spins relax back to the lower energy level. There are two general types of relaxation. One is where energy is transferred to other nuclei within the system through dipolar interaction mechanisms; examples include T_2 and NOE effects that can operate either through bonds or through space (nm or less). This transfer of energy does not change the total magnetization in the system, in contrast to T_1 mechanisms that involve loss of energy to the environment. Relaxation processes are modulated by both the electronic environment of nuclei and local molecular motions. Depending on the type of NMR experiment, the dominant mechanism of relaxation varies, as does the motional timescale that can be probed. An overview of relevant timescales is shown in Figure 9.2. For ^{1}H signals observed in solutions, relaxation effects are sensitive to very fast motions typical of small molecules tumbling in solution. More rigid molecules (e.g., gels or solids) relax so rapidly through ^{1}H—^{1}H dipolar interactions that conventional ^{1}H NMR spectra cannot be observed. Recent advances in magic-angle spinning (MAS) probes (see later) are now allowing high-resolution ^{1}H spectra to be obtained (Brown, 2009; Diez-Pena, Quijada-Garrido, Barrales-Rienda, Schnell, & Spiess, 2005), but information is still limited.

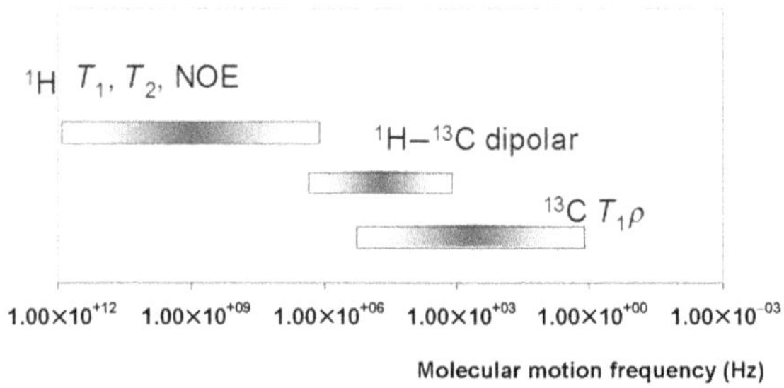

FIGURE 9.2 Schematic representation of the timescales of molecular motion that can be probed through different NMR relaxation mechanisms.

Molecular motions in the MHz and kHz frequency regions are more characteristic of (semi)solid systems and can be probed via $^{1}H—^{13}C$ dipolar interactions and ^{13}C $T_{1\rho}$ effects, respectively, using ^{13}C "solid-state" spectroscopy. For example, variable contact time experiments probe both the buildup of magnetization by cross-polarization from ^{1}H to ^{13}C via dipolar interactions and the subsequent decay of intensity via ^{13}C $T_{1\rho}$ effects, as will be discussed later.

2.1 High-Resolution NMR Spectra from Solids

The traditional application of NMR spectroscopy has been to molecules in solution where motion is fast enough to average any effects of orientation with respect to the magnetic field on observed chemical shifts, i.e., there is chemical shift isotropy. In addition, T_1 relaxation times are sufficiently short that the system returns to equilibrium rapidly and the next pulse of energy can be applied relatively quickly (typically of the order of 1 s) without saturating the system. In solids, T_1 values are longer, meaning that unreasonably long times are needed between pulses for full signal detection (typically of the order of minutes). In addition, dipolar interactions disperse energy across the system rapidly, leading to unmeasurably broad signals. One advantage of the dilute nature of the ^{13}C nucleus (~1% natural abundance) however is that $^{13}C—^{13}C$ dipolar interactions are infrequent and hence weak. For ^{13}C NMR of solids, the strong $^{13}C—^{1}H$ dipolar interaction can also be used to advantage as a route to selectively energizing ^{13}C nuclei via adjacent protons and allowing relaxation to occur through the ^{13}C $T_{1\rho}$ mechanism that typically occurs on millisecond timescales. Over the last 30 years, the applications of NMR spectroscopy to solids and semisolids such as found in many starch-based systems have become more frequent through the combined use of (a) cross-polarization (CP), (b) MAS, and (c) high-power proton dipolar decoupling (DD). CP involves the energizing of ^{13}C nuclei via ^{1}H nuclei and is most effective for carbons with directly attached hydrogens. The rate at which cross-polarization occurs depends on the strength of the $^{13}C—^{1}H$ dipolar interaction, which in turn is dependent on molecular mobility. In practical terms, only relatively rigid (segments of) molecules undergo CP, effectively discriminating against mobile components within a sample. MAS overcomes the effects of chemical shift anisotropy (i.e., the fact that each molecule has a different orientation with respect to the magnetic field resulting in different chemical shifts) inherent in solid samples because the magnitude of the anisotropy contains the term $3\cos^2\Theta - 1$. For the "magic" angle of 57° 44′, $3\cos^2\Theta = 1$ so the anisotropy term becomes zero. Solid sample rotors in NMR machines are therefore inclined at the magic angle with respect to the magnetic field and are spun sufficiently fast that "solution-like" averaging of molecular orientation is achieved. Finally, high-power proton decoupling is used to repress dipolar spreading of energy and to channel relaxation through the ^{13}C $T_{1\rho}$ pathway. The combined CP/MAS/DD approach (often abbreviated to CPMAS) has become the standard method for obtaining high-resolution ^{13}C spectra from solid starch samples (Gidley & Bociek, 1985, 1988; Paris, Bizot, Emery, Buzare, & Buleon, 2001; Tan, Flanagan, Halley, Whittaker, & Gidley, 2007).

One particularly powerful application of solid-state NMR spectroscopy is to quantify subcrystalline order in solid samples. The characteristic length scale for NMR is very short (nm or less), whereas the benchmark method for determining solid-state structure (X-ray diffraction) requires structural register on at least a 10 nm length scale. Many natural and fabricated structures containing starch have limited, if any, crystallinity but are thought to contain molecular order. Examples include cooked foods and biodegradable films based on starch. Figure 9.3 illustrates the analysis

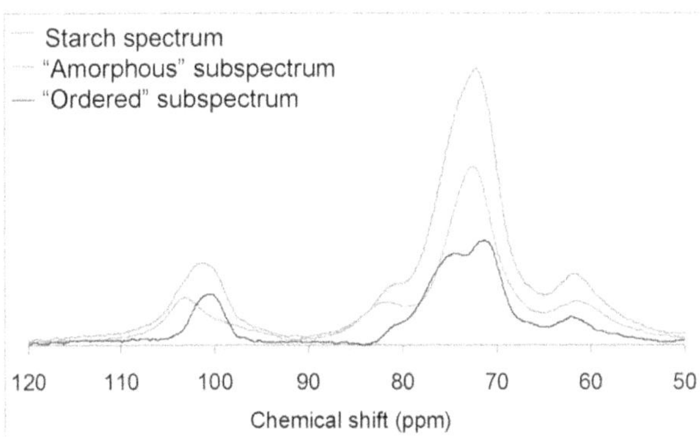

FIGURE 9.3 Solid-state spectrum of a high-amylose maize starch and computed "ordered" subspectrum obtained by subtracting a model amorphous subspectrum such that intensity is zero at 84 ppm. (For color version of this figure, the reader is referred to the online version of this chapter.)

of starch-based samples. The observed spectrum is split into subspectra representing "crystalline" and "amorphous" components based on the finding that model crystalline materials (Figure 9.4) have no detectable intensity at 84 ppm (Tan et al., 2007). The spectral features of a model amorphous material (e.g., after thorough cooking and drying—Figure 9.4) are subtracted from the observed spectrum until intensity at 84 ppm is zero. The spectral intensity required for subtraction quantifies the "amorphous"/"crystalline" ratio, which has been shown to equate to X-ray quantification for most uncooked starch granules provided an imperfect crystallite model is used (Lopez-Rubio, Flanagan, Gilbert, & Gidley, 2008). Following

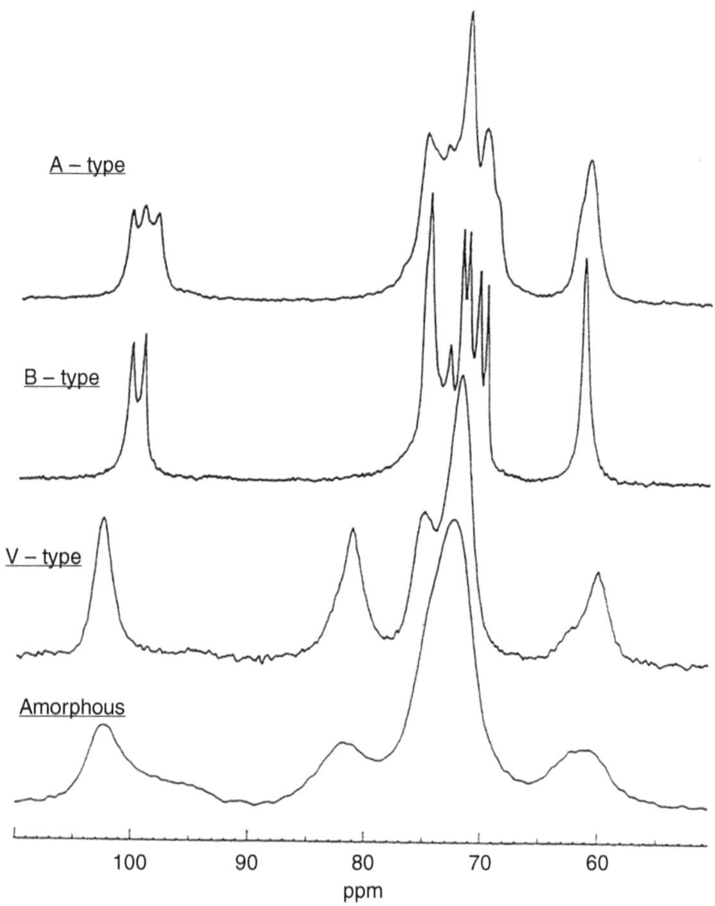

FIGURE 9.4 ^{13}C CPMAS spectra for highly crystalline (by X-ray diffraction) solid starch samples with A-, B-, and V-type crystallinity, compared with amorphous starch.

subtraction of amorphous features from starch spectra, a "crystalline" sub-spectrum is generated. The chemical shifts for this spectrum—particularly for the signal at ca. 100 ppm—are characteristic of the crystal polymorph (Figure 9.4); the relatively broad signals (Figure 9.3) are consistent with a low level of crystalline register as also observed by X-ray diffraction. Similar polymorph identification and quantification has also been achieved for cellulose in isolated form (Atalla & VanderHart, 1984), in composites with other polysaccharides (Whitney, Brigham, Darke, Reid, & Gidley, 1995), and as found in, e.g., fruit or vegetable cell walls (Foster, Ablett, McCann, & Gidley, 1996).

The relatively broad signals in amorphous starch solid-state spectra have been suggested to be due to the trapping of many different local polymer chain conformations (Gidley & Bociek, 1988). This is in contrast to the relatively sharp signals obtained for the same molecules in solution (Figure 9.5a). However, when the same solution is frozen (Figure 9.5b), the spectrum becomes very similar to that of dried amorphous starch (Figure 9.5c) apart from a lower signal-to-noise ratio due to the lower amylopectin concentration in solution. There is still scope for more information on the chain conformations in noncrystalline starches to be discovered if the structural basis for the wide range of chemical shifts could be better defined (Durran, Howlin, Webb, & Gidley, 1995).

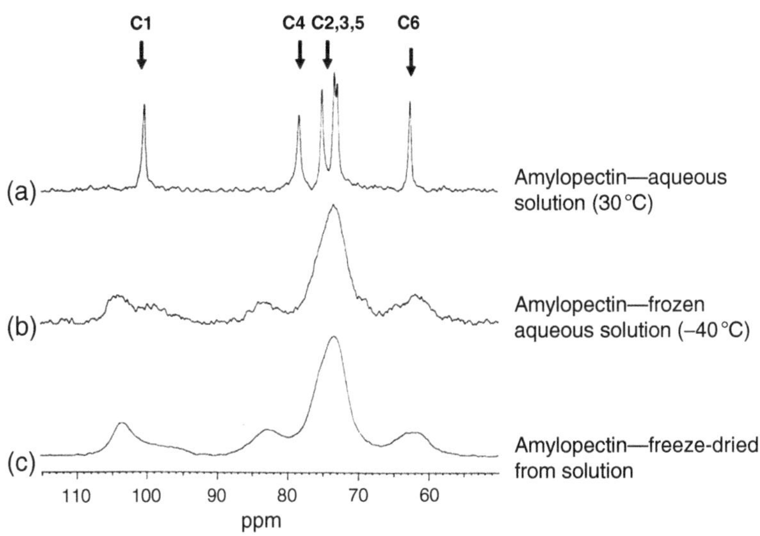

FIGURE 9.5 ^{13}C spectra for amylopectin in solution at 30 °C (using direct polarization) and −40 °C (using CPMAS) compared with solid "amorphous" amylopectin (CPMAS).

2.2 Mobility-Resolved NMR Spectroscopy

A major advantage of NMR spectroscopy is that the operator has control over how and when energy is transferred to the sample, which mechanisms are available for energy evolution, and how energy is lost to the environment. This is possible through a wide range of sophisticated "pulse sequences" that control the timing, power, and selectivity of energy inputs into the system. A practical consequence is that a single sample can be analyzed by a number of complementary techniques to build a more complete picture of the molecular basis for the material's observed macroscopic behavior. One example of particular value to the semisolid systems frequently encountered in food is that of mobility-resolved spectroscopy (Gidley, 1992) in which a series of spectra—typically ^{13}C—are acquired under conditions that detect signals with characteristic molecular mobility properties. This allows the association of specific molecules or parts of molecules with defined molecular mobility regimes within intact samples. Examples of applications include a number of food gels (Gidley, 1989, 1992), plant cell walls (Foster et al., 1996), and proteins in limited water (Kealley et al., 2008). For plant cell walls, rigid cellulose fibers are only observed in CPMAS spectra, whereas the more flexible pectin network has a sufficiently short T_1 value to also be detected by conventional direct polarization (DP), provided that high-power proton decoupling is also applied to prevent energy spreading via T_2 mechanisms. One example of this approach in the starch area is the molecular characterization of amylose gelation (Gidley, 1989). This showed that the entire "solid" component within amylose gels (defined as being detectable under standard CPMAS acquisition conditions) had a spectrum identical with that of B-type crystalline material (Figure 9.4), although there was only very limited X-ray diffraction from the sample. In contrast, the "mobile" component within amylose gels (defined as being detectable under standard direct polarization conditions) had the same spectrum as soluble starch polysaccharides (Figure 9.5). The ability to trap soluble polysaccharides as an amorphous solid (Figure 9.5) allowed the two conformational features within amylose gels to be quantified by acquiring a CPMAS spectrum in the frozen state. This spectrum (Gidley, 1989) contained contributions from amorphous and B-type crystalline solid forms of starch and could thus be quantified by integration. Depending on the amylose sample, 10% (w/v) gels were shown to contain between 67% and 83% "solid" B-type conformations, with the residual 33-17% being due to "mobile" amorphous conformations (Gidley, 1989). There are many other possibilities for a more precise definition of structure/rheology relationships in starch-based systems using this type of approach. One example is the characterization of the response of a range of granular starches to hydration (Tang & Hills, 2003), which showed a richness of local environment and raw material dependence that shed considerable light on structure and mobility within hydrated granules.

2.3 Probing Water "Pool" Sizes Using ^{1}H T_2 Properties

Water is often a major component in starch-based systems—even in solids. Techniques for characterizing the state of water within intact materials are important in understanding microenvironments that contribute to bulk material properties. The fact that water (as a very small molecule) has different relaxation properties to other components has been exploited to develop methods for characterizing the "pool" size of water within complex materials. The principle involved is that liquid water has a comparatively long ^{1}H T_2 (several seconds), whereas materials such as emulsifiers/oils found at the droplet surface of water-in-oil emulsions or biopolymers such as starch have much shorter values (milliseconds or microseconds). Water protons are mobile and in most cases will be present at the interface/surface for a fraction of time that is dependent on the water pool size and potentially the local viscosity. When water protons visit a surface, they can exchange energy with the much faster relaxing molecules present there. Thus, measurement of water ^{1}H T_2 values in the presence of a surface in comparison with the absence of the surface gives a measure of the proportion of time that a water proton spends adjacent to the surface and hence the size of the "pool" in which it is contained (Figure 9.6).

This method has been found to be so useful for characterizing water-in-oil emulsion droplet sizes (Johns & Hollingsworth, 2007) that it has become the industry standard method for defining droplet size distributions of critical importance to material properties and is one of the few examples where NMR equipment is used routinely within industrial processing operations. In the area of starch, ^{1}H relaxation studies have been used for the detailed characterization of intragranular water (Tang, Godward, & Hills, 2000) and a number of biopolymers at low moisture

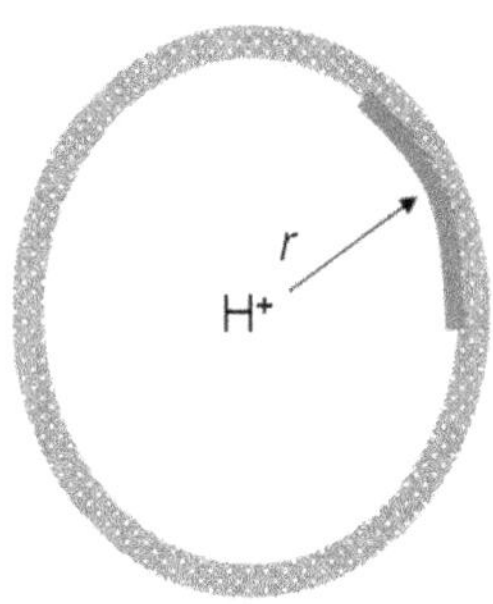

FIGURE 9.6 Schematic illustration of one mechanism that determines water ^{1}H T_2 values. The distance "r" that a ^{1}H nucleus is from a protein surface defines the fractional time spent within the sphere of influence of fast-relaxing protein compared with slow-relaxing bulk water. (For color version of this figure, the reader is referred to the online version of this chapter.)

level, although full understanding of the structural basis for some of the observed relaxation behaviors remains elusive (Belton, 2012).

3 CONCLUSIONS AND PERSPECTIVES

The wide diversity of experiments that can be performed using magnetic resonance principles and the ability to use multiple NMR techniques to characterize a single sample gives many possibilities for application of the suite of techniques to starch-based systems. The (deceptively) simple molecular structure allows a level of understanding that would not be possible with more complex monomer types as found, e.g., in proteins. The ability to measure quantitatively the degree of branching and levels of substitution introduced either naturally or by chemical modification provides very useful information, but the real power of NMR comes in connecting molecular structures and conformations to physical and rheological properties for (semi)solids, e.g., by identifying the structural basis for segmental flexibility. In addition, the toolkit of MRI allows at least some of these experiments to be carried out on defined subcentimeter regions within samples, thereby further broadening the utility of magnetic resonance. The information that will be deduced from these methods is likely to be of major importance in the beneficial utilization of starch, particularly related to its status as the dominant processable biopolymer produced by plants.

References

Ao, Z. H., Simsek, S., Zhang, G. Y., Venkatachalam, M., Reuhs, B. L., & Hamaker, B. R. (2007). Starch with a slow digestion property produced by altering its chain length, branch density, and crystalline structure. *Journal of Agricultural and Food Chemistry, 55*, 4540–4547.

Atalla, R. H., & VanderHart, D. L. (1984). Native cellulose—A composite of 2 distinct crystalline forms. *Science, 223*, 283–285.

Belton, P. S. (2012). NMR studies of hydration in low water content biopolymer systems. *Magnetic Resonance in Chemistry, 49*, S127–S132.

Brown, S. P. (2009). Recent advances in solid-state MAS NMR methodology for probing structure and dynamics in polymeric and supramolecular systems. *Macromolecular Rapid Communications, 30*, 688–716.

Diez-Pena, E., Quijada-Garrido, I., Barrales-Rienda, J. M., Schnell, I., & Spiess, H. W. (2005). Advanced H-1 solid-state NMR spectroscopy on hydrogels, 1—The effect of hydrogen bonding in the collapse of poly(methacrylic acid) hydrogels. *Macromolecular Chemistry and Physics, 205*, 430–437.

Durran, D. M., Howlin, B. J., Webb, G. A., & Gidley, M. J. (1995). Ab initio nuclear shielding calculations of a model α-(1→4) glucan. *Carbohydrate Research, 271*, C1–C5.

Foster, T. J., Ablett, S., McCann, M. C., & Gidley, M. J. (1996). Mobility—Resolved 13C NMR spectroscopy of primary plant cell walls. *Biopolymers, 39*, 51–66.

Gidley, M. J. (1985). Quantification of the structural features of starch polysaccharides by NMR spectroscopy. *Carbohydrate Research, 139*, 85–94.

Gidley, M. J. (1989). Molecular mechanisms underlying amylose aggregation and gelation. *Macromolecules, 22*, 351–358.
Gidley, M. J. (1992). High resolution solid state NMR of food materials. *Trends in Food Science and Technology, 3*, 231–236.
Gidley, M. J., & Bociek, S. M. (1985). Molecular organisation in starches: A 13C CP/MAS NMR study. *Journal of the American Chemical Society, 107*, 7040–7044.
Gidley, M. J., & Bociek, S. M. (1988). 13C CP/MAS NMR studies of amylose inclusion complexes, cyclodextrins and the amorphous phase of starch granules: relationships between glycosidic linkage conformation and solid state 13C chemical shifts. *Journal of the American Chemical Society, 110*, 3820–3829.
Johns, M. L., & Hollingsworth, K. G. (2007). Characterisation of emulsion systems using NMR and MRI. *Progress in Nuclear Magnetic Resonance Spectroscopy, 50*, 51–70.
Kasemsuwan, T., & Jane, J. L. (1996). Quantitative method for the survey of starch phosphate derivatives and starch phospholipids by P-31 nuclear magnetic resonance spectroscopy. *Cereal Chemistry, 73*, 702–707.
Kealley, C. S., Rout, M. K., Dezfouli, M. R., Strounina, E., Whittaker, A. K., Appelqvist, I. A. M., et al. (2008). Solid state structure and molecular mobility of soy glycinin as a function of moisture content. *Biomacromolecules, 9*, 2937–2946.
Lopez-Rubio, A., Flanagan, B. M., Gilbert, E. P., & Gidley, M. J. (2008). A novel approach for calculating starch crystallinity and its correlation with double helix content: A combined XRD and NMR study. *Biopolymers, 89*, 761–768.
Mette Bay-Smidt, A., Wischmann, B., Olsen, C. E., & Nielsen, T. H. (1994). Starch-bound phosphate in potato as studied by a simple method for determination of organic phosphate and ^{31}P NMR. *Starch, 46*, 167–172.
Paris, M., Bizot, H., Emery, J., Buzare, J. Y., & Buleon, A. (2001). NMR local range investigations in amorphous starchy substrates I. Structural heterogeneity probed by C-13 CP-MAS NMR. *International Journal of Biological Macromolecules, 29*, 127–136.
Safford, R., Jobling, S. A., Sidebottom, C. M., Westcott, R. J., Cooke, D., Tober, K. J., et al. (1998). Consequences of antisense RNA inhibition of starch branching enzyme activity on properties of potato starch. *Carbohydrate Polymers, 35*, 155–168.
Tan, I., Flanagan, B. M., Halley, P. J., Whittaker, A. K., & Gidley, M. J. (2007). A method for estimating the nature and relative proportions of amorphous, single, and double-helical components in starch granules by 13C CP/MAS NMR. *Biomacromolecules, 8*, 885–891.
Tang, H.-R., Godward, J., & Hills, B. (2000). The distribution of water in native starch granules—A multinuclear NMR study. *Carbohydrate Polymers, 43*, 375–387.
Tang, H., & Hills, B. P. (2003). Use of 13C MAS NMR to study domain structure and dynamics of polysaccharides in the native starch granules. *Biomacromolecules, 4*, 1269–1276.
Tizzotti, M., Sweedman, M. C., Tang, D., Schaefer, C., & Gilbert, R. G. (2011). New 1H NMR procedure for the characterization of native and modified food-grade starches. *Journal of Agricultural and Food Chemistry, 59*, 6913–6919.
Whitney, S. E. C., Brigham, J. E., Darke, A. H., Reid, J. S. G., & Gidley, M. J. (1995). In vitro assembly of cellulose/xyloglucan networks: Ultrastructural and molecular aspects. *The Plant Journal, 8*, 491–504.

PART VII

NOVEL PROCESSING

CHAPTER

10

Processing of Plasticized Starch-Based Materials: State of the Art and Perspectives

Fengwei Xie[1], *Peng Liu*[2], *Long Yu*[3,4]

[1]Australian Institute for Bioengineering and Nanotechnology, The University of Queensland, Brisbane, Queensland, Australia

[2]School of Chemistry and Chemical Engineering, Guangzhou University, Guangzhou, China

[3]Centre for Polymer from Renewable Resources, ERCSPP, College of Light Industry and Food Sciences, South China University of Technology, Guangzhou, China

[4]CSIRO Materials Science and Engineering, Melbourne, Victoria, Australia

1 INTRODUCTION

Due to the environmental concerns and the shortage of oil, the use of starch resources in nonfood applications has experienced considerable development in the past decades in order to find substitutes to petroleum-based plastics (Liu, Xie, Yu, Chen, & Li, 2009; Yu, Dean, & Li, 2006). Starch has advantages such as low cost, wide availability, and total compostability without toxic residues. And it can be processed by traditional polymer-processing techniques such as extrusion without significant modification. However, since raw starch exists in granules (as a form of energy storage in nature), it is not truly thermoplastic. In order to produce starch-based materials, starch has to be thermally processed with a plasticizer (usually water) where the phase transition of starch occurs, which is considered as the plasticization of starch (which is more complex than that of standard synthetic polymers). By plasticization, raw starch can be transformed into a homogeneous polymeric state, which is known as "thermoplastic starch (TPS)." The phase transition of starch plays a critical role in the plasticization of starch. The degree of phase transition also determines the processability and final product properties.

During the processing of starch, it is important to control the rheological behavior of plasticized starch to avoid fundamental flow-related engineering problems and to meet the quality requirements of the final products (Xie, Halley, & Avérous, 2012). Different conditions may be used for different processing techniques. For example, starch/polyester blends need to be processed in melt state under high shear, 10^2-10^3 s^{-1} in extrusion and 10^3-10^4 s^{-1} in injection molding (Schwach & Avérous, 2004). However, the rheological characterization of plasticized starch is not a simple issue. Not only we need a rheometer to be capable to prevent the loss of water (as the most commonly used plasticizer for starch), to impose enough thermomechanical input and cause it to flow prior to measurement, and to handle the high melt viscosity of plasticized starch, but also we have to link the rheological properties to the structural changes of starch that are determined by different processing conditions.

This chapter recognizes the phase transition and rheology as two most important aspects in the processing of plasticized starch-based materials and reviews the state of the art of the processing strategies and techniques for producing different kinds of plasticized starch-based products such as films, foams, and shaped articles. Though solution casting is the simplest and cheapest way of producing plasticized starch-based films/sheets as has been used in numerous studies, this technique has limited potential for practical industrial production due to its low efficiency and high energy consumption and thus will not be involved in the current chapter. While a comprehensive review on thermal processing of starch has been done before (Liu et al., 2009), this chapter will more focus on the extrusion and related techniques, which are considered as the core of ongoing development in this field.

2 PLASTICIZATION OF STARCH

2.1 Phase Transition of Starch

2.1.1 Phase Transition of Starch Under Shearless Conditions

When native starch granules are heated in water, their semicrystalline nature and 3D architecture are gradually disrupted, resulting in the phase transition from an ordered granular structure into a disordered state in water, which is known as "gelatinization" (Atwell, Hood, Lineback, Varriano-Marston, & Zobel, 1988; Lelievre, 1974; Ratnayake, Jackson, & Steve, 2008). Gelatinization is an irreversible process that includes, in a broad sense and in time-temperature sequence, granular swelling, native crystalline melting (loss of birefringence), and molecular solubilization (Sullivan & Johnson, 1964). If the water concentration is too high, the crystallites in starch might be pulled apart by swelling, leaving none to be melted at higher temperatures. If the water concentration is limited, the swelling forces will be much less significant and the steric hindrance is high; thus, complete gelatinization will not occur in the usual temperature range (Lai & Kokini, 1991; Liu et al., 2005; Liu, Yu, Xie, & Chen, 2006). However, as the temperature increases, starch molecules will become progressively more mobile and eventually the crystalline regions will be destructured (Donovan, 1979). The process of gelatinization with low water content could more accurately be defined as the "melting" of starch (Lai & Kokini, 1991).

Different techniques have been developed to characterize the phase transition of starch, such as microscope with hot-stage, differential scanning calorimetry (DSC), X-ray diffraction, and nuclear magnetic resonance (Liu et al., 2009; Xie, Liu, Chen, et al., 2006). Among them, DSC has gained most popularity due to its convenience in use and accuracy in results. By using a stainless steel high-pressure pan (cf. Figure 10.1a), Liu et al. (2006) systematically studied the phase transition behaviors of maize starches with different amylose/amylopectin ratios and found that the temperature positions, number, and enthalpies of the endotherms closely depended on amylose/amylopectin ratio and moisture content (cf. Figure 10.1b). The total enthalpy of gelatinization was increased with increasing amylopectin and water content in general. In another study where the same starch samples were used and both water and glycerol were used as plasticizers, Liu, Xie, Li, et al. (2011) revealed that the distinct pattern of multiphase transitions could be greatly affected by glycerol/water mixture content and glycerol/water ratio (cf. Figure 10.1c and d). Higher glycerol/water ratio tended to shift upward and even merged the multiple thermal transitions. Glycerol behaved as a coagent with water for starch phase transition. Higher temperature enabled glycerol to facilitate phase transitions of maize starch to a greater extent if abundant plasticizer was present. For maize starch with higher amylose content, this "abundant" level of plasticizer content could be higher. All these results could be instructive to the thermal processing of starch-based foods and thermoplastics.

2.1.2 Phase Transition of Starch Under Shear Conditions

The gelatinization/melting behavior of starch is quite different when shear treatment is imposed (Xie, Liu, Chen, et al., 2006). The significance of such studies is that most processing techniques for starch polymers involve shear treatment. In extrusion processing, shear forces can physically tear apart the starch granules, allowing faster transfer of water into

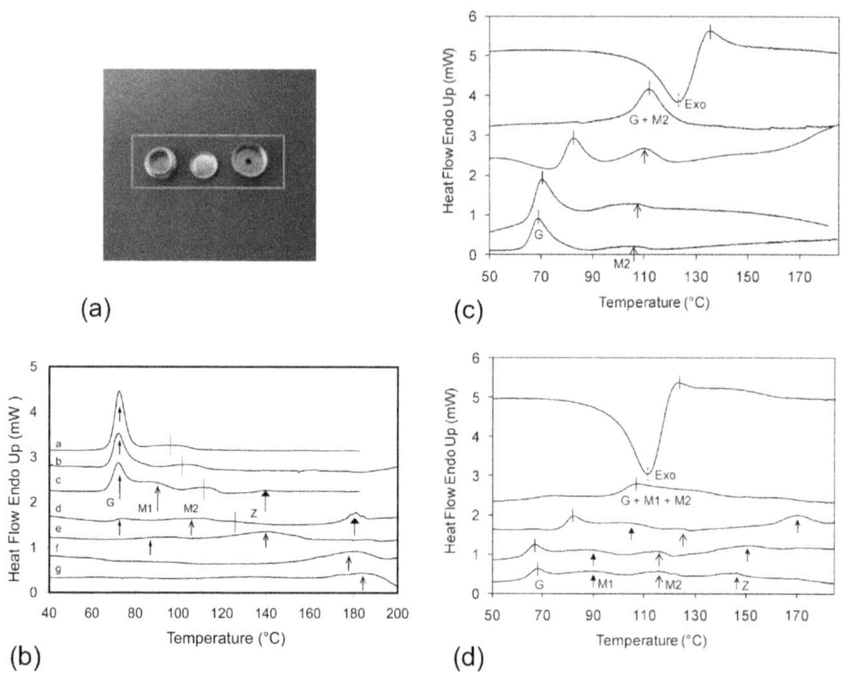

FIGURE 10.1 Application of DSC in the studies of phase transition of starch: (a) high-pressure stainless steel pan of DSC; (b) DSC results of regular maize starch with different water contents, 9%, 16.15%, 29.9%, 40%, 51.8%, 65.3%, and 74.57%, from bottom to top; and (c and d) DSC results of regular maize starch with 70% and 50% glycerol/water mixture content, respectively (glycerol/water ratio, 0/100, 10/90, 50/50, 90/10, and 100/0, from bottom to top). (For color version of this figure, the reader is referred to the online version of this chapter.) *Panel (b): Reprinted from Liu et al. (2006), Copyright (2006), with permission from Elsevier. Panels (c and d): Reprinted from Liu et al. (2011), Copyright (2011), with permission from Elsevier.*

the interior molecules (Burros, Young, & Carroad, 1987). Therefore, during extrusion, the loss of crystallinity is caused not only by water penetration but also by the mechanical disruption of molecular bonds due to the intense shear fields within the extruder (Barron, Bouchet, Della Valle, Gallant, & Planchot, 2001; Wang & Zheng, 1995; Zheng, Chiang, & Wang, 1995; Zheng & Wang, 1994).

An attempt was made by Xie, Yu, Chen, and Li (2008) to study the phase transition of starch with low moisture content (45%) and under shear stress by using dynamic mechanical analysis (DMA) with a shear sandwich mode (cf. Figure 10.2a). The results showed that the variations in physical properties (indicated by damp tanδ in DMA) occurred both before thermal transition (indicated by heat flow in DSC) started and after thermal transition completed (cf. Figure 10.2b). However, the shear treatment in DMA was quite small. By using a twin-rotor mixer

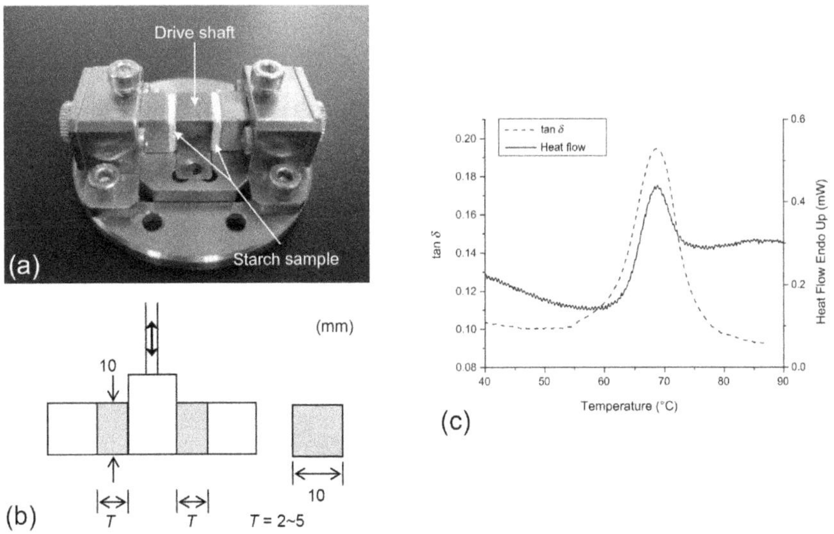

FIGURE 10.2 Application of DMA in the study of phase transition of starch under shear stress: (a) and (b) starch samples in the sandwich mode sample holder of DMA and (c) phase transition process measured by DMA (sample thickness, 3 mm; frequency, 1.0 Hz; heating rate, 2 °C min^{-1}) and DSC (heating rate, 2 °C min^{-1}) for rice starch with 45.5% water content. (For color version of this figure, the reader is referred to the online version of this chapter.) *Reprinted from Xie et al. (2008), Copyright (2008), with permission from Elsevier.*

(Haake Rheomix®) (cf. Figure 10.3a and b), Xue, Yu, Xie, Chen, and Li (2008) and Wang et al. (2010) investigated the phase transition of starch under strong shear treatment, just like in an extruder. The torque curve could be used to reflect the viscosity change and to estimate the time needed to achieve the steady molten starch phase (cf. Figure 10.3), which could provide useful information for practical extrusion processing.

2.2 Rheology of Plasticized Starch Melts

2.2.1 Viscous Properties of Plasticized Starch Melts

Steady-shear viscosity is the most prevalently measured rheological property of plasticized starch melts. This is because practical production requires the understanding of melt flow curve of the material. The most common method for studying the steady-shear viscosity of plasticized starch is to use a capillary/slit die viscometer, which is incorporated to an extruder. This can be practiced either in one step where the viscometer is directly attached to the extruder where the starch is plasticized (Martin, Avérous, & Della Valle, 2003) or in two steps where the starch is firstly plasticized in an extrusion process and then remelted and measured through another extrusion run with die viscometer (Xie, Yu,

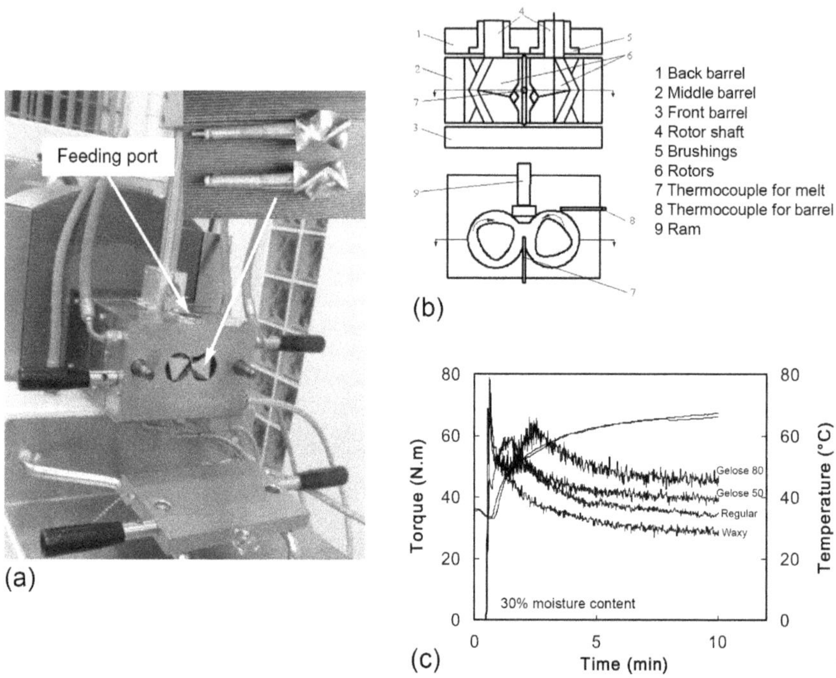

FIGURE 10.3 Application of Haake Rheomix® in the study of phase transition of starch under shear stress: (a) and (b) Haake Rheomix® with twin rotors and (c) torque variations with time for different maize starches with different amylose contents (30% water content, 30 rpm, 40 °C initial temperature). (For color version of this figure, the reader is referred to the online version of this chapter.) *Panel (a): Reprinted from Xue et al. (2008), Copyright (2008), with permission from Elsevier. Panels (b and c): Reprinted from Wang et al. (2010), Copyright (2010), with permission from John Wiley and Sons.*

Su, et al., 2009). In the former case, "in-line" measurements can be done with thermomechanical conditions (shear rate, SME, and temperature) comparable to those used during the practical extrusion processing. Besides, some interesting rheometers have been used for understanding the viscous properties of plasticized starch, including preshearing rheometer (Rheoplast®) (Barron, Valle, Colonna, & Vergnes, 2002; Della Valle, Vergnes, & Lourdin, 2007; Vergnes & Villemaire, 1987; Vergnes, Villemaire, Colonna, & Tayeb, 1987), mixer rheometer (Haake Rheomix®) (Brouillet-Fourmann, Carrot, Mignard, & Prochazka, 2002), and multipass rheometer (Tajuddin, Xie, Nicholson, Liu, & Halley, 2011). It has been generally considered that plasticized starch melts have a shear-thinning behavior, which can be expressed in the power-law equation (Della Valle, Colonna, Patria, & Vergnes, 1996; Martin et al., 2003; Tajuddin et al., 2011; Xie et al., 2009):

$$\eta_s = K\dot{\gamma}^{n-1} \tag{10.1}$$

where $\dot{\gamma}$ is the shear rate, K is the consistency coefficient, and n is the power-law index (<1). The K value is dependent on thermomechanical treatment, temperature, and plasticizer content, which can be expressed as (Martin et al., 2003; Tajuddin et al., 2011)

$$K = K_0 \exp\left(\frac{E}{RT} - \alpha MC - \alpha' GC - \beta SME\right) \tag{10.2}$$

where MC is the moisture content, GC is the glycerol content, SME is the specific mechanical energy, T is the temperature, E/R is the reduced flow activation energy, and K_0, α, α', and β are coefficients.

During extrusion processing, the structural modification of starch (which influences the viscosity of the product) depends on a number of factors such as specific mechanical energy (SME), feeding rate, screw speed, screw profile, and barrel pressure in an extruder. However, SME has been preferably used to evaluate the effect of thermomechanical treatment on the structural changes (granular transformation and macromolecular degradation) in starch (Martin et al., 2003) because SME is a combined effect manipulated by the screw speed and feed rate and it is independent of machine characteristics. It has been unanimously agreed that higher SME input can result in a greater starch structural changes (Aichholzer & Fritz, 1998; Berzin, Tara, & Tighzert, 2007; Della Valle, Boché, Colonna, & Vergnes, 1995; Della Valle et al., 1996; Vergnes, Della Valle, & Tayeb, 1993), which lead to the decreased viscosity of plasticized starch melts (cf. Equation 10.2).

It is well known that an increase in temperature or plasticizer content would result in a lower viscosity of plasticized starch melts, which has been widely reported in the literature (Della Valle et al., 1996; Martin et al., 2003; Tajuddin et al., 2011; Xie et al., 2009). This is because a higher temperature could increase the mobility of polymer molecular chains and reduce the flow resistance of polymer melts, whereas a higher plasticizer content could decrease the polymer entanglement density and increase the ease of disentanglement. It is worth noting that, though both higher amounts of water and glycerol would result in a lower viscosity, they may have different plasticizing effects on starch. According to a recently study by Tajuddin et al. (2011), an increase in glycerol/water ratio (1:4, 2:3, and 3:2) resulted in a stronger shear-thinning behavior (as reflected by a lower n value), while total plasticizer content (70-110%) had little influence on n (cf. Figure 10.4). It was thus proposed that, when plasticized mainly by water, amylopectin molecule (mainly having clusters composed of short-chain branches (DP $\approx$ 15)) has a compact ellipsoidal conformation by greatly unfolding their branches, and these "ellipsoids" are easy to move; thus, more Newtonian-like behavior is shown; the "ellipsoids," however, are largely absent when glycerol is the main plasticizer,

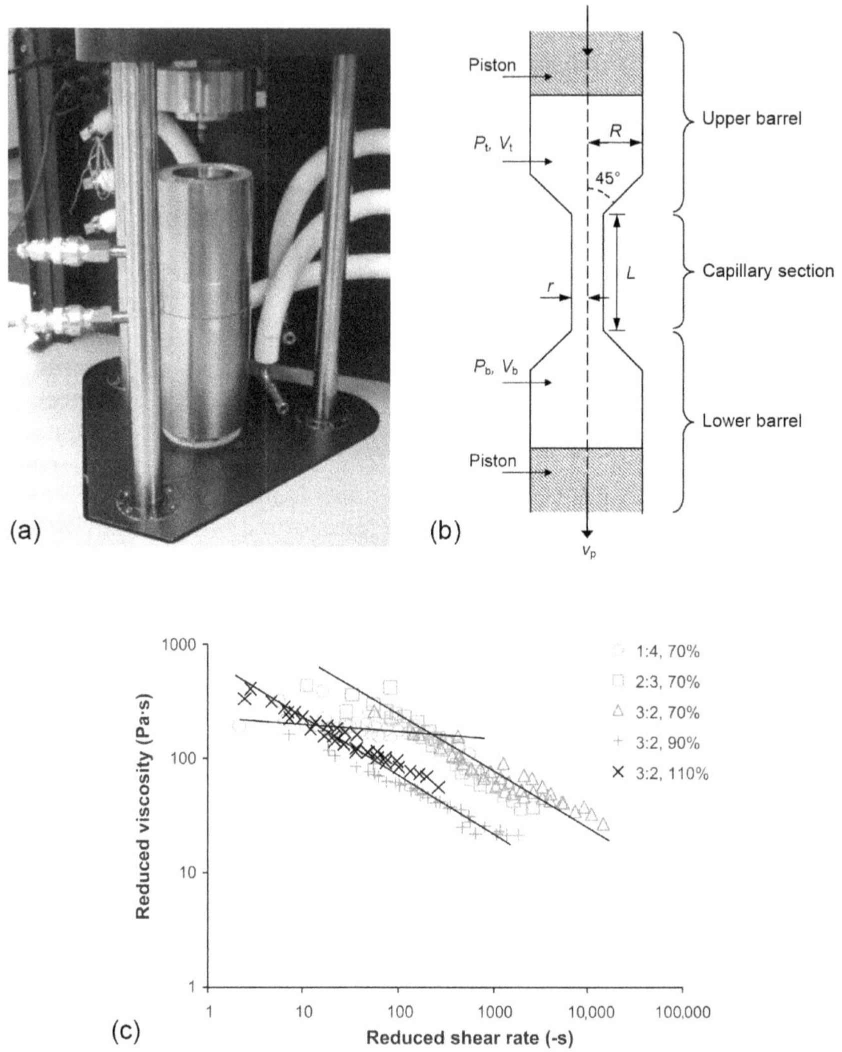

FIGURE 10.4 Application of multipass rheometer in the study of rheological properties of starch: (a) and (b) multipass rheometer ($R=5$ mm; $r=2$ mm; $L=1$, 10, or 40 mm) and (c) master curve of rheological results of plasticized waxy starch melt at temperature (T_0) of 110 °C, glycerol content (GC_0) of 54%, and moisture content (MC_0) of 36%, obtained by time-temperature-moisture content-glycerol content superposition (plotting η/a_{TMCGC} as a function of γa_{TMCGC} where a_{TMCGC} is the shift factor defined as $a_{TGCMC} = \exp\left[\frac{E}{R}\left(\frac{1}{T_i} - \frac{1}{T_0}\right) - \alpha(GC_i - GC_0) - \alpha'(MC_i - MC_0)\right]$) of the results measured at various T_i, GC_i, and MC_i conditions. (For color version of this figure, the reader is referred to the online version of this chapter.) *Panels (b) and (c): Reprinted from Tajuddin et al. (2011), Copyright (2011), with permission from Elsevier.*

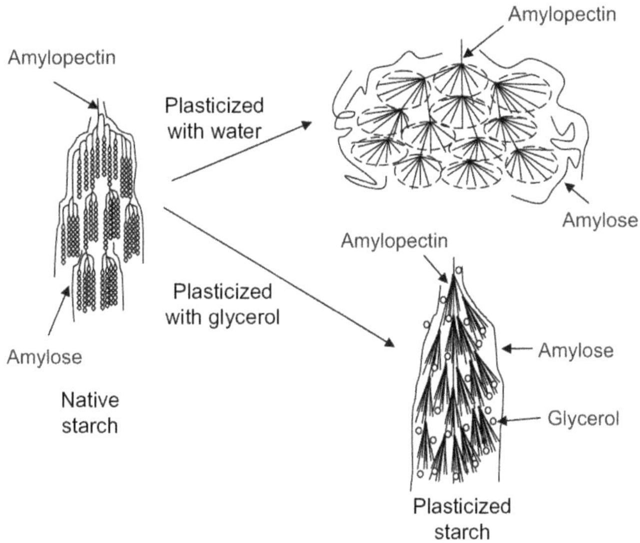

FIGURE 10.5 Schematic representation of microstructural changes of starch under processing with different plasticizers. *Reprinted from Tajuddin et al. (2011), Copyright (2011), with permission from Elsevier.*

and the amylopectin molecules behave more like a standard polymer and show stronger shear-thinning behavior (cf. Figure 10.5).

Starches with different amylose/amylopectin ratios but from the same botanical origin have attracted particular scientific interests as they display different processing and product properties. By using a slit die viscometer attached to a single-screw extruder (SSE) (cf. Figure 10.6a), Xie et al. (2009) systematically investigated the shear viscous properties of maize starches with different amylose/amylopectin ratios and found that the higher the amylose content, the higher is the shear viscosity and the more pronounced is the shear-thinning behavior, under the same shear rate range (cf. Figure 10.6). This is in agreement with a previous study by Della Valle et al. (1996). These results can be ascribed to the special molecular structures of amylopectin and amylose. As discussed before, the compact ellipsoidal conformation of amylopectin reduced its ability to create entanglements, resulting in a lower viscosity (Della Valle et al., 1996; Xie et al., 2009). In contrast, the long linear chains of amylose cause more entanglements between macromolecules, which increases zero shear viscosity and enhanced non-Newtonian behavior. In addition, the high viscosity of high-amylose starch can also be attributed to its lower tendency to undergo granular transformation (Liu et al., 2006) and macromolecular degradation (Liu, Halley, & Gilbert, 2010).

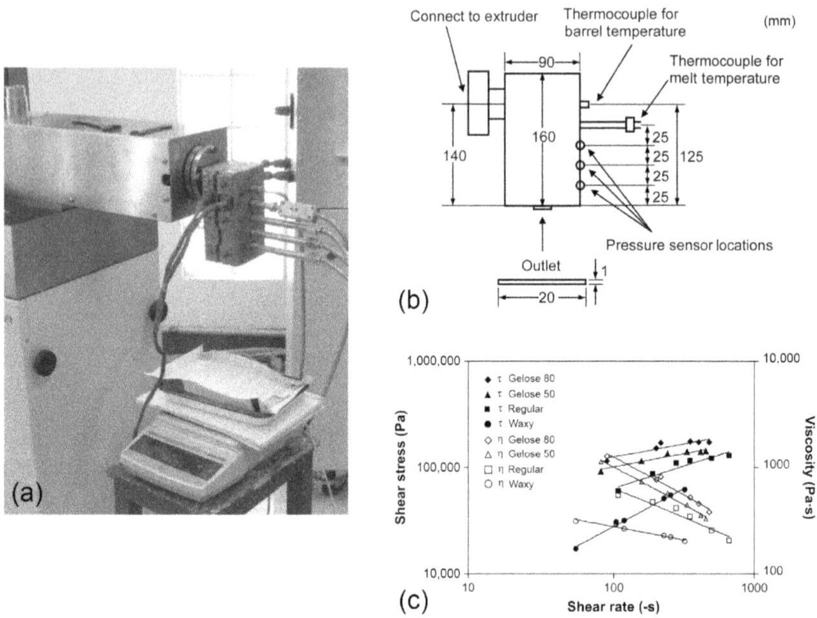

FIGURE 10.6 Application of a slit die rheometer in the study of rheological properties of starch: (a) the slit die rheometer attached to a single-screw extruder; (b) schematic representation of the slit die rheometer; and (c) rheological results of maize starches with different amylose contents (moisture content 23%; extrusion temperature 130 °C). (For color version of this figure, the reader is referred to the online version of this chapter.) *Panel (c): Reprinted from Xie et al. (2009), Copyright (2009), with permission from Elsevier.*

2.2.2 *Elastic Properties of Plasticized Starch Melts*

Though plasticized starch melts are commonly considered to exhibit a viscoelastic behavior, the elastic properties of them have received much less attention than viscous properties, partly due to the difficulty in measurements (Dealy & Wissbrun, 1999). The characterization of elastic component of plasticized starch melts can be associated with the extensional viscosity (η_e) and the first normal stress difference (N_1), which can be measured by the analysis of entrance and exit pressures data in a die viscometer (Martin et al., 2003; Senouci & Smith, 1988). According to the η_e and N_1 results, Martin et al. (2003) found a strong elastic behavior displayed by plasticized starch melts. The Trouton ratio, T_r, which is the ratio of the extensional viscosity to the shear viscosity, is commonly used to evaluate the elasticity of a material (Jones, Walters, & Williams, 1987). In uniaxial extensional flow, $T_r=3$ is displayed for inelastic non-Newtonian fluids, and $T_r>3$ for viscoelastic fluids. As reported by Bhattacharya, Padmanabhan, and Seethamraju (1994) for maize meal in uniaxial extensional flow, the T_r values ranging from 25 to 50, as affected by moisture content, barrel temperature, and deformation rate, clearly demonstrated that the extensional viscosity was much higher than the shear viscosity; thus, a significant elastic behavior was confirmed. The significance of understanding the elastic properties of plasticized starch is that they have a close relationship with some processing techniques such as film blowing and foaming (Xie et al., 2012).

3 EXTRUSION PROCESSING OF PLASTICIZED STARCH-BASED MATERIALS

3.1 Extrusion Facilities

The main extrusion facility for starch extrusion processing can be an SSE or a twin-screw extruder (TSE). An SSE can handle the high viscosity of starch and provide a high processing pressure for continuous metering of starch through die shapes, while a TSE has a large operational flexibility (individual barrel zone temperature control, multiple feeding/injection, and screw configuration for a different degree of mixing/kneading) and is useful for intensive mixing and compounding of components into plasticized starch. In both SSE and TSE, residence times and SME inputs can be controlled, and high efficiency production can be achieved. It is worth noting that a TSE is more suitable to process raw starch powder since its twin screws have a self-wiping ability, whereas an SSE may have the problem with conveying the starch powder at the feeding port.

Figure 10.7a gives a schematic representation of the TSE (ThermoHaake Rheomex PTW 24/40p, Ø30, screw diameter $D=24$ mm, and screw length $L=28D$) used in a number of studies by Yu and coworkers (Li, Liu, Zou, et al., 2011; Su, Xie, Li, et al., 2009; Xie, Yu, Liu, & Dean, 2006; Xie et al., 2009). There are eight temperature controlling zones (with water cooling) along the barrels of the extruder. The water cooling is necessary since the massive viscous heat dissipation during extrusion may increase the temperature of plasticized starch by up to 50 °C (Xie et al., 2012). Though the barrels have a longitudinal split design for sample collection and cleaning purposes, this extruder is well capable to prevent the moisture loss during extrusion. By coupling with different kinds of dies, the extruder can be used to produce threads that are then blade cut into pellets or films/sheets, which are then collected by a takeoff device.

Due to the different characteristics of SSE and TSE as mentioned before, a TSE is normally used for the first run to produce plasticized starch pellets and an SSE is used for the second run to produce a product like films/sheets or foams (Fishman, Coffin, Onwulata, & Willett, 2006). This two-stage technique can offer an easier and more stable extrusion process and higher quality of starch products. Technically, the use of a complex

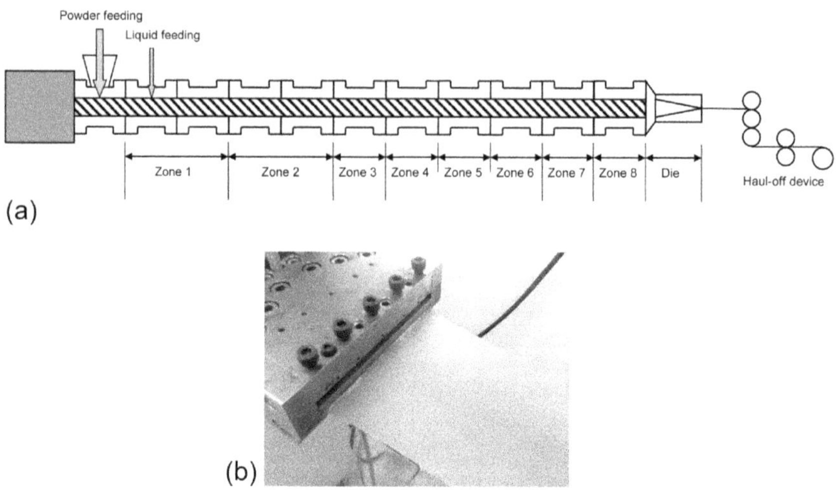

FIGURE 10.7 (a) Schematic representation of the extrusion system for starch film production and (b) photo of a typical extruded corn starch film leaving the die. (For color version of this figure, the reader is referred to the online version of this chapter.) *Panels (a) and (b): Reprinted from Li et al. (2011), Copyright (2011), with permission from Elsevier.*

extruder consisting of a TSE in the first half and an SSE in the second half in a single run could also be a solution; this, however, requires a much higher facility cost.

3.2 Extrusion Strategies and Issues

3.2.1 *Plasticizer and Additives*

Starch cannot be thermally processed without water since the melting temperature of dry starch is often higher than its decomposition temperature (as extrapolated by Flory law) (Liu, Yu, Liu, Chen, & Li, 2008; Russell, 1987). Water functions by lowering the melting temperature and plasticizing the starch polymer. As a result, in practical processing, water acts as a "plasticizer." It has been suggested that preblending starch with water and subsequent equilibration are favorable for a stable and easy extrusion processing (Li et al., 2011; Su et al., 2009). This is because premixing and equilibration allow water to diffuse into the starch granules and to distribute well in a system, resulting in a homogenous process of melting and fragmentation of the starch during extrusion (Lai & Kokini, 1991).

When water is used as the sole plasticizer, unstable processing may be caused due to the water evaporation. Further, the final products based on starch containing only water usually have poor mechanical properties especially due to the brittleness since its final temperature is lower than its glass transition temperature (T_g). To overcome these issues, nonvolatile (at the processing temperature) plasticizers such as polyols (glycerol, glycol, sorbitol, etc.) (Follain, Joly, Dole, Roge, & Mathlouthi, 2006; Forssell, Mikkilä, Moates, & Parker, 1997; Gaudin, Lourdin, Le Botlan, Ilari, & Colonna, 1999; Lourdin, Coignard, Bizot, & Colonna, 1997; Mathew & Dufresne, 2002; Qiao, Tang, & Sun, 2011; Rodriguez-Gonzalez, Ramsay, & Favis, 2004; Róz, Carvalho, Gandini, & Curvelo, 2006; Smits et al., 2001; Teixeira, Da Róz, Carvalho, & Curvelo, 2007; Wang, Yu, Chang, & Ma, 2007; Yu, Wang, & Ma, 2005), compounds containing nitrogen (urea, ammonium derived, and amines) (Ma & Yu, 2004a, 2004b, 2004c; Ma, Yu, & Feng, 2004; Ma, Yu, & Ma, 2005; Ma, Yu, & Wan, 2006; Shogren, Swanson, & Thompson, 1992; Wang, Yu, Chang, & Ma, 2008; Yang, Yu, & Ma, 2006a, 2006b, 2006c), and citric acid (Shi, Zhang, Liu, et al., 2007; Wang, Yu, Chang, et al., 2007; Yu et al., 2005) are utilized.

Lubricants can be added to improve the processability of plasticized starch-based materials, such as magnesium stearate (Onteniente, Abbès, & Safa, 2000), calcium stearate (Keszei, Szabó, Marosi, Anna, & Nagy, 2006), and fluoroelastomers (Thuwall, Boldizar, & Rigdahl, 2006). The use of a lubricant during processing can reduce the tendency of the material to stick to the die and clog it.

3.2.2 Starch Type and Chemical Modification

Starches with different amylose/amylopectin ratios but from the same botanical origin have attracted particular scientific interests as they display different processing and product properties. Though materials have shown that polymeric materials based on high-amylose starch exhibit superior strength and toughness (Li et al., 2011; Yu & Christie, 2005), the extrusion processing of high-amylose starch is much more difficult than that of low-amylose starch, partly due to the higher die pressure and torque requirements (Li et al., 2011; Shogren, 1992; Shogren & Jasberg, 1994; Su et al., 2009) as results of the higher melting temperature and viscosity of the former as already discussed. By systematically investigating the effects of various extrusion processing conditions such as temperature, screw speed, feeding rate, and water content on the processability of maize starches with different amylose contents (4.3-77.4%), Li et al. (2011) suggested that the difficulty in the extrusion processing of high-amylose starch could be alleviated by increasing the processing temperature, moisture content, screw speed, and equilibration time (cf. Figure 10.8).

Chemical modification of starch by substituting ester or ether groups for the hydroxyls is an effectively way to improve the processing and product properties. Chaudhary, Miler, Torley, Sopade, and Halley (2008) found that the torque, die pressure, and SME for the extrusion processing of hydroxypropylated high-amylose starch (about 80% amylose

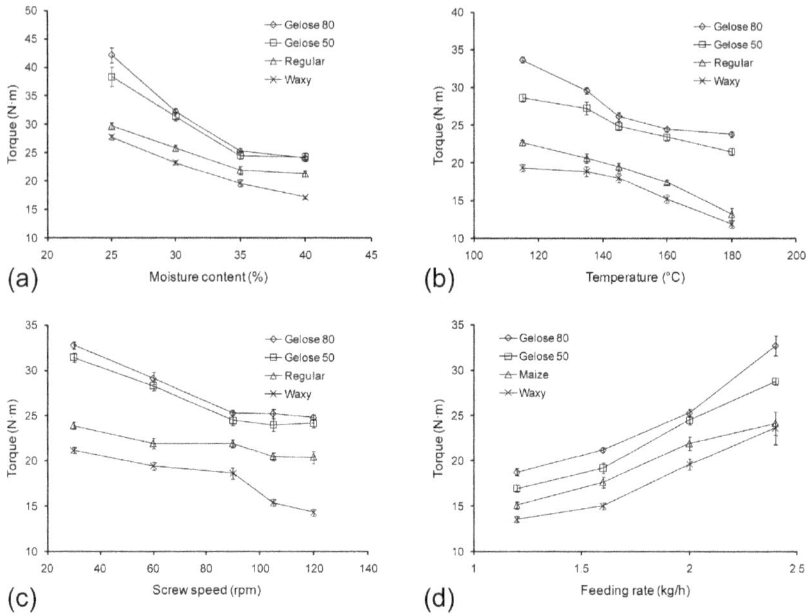

FIGURE 10.8 Effects on extrusion torque of (a) starch moisture content, (b) extrusion temperature, (c) screw speed, and (d) feeding rate (fixed conditions were moisture content 35%, extrusion temperature 145 °C, screw speed 90 rpm, feeding rate 2 kg h^{-1}). *Reprinted from Li et al. (2011), Copyright (2011), with permission from Elsevier.*

content) were all much lower than those for the processing of unmodified counterpart. They explained that modification due to hydroxypropylation of the starch affected its gelatinization behavior (increased interactions with water molecules), which in turn could result in a lower viscosity (Chaudhary et al., 2008).

3.2.3 Extrusion Blending

To improve the performances such as moisture resistibility, mechanical properties, and long-term stability, various starch-based polymeric blends and (nano)composites have been widely examined (Avérous, 2004; Avérous & Halley, 2009; Wang, Yang, & Wang, 2003; Xie, Halley, & Avérous, 2011; Yu et al., 2006). Starch is often blended with other polymers especially biodegradable aliphatic polyesters such as polylactide (PLA) (Guan, Eskridge, & Hanna, 2005; Sarazin, Li, Orts, & Favis, 2008; Schwach & Avérous, 2004; Schwach, Six, & Avérous, 2008; Wang, Yu, Chang, et al., 2007; Wang et al., 2008; Wang, Yu, & Ma, 2007; Willett & Shogren, 2002; Xie et al., 2007), polycaprolactone (PCL) (Bélard, Dole, & Avérous, 2009; Kalambur & Rizvi, 2004; Sarazin et al., 2008; Schwach & Avérous, 2004; Xie, Yu, Liu, & Dean, 2006; Zeng, Huang, Lu,

Fan, & Lourdin, 2011), poly(butylene succinate-*co*-adipate) (PBSA) (Ratto, Stenhouse, Auerbach, Mitchell, & Farrell, 1999; Schwach & Avérous, 2004; Xie, Yu, Liu, & Dean, 2006), and polyhydroxyalkanoates (PHA) (Willett, Kotnis, O'Brien, Fanta, & Gordon, 1998; Willett & Shogren, 2002; Zhang, Deng, Zhao, & Huang, 1997). However, since hydrophilic starch and hydrophobic polyesters are thermodynamically immiscible, a compatibilizer is always used to enhance the compatibility between starch and polyester. The compatibilizers are normally produced in two ways: (1) the graft polymerization of a polyester monomer onto starch backbone and (2) the modification of polyester by maleic anhydride (MA). As an example, Huneault and Li (2007) investigated the interfacial modification of starch/PLA blends by a twin-screw extrusion process with the side feeding of MA-grafted PLA. The first half of twin-screw extrusion line was used to prepare the plasticized starch, and the melted PLA was added at mid-extruder using an SSE as a side feeder to the twin-screw line. The water in plasticized starch was devolatilized before PLA was fed to minimize the decomposition of PLA in the presence of water. The authors claimed that the interface of starch and PLA could be improved (as reflected by much finer-dispersed phase size, in the 1-3 μm range) by this technique.

Starch/polyester blends can also be prepared by a reactive extrusion (REX) process, where a reagent like MA (Maliger, McGlashan, Halley, & Matthew, 2006; Raquez, Nabar, Narayan, & Dubois, 2011; Wang, Yu, & Ma, 2007) or dicumyl peroxide (Maliger et al., 2006) is added for *in situ* formation of compatibilizer in the blends. Sugih, Drijfhout, Picchioni, Janssen, and Heeres (2009) showed that PCL-*g*-glycidyl methacrylate and PCL-*g*-diethyl maleate could be used to compatibilize starch and PCL by REX. Kalambur and Rizvi (2005, 2006b, 2004) developed a novel REX process to prepare starch/PCL/nanoclay composite, where Fenton's reagent (hydrogen peroxide with iron and copper catalysts) as a single oxidizing/cross-linking agent was added. The cross-linking reaction was catalyzed by the organomodified nanoclay. An overview of starch-based blends by REX can be read from another paper (Kalambur & Rizvi, 2006a).

While methylene diphenyl diisocyanate (MDI) is one of the well-studied compatibilizers for starch/polyester blends, the highly reactive isocyanate group in MDI could be consumed by the water absorbed in starch granules, which results in the weakening of the coupling function at the interface between starch and polyester. Yu and coworkers (Xie, Yu, Liu, & Dean, 2006; Yu, Dean, Yuan, Chen, & Zhang, 2007; Yu, Petinakis, Dean, Liu, & Yuan, 2011) discovered that controlling the distribution of the compatibilizer, MDI, by multiple extrusion runs is an effective way to improve the compatibility between starch and polyester and the final product properties of the starch/polyester blends. As an example, the

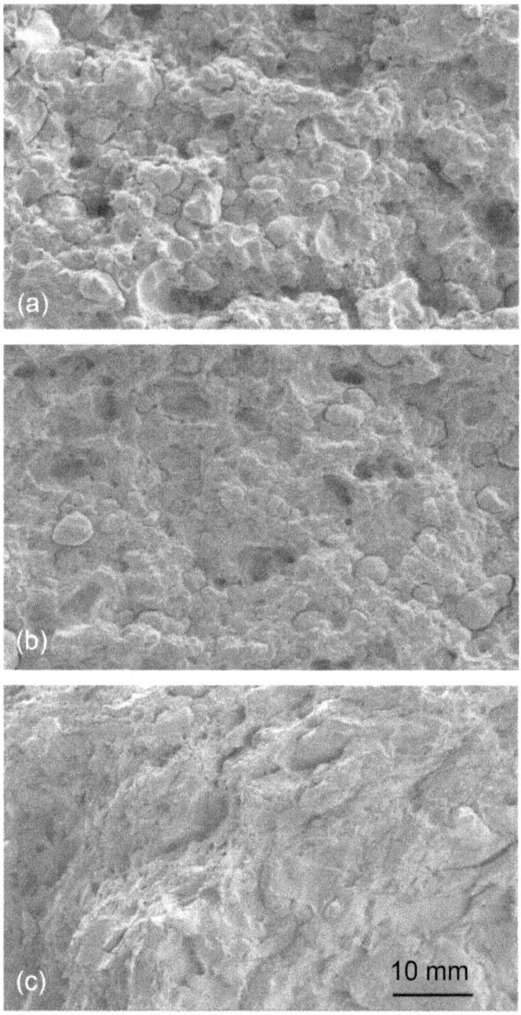

FIGURE 10.9 SEM images of the fracture surfaces of starch 50/PCL 50 blends by controlling the distribution of compatibilizer, MDI: (a) starch/PCL without MDI, (b) (starch/MDI-1.0)/PCL, and (c) starch/(PCL/MDI-1.0). *Reprinted from Yu et al. (2007), Copyright (2007), with permission from John Wiley and Sons.*

morphologies of starch/PCL blends by controlling the compatibilizer distribution were shown in Figure 10.9.

3.2.4 Molecular Degradation of Starch During Extrusion

Molecular degradation of starch is an important issue during extrusion processing since it would influence the final performance of the products. A number of studies have shown that amylopectin is much more susceptible to shear degradation than amylose during extrusion (Brümmer, Meuser, van Lengerich, & Niemann, 2002; Chinnaswamy & Hanna, 1990; Liu et al., 2010; Politz, Timpa, & Wasserman, 1994; Politz, Timpa, White, & Wasserman, 1994; Pushpadass et al., 2009; Rodis, Wen, & Wasserman, 1993; Yamada, Suzuki, Katuzaki, Hisamatsu, & Komiya, 1990). Liu et al. (2010) recently pointed out that there are two reasons for these phenomena. One reason could be the preferential degradation of large molecules of amylopectin; the other reason is that amylopectin starch, with many relatively short branches, has a relatively inflexible structure and consequently is less able to withstand deformation without incurring the chain-breaking events that result in degradation (Liu et al., 2010).

3.3 Extrusion Techniques

3.3.1 Extrusion Film Casting

Films/sheets can be produced by an extruder with a slit or a flat film die attached to the end of it, followed by a takeoff device for sample collection and orientation process (Dean, Yu, & Wu, 2007; Fishman, Coffin, Onwulata, & Konstance, 2004; Van Soest & Knooren, 1997; Walenta, Fink, Weigel, & Ganster, 2001; Walenta, Fink, Weigel, Ganster, & Schaaf, 2001). Figure 10.7b shows a photo of a typical film extrusion process of plasticized starch. Orientation, uniaxial or biaxial orientation, can help in enhancing the properties of the films/sheets, ranging from mechanical, to optical, to barrier characteristics. Yu and Christie (2005) investigated the effect of uniaxial orientation on microstructure and mechanical properties of plasticized maize starches with different amylose/amylopectin ratios. They found that the hydrogen bonds and highly branched microstructure in amylopectin resisted the orientation of the polymer chains. For different starches, orientation increased both modulus and yield stress in the drawing direction but decreased elongation (cf. Table 10.1), which was mainly contributed to the orientation of amorphous phase.

In order to decrease the water sensitivity and improve the mechanical and gas barrier properties of extruded films/sheets, multilayer coextrusion technique has been introduced. The middle layer consists of plasticized starch and the two external layers are usually made of biodegradable polyesters such as PLA (Fang et al., 2005; Gattin, Copinet, Bertrand, & Couturier, 2002; Martin, Schwach, Avérous, & Couturier, 2001), PCL (Martin et al., 2001; Wang, Shogren, & Carriere, 2000), polyesteramide (Martin & Avérous, 2002; Martin et al., 2001), PBSA (Martin et al., 2001), and poly(hydroxybutyrate-*co*-valerate) (Martin et al., 2001). In the studies by Avérous and coworkers (Martin & Avérous, 2002;

TABLE 10.1 Effect of Orientation on Mechanical Properties of Extruded Sheets from Maize Starches with Different Amylose Contents

Materials	Speed (cm min^{-1})	Thickness (mm)	Direction (Extrusion)	Modulus (Mpa)	Yield Strength (Mpa)	Elongation at Break (%)
Waxy	50	0.33	Parallel	1251 ± 72	10.6 ± 0.87	10.3 ± 1.14
			Cross	1172 ± 68	10.5 ± 0.43	2.5 ± 0.56
	90	0.27	Parallel	111 ± 58	14.3 ± 0.53	8.1 ± 0.71
			Cross	2062 ± 81	13.7 ± 0.79	1.7 ± 0.23
	120	0.21	Parallel	2332 ± 132	17.2 ± 0.88	4.6 ± 0.62
			Cross	2222 ± 121	16.5 ± 1.10	2.0 ± 0.19
	150	0.17	Parallel	2404 ± 146	19.8 ± 1.31	3.9 ± 0.31
			Cross	2320 ± 142	19.6 ± 1.27	1.9 ± 0.19
Regular	50	0.32	Parallel	1305 ± 52	11.1 ± 0.54	12.8 ± 0.98
			Cross	1290 ± 61	10.8 ± 0.62	7.5 ± 0.67
	90	0.27	Parallel	2245 ± 88	19.2 ± 0.97	8.6 ± 0.34
			Cross	2125 ± 67	18.8 ± 1.32	2.1 ± 0.11
	120	0.22	Parallel	2245 ± 89	19.2 ± 1.71	5.2 ± 0.29
			Cross	2125 ± 94	18.8 ± 2.11	2.2 ± 0.38
	150	0.17	Parallel	2249 ± 178	19.3 ± 2.81	3.9 ± 0.52
			Cross	2119 ± 290	18.7 ± 2.99	1.7 ± 0.33

Gelose 50	50	0.32	Parallel	1814 ± 92	10.9 ± 1.29	12.6 ± 0.49
			Cross	1332 ± 87	10.6 ± 1.12	12.0 ± 0.62
	90	0.26	Parallel	2322 ± 172	11.8 ± 1.35	11.6 ± 1.25
			Cross	2023 ± 224	10.9 ± 1.51	7.5 ± 1.32
	120	0.21	Parallel	2558 ± 541	18.1 ± 1.22	7.5 ± 2.13
			Cross	2246 ± 394	17.5 ± 1.94	7.2 ± 1.92
Gelose 80	50	0.33	Parallel	1881 ± 139	14.5 ± 2.30	15.2 ± 1.73
			Cross	1200 ± 119	7.3 ± 2.81	11.9 ± 1.39
	90	0.26	Parallel	2550 ± 392	17.1 ± 3.47	11.8 ± 2.20
			Cross	1503 ± 620	10.6 ± 2.78	7.9 ± 2.28

With kind permission from Springer Science+Business Media: Yu & Christie (2005), Table III.

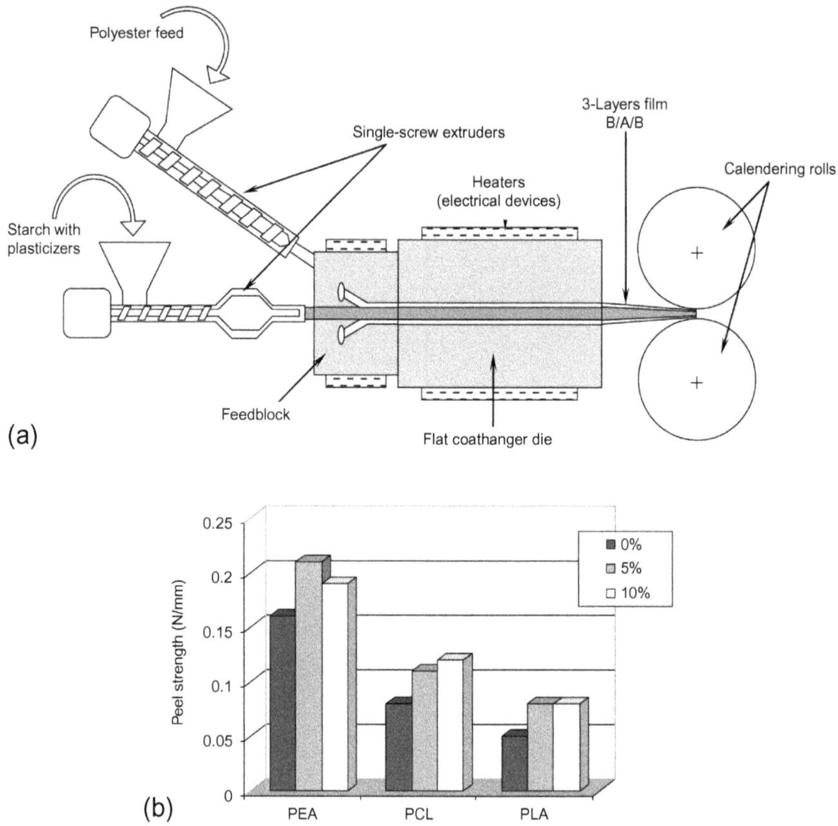

FIGURE 10.10 (a) Schematic of the labscale three-layer coextrusion line (the gray stream is the inner layer, and the white polymer stream constitutes the outer layers); (b) effect of the added polyester (0%, 5%, and 10% w/w) in the plasticized starch phase on the total peel strength of subsequent coextruded films. *Reprinted from Martin et al. (2001), Copyright (2001), with permission from John Wiley and Sons.*

Martin et al., 2001), the preparation processing line consisted of two SSE (one for the inner starch layer and the other for the outer polymer layers), a feed block, a coat hanger-type sheet die, and a three-roll calendering system (cf. Figure 10.10a). They found that the key parameters were the skin-layer viscosity and thickness, the global extrusion rate, and the die geometry. It has been suggested that the interfacial instabilities (the wave-like distortions of the interface between two polymers across the width of the film), which were strongly related to the shear stress at the interface, were more detrimental to the coextruded and final product than the encapsulation phenomena (the less viscous outer layers starting to surround the other layer) (Martin & Avérous, 2002; Martin et al., 2001). On the other hand, since no specific compatibilizer or tie layer was added, the properties of subsequent films relied on the compatibility between the plasticized starch and the polyester, which varied according to the starch plasticizer content and the polar/apolar nature of the polyester. A strategy was tested involving the blending of a small amount (5 or 10 wt%) of the polyester with plasticized starch for the inner layer (Martin et al., 2001). This strategy yielded satisfactory results (cf. Figure 10.10b).

Without using a complex multilayer coextrusion system, Bélard et al. (2009) studied the phase separation phenomenon during the extrusion processing of TPS/PCL blends and found that, in certain conditions, a pseudo-multilayer structure could be generated with a polyester-rich skin. In the range of available shear rates, the phase separation was mainly driven by the molecular weight of PCL, linked to its molten state viscosity. Above 60,000 g mol^{-1}, no surface enrichment could be detected; below 37,000 g mol^{-1}, the phase separation occurred with no dependence on the processing conditions; and, between these two limits, the phase separation depended on the residence time inside the die.

3.3.2 *Extrusion Film Blowing*

Film blowing can be practiced by equipping the extruder with a temperature-controlled film-blowing die, followed by a film-blowing tower with a calendering nip and takeoff rolls. The velocity of the takeoff rolls and the pressure inside the film "bubble" are adjusted in order to achieve as good a result as possible, i.e., a stable process with a blow-up ratio greater than two (Thunwall, Kuthanova, Boldizar, & Rigdahl, 2008).

Film blowing requires a high η_e due to a sufficient extensional viscosity required for the initial form of the bubble from the die in extension (Zullo & Iannace, 2009). Further, it was reported that the occurrence of strain-hardening behavior (an increase of the slope of η_e versus. time curve above a specific strain rate) in urea-formamide-plasticized starch samples leads to a positive effect on the film-blowing process (Zullo & Iannace, 2009).

Thunwall et al. (2008) has showed that, when the shear viscosity was higher than a critical value, TPS becomes too stiff to expand during the blowing and the material failed in a brittle manner at the calendering nip. Moreover, the significant increase of the torque required from the extruder and the die pressure as a result of high viscosity also made the film-blowing process difficult (Thunwall et al., 2008). In contrast, when the viscosity was too low, the surface of the extruded material became sticky, resulting in a double-walled film impossible to separate through the calendering nip. A low viscosity accompanied by high temperature and high plasticizer content might also cause the bubble formation in the extrudate and the subsequent rupture of the stretched melt (Thunwall et al., 2008). Based on these results, Thunwall et al. (2008) proposed an illustration of the processing parameters (glycerol content,

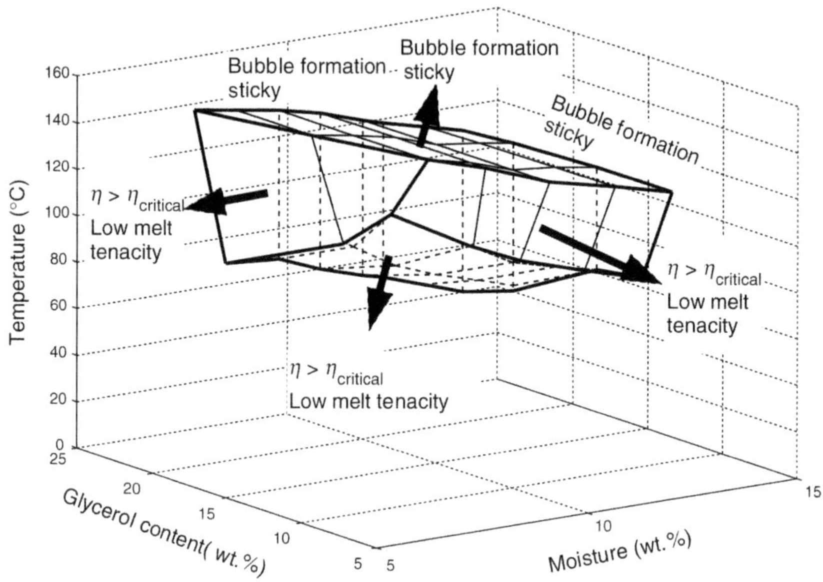

FIGURE 10.11 Illustration of the limiting processing parameters (temperature, moisture content, and glycerol content) in film blowing of TPS. *Reprinted from Thunwall et al. (2008), Copyright (2008), with permission from Elsevier.*

moisture content, and temperature) that influence the viscosity and the film-blowing properties of TPS (cf. Figure 10.11).

Orientation is also an important process during film blowing. Fishman et al. (2006) found that, unlike uniaxial orientation process (which results in very large difference between drawing and transverse directions) as mentioned before, simultaneous biaxial drawing (biaxial orientation) during film blowing produced nearly equivalent mechanical properties in both two directions.

Halley and coworkers (Halley, Mcglashan, & Gralton, 2006; McGlashan & Halley, 2003) were among the first to successfully prepare high-performance starch/polyester/nanoclay nanocomposite blown films by conventional extrusion and film-blowing facilities. They found that the addition of organomodified montmorillonite clay to starch/polyester blends can result in easier processing—less work required and more stable process. The nanocomposites did not foam when temperatures as high as 150 °C were used at the die and die lip; and higher processing temperatures lowered the viscosity of the predominantly starch-based nanocomposite to a level that alleviated a lot of the problems associated with processing the base blends (i.e., high melt viscosity and trying to stretch a nearly solid extrudate). The nanocomposite films produced were more transparent and homogenous (cf. Figure 10.12), which could be attributed

FIGURE 10.12 Photomicrographs of 50 wt% starch blends: (a) nanocomposite (1.5 wt% MMT); (b) base blend; (c) nanocomposite (5 wt% MMT). *Reprinted from McGlashan and Halley (2003), Copyright (2003), with permission from John Wiley and Sons.*

to the dual effect of platelets trapping the volatile plasticizer in place and retarding crystal growth in the cooling phase of the process.

3.3.3 *Extrusion Foaming*

The extrusion-foaming technique has been mainly used to produce loose-fill packaging materials, in a similar way for the production of extruded expanded snack foods (Chinnaswamy, 1993; Ilo, Liu, & Berghofer, 1999). Water can serve as the blowing agent as it expands upon exiting the extruder (Shogren, Lawton, Doane, & Tiefenbacher, 1998). Fan, Mitchell, and Blanshard (1994) developed a model to describe the dynamics of bubble growth and shrinkage in starchy extrudate: a bubble cell first grows by rapid vaporization of the superheated moisture and subsequently shrinks by the cooling of vapor, which produces a negative pressure difference. Different processing conditions such as moisture content, temperature, screw speed and configuration, feeding rate, and die geometry can influence the quality (cell size, cell number density, etc.) of foams since they influence the rheological properties and T_g of plasticized starch melt, which have a close relationship with its macroscopic expansion behavior. The details have been summarized in a recent review paper (Xie et al., 2012).

When acetate starch is used to produce foams (which can improve water resistance and mechanical properties Liu et al., 2009), water may not be an effective blowing agent because of the reduced solubility of acetate starch in water. Considering the different requirements of the physical and chemical properties of the blowing agent in various stages of extrusion foaming, Zhou and Hanna (2004, 2005) investigated secondary blowing agents such as pentane, ethyl acetate, and sodium bicarbonate, which were used in water- and ethanol-based foaming processes. They found that a secondary blowing agent with higher evaporation rates and/or solvency for starch acetate than the primary blowing agent can increase cell density at low concentrations of primary and secondary blowing agents. High blowing agent concentrations eventually lead to foams with low cell number density (Zhou & Hanna, 2004, 2005).

Rizvi and coworkers (Alavi, Gogoi, Khan, Bowman, & Rizvi, 1999; Alavi & Rizvi, 2005; Chen & Rizvi, 2006; Cho & Rizvi, 2008; Rizvi & Mulvaney, 1992) introduced a supercritical fluid extrusion (SCFX) technique in which supercritical carbon dioxide was used as a blowing agent and protein was added as a stabilizing agent to produce starch microcellular foams. It was claimed that a greater uniformity in the cellular structure of starch extrudates can be obtained by the SCFX process as compared to steam extrusion (Alavi et al., 1999).

3.3.4 *Reactive Extrusion*

Unlike the techniques mentioned earlier, REX involves concurrent reaction and extrusion. This technique has received more and more focus in recent years due to its advantages such as high conversion efficiency and a rapid rate of production. It can be used to produce either chemical-modified starch or starch-based blends, which has already been well reviewed in several papers (Kalambur & Rizvi, 2006a; Moad, 2011; Xie, Yu, Liu, & Chen, 2006). This technique has recently been expanded into wider applications. Frost et al. (2011) created plasticized starch-silica (SiO_2) poly(vinyl alcohol) composite films by REX using tetraethyl orthosilicate (TEOS) as a precursor. The TEOS-SiO_2 conversion efficiencies of up to 41.3% were achieved, and the resulting films had well-dispersed SiO_2 particles of size ranging from 20 to <1 μm, which are responsible for the increased tensile strength and Young's modulus. Song, Thio, and Deng (2011) discovered that starch nanoparticles could be formed by REX with the use of glyoxal as the cross-linker. They found that starch particle size could be reduced to around 160 nm, even at a lower extrusion temperature of 75 °C, which could be attributed to the mechanical and thermal treatment during processing.

4 OTHER PROCESSING TECHNIQUES FOR PLASTICIZED STARCH-BASED MATERIALS

After extrusion, plasticized starch can be further processed by injection molding or compression molding to produce more specific products.

A number of researchers (Abbégs, Ayad, Prudhomme, & Onteniente, 1998; Onteniente et al., 2000; Stepto, 2009; Tabi & Kovacs, 2007) have discussed the injection molding behavior and processing parameters of starch injection. The main difficulties in injection molding of plasticized starch are usually associated with its high-viscosity and poor-flow properties. However, at relatively lower viscosity, the adhesion of TPS part to the mold can occur (though the adhesion can be minimized by using an appropriate lubricant) (Onteniente et al., 2000). There are also problems with the serious distortion and shrinkage of injection molded parts especially when high content of water is used as plasticizer since the equilibrium water content is naturally achieved after processing (Stepto, 2009). Other drawbacks are related to the product performances such as poor mechanical properties, high hydrophilic properties, and poor long-term stability, which may be overcome by reinforcing with natural fibers or by blending with polyester in the formulations (Liu et al., 2009).

Compression molding can be used to process plasticized starch-based materials into foams and containers, such as endcaps, edges, or corner cushions for protective packaging, which are difficult to produce with extrusion-foaming technology (Zhou, Song, & Parker, 2006). Besides the simple traditional techniques from food processing, such as baking technique and explosion-puffing technique, microwave has been used for producing starch-based foams recently (Sjöqvist & Gatenholm, 2007; Zhou et al., 2006; Zhou, Song, & Parker, 2007). When compression, selection of adequate mold material, pretreatment of starch, and the initial loading of samples in mold cavity are the key issues to achieve a uniformly foamed block with good integrity (Zhou et al., 2007). And apart from gelatinization agents, mold-releasing agents such as magnesium stearate and steric acid are often used in formulations to prevent the starch from sticking to the mold (Liu et al., 2009).

5 SUMMARY AND FUTURE PERSPECTIVE

The traditional polymer-processing techniques such as extrusion can be readily used to process starch from native granular starch into a homogeneous molten phase and then into different product forms such as films/sheets, foams, and shaped articles. During the processing, the phase transition of starch and the rheology of plasticized starch melts have been recognized as two of the most important aspects, which determine the processability and product performance.

A plasticizer is indispensable for the plasticization of starch, which involves the phase transition of starch. The phase transition of starch can be largely influenced by plasticizer type and content and starch type and structure (amylose/amylopectin ratios). In addition, shear treatment, which often exists in processing, can play an important role in the phase transition of starch. Shear enhances the destructuration of starch granules and the melting of crystallites during the phase transition process. Measures need to be ensured to avoid the loss of water (which is the most commonly used plasticizer for starch) while studying the phase transition of starch.

The rheological characterization of plasticized starch is not a simple issue because that moisture loss has to be avoided and that thermomechanical input has to be imposed for starch to remelt before measurement. In addition, the macromolecular degradation, which is determined by the thermomechanical input, has to be considered since it has a close relationship with the rheological properties. Both viscous and elastic properties of plasticized starch play critical roles in the processing of plasticized starch-based materials (of which the details can be read in a recent review paper; Xie, Halley, & Avérous, 2012).

Since SSE and TSE have different advantages in processing, it is important to choose the right extruder at a different stage of starch processing. A TSE is preferably used in the first stage and an SSE is usually used in the latter stage. Premixing starch with plasticizer before extrusion is a good way to improve the stability of extrusion. Besides, it is important to employ the way of multiple extrusion runs to produce starch/biodegradable aliphatic polyester blend products with greater compatibility. The choice of die and subsequent takeoff and collecting device(s) determines the specific product to be produced. For film/sheets production, orientation is an important process to enhance the mechanical properties of the products.

The future development of plasticized starch-based materials still lies in the improvement of processability and product performance. Novel processing techniques based on extrusion need to be developed to produce plasticized starch with maximum phase transition but minimum macromolecular shear degradation. Rheology (both viscosity and elasticity) should be more emphasized in the processing of plasticized starch-based materials. Starch-based blends and (nano)composites will still be the core solutions to maximally utilize starch as a renewable resource. Their developments, however, need to be integrated with the development of extrusion techniques for further industrialization. In addition, REX is a promising way to produce high-performance starch-based blend and (nano)composite materials.

References

Abbégs, B., Ayad, R., Prudhomme, J. C., & Onteniente, J. P. (1998). Numerical simulation of thermoplastic wheat starch injection molding process. *Polymer Engineering & Science, 38*, 2029–2038.

Aichholzer, W., & Fritz, H.-G. (1998). Rheological characterization of thermoplastic starch materials. *Starch-Stärke, 50*, 77–83.

Alavi, S. H., Gogoi, B. K., Khan, M., Bowman, B. J., & Rizvi, S. S. H. (1999). Structural properties of protein-stabilized starch-based supercritical fluid extrudates. *Food Research International, 32*, 107–118.

Alavi, S., & Rizvi, S. S. H. (2005). Strategies for enhancing expansion in starch-based microcellular foams produced by supercritical fluid extrusion. *International Journal of Food Properties, 8*, 23–34.

Atwell, W. A., Hood, L. F., Lineback, D. R., Varriano-Marston, E., & Zobel, H. F. (1988). The terminology and methodology associated with basic starch phenomena. *Cereal Food World, 33*, 306–311.

Avérous, L. (2004). Biodegradable multiphase systems based on plasticized starch: A review. *Polymer Reviews, 44*, 231–274.

Avérous, L., & Halley, P. J. (2009). Biocomposites based on plasticized starch. *Biofuels, Bioproducts and Biorefining, 3*, 329–343.

Barron, C., Bouchet, B., Della Valle, G., Gallant, D. J., & Planchot, V. (2001). Microscopical study of the destructuring of waxy maize and smooth pea starches by shear and heat at low hydration. *Journal of Cereal Science, 33*, 289–300.

Barron, C., Valle, G. D., Colonna, P., & Vergnes, B. (2002). Energy balance of low hydrated starches transition under shear. *Journal of Food Science, 67*, 1426–1437.

Bélard, L., Dole, P., & Avérous, L. (2009). Study of pseudo-multilayer structures based on starch-polycaprolactone extruded blends. *Polymer Engineering & Science, 49*, 1177–1186.

Berzin, F., Tara, A., & Tighzert, L. (2007). In-line measurement of the viscous behaviour of wheat starch during extrusion. Application to starch cationisation. *Applied Rheology, 17* (21222), 1–7.

Bhattacharya, M., Padmanabhan, M., & Seethamraju, K. (1994). Uniaxial extensional viscosity during extrusion cooking from entrance pressure drop method. *Journal of Food Science, 59*, 221–226.

Brouillet-Fourmann, S., Carrot, C., Mignard, N., & Prochazka, F. (2002). On the use of an internal mixer for the rheological characterization of maize starch. *Applied Rheology, 12*, 192–199.

Brümmer, T., Meuser, F., van Lengerich, B., & Niemann, C. (2002). Effect of extrusion cooking on molecular parameters of corn starch. *Starch-Starke, 54*, 1–8.

Burros, B. C., Young, L. A., & Carroad, P. A. (1987). Kinetics of corn meal gelatinization at high temperature and low moisture. *Journal of Food Science, 52*, 1372–1376.

Chaudhary, A. L., Miler, M., Torley, P. J., Sopade, P. A., & Halley, P. J. (2008). Amylose content and chemical modification effects on the extrusion of thermoplastic starch from maize. *Carbohydrate Polymers, 74*, 907–913.

Chen, K. H., & Rizvi, S. S. H. (2006). Rheology and expansion of starch-water-CO_2 mixtures with controlled gelatinization by supercritical fluid extrusion. *International Journal of Food Properties, 9*, 863–876.

Chinnaswamy, R. (1993). Basis of cereal starch expansion. *Carbohydrate Polymers, 21*, 157–167.

Chinnaswamy, R., & Hanna, M. A. (1990). Macromolecular and functional properties of native and extrusion-cooked corn starch. *Cereal Chemistry, 67*, 490–499.

Cho, K. Y., & Rizvi, S. S. H. (2008). The time-delayed expansion profile of supercritical fluid extrudates. *Food Research International, 41*, 31–42.

Dealy, J., & Wissbrun, K. (1999). *Melt rheology and its role in plastics processing: Theory and applications.* Dordrecht, The Netherlands: Kluwer Academic Publishers.

Dean, K., Yu, L., & Wu, D. Y. (2007). Preparation and characterization of melt-extruded thermoplastic starch/clay nanocomposites. *Composites Science and Technology, 67*, 413–421.

Della Valle, G., Boché, Y., Colonna, P., & Vergnes, B. (1995). The extrusion behaviour of potato starch. *Carbohydrate Polymers, 28*, 255–264.

Della Valle, G., Colonna, P., Patria, A., & Vergnes, B. (1996). Influence of amylose content on the viscous behavior of low hydrated molten starches. *Journal of Rheology, 40*, 347–362.

Della Valle, G., Vergnes, B., & Lourdin, D. (2007). Viscous properties of thermoplastic starches from different botanical origin. *International Polymer Processing, 22*, 471–479.

Donovan, J. W. (1979). Phase transitions of the starch-water system. *Biopolymers, 18*, 263–275.

Fan, J., Mitchell, J. R., & Blanshard, J. M. V. (1994). A computer simulation of the dynamics of bubble growth and shrinkage during extrudate expansion. *Journal of Food Engineering, 23*, 337–356.

Fang, J. M., Fowler, P. A., Escrig, C., Gonzalez, R., Costa, J. A., & Chamudis, L. (2005). Development of biodegradable laminate films derived from naturally occurring carbohydrate polymers. *Carbohydrate Polymers, 60*, 39–42.

Fishman, M. L., Coffin, D. R., Onwulata, C. I., & Konstance, R. P. (2004). Extrusion of pectin and glycerol with various combinations of orange albedo and starch. *Carbohydrate Polymers, 57*, 401–413.

Fishman, M. L., Coffin, D. R., Onwulata, C. I., & Willett, J. L. (2006). Two stage extrusion of plasticized pectin/poly(vinyl alcohol) blends. *Carbohydrate Polymers, 65*, 421–429.

Follain, N., Joly, C., Dole, P., Roge, B., & Mathlouthi, M. (2006). Quaternary starch based blends: Influence of a fourth component addition to the starch/water/glycerol system. *Carbohydrate Polymers, 63*, 400–407.

Forssell, P. M., Mikkilä, J. M., Moates, G. K., & Parker, R. (1997). Phase and glass transition behaviour of concentrated barley starch-glycerol-water mixtures, a model for thermoplastic starch. *Carbohydrate Polymers, 34*, 275–282.

Frost, K., Barthes, J., Kaminski, D., Lascaris, E., Niere, J., & Shanks, R. (2011). Thermoplastic starch-silica-polyvinyl alcohol composites by reactive extrusion. *Carbohydrate Polymers, 84*, 343–350.

Gattin, R., Copinet, A., Bertrand, C., & Couturier, Y. (2002). Biodegradation study of a starch and poly(lactic acid) co-extruded material in liquid, composting and inert mineral media. *International Biodeterioration & Biodegradation, 50*, 25–31.

Gaudin, S., Lourdin, D., Le Botlan, D., Ilari, J. L., & Colonna, P. (1999). Plasticisation and mobility in starch-sorbitol films. *Journal of Cereal Science, 29*, 273–284.

Guan, J., Eskridge, K. M., & Hanna, M. A. (2005). Acetylated starch-polylactic acid loose-fill packaging materials. *Industrial Crops and Products, 22*, 109–123.

Halley, P., Mcglashan, S., & Gralton, J. (2006). *Biodegradable polymer*. United States patent application 10/474,907.

Huneault, M. A., & Li, H. (2007). Morphology and properties of compatibilized polylactide/thermoplastic starch blends. *Polymer, 48*, 270–280.

Ilo, S., Liu, Y., & Berghofer, E. (1999). Extrusion cooking of rice flour and amaranth blends. *Lebensmittel-Wissenschaft und-Technologie, 32*, 79–88.

Jones, D. M., Walters, K., & Williams, P. R. (1987). On the extensional viscosity of mobile polymer solutions. *Rheologica Acta, 26*, 20–30.

Kalambur, S. B., & Rizvi, S. S. (2004). Starch-based nanocomposites by reactive extrusion processing. *Polymer International, 53*, 1413–1416.

Kalambur, S., & Rizvi, S. S. H. (2005). Biodegradable and functionally superior starch-polyester nanocomposites from reactive extrusion. *Journal of Applied Polymer Science, 96*, 1072–1082.

Kalambur, S., & Rizvi, S. S. H. (2006a). An overview of starch-based plastic blends from reactive extrusion. *Journal of Plastic Film and Sheeting, 22*, 39–58.

Kalambur, S., & Rizvi, S. S. H. (2006b). Rheological behavior of starch-polycaprolactone (PCL) nanocomposite melts synthesized by reactive extrusion. *Polymer Engineering & Science, 46*, 650–658.

Keszei, S., Szabó, A., Marosi, G., Anna, P., & Nagy, S. (2006). Use of thermoplastic starch in continuous pharmaceutical process. *Macromolecular Symposia, 239*, 101–104.

Lai, L. S., & Kokini, J. L. (1991). Physicochemical changes and rheological properties of starch during extrusion. *Biotechnology Progress, 7*, 251–266.

Lelievre, J. (1974). Starch gelatinization. *Journal of Applied Polymer Science, 18*, 293–296.
Li, M., Liu, P., Zou, W., Yu, L., Xie, F., Pu, H., et al. (2011). Extrusion processing and characterization of edible starch films with different amylose contents. *Journal of Food Engineering, 106*, 95–101.
Liu, W.-C., Halley, P. J., & Gilbert, R. G. (2010). Mechanism of degradation of starch, a highly branched polymer, during extrusion. *Macromolecules, 43*, 2855–2864.
Liu, H., Xie, F., Chen, L., Yu, L., Dean, K., & Bateman, S. (2005). Thermal behaviour of high amylose cornstarch studied by DSC. *International Journal of Food Engineering, 1*(1), 1–6, Article 3.
Liu, P., Xie, F., Li, M., Liu, X., Yu, L., Halley, P. J., et al. (2011). Phase transitions of maize starches with different amylose contents in glycerol-water systems. *Carbohydrate Polymers, 85*, 180–187.
Liu, H., Xie, F., Yu, L., Chen, L., & Li, L. (2009). Thermal processing of starch-based polymers. *Progress in Polymer Science, 34*, 1348–1368.
Liu, X., Yu, L., Liu, H., Chen, L., & Li, L. (2008). In situ thermal decomposition of starch with constant moisture in a sealed system. *Polymer Degradation and Stability, 93*, 260–262.
Liu, H., Yu, L., Xie, F., & Chen, L. (2006). Gelatinization of cornstarch with different amylose/amylopectin content. *Carbohydrate Polymers, 65*, 357–363.
Lourdin, D., Coignard, L., Bizot, H., & Colonna, P. (1997). Influence of equilibrium relative humidity and plasticizer concentration on the water content and glass transition of starch materials. *Polymer, 38*, 5401–5406.
Ma, X., & Yu, J. (2004a). The plasticizers containing amide groups for thermoplastic starch. *Carbohydrate Polymers, 57*, 197–203.
Ma, X., & Yu, J. (2004b). Formamide as the plasticizer for thermoplastic starch. *Journal of Applied Polymer Science, 93*, 1769–1773.
Ma, X., & Yu, J. (2004c). The effects of plasticizers containing amide groups on the properties of thermoplastic starch. *Starch-Stärke, 56*, 545–551.
Ma, X., Yu, J., & Feng, J. (2004). Urea and formamide as a mixed plasticizer for thermoplastic starch. *Polymer International, 53*, 1780–1785.
Ma, X. F., Yu, J. G., & Ma, Y. B. (2005). Urea and formamide as a mixed plasticizer for thermoplastic wheat flour. *Carbohydrate Polymers, 60*, 111–116.
Ma, X. F., Yu, J. G., & Wan, J. J. (2006). Urea and ethanolamine as a mixed plasticizer for thermoplastic starch. *Carbohydrate Polymers, 64*, 267–273.
Maliger, R. B., McGlashan, S. A., Halley, P. J., & Matthew, L. G. (2006). Compatibilization of starch-polyester blends using reactive extrusion. *Polymer Engineering & Science, 46*, 248–263.
Martin, O., & Avérous, L. (2002). Comprehensive experimental study of a starch/polyesteramide coextrusion. *Journal of Applied Polymer Science, 86*, 2586–2600.
Martin, O., Avérous, L., & Della Valle, G. (2003). In-line determination of plasticized wheat starch viscoelastic behavior: Impact of processing. *Carbohydrate Polymers, 53*, 169–182.
Martin, O., Schwach, E., Avérous, L., & Couturier, Y. (2001). Properties of biodegradable multilayer films based on plasticized wheat starch. *Starch-Stärke, 53*, 372–380.
Mathew, A. P., & Dufresne, A. (2002). Plasticized waxy maize starch: Effect of polyols and relative humidity on material properties. *Biomacromolecules, 3*, 1101–1108.
McGlashan, S. A., & Halley, P. J. (2003). Preparation and characterisation of biodegradable starch-based nanocomposite materials. *Polymer International, 52*, 1767–1773.
Moad, G. (2011). Chemical modification of starch by reactive extrusion. *Progress in Polymer Science, 36*, 218–237.
Onteniente, J.-P., Abbès, B., & Safa, L. H. (2000). Fully biodegradable lubricated thermoplastic wheat starch: Mechanical and rheological properties of an injection grade. *Starch-Stärke, 52*, 112–117.

Politz, M. L., Timpa, J. D., & Wasserman, B. P. (1994). Quantitative measurement of extrusion-induced starch fragmentation products in maize flour using nonaqueous automated gel-permeation chromatography. *Cereal Chemistry, 71*, 532–536.

Politz, M. L., Timpa, J. D., White, A. R., & Wasserman, B. P. (1994). Non-aqueous gel permeation chromatography of wheat starch in dimethylacetamide (DMAC) and LiCl: Extrusion-induced fragmentation. *Carbohydrate Polymers, 24*, 91–99.

Pushpadass, H. A., Kumar, A., Jackson, D. S., Wehling, R. L., Dumais, J. J., & Hanna, M. A. (2009). Macromolecular changes in extruded starch-films plasticized with glycerol, water and stearic acid. *Starch-Stärke, 61*, 256–266.

Qiao, X., Tang, Z., & Sun, K. (2011). Plasticization of corn starch by polyol mixtures. *Carbohydrate Polymers, 83*, 659–664.

Raquez, J.-M., Nabar, Y., Narayan, R., & Dubois, P. (2011). Preparation and characterization of maleated thermoplastic starch-based nanocomposites. *Journal of Applied Polymer Science, 122*, 639–647.

Ratnayake, W. S., Jackson, D. S., & Steve, L. T. (2008). Starch gelatinization. *Advances in food and nutrition research, 55*, 221–268.

Ratto, J. A., Stenhouse, P. J., Auerbach, M., Mitchell, J., & Farrell, R. (1999). Processing, performance and biodegradability of a thermoplastic aliphatic polyester/starch system. *Polymer, 40*, 6777–6788.

Rizvi, S. S. H., & Mulvaney, S. (1992). Extrusion processing with supercritical fluids. *United States Patent Application, 770*, 345.

Rodis, P., Wen, L. F., & Wasserman, B. P. (1993). Assessment of extrusion induced starch fragmentation by gel permeation chromatography and methylation analysis. *Cereal Chemistry, 70*, 152–157.

Rodriguez-Gonzalez, F. J., Ramsay, B. A., & Favis, B. D. (2004). Rheological and thermal properties of thermoplastic starch with high glycerol content. *Carbohydrate Polymers, 58*, 139–147.

Róz, A. L. D., Carvalho, A. J. F., Gandini, A., & Curvelo, A. A. S. (2006). The effect of plasticizers on thermoplastic starch compositions obtained by melt processing. *Carbohydrate Polymers, 63*, 417–424.

Russell, P. L. (1987). Gelatinisation of starches of different amylose/amylopectin content. A study by differential scanning calorimetry. *Journal of Cereal Science, 6*, 133–145.

Sarazin, P., Li, G., Orts, W. J., & Favis, B. D. (2008). Binary and ternary blends of polylactide, polycaprolactone and thermoplastic starch. *Polymer, 49*, 599–609.

Schwach, E., & Avérous, L. (2004). Starch-based biodegradable blends: Morphology and interface properties. *Polymer International, 53*, 2115–2124.

Schwach, E., Six, J.-L., & Avérous, L. (2008). Biodegradable blends based on starch and poly (lactic acid): Comparison of different strategies and estimate of compatibilization. *Journal of Polymers and the Environment, 16*, 286–297.

Senouci, A., & Smith, A. C. (1988). An experimental study of food melt rheology. II. End pressure effects. *Rheologica Acta, 27*, 649–655.

Shi, R., Zhang, Z., Liu, Q., Han, Y., Zhang, L., Chen, D., et al. (2007). Characterization of citric acid/glycerol co-plasticized thermoplastic starch prepared by melt blending. *Carbohydrate Polymers, 69*, 748–755.

Shogren, R. L. (1992). Effect of moisture content on the melting and subsequent physical aging of cornstarch. *Carbohydrate Polymers, 19*, 83–90.

Shogren, R. L., & Jasberg, B. K. (1994). Aging properties of extruded high-amylose starch. *Journal of Polymers and the Environment, 2*, 99–109.

Shogren, R. L., Lawton, J. W., Doane, W. M., & Tiefenbacher, K. F. (1998). Structure and morphology of baked starch foams. *Polymer, 39*, 6649–6655.

Shogren, R. L., Swanson, C. L., & Thompson, A. R. (1992). Extrudates of cornstarch with urea and glycols: Structure/mechanical property relations. *Starch-Stärke, 44*, 335–338.

Sjöqvist, M., & Gatenholm, P. (2007). Effect of water content in potato amylopectin starch on microwave foaming process. *Journal of Polymers and the Environment, 15*, 43–50.

Smits, A. L. M., Wübbenhorst, M., Kruiskamp, P. H., van Soest, J. J. G., Vliegenthart, J. F. G., & van Turnhout, J. (2001). Structure evolution in amylopectin/ethylene glycol mixtures by H-bond formation and phase separation studied with dielectric relaxation spectroscopy. *The Journal of Physical Chemistry B, 105*, 5630–5636.

Song, D., Thio, Y. S., & Deng, Y. (2011). Starch nanoparticle formation via reactive extrusion and related mechanism study. *Carbohydrate Polymers, 85*, 208–214.

Stepto, R. F. T. (2009). Thermoplastic starch. *Macromolecular Symposia, 279*, 163–168.

Su, B., Xie, F., Li, M., Corrigan, P. A., Yu, L., Li, X., et al. (2009). Extrusion processing of starch film. *International Journal of Food Engineering, 5*, Article 7/1–12.

Sugih, A. K., Drijfhout, J. P., Picchioni, F., Janssen, L. P. B. M., & Heeres, H. J. (2009). Synthesis and properties of reactive interfacial agents for polycaprolactone-starch blends. *Journal of Applied Polymer Science, 9999*.

Sullivan, J. W., & Johnson, J. A. (1964). Measurement of starch gelatinization by enzyme susceptibility. *Cereal Chemistry, 41*, 73–77.

Tabi, T., & Kovacs, J. G. (2007). Examination of injection moulded thermoplastic maize starch. *Express Polymer Letters, 1*, 804–809.

Tajuddin, S., Xie, F., Nicholson, T. M., Liu, P., & Halley, P. J. (2011). Rheological properties of thermoplastic starch studied by multipass rheometer. *Carbohydrate Polymers, 83*, 914–919.

Teixeira, E. M., Da Róz, A. L., Carvalho, A. J. F., & Curvelo, A. A. S. (2007). The effect of glycerol/sugar/water and sugar/water mixtures on the plasticization of thermoplastic cassava starch. *Carbohydrate Polymers, 69*, 619–624.

Thunwall, M., Kuthanova, V., Boldizar, A., & Rigdahl, M. (2008). Film blowing of thermoplastic starch. *Carbohydrate Polymers, 71*, 583–590.

Thuwall, M., Boldizar, A., & Rigdahl, M. (2006). Extrusion processing of high amylose potato starch materials. *Carbohydrate Polymers, 65*, 441–446.

Van Soest, J. J. G., & Knooren, N. (1997). Influence of glycerol and water content on the structure and properties of extruded starch plastic sheets during aging. *Journal of Applied Polymer Science, 64*, 1411–1422.

Vergnes, B., Della Valle, G., & Tayeb, J. (1993). A specific slit die rheometer for extruded starchy products. Design, validation and application to maize starch. *Rheologica Acta, 32*, 465–476.

Vergnes, B., & Villemaire, J. P. (1987). Rheological behaviour of low moisture molten maize starch. *Rheologica Acta, 26*, 570–576.

Vergnes, B., Villemaire, J. P., Colonna, P., & Tayeb, J. (1987). Interrelationships between thermomechanical treatment and macromolecular degradation of maize starch in a novel rheometer with preshearing. *Journal of Cereal Science, 5*, 189–202.

Walenta, E., Fink, H.-P., Weigel, P., & Ganster, J. (2001). Structure-property relationships in extruded starch, 1 supermolecular structure of pea amylose and extruded pea amylose. *Macromolecular Materials and Engineering, 286*, 456–461.

Walenta, E., Fink, H.-P., Weigel, P., Ganster, J., & Schaaf, E. (2001). Structure-property relationships of extruded starch, 2 extrusion products from native starch. *Macromolecular Materials and Engineering, 286*, 462–471.

Wang, L., Shogren, R. L., & Carriere, C. (2000). Preparation and properties of thermoplastic starch-polyester laminate sheets by coextrusion. *Polymer Engineering & Science, 40*, 499–506.

Wang, X.-L., Yang, K.-K., & Wang, Y.-Z. (2003). Properties of starch blends with biodegradable polymers. *Polymer Reviews, 43*, 385–409.

Wang, N., Yu, J., Chang, P. R., & Ma, X. (2007). Influence of citric acid on the properties of glycerol-plasticized dry starch (DTPS) and DTPS/poly(lactic acid) blends. *Starch-Stärke, 59*, 409–417.

Wang, N., Yu, J., Chang, P. R., & Ma, X. (2008). Influence of formamide and water on the properties of thermoplastic starch/poly(lactic acid) blends. *Carbohydrate Polymers, 71*, 109–118.

Wang, N., Yu, J., & Ma, X. (2007). Preparation and characterization of thermoplastic starch/PLA blends by one-step reactive extrusion. *Polymer International, 56*, 1440–1447.

Wang, J., Yu, L., Xie, F., Chen, L., Li, X., & Liu, H. (2010). Rheological properties and phase transition of cornstarches with different amylose/amylopectin ratios under shear stress. *Starch-Stärke, 62*, 667–675.

Wang, S. S., & Zheng, X. (1995). Tribological shear conversion of starch. *Journal of Food Science, 60*, 520–522.

Willett, J. L., Kotnis, M. A., O'Brien, G. S., Fanta, G. F., & Gordon, S. H. (1998). Properties of starch-graft-poly(glycidyl methacrylate)–PHBV composites. *Journal of Applied Polymer Science, 70*, 1121–1127.

Willett, J. L., & Shogren, R. L. (2002). Processing and properties of extruded starch/polymer foams. *Polymer, 43*, 5935–5947.

Xie, F., Halley, P. J., & Avérous, L. (2011). Bio-nanocomposites based on starch. In V. Mittal (Ed.), *Nanocomposites with biodegradable polymers: Synthesis, properties and future perspectives* (pp. 234–260). Oxford, UK: Oxford University Press.

Xie, F., Halley, P. J., & Avérous, L. (2012). Rheology to understand and optimize processibility, structures and properties of starch polymeric materials. *Progress in Polymer Science, 37*, 595–623.

Xie, F., Liu, H., Chen, P., et al. (2006). Starch gelatinization under shearless and shear conditions. *International Journal of Food Engineering*, 2(5), 1–29. Article 6.

Xie, F., Xue, T., Yu, L., Chen, L., Li, X., & Zhang, X. (2007). Rheological properties of starch-based materials and starch/poly(lactic acid) blends. *Macromolecular Symposia, 249–250*, 529–534.

Xie, F., Yu, L., Chen, L., & Li, L. (2008). A new study of starch gelatinization under shear stress using dynamic mechanical analysis. *Carbohydrate Polymers, 72*, 229–234.

Xie, F., Yu, L., Liu, H., & Chen, L. (2006). Starch modification using reactive extrusion. *Starch-Stärke, 58*, 131–139.

Xie, F., Yu, L., Liu, H., & Dean, K. (2006). Effect of compatibilizer distribution on thermal and rheological properties of gelatinized starch/biodegradable polyesters blends. *International Polymer Processing, 21*, 379–385.

Xie, F., Yu, L., Su, B., Liu, P., Wang, J., Liu, H., et al. (2009). Rheological properties of starches with different amylose/amylopectin ratios. *Journal of Cereal Science, 49*, 371–377.

Xue, T., Yu, L., Xie, F., Chen, L., & Li, L. (2008). Rheological properties and phase transition of starch under shear stress. *Food Hydrocolloids, 22*, 973–978.

Yamada, T., Suzuki, K., Katuzaki, H., Hisamatsu, M., & Komiya, T. (1990). GPC profile change of potato starch with extrusion processing. *Starch-Starke, 42*, 217–223.

Yang, J.-H., Yu, J.-G., & Ma, X.-F. (2006a). Preparation and properties of ethylenebisformamide plasticized potato starch (EPTPS). *Carbohydrate Polymers, 63*, 218–223.

Yang, J.-h., Yu, J.-g., & Ma, X.-f. (2006b). Preparation of a novel thermoplastic starch (TPS) material using ethylenebisformamide as the plasticizer. *Starch-Stärke, 58*, 330–337.

Yang, J.-h., Yu, J.-g., & Ma, X.-f. (2006c). Study on the properties of ethylenebisformamide and sorbitol plasticized corn starch (ESPTPS). *Carbohydrate Polymers, 66*, 110–116.

Yu, L., & Christie, G. (2005). Microstructure and mechanical properties of orientated thermoplastic starches. *Journal of Materials Science, 40*, 111–116.

Yu, L., Dean, K., & Li, L. (2006). Polymer blends and composites from renewable resources. *Progress in Polymer Science, 31*, 576–602.

Yu, L., Dean, K., Yuan, Q., Chen, L., & Zhang, X. (2007). Effect of compatibilizer distribution on the blends of starch/biodegradable polyesters. *Journal of Applied Polymer Science, 103*, 812–818.

Yu, L., Petinakis, E., Dean, K., Liu, H., & Yuan, Q. (2011). Enhancing compatibilizer function by controlled distribution in hydrophobic polylactic acid/hydrophilic starch blends. *Journal of Applied Polymer Science, 119*, 2189–2195.

Yu, J., Wang, N., & Ma, X. (2005). The effects of citric acid on the properties of thermoplastic starch plasticized by glycerol. *Starch-Stärke, 57*, 494–504.

Zeng, M., Huang, Y., Lu, L., Fan, L., & Lourdin, D. (2011). Effects of filler-matrix morphology on mechanical properties of corn starch-zein thermo-moulded films. *Carbohydrate Polymers, 84*, 323–328.

Zhang, L., Deng, X., Zhao, S., & Huang, Z. (1997). Biodegradable polymer blends of poly(3-hydroxybutyrate) and starch acetate. *Polymer International, 44*, 104–110.

Zheng, X., Chiang, W.-C., & Wang, S. S. (1995). Effect of shear energy on size reduction of starch granules in extrusion. *Starch-Stärke, 47*, 146–151.

Zheng, X., & Wang, S. S. (1994). Shear induced starch conversion during extrusion. *Journal of Food Science, 59*, 1137–1143.

Zhou, J., & Hanna, M. A. (2004). Extrusion of starch acetate with mixed blowing agents. *Starch-Stärke, 56*, 484–494.

Zhou, J., & Hanna, M. A. (2005). Effects of the properties of blowing agents on the processing and performance of extruded starch acetate. *Journal of Applied Polymer Science, 97*, 1880–1890.

Zhou, J., Song, J., & Parker, R. (2006). Structure and properties of starch-based foams prepared by microwave heating from extruded pellets. *Carbohydrate Polymers, 63*, 466–475.

Zhou, J., Song, J., & Parker, R. (2007). Microwave-assisted moulding using expandable extruded pellets from wheat flours and starch. *Carbohydrate Polymers, 69*, 445–454.

Zullo, R., & Iannace, S. (2009). The effects of different starch sources and plasticizers on film blowing of thermoplastic starch: Correlation among process, elongational properties and macromolecular structure. *Carbohydrate Polymers, 77*, 376–383.

CHAPTER

11

Reactive Extrusion for Thermoplastic Starch-Polymer Blends

R.B. Maliger, Peter J. Halley

School of Chemical Engineering, The University of Queensland, Brisbane, Queensland, Australia

1 INTRODUCTION

Plastics obtained from petrochemical sources generally have lifetimes of hundreds of years when buried in typical solid waste sites and thus have posed significant environmental threats (Chandra & Rustgi, 1997). These plastics have contributed 17-25% of the volume of waste being landfilled (Doane, 1994). Hence, recently, much research has concentrated on the development of environmental-friendly biodegradable plastics obtained from various natural renewable resources including starch, cellulose, polyhydroxyalkanoates, and poly(lactic acid) (PLA).

Three main classes of biodegradable polymers have been recognized. The first class consists of synthetic polymers sourced from nonrenewable resources with vulnerable groups susceptible to hydrolysis attack by microbes. The second class of materials is composed of naturally occurring bacterial polymers like polyhydroxybutyrates, polyhydroxyvalerates, and polyhydroxybutyrate-co-valerates. Although polyhydroxyalkanoates are truly renewable and biodegradable, the determination of processing parameters and high production costs are the concerning factors (Chandra & Rustgi, 1997). The third class of materials is biodegradable polymers from renewable resources such as starch, cellulose, and PLA. Starch-based thermoplastics are inexpensive but generally possess high viscosities and poor melt properties that make them difficult to process. As a result, products made from starch are often brittle and water-sensitive (Kalambur & Rizvi, 2006). In order to alleviate these problems, starch polymers are often blended with more high-performance synthetic polymers. Of course, in the case of starch-polyester blend systems, one must also factor in the higher cost of the polyester and aim to increase the percentage of the lower-cost starch in the blend while maintaining the desired mechanical properties and performance of the polymer product. Thus, to achieve this cost-performance optimization, much must be understood and optimized about the individual polymers and the blending process.

1.1 Structure, Morphology, Composition, and Properties of Starch

Recent NMR studies suggest that there are at least three distinct components in wheat starch granules (Donald, 1997). They are (i) highly crystalline regions formed from double-helical starch chains, (ii) solid-like regions formed from amylose-lipid inclusion complexes, and (iii) completely amorphous regions associated with branching regions of the amylopectin components of starch and possibly the lipid-free amylose. Starch granules essentially consist of linear poly-(1-4)-α-D-glucose, amylose, and the branched molecule, amylopectin, where the linear chains are connected through (1-6)-α linkages. When starch granules are observed under polarized light, a characteristic dark cross is seen that has led to regard the granules as distorted spherocrystals (Galliard, 1987). The proportion of amylopectin ranges from 95% in the low-amylose or waxy starches, to 70-75% in normal starches, to 30% in some high-amylose starches (Galliard, 1987).

Hydroxyl groups cause starch to generally behave as an alcohol during chemical reactions. This property of starch is important when considering reactive blending of starch with synthetic polymers. The presence of such large number of —OH groups affords starch hydrophilic properties and therefore adds affinity for moisture and dispersibility in water (Imberty, Buleon, Tran, & Perez, 1991). However, hydrophilicity is undesirable in many plastic packaging applications, and hence, it is a major limitation in using starch as a homopolymer.

1.2 Starch-Polymer Blends

Most synthetic polymers are incompatible with starch and it is difficult to add starch to synthetic polymer systems in large quantities without dramatically reducing performance properties (Ramkumar, Bhattacharya, & Zhang, 1997). To enhance the compatibility of starch and synthetic polymers, a reactive functional group such as maleic anhydride (MA) can be added to the synthetic polymer (Ramkumar, Yang, & Bhattacharya, 1997). In these blends, the interfacial tension is lowered to generate a small phase size and strong interfacial adhesion to transmit applied force effectively between the component phases (Barlow & Paul, 1984). Thus, compatibilization through *in situ* formation of compatibilizer in polymer blends has become increasingly important and an alternative to replace the method of adding block or graft copolymers separately (Mani, Bhattacharya, & Tang, 1999). Synthetic polymers having functional groups such as carboxylic acid, anhydride, epoxy, urethane, or oxazoline can react with the —OH group in starch to form a blend with stable morphology. These reactive groups form a chemical or physical bond with hydroxyl or carboxyl groups in natural polymers such as starch (John, Tang, Yang, & Bhattacharya, 1997).

The role of a compatibilizer is to control the properties of multiphase blends not only by converting immiscible blends into fully miscible blends but also by controlling the size of the phase domains of immiscible blends. Effective compatibilizers must be located at the interface between the phase domains of the immiscible blend. Most importantly, the degree of compatibilization in a particular system depends on the reactivity of the compatibilizer used. It has been found that a compatibilizer is most effective when its sections are of higher molecular weight than the corresponding blend components (Lohse & Datta, 1996). Several theories have been proposed to explain the role of compatibilizers, of which two mechanisms are considered plausible. The first mechanism is thermodynamic in nature in that the compatibilizer reduces the interfacial tension between the phases. The second mechanism is kinetic in nature in that the presence of the compatibilizer at the interface reduces the agglomeration of domains by steric stabilization. However, it is often unclear which of these mechanisms dominate the reduction in the particle size (Lohse & Datta, 1996), and hence, the mechanism of compatibilization is still a debatable issue.

Several researchers have showed that dissimilar blends could be compatibilized using different techniques. Bacon and Farmer (1939) showed that MA reacts with natural rubber in the presence of benzoyl peroxide and results in the addition of MA to the double bonds and to the α-methylene groups of the polymer. Gaylord (1996) had confirmed the addition reaction of MA with rubber in solid phase and in solution. Mani et al. (1999) had used initiators like dicumyl peroxide (DCP), benzoyl peroxide, and di-*tert*-butyl peroxide for grafting biodegradable polybutylene succinate (PBS) with starch in the presence of MA as a cross-linking agent. Sailaja and Chanda (2000) had used MA-grafted polyethylene (PE) for preparing PE-starch blends in the presence of benzoyl peroxide as initiator. Ceric ammonium nitrate (CAN) is another initiator commonly used during reactive grafting. Lutfor et al. (2001) studied the kinetics of graft polymerization of acrylonitrile onto sago starch using CAN as initiator. These works suggested the use of other cross-linking agents such as phthalic anhydride and dodecyl succinic anhydride in the presence of initiators to facilitate grafting reactions.

Of all the methods used to achieve compatibilization in starch-polyester blends, REX is a well-utilized technique as it combines the traditionally separated chemical processes (polymer synthesis and modification) and extrusion (melting, blending, structuring, devolatilization, and shaping) into a single process carried out in the extruder (Kalambur & Rizvi, 2006). Using REX, modified polymer can be obtained in a ready-to-use form at the die. Specifically, Kalambur and Rizvi (2006) reviewed some of the most promising chemistries available for the REX of starch-based polymer blends to overcome the thermodynamic immiscibility and non-wetting of starch with other polymers, which leads to serious deterioration of mechanical properties at >25-30 wt% starch. They identified the primary challenge was to develop fast reaction chemistries that can be transformed into viable processes and integrated into existing process lines with economically viable formulations. Key results indicate that graft copolymer properties can be controlled through the monomer/starch ratio and moisture content during REX. Xie, Yu, Liu, and Chen (2006) reviewed REX and their application in starch modification and preparation of starch derivatives, such as starch graft copolymers, glycosides, cationic starch, and oxidized starch. Moad (2011) recently reviewed processes for chemical transformation of starch by extrusion processing (REX). The review included focusing on the grafting of monomers from starch such as ring opening of epoxides, esterification (with lactones, anhydrides, acids, halides, or vinyl esters), phosphorylation, and silylation; graft polymerization from starch by radical-induced grafting or the ring-opening polymerization of lactones; reactive compatibilization with polyesters and polyolefins by grafting to or from starch; cross-linking of starch with epichlorohydrin or by phosphorylation; and the degradation of starch thermally or catalyzed by acid or enzymes.

Recent work by Zeng et al. (2011) examined blending PBS with thermoplastic starch (TPS) to improve the mechanical properties and reduce water absorption of the resulting starch-based plastics. In order to enhance the miscibility, reactive PBS (RPBS) with terminal NCO group was synthesized first and then blended with TPS. The mechanical properties of the TPS were greatly improved after blending with RPBS and the tensile strength was increased to 10 times more than TPS at 10 wt% RPBS. The water absorption of the blend was significantly reduced with increasing RPBS content and hydrophobicity of the blend was enhanced.

Shin, Jang, and Kim (2011) examined PLA blended with chemically modified thermoplastic starch (CMPS) in a twin-screw extruder. Differential scanning calorimetry showed that the PLA/CMPS was thermodynamically immiscible; however, scanning electron microscopy and Fourier transform infrared studies revealed that the interfacial adhesion was improved by the PLA-g-starch copolymers that were formed at the interface through a transesterification reaction between PLA and CMPS. The crystallinity of the PLA component in the blend was increased by the addition of the CMPS and was highly affected by the morphology of the blend. The tensile strength and elongation were found to decrease in a linear trend with increasing CMPS content. The biodegradability of the blends increased with increasing CMPS content.

Frost et al. (2011) examined TPS-silica (SiO_2) polyvinyl alcohol (PVOH) composite films created via an REX process using tetraethyl orthosilicate as a precursor. Tensile strength and Young's modulus increased, while elongation at break decreased with increasing SiO_2 content. Dynamic mechanical analysis results showed that the starch-silica composite storage modulus increased and the loss modulus decreased with increasing SiO_2 content.

Nayak (2010) examined starch-based biodegradable nanocomposites of poly(butylene adipate-co-terephthalate) (PBAT) and organically modified nanoclays using melt intercalation technique in a Haake Rheocord 9000 torque rheometer where the interfacial region between the biodegradable polymer matrix and the clays was also modified with grafting of MA with PBAT chains, during melt blending through two-stage REX process. Mechanical tests revealed an increase in tensile modulus and elongation at break with the incorporation of compatibilizer and nanoclays.

Manoi and Rizvi (2010) recently examined cross-linking of starch blends by phosphorylation using reactive supercritical fluid extrusion. A starch blend was mixed with sodium trimetaphosphate and extruded NaOH solution and supercritical carbon dioxide (SC-CO_2). Cross-linking of starch at highest reaction pH was also evidenced by its lowest water solubility and highest proportion of distarch monophosphate. Introduction of phosphate cross-links into the starch restricted the molecular structure mobility, leading to an increase in the gelatinization temperature. SC-CO_2-based expansion provided lightweight and nonporous skin starch foams with excellent water resistance.

Willett and Finkenstadt (2009) examined graft copolymers of starch and polyacrylamide (PAAm) prepared using REX. The effect of cationic starch modification was examined using unmodified and cationic starch. For a fixed acrylamide/starch feed ratio, conversion, graft content, and grafting efficiency were essentially unaffected by starch type or modification. Cationic starch graft copolymers generally had lower-molecular-weight PAAm grafts compared to the unmodified starch, whereas amylose content had little or no significant effect. In addition, the frequency of grafting was higher with cationic starch. These results indicate that cationic modification of starch enhances the formation of grafting sites, resulting in graft copolymers with more grafts of lower molecular weight compared to unmodified starch.

O'Brien, Wang, Vervaet, and Remon (2009) and O'brien and Wang (2009) investigated extruded starch phosphates as a potential material for sustained release applications because of its nontoxicity and biodegradability. Starches from various botanical sources with different amylose contents were used to prepare starch phosphates using an REX method. REX produced starch extrudates that upon forming hydrogels were capable of sustaining release of metoprolol tartrate. The structural features of the hydrogel as modified by the phosphorylation reaction were found to alter the kinetics of drug release from the swellable matrices. The unmodified extrudates formed weaker gels as evidenced by their rheological properties and showed faster drug release. Waxy cornstarch phosphorylated at pH 9.0 exhibited the lowest degree of drug release. The results demonstrate that the reaction efficiency of phosphorylation by REX and subsequent drug release can be affected by the shear rate and pH for different starch types.

Murua-Pagola, Beristain-Guevara, and Martinez-Bustos (2009) prepared acetylated, *n*-octenylsuccinylated, and phosphorylated waxy maize starches using REX in a single-screw extruder. The viscosity of the modified starches was reduced after extrusion and the water solubility index, and water absorption index increased. The modified starches prepared using REX showed good characteristics as shell materials for encapsulation by spray drying.

Raquez, Nabar, Narayan, and Dubois (2008) examined maleated thermoplastic starch (MTPS) in the REX melt blending with PBAT for blown film applications. For high polyester fractions, PBAT-g-MTPS graft copolymers were obtained through transesterification reactions, whereas at lower polyester content, no significant reaction occurred, more likely due to an inversion in the phase morphology between both components. Tensile properties of PBAT-g-MTPS graft copolymer containing 70 wt% polyester were much higher as the TPS/PBAT melt blend modified with MA. This was explained by a finer morphology of the dispersed phase in the continuous PBAT matrix and an increased interfacial area for the grafting reaction.

Willett and Finkenstadt (2006) and Finkenstadt and Willett (2005) examined graft copolymers of PAAm and various substrates including unmodified starches (corn, waxy maize, wheat, and potato), cationic starches, dextrin, dextran, and PVOH by REX in a twin-screw extruder using ammonium persulfate as initiator. Conversions and absorbances were monitored.

1.3 Case Study: Thermoplastics Starch/Biopolyester/ Nanocomposite

Unmodified starch-based thermoplastics generally have higher viscosities and poor melt properties than traditional synthetic polymers that make them difficult to process. Also, starch and synthetic polymers are generally thermodynamically dissimilar in nature and hence are incompatible unless a compatibilizer is used. In REX of cross-linking dissimilar polymers, free-radical initiation plays a predominant role. However, in the extruder, mechanochemistry, by itself, is not powerful enough for free-radical generation in such grafting reactions. Hence, in this study, we have attempted to graft MA to a biodegradable polyester in the presence of a free-radical initiator during stage one of extrusion and further cross-linked MA-grafted polyester to starch during a second extrusion stage. We used different temperature profiles during the REX process. Should the compatibilized blends exhibit improved mechanical properties, determination of interfacial tension and controlling the interfacial properties of such blends will lead us on the path to developing well-controlled biodegradable blends with tailorable mechanical properties.

2 MATERIALS

EnPol®, a biodegradable thermoplastic polyester, was obtained from IRe Chemical Ltd. (Korea). Two types of starch with different levels of amylopectin-to-amylose ratios were used in a preset blend ratio in all formulations. The first was a low-amylose common waxy wheat starch and the other a high-amylose, chemically modified maize starch (hydroxypropylated starch). The second has a lower gelatinization temperature than the first and is able to withstand higher processing temperatures. Both starches were sourced from Penford (Australia). These starches were blended together with plasticizers in accordance with the patents (Buehler & Schultze, 1998; Buehler, Ernst, & Schultze, 1998). MA (98.06%), obtained from ICN Biomedicals Inc. (United States), was used as a cross-linking agent, and DCP (99%), obtained from Sigma-Aldrich Company, Pty Ltd., was used as a free-radical initiator. Inert nitrogen gas was used to prevent the effect of moisture on grafting during the first stage of extrusion.

2.1 Blend Preparation

A batch of starch patent formulation (SPF) was prepared using a granulator. The composition of SPF to EnPol® was maintained at 40:60 (wt%) in all the blends. The different blends prepared using a laboratory-scale PRISM corotating twin-screw extruder (length-to-diameter ratio of 40:1 and screw diameter of 16 mm) are shown in Table 11.1.

TABLE 11.1 Blends Prepared Using Twin-Screw Extruder

S (wt%)	PEst (wt%)	MA (wt%)	DCP (wt%)
40	60	0.0	0.0
40	60	0.5	0.3
40	60	0.5	0.5
40	60	0.5	0.8
40	60	0.5	1.2
40	60	1.0	0.0
40	60	1.0	0.3
40	60	1.0	0.5
40	60	1.0	0.8
40	60	1.0	1.2
40	60	1.5	0.0
40	60	1.5	0.3
40	60	1.5	0.5
40	60	1.5	0.8
40	60	1.5	1.2

2.2 Grafting Procedure

The grafting reactions were carried out in a laboratory-scale PRISM corotating twin-screw extruder with a barrel length-to-diameter ratio of 40:1, a screw diameter of 16 mm, and eight heating zones. The composition of SPF and the polyester was maintained at 40:60 (wt%) in all the blends.

The first step was the preparation of maleated polyester. From initial trials, we found that the color of the extrudate varied as the composition of MA was increased (0.5% MA: white; 1% MA: light pink; 1.5% MA: dark pink; 2% MA: brown; 3% MA: brownish black; 5% MA: black). Maleated polyesters with MA concentration of more than 2% were found to be unsuitable for compounding with starch. Therefore, after optimizing the process, three different compositions of MA (0.5%, 1%, and 1.5%) and four different compositions of DCP (0.3%, 0.5%, 0.8%, and 1.2%) were selected. The temperature profiles for stages 1 and 2 of extrusion are shown in Table 11.2.

EnPol® was dried under vacuum for 24 h prior to the day of extrusion. MA and DCP were used in their powder form. A mixture of polyester, MA, and DCP was introduced using a mechanical feeder at a feed rate

TABLE 11.2 Temperature Profiles for Stages 1 and 2 of Extrusion

Die (°C)	Zone9 (°C)	Zone8 (°C)	Zone7 (°C)	Zone6 (°C)	Zone5 (°C)	Zone4 (°C)	Zone3 (°C)	Zone2 (°C)	Zone1 (°C)
100	120	130	140	150	150	140	130	120	100
100	120	160	180	180	180	180	180	150	120

of 0.38 kg/h. Screw speed of 45 rpm was used here. A continuous flow of nitrogen was maintained with the help of an inlet device during the first stage of extrusion. The torque and die pressure were monitored using a torque meter and pressure transducer, respectively. The temperature between mixing and transportation zones was maintained at 180 °C to facilitate peroxide-initiated free-radical generation. The extrudate obtained from the first stage of extrusion was pelletized using a pelletizer and stored in a humidifier at 60% RH until further use.

The second step was compounding of the pelletized MA-grafted polyester with SPF. Here, a screw speed of 70 rpm and a feed rate of 0.45 kg/h were used. The temperature between mixing and transportation zones was maintained at 145 °C. The extruded strands were stored in a humidifier at 60% RH.

3 RESULTS AND DISCUSSION

3.1 Twin-Screw Extrusion

The twin-screw extruder here was used as a reactive extruder to combine peroxide-initiated grafting reaction and conventional extrusion into a single process. The screw configuration, torque, rotational speed, and mass flow rate are the important terms in determining specific mechanical energy (SME), which is given as

$$\mathrm{SME} = M_{\mathrm{d}}\omega/m \tag{11.1}$$

where SME = specific mechanical energy (J/kg), M_{d} = torque (N m), ω = rotational speed of screw (s^{-1}), and m = mass flow rate (kg/s).

The greater the torque in the extruder, the higher the bulk viscosity of the system. Thus, for cross-linked systems, the SME required should be more. The calculated SMEs for different blends are given in Table 11.3.

During blending of starch with polyesters in the presence of a cross-linking agent and initiator, the anhydride functional group could react with —OH group of starch to form ester linkages (23). Hence, in this REX, inter-cross-linked polymer chains are expected, and as a result, the torque is generated and hence, the SME of a compatibilized blend

TABLE 11.3 Specific Mechanical Energy of Different Blends

Blend	SME (kJ/kG)
40S, 60PEst	152.98
40S, 60PEst, 0.5MA	198.53
40S, 60PEst, 0.5MA, 0.3DCP	535.23
40S, 60PEst, 0.5MA, 0.5DCP	560.42
40S, 60PEst, 0.5MA, 0.8DCP	579.31
40S, 60PEst, 0.5MA, 1.2DCP	540.81
40S, 60PEst, 1MA	214.35
40S, 60PEst, 1MA, 0.3DCP	271.66
40S, 60PEst, 1MA, 0.5DCP	353.23
40S, 60PEst, 1MA, 0.8DCP	413.23
40S, 60PEst, 1MA, 1.2DCP	598.97
40S, 60PEst, 1.5MA	348.94
40S, 60PEst, 1.5MA, 0.3DCP	453.37
40S, 60PEst, 1.5MA, 0.5DCP	454.46
40S, 60PEst, 1.5MA, 0.8DCP	452.53
40S, 60PEst, 1.5MA, 1.2DCP	458.9

should be higher than the SME of uncompatibilized blends. It is evident from Table 11.3 that the SME of each compatibilized blend is higher than the SME of the uncompatibilized blend (40S, 60PEst). It can be observed that the SME variation in blends containing 1% MA is greater than in other blends. In the case of blends containing 0.5% MA and 1.5% MA, the variation in SME was negligible. This indicates that blends containing 1% MA had higher microradical generation and better cross-linking than in other blends. This could have increased the bulk viscosity and SME of the system.

3.2 Tensile Testing Analysis

Three important parameters were considered for tensile testing analysis, namely, tensile stress at maximum load, stress at break, and Young's modulus (tangent 5%). The effect of the concentration of MA and DCP on these parameters was determined and is shown in Table 11.4.

The blends at 0.5% MA exhibited better tensile properties in terms of Young's modulus and stress at break than the unmodified blend but

TABLE 11.4 Tensile Properties of Different Blends

Blend	Tensile Stress at Maximum Load (MPa)	Strain at Break (%)	Young's Modulus (5% Tangent) (MPa)
40S, 60PEst	14.68 ± 0.95	78.68 ± 9.58	36.77 ± 6.02
40S, 60PEst, 0.5MA, 0DCP	15.70 ± 1.22	80.82 ± 4.38	38.57 ± 12.48
40S, 60PEst, 0.5MA, 0.3DCP	20.11 ± 0.9	75.18 ± 8.74	86.87 ± 13.22
40S, 60PEst, 0.5MA, 0.5DCP	19.89 ± 1.78	78.68 ± 11.52	79.87 ± 21.10
40S, 60PEst, 0.5MA, 0.8DCP	18.63 ± 1.75	92.38 ± 8.65	62.87 ± 15.55
40S, 60PEst, 0.5MA, 1.2DCP	17.55 ± 1.55	50.24 ± 10.19	48.38 ± 19.83
40S, 60PEst, 1MA	12.54 ± 4.42	31.06 ± 2.42	51.74 ± 21.79
40S, 60PEst, 1MA, 0.3DCP	10.5 ± 3.33	150.12 ± 15.15	30.88 ± 9.94
40S, 60PEst, 1MA, 0.5DCP	15.71 ± 0.96	111.24 ± 12.98	43.07 ± 9.09
40S, 60PEst, 1MA, 0.8DCP	18.92 ± 1.04	257.35 ± 37.69	51.59 ± 6.89
40S, 60PEst, 1MA, 1.2DCP	16.05 ± 1.71	165.73 ± 38.61	44.07 ± 5.88
40S, 60PEst, 1.5MA	17.61 ± 0.54	35.34 ± 3.62	82.64 ± 20.33
40S, 60PEst, 1.5MA, 0.3DCP	19.8 ± 2.78	40.86 ± 4.81	88.31 ± 11.63
40S, 60PEst, 1.5MA, 0.5DCP	18.27 ± 1.56	36.06 ± 5.51	80.86 ± 5.18
40S, 60PEst, 1.5MA, 0.8DCP	19.35 ± 1.61	44.08 ± 7.85	74.78 ± 27.36
40S, 60PEst, 1.5MA, 1.2DCP	18.98 ± 1.58	43.02 ± 4.96	79.99 ± 20.28

showed little or no improvement in elongation at break. The blends with 1% MA had moderately higher Young's modulus and stress at break values than the unmodified blend. They also showed much higher elongation at break than the uncompatibilized blend and indeed the other reactively extruded blends. All blends containing 1.5% MA had lower elongation at break than the uncompatibilized blend. The Young's modulus and stress at break were increased over the uncompatibilized system. In comparing between the compatibilized blends, blends containing 0.5% MA and 1.5% MA had higher Young's modulus than the blends containing 1% MA. However, the blends at 1% MA had greatly improved elongation at break values than 0.5% MA and 1.5% MA blends.

Mani and Bhattacharya (1998) observed that starch blends that encountered higher SME had high tensile strength. Our results indicated that only few starch blends with higher SME had better tensile strength. It is hence important to consider the effects of interfacial properties and maleation on the tensile properties of starch-polyester blends.

Bhattacharya, Vaidya, Zhang, and Narayan (1995) noticed that the addition of compatibilizers (styrene MA copolymer and ethylene-propylene-g-maleic anhydride copolymer) had a profound effect on the tensile properties of the starch blend (60% starch, 40% compatibilizer). However, those blends exhibited poor elongation at break. Avella et al. (2000) had shown that an increase in the composition of starch and pre-compatibilizer decreases both tensile strength and elongation at break but increases Young's modulus. In synthetic polymer blends, the addition of a second phase to the polymer matrix usually diminishes the elongation properties at break (Stpierre, Favis, Ramsay, Ramsay, & Verhoogt, 1997), and in many cases, when 20% of the dispersed minor phase has been added, highly deformable matrix materials are transformed into brittle materials (Paul, Vinson, & Locke, 1972). The elongation at break in synthetic polymer blends is therefore considered to be highly sensitive to the state of the interface. However, we observed a notable increase in elongation at break and overall decrease in Young's modulus in blends containing 1% MA. Krishnan and Narayan (2002) had described in their patent that hydroxyl groups of plasticizers and starch molecules could interact with compatibilizers promoting interfacial adhesion. They also indicate an acute possibility of plasticizers acting as stretching agents. If it is assumed that maleation did not occur, then starch plasticizers should have increased elongation at break and interfacial adhesion in all blends, irrespective of the composition of the cross-linking agent and the initiator. The fact that only few maleated starch blends showed improved mechanical properties indicates the importance of optimizing the composition of the cross-linking agent and the initiator. In particular, the blend (40S, 60PEst, 1MA, 0.8DCP) showed the highest elongation at break (257%). This could have resulted from the strong interfacial adhesion due to cross-linking of the two phases and the greater ability of this interface to withstand higher extension to break. Whereas, blends containing 1.5% MA had relatively higher Young's modulus and lower elongation at break. Ideally, an optimized compatibilized blend here is a compromise between desired mechanical properties and starch composition.

3.3 Dynamic Mechanical and Thermal Analysis

Dynamic mechanical and thermal analysis (DMTA) experiments are used to investigate the mechanical behavior of materials and to obtain information about the relaxation mechanisms that may be correlated with the dynamics and the microstructure of the material (Demirgoz et al., 2000). The effect of temperature on storage modulus (E'), loss modulus (E''), and loss factor ($\tan\delta$) at a fixed strain rate (0.1%) was studied here. The storage modulus E' is related with the mechanical energy stored during each load cycle and per unit volume. Loss modulus E'' signifies the dissipation of energy as heat during the deformation. The loss factor $\tan\delta$ is equal to E''/E' and is thus sensitive to balance of the dissipated and stored energy of the system and is useful to detect thermomechanical relaxations (Vaidya, Bhattacharya, & Zhang, 1995).

E' for the uncompatibilized blend (40S, 60PEst) was found to decrease gradually with increase in temperature, indicating a stiffness loss. Here, a definite glass transition was observed for E'' and $\tan\delta$ curves between 35 and 50 °C, which should be due to starch. The graphs for E' and $\tan\delta$ are shown in Figure 11.1.

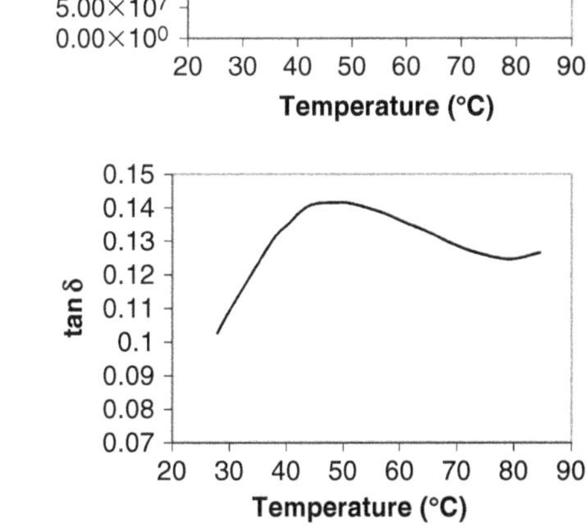

FIGURE 11.1 Graph of E' and E'' versus temperature for uncompatibilized blend.

3.4 Analysis of Compatibilized Blends

Graphs of storage modulus versus temperature of different blends containing 0.5% MA, 1% MA, and 1.5% MA and various compositions of DCP are shown in Figures 11.2–11.4.

While there appears to be no specific trend, the blend (40S, 60PEst, 0.5MA, 0.8DCP) exhibited the highest storage modulus (4.64×10^8 Pa) at 30 °C, followed by blends containing 0.5% DCP, 0.3% DCP, and 1.2% DCP (Figure 11.2). It was also observed that at higher temperatures (>80 °C), blends containing higher compositions of DCP lost their stiffness more quickly than the blends containing lower compositions of DCP. For blend (40S, 60PEst, 0.5MA, 0.3DCP), two distinct peaks were observed at 55 and 68 °C. The first peak at 55 °C is due to the addition of the interface modifier (compatibilizer). The other peak at 68 °C could

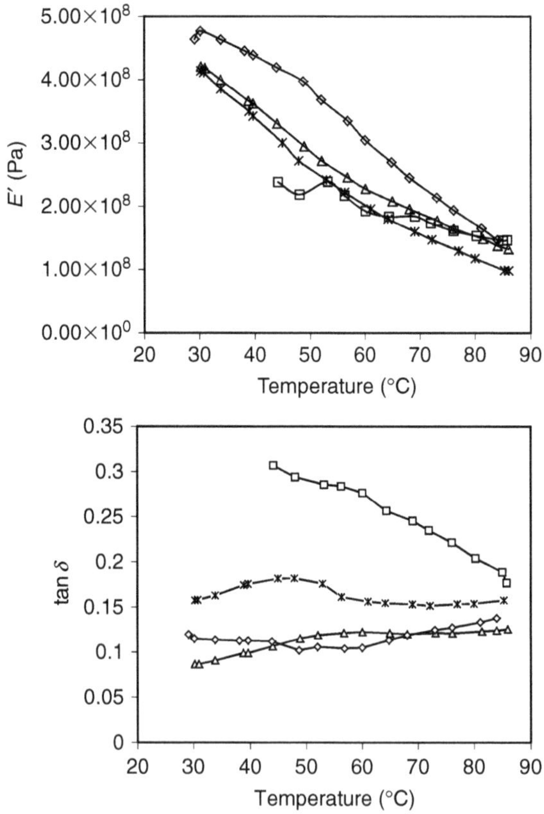

FIGURE 11.2 Graph of E' and tan δ versus temperature for blends containing 0.5% MA {□ (40S, 60PEst, 0.5MA, 0.3DCP); Δ (40S, 60PEst, 0.5MA, 0.5DCP); ◇ (40S, 60PEst, 0.5MA, 0.8DCP); * (40S, 60PEst, 0.5MA, 1.2DCP)}.

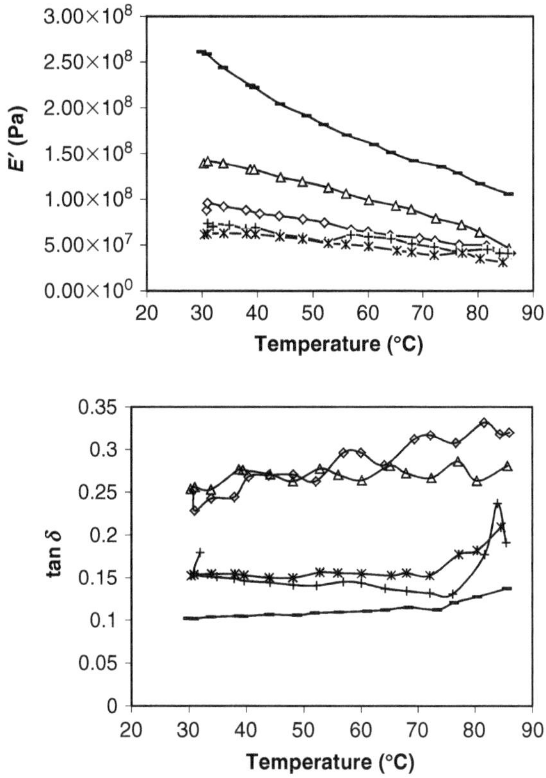

FIGURE 11.3 Graph of E' and tan δ versus temperature for blends containing 1% MA {– (40S, 60PEst, 1MA); ◇ (40S, 60PEst, 1MA, 0.3DCP); Δ (40S, 60PEst, 1MA, 0.5DCP); + (40S, 60PEst, 1MA, 0.8DCP); * (40S, 60PEst, 1MA, 1.2DCP)}.

be due to the molecular motions within the starch phase. At higher concentrations of DCP, these peaks were found to disappear. When compared with the transition peak in the uncompatibilized blend, it is found that with the addition of compatibilizer, the transition peak in compatibilized blends shifted towards higher temperatures. It can be observed from Figure 11.3 that in all the blends containing 1% MA, a drastic decrease in storage modulus (E') is observed. This indicates that as the concentration of MA in the blend is increased from 0.5% to 1%, there is a sharp decrease in stiffness. This is in agreement with the tensile testing results. In all the blends containing 1% MA, a transition peak is observed between 70 and 85 °C, which is a shift by about 10 °C compared to the transition peak in the blend (40S, 60PEst, 0.5MA, 0.3DCP). This indicates that the cross-linked points could have been obstructing the conformational mobility of the segments of EnPol®, resulting in further shift of the transition peak. A series of transitions observed in the blends containing 0.3%,

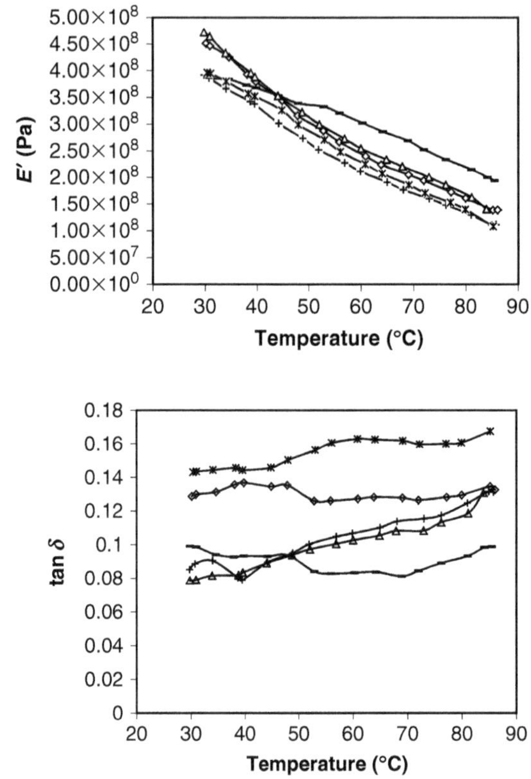

FIGURE 11.4 Graph of E' and tan δ versus temperature for blends containing 1.5% MA {– (40S, 60PEst, 1.5MA); ◇ (40S, 60PEst, 1.5MA, 0.3DCP); Δ (40S, 60PEst, 1.5MA, 0.5DCP); + (40S, 60PEst, 1.5MA, 0.8DCP); * (40S, 60PEst, 1.5MA, 1.2DCP)}.

0.5%, and 0.8% DCP could be due to the side groups and segmental motions of the cross-linked chains.

It is noticeable that all the blends containing 1.5% MA have storage modulus between that of blends containing 1% MA and 0.5% MA. These blends are found to exhibit similar tensile properties to the blends containing 0.5% MA. Most of these blends exhibit a transition between 50 and 60 °C that could again be due to the presence of interface modifier. These blends also showed transition peak between 60 and 70 °C, signifying the molecular motion within starch.

3.5 X-ray Photoelectron Spectroscopy

Although x-ray photoelectron spectroscopy (XPS) is a surface analysis technique, it still gives information about the structure of cross-linked polymers. Compression-molded films of six blends ((40S, 60PEst), (40S, 60PEst, 0.5MA, 0.5DCP), (40S, 60PEst, 1MA), (40S, 60PEst, 1MA,

0.8DCP), (40S, 60PEst, 1.5MA), and (40S, 60PEst, 1.5MA, 0.5DCP)) were washed using hexane to remove the surface impurities and then examined with XPS. In the "survey scans" of all the graphs, Si (2p) and Si (2s) peaks were observed in the range of 102 and 106 eV. Peaks at binding energies of 102 and 103.4 eV correspond to Si and methyl (CH_3) group, respectively. Hence, it is assumed that the active peaks between 102 and 106 eV were due to the presence of poly(dimethylsiloxane) (PDMS), which could have been present in polyester in the form of crystal impurity. In the survey scan, a graph of electron count versus binding energy (eV) was plotted. In all the graphs, distinct peaks corresponding to C (1s) and O (1s) were observed in the range 250-300 and 500-550 eV, respectively.

Then, a high-resolution scan (called multiplex) of the C(1s) and O(1s) energy levels of the mentioned blends was carried out at a pass energy of 25 eV. It was observed that the ratio of C/O of the blends was approximately 3. According to the proposed structure, the —OH group of C6 of starch attaches itself to the C=O group of MA resulting in the formation of a carboxylic group (—COOH). We determined the compositions of —CO and —COO groups in different blends using curve-fitting software "Resident PHI V6.0." All the data were charge-corrected at 101.8 eV using Si (2p) as reference for PDMS and are shown in Figure 11.5. The curve-fitting results of C (1s) energy levels of different blends are shown in Table 11.5.

For six different blends, the factor A_1/A_2 (area of —COO/area of —CO) was determined based on the area of their respective peaks in the curve-fitted graphs. It is to be noted here that XPS does not detect H. It was observed that the composition of —COO groups in the blend (40S, 60PEst, 1MA, 0.8DCP) was higher than the composition of —COO groups of other blends. The —COO peak in the curve-fitted graph is due to the presence of —COO groups of EnPol® and —COOH groups, which are formed due to cross-linking (refer to mechanism). As cross-linking increases, the percentage of —COOH groups in the compatibilized blends also increases, which is due to the transfer of —OH group of C6 of starch to the C=O group of MA to form —COOH group. The number of C=O groups in the compatibilized blends varies with the degree of cross-linking. For example, even in compatibilized blends, the presence of unreacted MA increases the number of C=O groups. Also, in the blends that did not contain DCP, XPS detected more C=O groups, indicating that there is a possibility of MA remaining in unreacted state or partially reacted state.

3.6 Differential Scanning Calorimetry

Differential scanning calorimetry was used to evaluate the thermal transition of the blends. To eliminate thermal history, the samples were equilibrated at 30 °C, heated to 220 °C at 10 °C min^{-1}, cooled to −70 °C

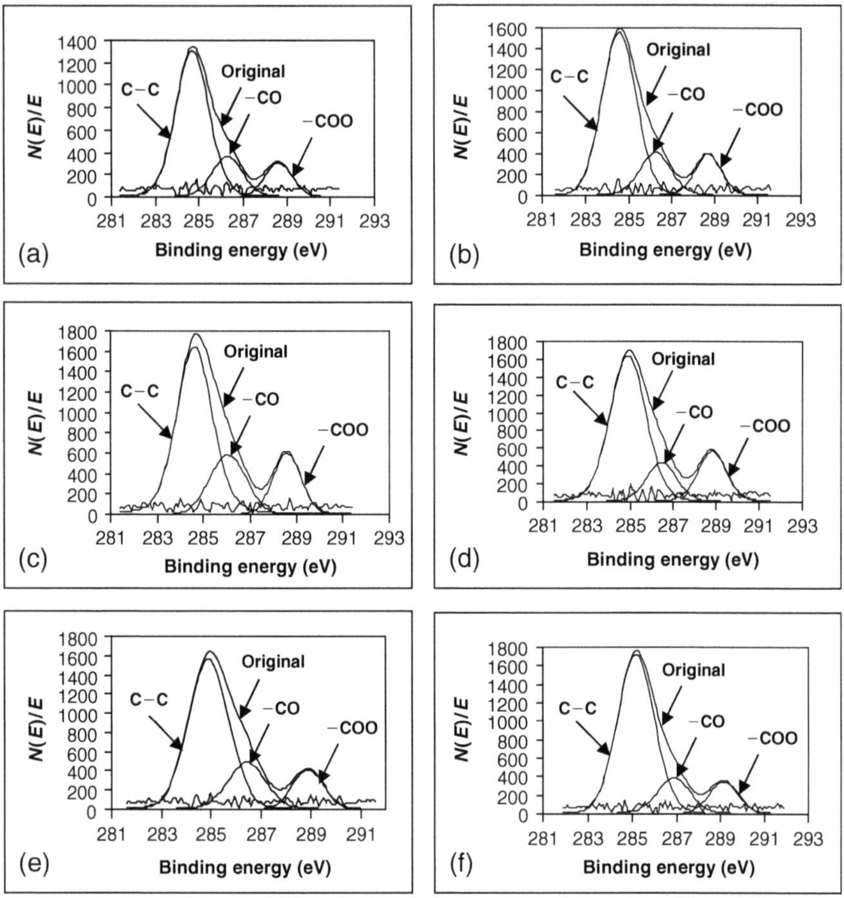

FIGURE 11.5 High-resolution scan of (a) uncompatibilized blend, (b) (40S, 60PEst, 0.5MA, 0.5DCP), (c) (40S, 60PEst, 1MA), (d) (40S, 60PEst, 1MA, 0.8DCP), (e) (40S, 60PEst, 1.5MA), and (f) (40S, 60PEst, 1.5MA, 0.5DCP).

TABLE 11.5 Curve-Fitting Results of C (1 s) Energy Levels of Different Blends

Blend	Area of —COO Groups (A_1)	Area of —CO Groups (A_2)	A_1/A_2
40S, 60PEst	983	1372	0.72
40S, 60PEst, 0.5MA, 0.5DCP	880	1074	0.82
40S, 60PEst, 1MA	1246	1508	0.83
40S, 60PEst, 1MA, 0.8DCP	1362	1125	1.21
40S, 60PEst, 1.5MA	872	1354	0.64
40S, 60PEst, 1.5MA, 0.8DCP	789	977	0.81

at 20 °C min^{-1}, maintained under isothermal conditions for 5 min, and heated to 220 °C min^{-1} at 10 °C min^{-1}.

Thermal transitions such as melting and crystallization are of high importance in polymer processing techniques. For the base blend containing 60% polyester and 40% starch-based plastic, the melting endotherm occurred at 98.8 °C, while the crystallization exotherm occurred at 63.8 °C (Table 11.6).

For compatibilized starch-polyester blends, no significant change in the endotherm (melting) temperature was observed, but the exotherm (crystallization) temperatures of all the compatibilized blends were lower than the crystallization temperature of the base blend (refer to Table 11.6). Hence, cross-linking agents do have significant effect on the crystallization temperature of starch-polyester blends. Due to β-scission of the polyester, the chains could form cross-linked regions resulting in the restriction of polymer chain mobility, which could significantly reduce the degree of crystallinity. However, it is interesting to note that the crystallization temperatures of the two blends at 1.0% MA addition (40S, 60PEst, 1MA, 0.8DCP) and (40S, 60PEst, 1MA, 1.2DCP) were higher than the crystallization temperatures of other compatibilized blends. In fact, the

TABLE 11.6 Thermal Transitions for Starch/Polyester Uncompatibilized and Compatibilized Blends

Blends	Endotherm T (°C)	Exotherm T (°C)
40S, 60PEst	98.8	63.8
40S, 60PEst, 0.5MA, 0.3DCP	99.1	49.6
40S, 60PEst, 0.5MA, 0.5DCP	98.4	45.9
40S, 60PEst, 0.5MA, 0.8DCP	98.4	51.6
40S, 60PEst, 0.5MA, 1.2DCP	97.1	52.6
40S, 60PEst, 1MA	99.9	48.4
40S, 60PEst, 1MA, 0.3DCP	98.5	50.7
40S, 60PEst, 1MA, 0.5DCP	98.2	48.4
40S, 60PEst, 1MA, 0.8DCP	97.5	59.2
40S, 60PEst, 1MA, 1.2DCP	97.5	64.8
40S, 60PEst, 1.5MA	99.0	50.2
40S, 60PEst, 1.5MA, 0.3DCP	98.9	55.9
40S, 60PEst, 1.5MA, 0.5DCP	98.1	45.3
40S, 60PEst, 1.5MA, 0.8DCP	98.1	44.8
40S, 60PEst, 1.5MA, 1.2DCP	98.2	45.1

greatest increase in exotherm temperature was found in (40S, 60PEst, 1MA, 1.2DCP). The presence of optimum concentration of the compatibilizer could have prevented appreciable micelle formation and hence reduced the interfacial energy of the blend. As a result, the degree of crystallization could have increased in both phases and at the interface further leading to an increase in the number of nucleation sites. In the two blends mentioned earlier, the compositions of MA and DCP could hence be in the proximity of their respective optimum concentrations. The crystallization temperatures of other compatibilized blends were low, indicating that compositions of the cross-linking agent and the initiator in these blends are not optimized. It could also be observed that the crystallization temperature for the uncompatibilized blend (40S, 60PEst) is quite high (63.8 °C), and yet, the blend exhibited poor mechanical properties. The absence of cross-linking agent could promote micelle formation at the interface of the hydrophobic polyester and the hydrophilic starch, thus contributing toward poor mechanical properties. At the same time, the two phases could crystallize in their domains resulting in an increase in the crystallization temperature of the blend.

3.7 Optical Microscopy

The optical micrographs of the uncompatibilized blend and blends that did not contain DCP are shown in Figure 11.6.

In the uncompatibilized blend (Figure 11.6a), due to interfacial tension between the starch and the polyester phases, the starch phase is not homogeneously distributed in the polymer matrix. In the blend (40S, 60PEst, 1MA), the starch phase could have remained as a separate phase (Figure 11.6b), whereas in the other blend (40S, 60PEst, 1.5MA), the starch phases are finer than the phases of the former blend. This could be due to hydrolysis of starch by MA that is present in excess concentration, leading to the breakdown of the starch phase (Figure 11.6c).

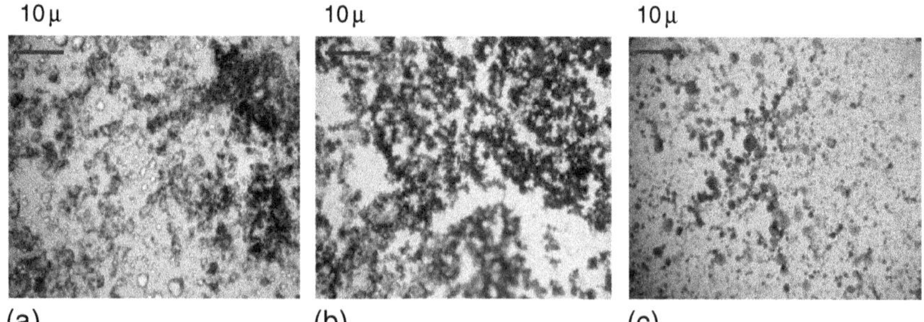

FIGURE 11.6 Optical micrograph of (a) uncompatibilized blend, (b) (40S, 60PEst, 1MA), and (c) (40S, 60PEst, 1.5MA).

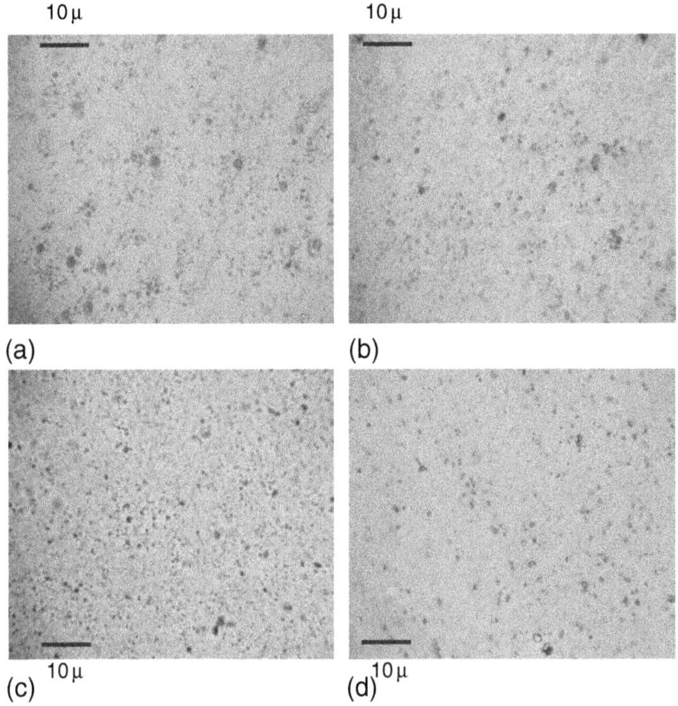

FIGURE 11.7 Optical micrograph of (a) (40S, 60PEst, 0.5MA, 0.3DCP), (b) (40S, 60PEst, 0.5MA, 0.5DCP), (c) (40S, 60PEst, 0.5MA, 0.8DCP), and (d) (40S, 60PEst, 0.5MA, 1.2DCP).

Analysis of blends containing 0.5% MA: The optical micrographs of the blends containing 0.5% MA and different compositions of DCP are shown in Figure 11.7.

It could be observed in the micrographs (Figure 11.7) that the starch phase is homogeneously distributed and has formed a cocontinuous phase with the polyester phase. These blends exhibited better mechanical properties than the uncompatibilized blend.

Analysis of blends containing 1% MA: The optical micrographs of blends containing 1% MA and different compositions of DCP are shown in Figure 11.8.

It can be observed that in the blends with higher DCP content (40S, 60PEst, 1MA, 0.8DCP) and (40S, 60PEst, 1MA, 1.2DCP) (Figure 11.8c and d), starch is evenly dispersed throughout the polymer matrix. This suggests optimum compositions of MA and DCP help generate polyester microradicals, promoting efficient cross-linking whereupon more and more starch reacts with the compatibilizer. This reduces the interfacial tension between the two dissimilar phases and promotes adhesion.

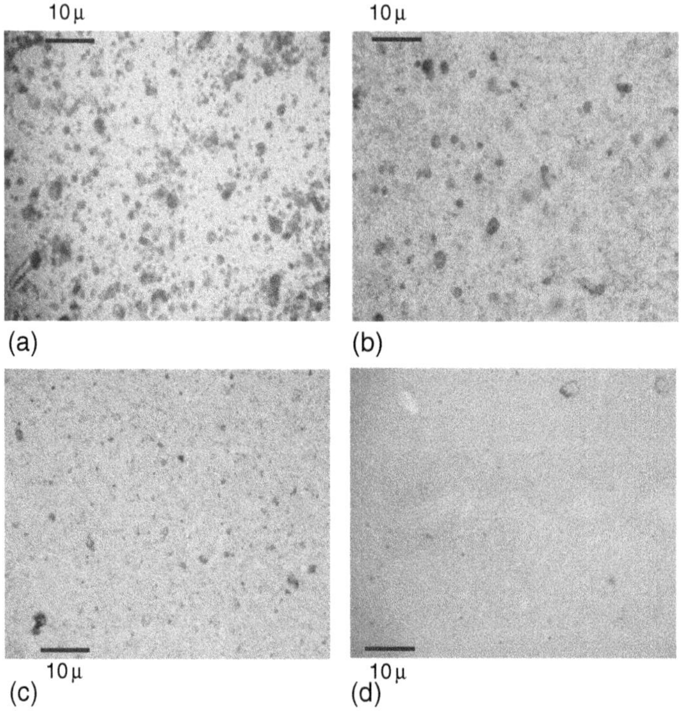

FIGURE 11.8 Optical micrograph of (a) (40S, 60PEst, 1MA, 0.3DCP), (b) (40S, 60PEst, 1MA, 0.5DCP), (c) (40S, 60PEst, 1MA, 0.8DCP), and (d) (40S, 60PEst, 1MA, 1.2DCP).

Analysis of blends containing 1.5% MA: The micrographs of the blends containing 1.5% MA and different compositions of DCP are shown in Figure 11.9.

It can be observed from micrographs of blends containing 1.5% MA and different compositions of DCP that starch phases are concentrated in certain regions of the polymer matrix and that the cocontinuous phases exhibited at 1% MA have now been lost. In these blends with higher composition of DCP (Figure 11.9b–d), numerous microradicals generated could react among themselves (disproportionation or coupling) and terminate the reaction. By this, the cross-linking of the two phases is hindered forcing some starch phase to remain in unreacted state.

4 DISCUSSION

It is apparent that mechanical and thermal properties of the compatibilized starch-based polymer-polyester polymers are affected by the addition of the compatibilization system. In summary, it appears that

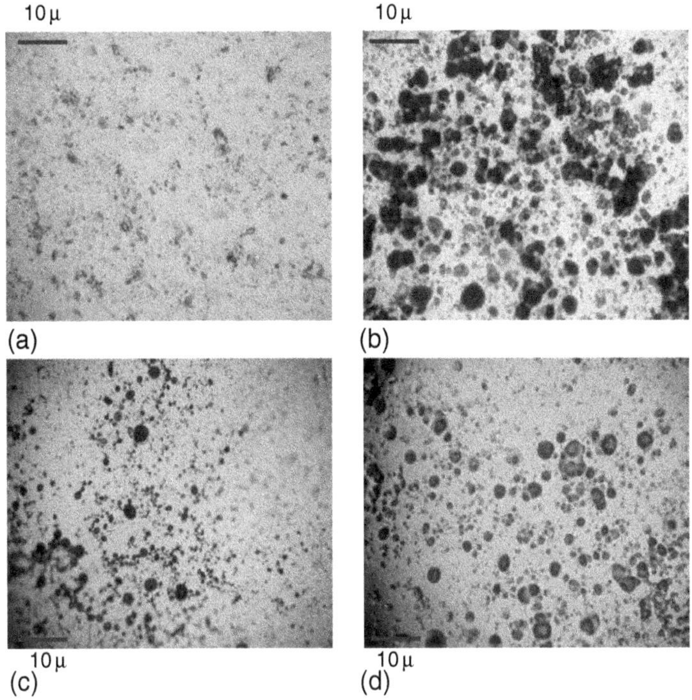

FIGURE 11.9 Optical micrograph of (a) (40S, 60PEst, 1.5MA, 0.3DCP), (b) (40S, 60PEst, 1.5MA, 0.5DCP), (c) (40S, 60PEst, 1.5MA, 0.8DCP), and (d) (40S, 60PEst, 1.5MA, 1.2DCP).

- all blends (0.5, 1.0, and 1.5 wt% MA) showed an improvement in Young's modulus and stress at break, and blends with 1.0% MA showed larger improvements in elongation at break;
- DMTA studies revealed that the blends containing 1% MA had lower stiffness (E') than blends containing 0.5% MA and 1.5% MA;
- XPS analysis indicated the presence of more —COO groups in the blend with 1% MA, thus supporting the proposed compatibilized structure for the blend;
- the DSC results indicated that the crystallization temperatures were reduced with increasing compatibilizers due to an increased hindrance to crystallization, and crystallization temperatures of the two blends at 1.0% MA addition (40S, 60PEst, 1MA, 0.8DCP) and (40S, 60PEst, 1MA, 1.2DCP) had higher than the crystallization temperatures than that of other compatibilized blends;
- optical micrographs of compatibilized blends with 1% MA revealed uniform distribution of starch in the polymer matrix, indicating compatibilization between starch and polyester phases.

Thus, optical, DSC, and XPS tests indicate a reactively compatibilized structure is optimized for the 1.0% MA samples. Interestingly, the 1.0% MA samples showed lower elastic modulus or stiffness during DMTA tests yet showed high Young's modulus, stress at break, and elongation at break during tensile testing. This is possibly due to the fact the DMTA tests are linear deformation tests and the tensile tests induce nonlinear deformation. That is, the linear DMTA tests determine a lower linear elastic modulus under small deformations, possibly due to the plasticizing addition of the MA to the system as previously noted (Vaidya et al., 1995), and the MA-induced cross-links have little effect. However, under nonlinear tensile testing, the deformation is larger and more destructive, and the compatibilized network structure is able to reduce the breaking of tie layers commonly described in basic polymer fracture studies. It is also interesting that the Young's modulus, stress at break, and elongation at break all increase, unlike normal composite systems, which sacrifice elongation (extensibility) for increases in strength.

5 CONCLUSIONS

The main conclusions from this work are summarized in the succeeding text. Firstly, all blends (0.5, 1.0, and 1.5 wt% MA) showed an improvement in Young's modulus and stress at break. Blends with 1.0% MA showed larger improvements in elongation at break. DMTA studies revealed that the blends containing 1% MA had lower stiffness than blends containing 0.5% MA and 1.5% MA. The compatibilized blends exhibited several transition peaks due to the presence of interface modifier, molecular motions within starch phase, and side groups and segmental motions of the cross-linked chains. XPS analysis indicated the presence of more —COO groups in the compatibilized blend with 1% MA, thus supporting the proposed structure for the blend. The DSC results indicated that the crystallization temperatures were reduced with increasing compatibilizers due to an increased hindrance to crystallization. Crystallization temperatures of the two blends at 1.0% MA addition {(40S, 60PEst, 1MA, 0.8DCP) and (40S, 60PEst, 1MA, 1.2DCP)} had higher than the crystallization temperatures than that of other compatibilized blends. We believe that the presence of optimum concentrations of MA and DCP in these blends reduces appreciable micelle formation and increases the number of nucleation sites resulting in an increase in crystallization temperature and interfacial adhesion. Optical micrographs of compatibilized blends with 1% MA revealed uniform distribution of starch in the polymer matrix, indicating compatibilization between starch and polyester phases.

In this study, compatibilization of starch and biodegradable polyester has been achieved using REX, and thus, it offers a new direction for enhancing the properties of low-cost base renewable biodegradable polymers. Also, for compatibilized blends, the study of interfacial tension (subject of future work) is a vital aspect with regard to improving the properties of biopolymers. This ability to tailor biodegradable polymer morphology and properties is crucial if low-cost biodegradable polymers are ever to be fully optimized for appropriate performance properties.

References

Avella, M., Errico, M. E., Laurienzo, P., Martuscelli, E., Raimo, M., & Rimedio, R. (2000). Preparation and characterisation of compatibilised polycaprolactone/starch composites. *Polymer, 41*, 3875–3881.

Bacon, R., & Farmer, E. H. (1939). The interaction of maleic anhydride with rubber. *Rubber Chemistry and Technology, 12*, 200.

Barlow, J. W., & Paul, D. R. (1984). Mechanical compatibilization of immiscible blends. *Polymer Engineering and Science, 24*, 525–534.

Bhattacharya, M., Vaidya, U. R., Zhang, D. M., & Narayan, R. (1995). Properties of blends of starch and synthetic-polymers containing anhydride groups. 2. Effect of amylopectin amylose ratio in starch. *Journal of Applied Polymer Science, 57*, 539–554.

Buehler, F.S., Casanova, J., Ernst, H., & Schultze, H. J. (1998a). *Process and apparatus for producing starch melt and products obtainable by this process.* USA patent application.

Buehler, F. S., Schmid, E., & Schultze, H. J. (1998b). *Starch/polymer mixture, process for the preparation of thereof, and products obtainable therefrom.* USA patent application.

Chandra, R., & Rustgi, R. (1997). Biodegradation of maleated linear low-density polyethylene and starch blends. *Polymer Degradation and Stability, 56*, 185–202.

Demirgoz, D., Elvira, C., Mano, J. F., Cunha, A. M., Piskin, E., & Reis, R. L. (2000). Chemical modification of starch based biodegradable polymeric blends: Effects on water uptake, degradation behaviour and mechanical properties. *Polymer Degradation and Stability, 70*, 161–170.

Doane, W. M. (1994). Biodegradable plastics. *Journal of Polymer Materials, 11*, 229–237.

Donald, A. (1997). *Starch structure and functionality.* London: Royal Society of Chemistry.

Finkenstadt, V. L., & Willett, J. L. (2005). Reactive extrusion of starch-polyacrylamide graft copolymers: Effects of monomer/starch ratio and moisture content. *Macromolecular Chemistry and Physics, 206*, 1648–1652.

Frost, K., Barthes, J., Kaminski, D., Lascaris, E., Niere, J., & Shanks, R. (2011). Thermoplastic starch-silica-polyvinyl alcohol composites by reactive extrusion. *Carbohydrate Polymers, 84*, 343–350.

Galliard, T. (1987). *Starch: Properties and potential.* New York: John Wiley & Sons.

Gaylord, N. (1996). *Maleic anhydride grafting.* New York: CRC press, Inc.

Imberty, A., Buleon, A., Tran, V., & Perez, S. (1991). Recent advances in knowledge of starch structure. *Starch-Starke, 43*, 375–384.

John, J., Tang, J., Yang, Z. H., & Bhattacharya, M. (1997). Synthesis and characterization of anhydride-functional polycaprolactone. *Journal of Polymer Science Part A—Polymer Chemistry, 35*, 1139–1148.

Kalambur, S., & Rizvi, S. S. H. (2006). An overview of starch-based plastic blends from reactive extrusion. *Journal of Plastic Film & Sheeting, 22*, 39–58.

Krishnan, M., Narayan, R. (2002). *Biodegradable multicomponent polymeric materials based on unmodified starch-like polysaccharides.*

Lohse, D. J., & Datta, S. (1996). *Polymeric materials encyclopedia.* Boca Raton, FL: CRC Press, Inc.

Lutfor, M. R., Rahman, M. Z. A., Sidik, S., Mansor, A., Haron, J., & Yunus, W. (2001). Kinetics of graft copolymerization of acrylonitrile onto sago starch using free radicals initiated by ceric ammonium nitrate. *Designed Monomers and Polymers, 4*, 253–260.

Mani, R., & Bhattacharya, M. (1998). Properties of injection mouldet starch synthetic polymer blends—III. Effect of amylopectin to amylose ratio in starch. *European Polymer Journal, 34*, 1467–1475.

Mani, R., Bhattacharya, M., & Tang, J. (1999). Functionalization of polyesters with maleic anhydride by reactive extrusion. *Journal of Polymer Science Part A—Polymer Chemistry, 37*, 1693–1702.

Manoi, K., & Rizvi, S. S. H. (2010). Physicochemical characteristics of phosphorylated cross-linked starch produced by reactive supercritical fluid extrusion. *Carbohydrate Polymers, 81*, 687–694.

Moad, G. (2011). Chemical modification of starch by reactive extrusion. *Progress in Polymer Science, 36*, 218–237.

Murua-Pagola, B., Beristain-Guevara, C. I., & Martinez-Bustos, F. (2009). Preparation of starch derivatives using reactive extrusion and evaluation of modified starches as shell materials for encapsulation of flavoring agents by spray drying. *Journal of Food Engineering, 91*, 380–386.

Nayak, S. K. (2010). Biodegradable PBAT/starch nanocomposites. *Polymer-Plastics Technology and Engineering, 49*, 1406–1418.

O'Brien, S., & Wang, Y. J. (2009). Effects of shear and pH on starch phosphates prepared by reactive extrusion as a sustained release agent. *Carbohydrate Polymers, 77*, 464–471.

O'Brien, S., Wang, Y. J., Vervaet, C., & REMON, J. P. (2009). Starch phosphates prepared by reactive extrusion as a sustained release agent. *Carbohydrate Polymers, 76*, 557–566.

Paul, D. R., Vinson, C. E., & Locke, C. E. (1972). Potential for reuse of plastics recovered from solid wastes. *Polymer Engineering and Science, 12*, 157–166.

Ramkumar, D. H. S., Bhattacharya, M., & ZHANG, D. (1997). High shear and transient viscosity of starch and maleated ethylene vinyl acetate copolymer blends. *Polymer Networks & Blends, 7*, 51–59.

Ramkumar, D. H. S., Yang, Z., & Bhattacharya, M. (1997). Linear viscoelastic properties of starch synthetic polymer blends. *Polymer Networks & Blends, 7*, 31–41.

Raquez, J. M., Nabar, Y., Narayan, R., & Dubois, P. (2008). In situ compatibilization of maleated thermoplastic starch/polyester melt-blends by reactive extrusion. *Polymer Engineering and Science, 48*, 1747–1754.

Sailaja, R. R. N., & Chanda, M. (2000). Use of maleic anhydride-grafted polyethylene as compatibilizer for polyethylene-starch blends: Effects on mechanical properties. *Journal of Polymer Materials, 17*, 165–176.

Shin, B. Y., Jang, S. H., & Kim, B. S. (2011). Thermal, morphological, and mechanical properties of biobased and biodegradable blends of poly(lactic acid) and chemically modified thermoplastic starch. *Polymer Engineering and Science, 51*, 826–834.

Stpierre, N., Favis, B. D., Ramsay, B. A., Ramsay, J. A., & Verhoogt, H. (1997). Processing and characterization of thermoplastic starch/polyethylene blends. *Polymer, 38*, 647–655.

Vaidya, U. R., Bhattacharya, M., & Zhang, D. M. (1995). Effect of processing conditions on the dynamic-mechanical properties of starch and anhydride functional polymer blends. *Polymer, 36*, 1179–1188.

Willett, J. L., & Finkenstadt, V. L. (2006). Reactive extrusion of starch-polyacrylamide graft copolymers using various starches. *Journal of Polymers and the Environment, 14*, 125–129.

Willett, J. L., & Finkenstadt, V. L. (2009). Comparison of cationic and unmodified starches in reactive extrusion of starch-polyacrylamide graft copolymers. *Journal of Polymers and the Environment, 17*, 248–253.

Xie, F. W., Yu, L., Liu, H. S., & Chen, L. (2006). Starch modification using reactive extrusion. *Starch-Starke, 58*, 131–139.

Zeng, J. B., Jiao, L., LI, Y. D., Srinivasan, M., Li, T., & Wang, Y. Z. (2011). Bio-based blends of starch and poly(butylene succinate) with improved miscibility, mechanical properties, and reduced water absorption. *Carbohydrate Polymers, 83*, 762–768.

PART VIII

DEGRADATION AND BIODEGRADATION

CHAPTER

12

Degradation and Biodegradation Standards for Starch-Based and Other Polymeric Materials

Joseph Greene

California State University, Chico, California, USA

1 INTRODUCTION

Biodegradable plastics are available throughout the world. Biodegradable plastics can be made with reduced carbon emissions, reduced waste, and reduced toxic pollution compared to traditional petroleum-based plastics. Typically, plastics account for 10% by weight or 20% of the volume of the landfill. Organic food waste can account for 20-30% of the landfill weight and 30-40% of the volume. Biodegradable plastics can reduce the amount of food waste and plastic waste if they are used to wrap or package the food items and then sent to industrial composting facilities rather than to landfills.

Biodegradation is an important feature of biodegradable plastics. Two essential components of biodegradation are that the material must be a food source for the bacteria in the disposal environment and that biodegradation must take place within a short time period, typically 6 months. Therefore, biodegradation can occur in an industrial compost environment for biodegradable plastics if they are used as food source for the bacteria in the compost and that they are consumed within a short time span. Likewise, biodegradation can occur in the marine environment if the bacteria in the seawater consume a major portion of the plastic within a short time span.

Biodegradation standards are created to capture these essential components. These standards define the environment of biodegradation and time of biodegradation. Thus, plastic materials can be defined as biodegradable in compost environment if they biodegrade in one growing season, or 6 months. Alternatively, materials can be properly defined as biodegradable in marine environment if they partially biodegrade within 6 months.

This chapter describes the worldwide biodegradation standards for biodegradable plastics, including starch-based plastics, in common disposal environments, including compost, marine, anaerobic digestion, soil, and landfill. Compost environments include aerobic conditions within hot industrial compost environments. Marine environments include cold aerobic conditions. Landfill disposal environments include aerobic and anaerobic conditions. Anaerobic-digestion environments include hot anaerobic conditions.

1.1 Biodegradation Standards

Biodegradation standards for plastic materials are established in two necessary categories for biodegradation: one for a biodegradation testing method and one for biodegradation performance specifications. The first standard is a test method that accurately simulates the intended environment and specifies a method for measuring biodegradation. The second standard is a specification standard that assigns a minimum value to establish biodegradation. Both types of standards are necessary and sufficient to adequately establish the biodegradation performance of plastic materials.

Solid waste disposal environments for plastic materials can include industrial compost, home compost, anaerobic digestion, landfill, litter, and ocean water. Only two disposal environments have both biodegradation standards for test methods and biodegradation performance standards, e.g., industrial compost and marine biodegradation environments. Test method standards are available for anaerobic-digestion and landfill environments. The second necessary performance specification standard for biodegradation performance is not available for anaerobic-digestion, home compost, or landfill environments. Therefore, plastic materials can claim meeting biodegradation performance standards for industrial compost and marine environments, but not for anaerobic-digestion, home compost, or landfill environments.

1.2 Worldwide Biodegradation Standards Agencies

Several worldwide organizations as listed in Table 12.1 are involved in setting standards for biodegradable and compostable plastics, including American Society for Testing and Materials (ASTM), International

TABLE 12.1 Biodegradation Standards Agencies

Standards Agency	Region	Web Address	Certification Agency
ASTM	The United States	www.astm.org	http://www.bpiworld.org/BPI-certification-requirements
CEN	International Union	www.cen.eu	www.din.de/cmd
ISO	Worldwide	www.iso.org	www.din.de/cmd
JIS	Japan	www.jsa.or.jp	http://www.jbpaweb.net/english/english.htm

Committee for Standardization (CEN), International Organization for Standardization (ISO), German Institute for Standardization (DIN), Japanese Industrial Standards (JIS), and British Plastics Federation. The standards from these organizations have helped the industry create biodegradable and compostable products that meet the increasing worldwide demand for more environmentally friendly plastics (Narayan & Pettigrew, 1999). International, American, and Japanese certification schemes are cooperating to enable international cross-certification of products so that a product certified in one of these countries would automatically be eligible for certification in another.

Biodegradation standards are provided in the following sections. The standards are organized based on disposal environment. Thus, the biodegradation standards from different worldwide standards organizations are presented in an industrial compost section, marine environment section, an anaerobic-digestion section, landfill section, and home compost section.

Standards for biobased materials are, also, presented in the following to establish the parameters of claims that plastic materials are biobased. Biodegradable plastics can be produced from organic materials or petroleum-based materials.

1.3 Certification

Certification is needed for biodegradable plastics to ensure that they meet the performance specification requirements in the biodegradation standards. Several certification companies, listed in Table 12.1, are available to certify the biodegradation performance of compostable plastics. In the United States, Biodegradable Products Institute (BPI) and the US Composting Council established the Compostable Plastics Certification

Program in the United States to certify compostable plastics as meeting the ASTM D6400 or ASTM D6868 compostability standards.

In Europe, the DIN certification organization was created to certify that biodegradable plastics (BPI Certified Compostable, 2013) meet the International Standard EN 13432 for compostability. Vinçotte is also a certification company for certification of compostable plastics in Europe. Vinçotte OK Biodegradable mark ensures that the plastic material will pass the performance biodegradable specification in a specific natural environment (soil, fresh water, seawater, etc.). Vinçotte OK Compost mark ensures that the plastic material meets the performance specification of compostable plastics in EN 13432.

In Japan, the BioPlastics Association (JBPA) was created to certify biodegradable plastics meet the International Standard EN 13432 for compostability. JBPA in Japan established GreenPla certification and labeling system based on international biodegradation standards.

2 BIOBASED STANDARD TEST METHOD

Many biodegradable plastics are made with biobased materials—those that are derived from organic carbon sources from cereal, corn, potato, rice, soy, sugar cane, wheat, vegetable oil, etc. Biobased plastics are made with a renewable resource that can have lower environmental impacts than petroleum-based plastics. Replacing petroleum-based carbon with organic carbon from today can reduce the carbon footprint of the plastic material (Narayan, 2011a, Rational, Drivers). Biological carbon content analysis can establish the carbon footprint of bioplastics with life cycle assessment methods (Narayan, 2011b, Carbon Footprint). The biobased content of the plastic material can be established by tests that measure carbon isotopes of the material.

2.1 US Biobased Standard

2.1.1 ASTM D6866-10 Standard Test Methods for Determining the Biobased Content of Solid, Liquid, and Gaseous Samples Using Radiocarbon Analysis

2.1.1.1 SUMMARY

This American test method establishes the procedures, equipment, materials, and conditions to measure the ^{14}carbon content of the plastic sample through radiocarbon analysis. This test method is applicable to carbon-based plastic products that can undergo combustion in the presence of oxygen to product carbon dioxide (CO_2) gas. This test method can be useful for companies and US federal agencies to promote the use of biobased products in commercial, nonfood products. The test method directly discriminates between 14C/12C contents of plastic samples made from contemporary carbon and those plastic samples made from petroleum sources that do not have 14C. Measurement of a plastic product's 14C/12C content is determined relative to the modern carbon-based oxalic acid radiocarbon Standard Reference Material 4990c.

2.1.1.2 PROCEDURE

The radiocarbon analysis can be performed with three methods, as listed in the following:

1. Method B: accelerator mass spectrometry (AMS) and isotope ratio mass spectrometry (IRMS) techniques to quantify the biobased content of a given product. Sample preparation methods are identical to Method A. Stored CO_2 is sent to an AMS facility for final processing and analysis. The maximum error is 1-2% for AMS and 0.1-0.5% for IRMS.
2. Method C: liquid scintillation counting (LSC) radiocarbon techniques to quantify the biobased content of a product. Method C uses LSC analysis of sample carbon that has been converted to benzene rather than CO_2 solutions of Method A. The maximum error is $\pm 3\%$.

2.1.1.3 EQUIPMENT

1. LSC
2. AMS
3. IRMS

2.2 International Biobased Standards

Currently, there are no ISO test methods that are equivalent to biobased standard test method.

3 INDUSTRIAL COMPOST ENVIRONMENT

Biodegradation standards for industrial compost environment require a biodegradation specification standard and a biodegradation test method standard. The specification standards in the United States are for compostable plastics and for coatings that can be used for packaging and containers. Both specification standards require the use of the same test method. Table 12.2 lists the biodegradation standards for industrial compost in several countries. The standards listed are the performance specification standard and the test method standards in the industrial compost environment. The standards are discussed in more detail in the following

TABLE 12.2 Biodegradation Standards for Industrial Composting Environment

Region	Performance Specification	Test Method	Measurement Method	Disintegration Test Method
Australia	ISO 17088	ISO 14855-2	CO_2 measurement	ISO 16929, ISO 20200
Europe	ISO 17088/ EN13432	ISO 14855-2	CO_2 measurement	ISO 16929, ISO 20200
Japan	ISO 17088	ISO 14855-2	CO_2 measurement	ISO 16929, ISO 20200
The United States	ASTM D6400, D6868	ASTM D5338	CO_2 measurement	ISO 16929[a]

[a] *Under review with the ASTM Subcommittee D20.96.*

section. For complete information on the standards, please consult the documentation from the standards agency.

3.1 US Biodegradation Standards for Industrial Compost Environment

3.1.1 Biodegradation Performance Specification Standard: ASTM D6400-04 (Standard Specification for Compostable Plastics)

3.1.1.1 SUMMARY

This American specification standard establishes the performance requirements for biodegradation of compostable plastic materials that are designed to biodegrade into CO_2, water, and biomass in an industrial compost environment. The industrial compost environment is one that maintains a temperature above 40 °C and results in thermophilic conditions. The performance specification standard requires the use of ASTM D5338-05 test method to measure the amount of CO_2 that is emitted from the degrading plastic sample.

ASTM D6400-04 performance specification standard requires the product must demonstrate each of the three characteristics as follows:

(1) Proper *disintegration* during composting
(2) Adequate level of inherent *biodegradation*
(3) No adverse impacts on the ability of composts to *support plant growth*

3.1.1.2 PROCEDURES

Three test procedures for the ASTM D6400-04 standard specify that three types of tests are performed on the plastic samples. The first test measures the percentage of disintegration of the plastic samples while under hot and moist compost conditions. The plastic samples are weighed prior to exposure to test conditions. The samples are placed in compost soil with the use of a sack, bag, or screened container. The plastic samples are exposed to industrial composting conditions of at least 58 °C and 50% moisture for 12 weeks. The mass of the plastic sample is measured after 12 weeks by passing the plastic sample and compost through a 2.0 mm sieve. ASTM D6400 specifies that.

The second test procedure for ASTM D6400-04 standard specifies a satisfactory rate of *biodegradation* of the plastic material that is under industrial composting conditions of at least 58 °C and 50% moisture for 180 days that more than 90% of the carbon in the original plastic sample is converted into CO_2 as measured by a CO_2 respirometer or wet chemistry methods. The details of the test procedure are listed in ASTM D5338-11 test method.

The third test procedure for ASTM D6400-04 standard specifies a satisfactory ability of the compost soil at the end of the biodegradation testing to *support plant growth* through phytotoxicity testing and very low-regulated heavy metal concentrations. Phytotoxicity testing is achieved through planting of tomato, cucumber, radish, rye, barley, or cress grass seeds in the tested compost soil. The growth of the plants after 10 days indicates positive soil conditions for plant growth. Plant biomass tests can reveal quality differences between composts and can indicate potential plant stress induced by the compost at the given level used in the test.

Regulated heavy metal testing measures the concentrations of regulated heavy metals in the compost soil at the end of the biodegradation testing. The level of regulated heavy metals can be measured with flame atomic absorption spectrophotometer using an air-acetylene flame and equipped with a Pb hollow cathode lamp. The compost samples must have regulated metal concentrations <50% of the acceptable levels of regulated heavy metals as prescribed in 40 CFR Part 503.13, i.e., lead (75 mg/kg), cadmium (8.5 mg/kg), chromium (not specified), copper (375 mg/kg), nickel (105 mg/kg), zinc (700 mg/kg), and mercury (4.25 mg/kg) (Title 40: Protection of the Environment, 2013).

3.1.1.3 SPECIFICATION

ASTM D6400-04 standard specifies that a plastic material is compostable if it meets the following specifications:

1. *Disintegration*: <10% of the original dry weight of the plastic material remains after 12 weeks of exposure to industrial composting conditions specified in ISO 16929 test method standard for disintegration.
2. *Biodegradation*: >90% of the carbon in the original plastic sample is converted into CO_2 after 180 days of exposure to industrial composting conditions specified in ASTM D5338 test method standard.

3. *Nontoxic to plants*: No measureable phytotoxicity or regulated heavy metals greater than allowable levels as described in 40 CFR Part 503.13.

3.1.2 Biodegradation Performance Specification Standard: ASTM D6868-03 (Standard Specification for Biodegradable Plastics Used as Coatings on Paper and Other Compostable Substrates)

3.1.2.1 SUMMARY

This American specification standard establishes the performance requirements for biodegradation of materials and products (including packaging, films, and coatings) that are attached (through either lamination or extrusion directly onto the paper) to compostable substrate plastic materials. The compostable plastics are designed to biodegrade in an industrial compost environment. The industrial compost environment is one that maintains a temperature above 40 °C and results in thermophilic conditions. The performance specification standard requires the use of ASTM D5338-05 test method to measure the amount of CO_2 that is emitted from the degrading plastic sample.

ASTM D6868-03 performance specification standard requires the product must demonstrate each of the three characteristics as follows:

(1) Proper *disintegration* during composting
(2) Adequate level of inherent *biodegradation*
(3) No adverse impacts on the ability of composts to *support plant growth*

3.1.2.2 PROCEDURES

The test procedures for ASTM D6868-03 standard specify that three types of tests are performed on the plastic samples. The first test measures the percentage of disintegration of the plastic samples while under hot and moist compost conditions. The plastic samples are weighed prior to exposure to test conditions. The samples are placed in compost soil with the use of a sack, bag, or screened container. The plastic samples are exposed to industrial composting conditions of at least 58 °C and 50% moisture for 12 weeks. The mass of the plastic sample is measured after 12 weeks by passing the plastic sample and compost through a 2.0 mm sieve. ASTM D6868 specifies that no more than 10% of the original dry weight of the plastic material remains.

The second test procedure for ASTM D6868-03 standard specifies a satisfactory rate of *biodegradation* of the plastic material that is under industrial composting conditions of at least 58 °C and 50% moisture for 180 days that more than 90% of the carbon in the original plastic sample is converted into CO_2 as measured by a CO_2 respirometer or wet chemistry methods. If the packaging or paper product is a combination of several materials, then each of the materials must individually pass the respirometry test. The details of the test procedure are listed in ASTM D5338-11 test method.

The third test procedure for ASTM D6868-03 standard specifies a satisfactory ability of the compost soil at the end of the biodegradation testing to *support plant growth* through phytotoxicity testing and very low-regulated heavy metal concentrations. Phytotoxicity testing is achieved through planting of tomato, cucumber, radish, rye, barley, or cress grass seeds in the tested compost soil. The growth of the plants after 10 days indicates positive soil conditions for plant growth.

Regulated heavy metal testing measures the concentrations of regulated metals in the compost soil at the end of the biodegradation testing.

3.1.2.3 SPECIFICATION

ASTM D6868-03 standard specifies that a plastic coating material is compostable if it meets the following specifications:

1. *Disintegration*: <10% of the original dry weight of the plastic material remains after 12 weeks of exposure to industrial composting conditions.
2. *Biodegradation*: >90% of the carbon in the original plastic sample is converted into CO_2 as measured by a CO_2 respirometer after 180 days of exposure to industrial composting conditions.
3. *Nontoxic to plants*: No measureable phytotoxicity or regulated heavy metals greater than allowable levels as prescribed in 40 CFR Part 503.13.

3.1.3 Biodegradation Test Method Standard: ASTM D5338-11 (Standard Test Method for Determining Aerobic Biodegradation of Plastic Materials Under Controlled Composting Conditions)

3.1.3.1 SUMMARY

This test method establishes the procedures, equipment, materials, and conditions to measure the degree and rate of biodegradation of plastic materials under aerobic thermophilic composting conditions. This test method is designed to produce repeatable and reproducible test results under controlled composting conditions that simulate industrial compost conditions. The plastic test samples are exposed to an inoculum that is derived from industrial compost. The aerobic thermophilic conditions of the test are provided in an environment where temperature, aeration, and humidity are closely monitored and controlled. The plastic samples are cut into small pieces and placed in a vessel with warm and moist compost soil. The test containers are at 58 °C (±2 °C) for 180 days. The biogas from the container is measured for CO_2 and O_2 over the testing period.

3.1.3.2 PROCEDURE

The test procedures for ASTM D5338-11 involve placing 500 g of industrial compost soil that has maturity level of 2-3 months into a container vessel >2 and <5 l. A minimum of 12 composting vessels are required for the test. The containers will be provided for the plastic sample and the following plastic samples, blank, positive control, and negative control. The testing is completed in triplicate. The plastic samples are added to the container in quantities of 100 g. The container must have 25% free air space above the samples. The pH, moisture content, C/N ratio, and percentage solids are measured for the compost soil. The sample containers must have moist air supplied to the container that is free of CO_2. The biogas from the sample containers is measured periodically for CO_2 and O_2. The O_2 level must be >6% during the duration of the test. The test containers should be rotated every 2 or 3 days to reduce the packing of the soil and to mix fresh air to the compost soil. An example of a test apparatus is provided in Figure 12.1.

3.1.3.3 MATERIALS

- Plastic samples
- Blank compost soil
- Positive control: cellulose powder or cellulose paper
- Negative control: polyethylene film
- Barium hydroxide solution

3.1.3.4 EQUIPMENT

- Composting vessels with size >2 and <5 l
- Water bath or other temperature control device
- Pressurized air system with CO_2-free and H_2O-saturated air
- Carbon dioxide trapping apparatus with $Ba(OH)_2$ solution
- Flexible tubing nonpermeable to CO_2

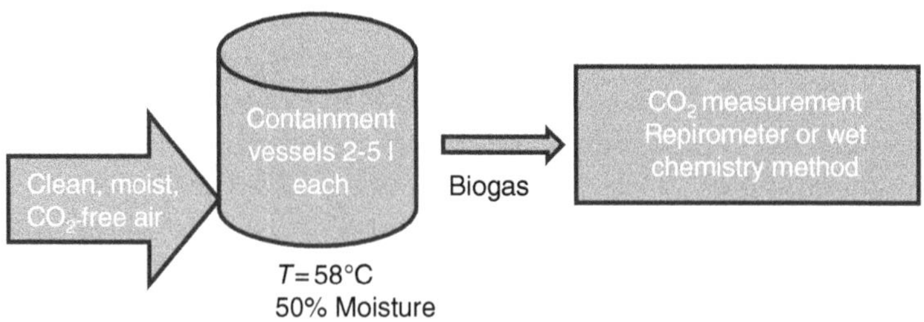

FIGURE 12.1 Experimental setup for laboratory environment of industrial compost per ASTM D5338 standards. (For color version of this figure, the reader is referred to the online version of this chapter.)

- Stoppers equipped with sampling ports
- Analytical balance (±0.1 mg)
- 100 ml burette
- 0.05 N HCl
- pH meter
- Respirometer devices for measuring CO_2 and O_2 periodically (optional)
- CO_2 trapping equipment with gas flow meter, gas chromatograph, or other suitable equipment

3.2 International Biodegradation Standards for Industrial Compost Environment

The ISO is the world's largest developer of standards. ISO is a network of the national standards institutes of 157 countries who agree on specifications and criteria to be applied consistently in the classification of materials, in the manufacture and supply of products, in testing and analysis, in terminology, and in the provision of services. The test method is designed to yield the percentage conversion of the carbon in the test material to evolved carbon dioxide as well as the rate of conversion.

In Europe, compostable plastics are being used in several applications. Compostable plastics must comply with the International Norm EN13432, which is the criterion for compostability. The EN13432 specifications standard is very similar to the ASTM D6400 specifications standard. Certification with ISO compostability standard is compatible with certification for ASTM D6400 standard. EN13432 requires a compostable plastic material to break down to the extent of at least 90% to H_2O and CO_2 and biomass within a period of 6 months. ISO14855 standard specifies a testing method to evaluate the ultimate aerobic biodegradability of plastics, based on organic compounds, under controlled composting conditions by measurement of the amount of carbon dioxide evolved and the degree of plastic disintegration at the end of test. ISO 14852 is the determination of the ultimate aerobic biodegradability of plastic materials in an aqueous medium. The test method measures the evolved carbon dioxide and is similar to ASTM D5338 test method standard.

DIN CERTCO is a very well-known and utilized certification system in Europe (Bastioli, 2005). Sample materials are tested for regulated metals, organic contaminants, complete biodegradation, disintegration under compost conditions, and phytotoxicity (plant toxicity). The regulated metals and organic chemical tests ensure that neither organic contaminants nor heavy metals such as lead, mercury, and cadmium can enter

TABLE 12.3 Regulated Heavy Metal Allowable Concentrations

Element	ASTM D6400 The United States (ppm)	ASTM D6400 Canada (ppm)	EN 13432 (ppm)	Japan (ppm)
Zn	1400	463	150	180
Cu	750	189	50	60
Ni	210	45	25	30
Cd	17	5	0.5	0.5
Pb	150	125	50	10
Hg	8.5	1	0.5	0.2
Cr	—	265	50	50
Mo	—	5	1	—
Se	50	4	0.75	—
As	20.5	19	5	5
F	—	—	100	—
Co	—	38	—	—

the soil via the biodegradable materials. The procedures for testing complete biodegradation in the laboratory and disintegration under compost conditions ensure that materials are completely degraded during one process cycle of a standard composting plant. The DIN compostability certification is very similar to BPI certification, which meets ASTM D6400 standards.

Table 12.3 illustrates the heavy metal limits in the international standard and the US standards. Heavy metal concentrations in the EN13432 standard allows a limited amount of metal, i.e., lead (30 mg/kg), cadmium (0.3 mg/kg), chromium (30 mg/kg), copper (22.5 mg/kg), nickel (15 mg/kg), zinc (100 mg/kg), and mercury (0.3 mg/kg). The US standard allows the following amounts: lead (150 mg/kg), cadmium (17 mg/kg), chromium (not specified), copper (750 mg/kg), nickel (210 mg/kg), zinc (1400 mg/kg), and mercury (8.5 mg/kg) (Rudnik, 2008). Acceptable levels of heavy metals in sewer sludge are provided per US EPA Subpart 503-13.

Biodegradation standards for industrial compost environment require a biodegradation specification standard and a biodegradation test method standard. The specification standards in Europe are for plastics and for packaging. Both specification standards require the use of the same test method.

3.2.1 *Biodegradation Performance Specification Standard: EN 13432-2000 (Packaging. Requirements for Packaging Recoverable Through Composting and Biodegradation Test Scheme and Evaluation Criteria for the Final Acceptance of Packaging)*

3.2.1.1 SUMMARY

This international specification establishes the performance requirements for biodegradation of plastic materials and products (including packaging, films, and other products). The compostable plastics are designed to biodegrade in an industrial compost environment. The industrial compost environment is one that maintains a temperature above 40 °C and results in thermophilic conditions. The performance specification standard requires the use of EN 14995 (ISO 17088) test method to measure the amount of CO_2 that is emitted from the degrading plastic sample. EN 13432-2000 performance specification standard requires the product must demonstrate each of the following characteristics:

1. Biodegradability
2. Disintegration during biological treatment
3. Minimal effect on the biological treatment process
4. Minimal effect on the quality of the resulting compost

3.2.1.2 PROCEDURES

The test procedures for EN 13432 standard specify that three types of tests are performed on the plastic samples. The first test measures the percentage of disintegration of the plastic samples while under hot and moist compost conditions. The plastic samples are weighed prior to exposure to test conditions. The samples are placed in compost soil with the use of a sack, bag, or screened container. The plastic samples are exposed to industrial composting conditions of at least 58 °C and 50% moisture for 12 weeks. The mass of the plastic sample is measured after 3 months by passing the plastic sample and compost through a 2.0 mm sieve. EN 13432 specifies that no more than 10% of the original dry weight of the plastic material remains.

The second test procedure for EN 13432 standard specifies a satisfactory rate of *biodegradation* of the plastic material that is under industrial composting conditions of at least 58 °C and 50% moisture for 26 weeks that more than 90% of the carbon in the original plastic sample is converted into CO_2 as measured by a CO_2 respirometer or wet chemistry methods. The details of the test procedures are listed in EN 14046 or ISO 14855 test method.

The third test procedure for EN 13432 standard specifies a satisfactory ability of the compost soil at the end of the biodegradation testing to *support plant growth* through phytotoxicity testing and very low-regulated heavy metal concentrations. Phytotoxicity testing is achieved through planting of two seed crops in the tested compost soil. The growth of the plants after 10 days indicates positive soil conditions for plant growth.

Plant biomass tests can reveal quality differences between composts and can indicate potential plant stress induced by the compost at the given level used in the test.

Regulated heavy metal testing measures the concentrations of regulated metals in the compost soil at the end of the biodegradation testing.

3.2.1.3 SPECIFICATIONS

EN 13432 standard specifies that a plastic material is compostable if it meets the following specifications:

- Satisfactory rate of *disintegration* of the plastic material that is under pilot-scale composting test of at least 58 °C and 50% moisture for 84 days that no more than 10% of the original dry weight of the plastic material remains after passing through a 2.0 mm sieve. Also, the biodegradation testing shall not have any adverse effects on the composting process. The standard requires the use of ISO 16929 or ISO 2020 test method standard.
- Satisfactory rate of *biodegradation* of the plastic material that is under industrial composting conditions of at least 58 °C and 50% moisture in 6 months that more than 90% of the carbon in the original plastic sample is converted into CO_2 as measured by measuring the amount of release CO_2 through wet chemistry methods. The standard requires the use of ISO 148455 or ASTM D5338 test method standard.
- Satisfactory ability of the compost soil at the end of the biodegradation testing to *support plant growth* through phytotoxicity testing with seedling emergence and seedling growth test (OECD 208), and very low-regulated heavy metal concentrations. Seedling emergence and seedling growth test – OECD 208.

3.2.2 *Biodegradation Performance Specification Standard: ISO 17088 (EN 13432) (Plastics. Evaluation of Compostability. Test Scheme and Specification)*

This international standard is very similar to EN 13432-2000 and establishes the performance requirements for biodegradation of plastic materials and products (including packaging, films, and other products). The compostable plastics are designed to biodegrade in an industrial compost environment. The industrial compost environment is one that maintains a temperature above 40 °C and results in thermophilic conditions. The performance specification standard requires the use of EN 14995 (ISO 17088) test method to measure the amount of CO_2 that is emitted from the degrading plastic sample. EN 13432-2000 performance specification standard requires the product must demonstrate each of the four characteristics:

1. Biodegradability
2. Disintegration during biological treatment

3. Minimal effect on the biological treatment process
4. Minimal effect on the quality of the resulting compost

ISO 17088 standard specification establishes the performance requirements for biodegradation of compostable plastic materials that are designed to biodegrade into CO_2, water, and biomass in an industrial compost environment according to test methods described in EN 14046 and ISO 14855. The industrial compost environment is one that maintains thermophilic temperatures. The performance specification standard requires the use of EN 14046 or ISO 14855 test method to measure the amount of CO_2 that is emitted from the degrading plastic sample.

3.2.2.1 PROCEDURES

The test procedures for ISO 17088 standard specify that three types of tests are performed on the plastic samples. The first test measures the percentage of disintegration of the plastic samples while under hot and moist compost conditions. The plastic samples are weighed prior to exposure to test conditions. The samples are placed in compost soil with the use of a sack, bag, or screened container. The plastic samples are exposed to industrial composting conditions of at least 58 °C and 50% moisture for 84 days. The mass of the plastic sample is measured after 3 months by passing the plastic sample and compost through a 2.0 mm sieve. EN 13432 specifies that no more than10% of the original dry weight of the plastic material remains.

The second test procedure for ISO 17088 standard specifies a satisfactory rate of *biodegradation* of the plastic material that is under industrial composting conditions of at least 58 °C and 50% moisture for 26 weeks that more than 90% of the carbon in the original plastic sample is converted into CO_2 as measured by a CO_2 respirometer. The details of the test procedures are listed in ISO 14855 or ASTM D5338 test method.

The third test procedure for ISO 17088 standard specifies a satisfactory ability of the compost soil at the end of the biodegradation testing to *support plant growth* through phytotoxicity testing and very low-regulated heavy metal concentrations. Phytotoxicity testing is achieved through planting of tomato, cucumber, radish, rye, barley, or cress grass seeds in the tested compost soil. The growth of the plants after 10 days indicates positive soil conditions for plant growth.

Regulated heavy metal testing measures the concentrations of regulated metals in the compost soil at the end of the biodegradation testing. The compost samples must have regulated metal concentrations less than prescribed levels, listed in Table 12.3.

3.2.2.2 SPECIFICATIONS

ISO 17088 standard specifies that a plastic material is compostable if it meets the following specifications:

- Satisfactory rate of *disintegration* of the plastic material that is under pilot-scale composting test of at least 58 °C and 50% moisture for 3 months that no more than10% of the original dry weight of the plastic material remains after passing through a 2.0 mm sieve. Also, the biodegradation testing shall not have any adverse effects on the composting process.
- Satisfactory rate of *biodegradation* of the plastic material that is under industrial composting conditions of at least 58 °C and 50% moisture for 6 months that more than 90% of the carbon in the original plastic sample is converted into CO_2 as measured by a CO_2 respirometer or wet chemistry methods.
- Satisfactory ability of the compost soil at the end of the biodegradation testing to *support plant growth* through phytotoxicity testing and very low-regulated heavy metal concentrations. Plant growth test OECD 208 test.

3.2.3 Biodegradation Test Method Standard: ISO 14855-2 (EN 14046) (Packaging. Evaluation of the Ultimate Aerobic Biodegradability and Disintegration of Packaging Materials Under Controlled Composting Conditions. Method by Analysis of Released Carbon Dioxide)

3.2.3.1 SUMMARY

This international standard specifies a test method for the evaluation of the ultimate aerobic biodegradability of packaging materials based on organic compounds under controlled composting conditions by measurement of released carbon dioxide at the end of the test. This method is designed to resemble typical aerobic composting conditions for the organic fraction of mixed municipal solid waste. The packaging material is exposed in a laboratory test to an inoculum that is derived from compost. The aerobic composting takes place in an environment where especially temperature, aeration, and humidity are closely monitored and controlled. The test method is designed to yield a percentage and rate of conversion of carbon of the test material to released carbon dioxide. The conditions described in this standard do not necessarily always correspond to the optimal conditions allowing the maximum degree of biodegradation to occur.

This test method establishes the procedures, equipment, materials, and conditions to measure the degree and rate of biodegradation of plastic materials under aerobic thermophilic composting conditions. This test method is designed to produce repeatable and reproducible test results under controlled composting conditions that simulate industrial compost conditions. The testing is completed in duplicate. The plastic test samples are exposed to an inoculum that is derived from industrial compost. The aerobic thermophilic conditions of the test are provided in an environment where temperature, aeration, and humidity are closely monitored and controlled. The plastic samples are cut into small pieces and placed in a vessel with warm and moist compost soil. The test containers are at 58 °C (±2 °C) for 180 days. The biogas from the container is measured for CO_2 and O_2 over the testing period.

3.2.3.2 PROCEDURE

The test procedure for ISO 14855 involves placing 120 g of moist (~50% moisture) industrial compost soil that has maturity level of 6 months and 50% total solids into a container vessel of approximately 500 ml. The containers will be provided for the plastic sample and the following samples, blank, positive control, and negative control. The testing is done in duplicate. The plastic samples are added to the container in quantities of 10 g. The pH, moisture content, C/N ratio, and percentage solids are measured for the compost soil. The sample containers must have moist air supplied to the container that is free of CO_2. The biogas from the sample containers is measured periodically for CO_2 and O_2. The test containers should be rotated every 2 or 3 days to reduce the packing of the soil and to mix fresh air to the compost soil. The tests are run in duplicate.

3.2.3.3 MATERIALS

- Plastic sample
- Blank compost soil
- Positive control reference: cellulose powder
- Negative control: polyethylene film

3.2.3.4 EQUIPMENT

- Composting vessels with size approximately 500 ml
- Water bath or other temperature control device
- Pressurized air system with CO_2-free and H_2O-saturated air
- Carbon dioxide trapping apparatus
- Flexible tubing nonpermeable to CO_2
- Stoppers equipped with sampling ports
- Analytical balance (±0.1 mg)
- 100 ml burette
- pH meter
- Respirometer devices for measuring CO_2 and O_2 periodically (optional)
- CO_2 trapping equipment with gas flow meter, gas chromatograph, or other suitable equipment

3.2.4 ISO 16929 (EN14045:2003) (Plastics. Determination of the Degree of Disintegration of Plastic Materials Under Simulated Composting Conditions in a Pilot-scale Test)

3.2.4.1 SUMMARY

This international standard specifies a method for the evaluation of the disintegration of plastic materials while under industrial composting conditions. The testing environment is pilot scale at an industrial compost facility. The mass of plastic samples is measured prior to the test. The samples are placed in a sack or aerated bag with compost soil. The sack is placed in an aerated pile at least 1 m deep at an industrial compost facility. The compost vessel should be a minimum of 140 l and a maximum of 250 l. The concentration is 1% by weight plastic sample and 99% compost. The temperature of the compost pile should be between 40 and 75 °C. The aerated compost pile should be turned twice per week.

3.2.4.2 PROCEDURE

The test procedure for ISO 16929:

1. Measure the mass of plastic samples
2. Place in sack or aerated container between 140 and 250 l
3. Add samples and compost to sack or container
4. Bury container in compost pile at least 1 m from the surface
5. Rotate the compost soil around container at least twice per week
6. Measure the temperature of the compost
7. After 12 weeks, remove compost and plastic
8. Sieve in 2 mm screen
9. Wash and dry plastic to remove compost soil
10. Measure the mass of the plastic samples
11. Measure the quality of the compost soil for maturity, % solid, pH, moisture content, carbon/nitrogen ratio, temperature

3.2.4.3 MATERIALS

- Plastic
- Mature compost
- Container minimum 140 l

3.2.4.4 EQUIPMENT

- Industrial compost facility

3.2.5 ISO 20200 (EN14806:2005) (Plastics. Determination of the Degree of Disintegration of Plastic Materials Under Simulated Composting Conditions in a Laboratory-Scale Test)

3.2.5.1 SUMMARY

This international standard specifies a method for the evaluation of the disintegration of plastic materials while under industrial composting conditions. The testing environment is pilot scale at an industrial compost facility. The mass of plastic samples is measured prior to the test. The samples are placed in a sack or aerated bag with compost soil. The sack is placed in an aerated pile at least 1 m deep at an industrial compost facility. The compost vessel should be a minimum of 5 l and a maximum of 20 l. The concentration is 1% by weight plastic sample and 99% compost. The temperature of the compost pile should be between 40 and 75 °C. The aerated compost pile should be turned twice per week.

3.2.5.2 PROCEDURE

The test procedure for ISO 20200:

1. Measure the mass of plastic samples
2. Place in sack or aerated container between 5 and 20 l
3. Add samples and compost to sack or container at concentrations of 1% (weight) plastic and 99% compost
4. Bury container in compost pile at least 1 m from the surface
5. Rotate the compost soil twice per week
6. Measure the temperature of the compost
7. After 12 weeks, remove compost and plastic
8. Sieve in 2 mm screen
9. Wash and dry plastic to remove compost soil
10. Measure the mass of the plastic samples
11. Measure the quality of the compost soil for maturity, % solid, pH, moisture content, carbon/nitrogen ratio, temperature

3.2.5.3 MATERIALS

- Plastic
- Mature compost
- Container minimum 5 l

3.2.5.4 EQUIPMENT

- Industrial compost facility

3.3 Australian Biodegradation Standards

The Australian standard for degradable plastics includes test methods that enable validation of biodegradation of degradable plastics and designation as "compostable" in Australia. The Australian standard for industrial compost is AS4736-2006 (AS4736-2013). The AS4736 standard is based on EN 13432-2000 standard for the determination of the ultimate aerobic biodegradability and disintegration of plastic materials under controlled composting conditions. AS4736-2006 requires that plastic materials meet the following requirements:

1. A minimum of 90% biodegradation of plastic material within 180 days while under industrial composting conditions of 58 °C and 50% moisture
2. A minimum of 90% disintegration of plastic material within 12 weeks while under industrial composting conditions of 58 °C and 50% moisture
3. No toxic effects on the resulting compost for plants and earthworms
4. Toxic levels of regulated metals are less than the maximum allowable levels.
5. Minimum of 50% organic material in the plastic

3.4 Japanese Biodegradation Standards

The JIS are met with a GreenPla certification system. The GreenPla system has very similar testing requirements as the US and international certification methods. In particular, the GreenPla certification assures biodegradability by measuring carbon dioxide evolution after microbial biodegradation, mineralization by the ability to disintegrate and not have visible fragments after composting, and organic compatibility by the ability of the compost to support plant growth. The same amount of carbon dioxide evolution (60%) in 45 days is required for certification. The same 11 regulated metals are monitored in GreenPla as EN 13432. However, several aspects of the certification are different than the US BPI and International DIN CERTCO certifications.

4 MARINE ENVIRONMENT

Plastic debris is accumulating in the oceans around the world, endangering animal life, releasing toxic chemicals, and collecting floating toxins that can enter the food stream through fish. The fate of plastics in the oceans can lead to fragmentation and result in slurry of plastic particles floating or suspended in the water column. The plastic fragments can degrade and release toxic chemicals into the ocean waters.

Biodegradation of biodegradable plastics in marine environment is based upon two levels of standards—the first for a test method standard

TABLE 12.4 Biodegradation Standards for Marine Environment

Region	Performance Specification	Test Method	Measurement Method	Other Requirements
Australia	None	ISO 14851, ISO 14852	CO_2 measurement	None
Europe	None	ISO 14851, ISO 14852	CO_2 measurement	None
Japan	None	ISO 14851, ISO 14852	CO_2 measurement	None
The United States	ASTM D7081	ASTM D6691	CO_2 measurement	ASTM D6400

and the second for a performance specification standard. Biodegradation standards for the marine environment are listed in Table 12.4. The marine biodegradation standards cover nonfloating products made from plastics that are designed to biodegrade in the aerobic marine environment. The standard applies to deep seawater, shallow seawater, and brackish inland waters. In the standards, plastic materials must demonstrate disintegration and inherent biodegradation during marine water exposure and not exhibit any adverse environmental impacts on the survival of marine organisms while in the marine environment.

The plastic sample also must pass several marine toxicity tests, including PolyTox (microbial oxygen absorption), Microtox (microbial bioluminescence), fish acute toxicity (static conditions) (OPPTS 850.1075), Daphnia acute toxicity (static conditions) (OPPTS 850.1010), and static algal toxicity (OPPTS 850.5400). The plastic samples must also have $<25\%$ of maximum allowable concentrations of regulated heavy metals.

Marine biodegradation standards require that the plastic samples also pass the ASTM D6400 standard for biodegradation under industrial aerobic compost conditions. The ASTM D6400 standard requires plastic samples to convert 90% of the carbon in the plastic sample to CO_2 after 180 days while at 58 °C.

4.1 US Biodegradation Standards for Marine Environment

4.1.1 Biodegradation Performance Specification Standard: ASTM D7081-05 (Nonfloating Biodegradable Plastic in the Marine Environment)

4.1.1.1 SUMMARY

This American specification establishes the performance requirements for biodegradation of plastic materials and products, including packaging, films, and coatings. The marine environment includes conditions of aerobic marine waters or anaerobic marine sediments, or both.

This specification establishes the requirements for biodegradation of plastic materials that have rates that are similar to known compostable materials. The specification also specifies that the degradation of the plastic materials will not cause any harm to sea life or habitat.

The performance specification standard requires the use of ASTM D6691-08 test method to measure the amount of CO_2 that is emitted from the degrading plastic sample while in the cool marine water.

ASTM D7081-05 performance specification standard requires the product must demonstrate each of the three characteristics as follows:

1. Disintegration during marine biodegradation
2. Inherent biodegradation of the plastic material
3. Minimal adverse effect on the marine environment

4.1.1.2 PROCEDURES

The test procedures for ASTM D7081-05 standard specify that three types of tests are performed on the plastic samples. The first test measures the percentage of disintegration of the plastic samples while exposed to cool marine water conditions, described in ASTM D6691-08 test conditions. The plastic samples are weighed prior to exposure to test conditions. The samples are placed in containers with ocean water. The plastic samples are exposed to aerobic marine water conditions of 30 °C for 12 weeks. The mass of the plastic sample is measured after 12 weeks by passing the plastic sample and marine water through a 2.0 mm sieve. ASTM D7081-05 specifies that no more than 30% of the original dry weight of the plastic material remains.

The second test procedure for ASTM D7081-05 standard specifies a satisfactory rate of *biodegradation* of the plastic material that is under marine conditions of 30 °C for 180 days that more than 30% of the carbon in the original plastic sample is converted into CO_2 as measured by a CO_2 respirometer or other measurement device. The details of the test procedures are listed in ASTM D6691-08 test method.

The third test procedure for ASTM D7081-05 standard specifies that the plastic product shall undergo toxicity testing with one of the following tests: PolyTox, Microtox, fish acute toxicity, Daphnia acute toxicity, or static algal toxicity.

The fourth test procedure for ASTM D7081-05 standard specifies that the plastic sample shall have <25% of the regulated heavy metal concentrations that is specified in the country where the product is sold.

4.1.1.3 SPECIFICATIONS

ASTM D7081-05 standard specifies that a plastic coating material is compostable if it meets the following specifications:

- Satisfactory *disintegration* of the plastic material that is under aerobic marine water test of 30 °C for 12 weeks that no more than 30% of the original dry weight of the plastic material remains, after passing through a 2.0-mm sieve
- Satisfactory rate of *biodegradation* of the plastic material that is under aerobic marine water test of 30 °C in 6 months that more than 30% of the carbon in the original plastic sample is converted into CO_2 as measured by measuring the amount of release CO_2
- Satisfactory passing of one of the several marine toxicity tests and have less that 25% of the regulated heavy metal concentrations that is specified in the country where the product is sold
- Satisfactory rate of *biodegradation* specified in ASTM D6400 standard for compostability

4.1.2 Biodegradation Test Method Standard: ASTM D6691-09 (Standard Test Method for Determining Aerobic Biodegradation of Plastic Materials in the Marine Environment by a Defined Microbial Consortium or Natural Sea Water Inoculum)

4.1.2.1 SUMMARY

ASTM D6691-09 test method provides a description of the testing procedures that best simulate the marine environment and a method by which to measure biodegradation. ASTM D6691-09 test method is used to determine the degree and rate of aerobic biodegradation of plastic materials exposed to the indigenous population of existing seawater or synthesized seawater with pregrown population of at least 10 aerobic marine microorganisms of known genera.

The test method consists of preparing a uniform inoculum of marine water, exposing the plastic samples to the marine water, measuring biodegradation with a carbon dioxide respirometer or equivalent measurement method, and assessing the percentage of carbon conversion in the plastic to carbon dioxide. The testing is carried out at 30 ± 2 °C under controlled laboratory conditions for 180 days. The standard recommends the use of 125 ml autoclave bottles, a recipe of marine organisms and nutrients, and seawater. The standard recommends adding 20 mg of plastic sample with 75 ml of marine water inoculum in a constant temperature environment of 30 ± 2 °C. The amount of biodegradation of the plastic sample is compared to the biodegradation of a positive control, e.g., cellulose powder.

This test method establishes the procedures, equipment, materials, and conditions to measure the degree and rate of biodegradation of plastic materials under aerobic mesophilic marine water conditions. This test method is designed to produce repeatable and reproducible test results under controlled test conditions that simulate the marine environment.

4.1.2.2 PROCEDURE

The test procedures for ASTM D6691-08 involve placing 20 mg of plastic sample with 75 ml of marine stock solution or natural seawater with inorganic nutrients placed in 125 ml bottles in a sterile environment as shown in Figure 12.2. A minimum of 12 composting vessels are required for the test. The containers will be provided for the plastic sample and the following samples, blank, positive control, and negative control. The testing is completed in triplicate. The marine inoculum is tested for marine ingredients.

4.1.2.3 MATERIALS

- Plastic samples
- Blank marine water
- Positive control: cellulose powder or paper
- Negative control: polyethylene film

4.1.2.4 EQUIPMENT

- Sterile jar with volume of 125 ml
- Water bath or other temperature control device
- Carbon dioxide trapping apparatus with $Ba(OH)_2$ solution
- Flexible tubing nonpermeable to CO_2
- Stoppers equipped with sampling ports
- Analytical balance (±0.1 mg)
- pH meter
- Respirometer devices for measuring CO_2 and O_2 periodically (optional)
- CO_2 trapping equipment with gas flow meter, gas chromatograph, or other suitable equipment

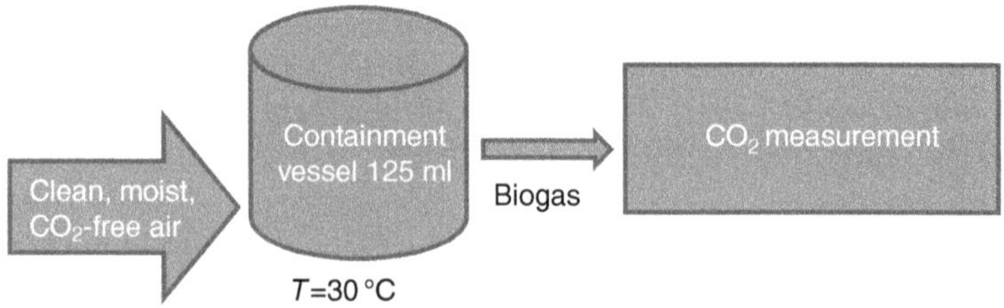

FIGURE 12.2 Experimental setup for laboratory environment of marine biodegradation per ASTM D6691 standards. (For color version of this figure, the reader is referred to the online version of this chapter.)

4.2 International Aqueous Biodegradation Standards

4.2.1 Biodegradation Test Method Standard: ISO 14852-1999 (EN14047) (Determination of Ultimate Aerobic Biodegradability of Plastic Materials in an Aqueous Medium. Method by Analysis of Evolved Carbon)

4.2.1.1 SUMMARY

This international standard specifies a method for the evaluation of the biodegradation of plastic materials while exposed to aqueous mineral medium. Biodegradation is determined by measuring the evolved carbon dioxide from the biodegrading sample. The test is used to measure the degree and rate of aerobic biodegradation of plastic materials on exposure to activated sludge biomass or a suspension of active soil under laboratory conditions at solid concentrations between 100 and 2000 mg/l. The temperature of the test environment should be between 20 and 25 °C. The tests are run in duplicate.

4.2.1.2 PROCEDURE

The test procedure for ISO 14852:

1. Measure the mass of plastic samples
2. Measure the quality of the inoculum soil for % solid, pH, moisture content, carbon/nitrogen ratio, temperature. Carbon/nitrogen ratio should be approximately 40:1
3. Add the plastic sample in the form of powder or small pieces to a concentration of 100-2000 mg/l of TOC with activated sludge or active soil in a test flask
4. Connect the sample flasks to CO_2-free air production
5. Maintain the test flasks at temperature between 20 and 25 °C
6. Bubble CO_2-free air through flasks at a rate of 50-100 ml/min
7. Measure CO_2 from the sample flasks

4.2.1.3 MATERIALS

- Plastic samples
- Blank-activated sludge or active soil
- Positive control: cellulose powder or cellulose paper
- Negative control: polyethylene film
- Barium hydroxide solution

4.2.1.4 EQUIPMENT

- Flasks
- Water bath or other temperature control device
- Pressurized air system with CO_2-free and H_2O-saturated air
- Carbon dioxide trapping apparatus

- Flexible tubing nonpermeable to CO_2
- Stoppers equipped with sampling ports
- Analytical balance (±0.1 mg)
- 100 ml burette
- 0.05 N HCl
- pH meter
- CO_2 trapping equipment with gas flow meter, gas chromatograph, or other suitable equipment

4.2.2 Biodegradation Test Method Standard: ISO 14851 (EN14048) (Determination of Ultimate Aerobic Biodegradability of Plastic Materials in an Aqueous Medium. Method by Measuring the Oxygen Demand in a Closed Respirometer)

4.2.2.1 SUMMARY

This international standard specifies a method for the evaluation of the biodegradation of plastic materials while exposed to aqueous mineral medium. Biodegradation is determined by measuring oxygen demand produced from the biodegrading sample. The percentage of biodegradation is calculated by comparing the biological oxygen demand to the chemical oxygen demand, determined by chemical oxidation of the test item or by stoichiometric formula. The principle and test procedures are very similar to ISO 14852 (Bastioli, 2005).

The test is used to measure the degree and rate of aerobic biodegradation of plastic materials on exposure to activated sludge biomass or a suspension of active soil under laboratory conditions at solid concentrations between 100 and 2000 mg/l. The temperature of the test environment should be between 20 and 25 °C. The tests are run in duplicate.

4.2.2.2 PROCEDURE

The test procedure for ISO 14852:

1. Measure the mass of plastic samples
2. Add the plastic sample in the form of powder or small pieces to a concentration of 100-2000 mg/l of TOC with activated sludge or active soil in a test flask
3. Connect the sample flasks to CO_2-free air production
4. Maintain the test flasks at temperature between 20 and 25 °C
5. Bubble CO_2-free air through flasks at a rate of 50-100 ml/min
6. Measure CO_2 from the sample flasks

4.2.2.3 MATERIALS

- Plastic samples
- Blank-activated sludge or active soil

- Positive control: cellulose powder or cellulose paper
- Negative control: polyethylene film

4.2.2.4 EQUIPMENT

- Flasks
- Water bath or other temperature control device
- Pressurized air system with CO_2-free and H_2O-saturated air
- Flexible tubing nonpermeable to CO_2
- Stoppers equipped with sampling ports
- Analytical balance (±0.1 mg)
- 100 ml burette
- pH meter
- O_2 trapping equipment with gas flow meter, gas chromatograph, or other suitable equipment

5 ANAEROBIC DIGESTION

Anaerobic digestion is a process where food and agricultural waste can be converted to methane and carbon dioxide in a thermophilic reactor. Anaerobic digestion is a series of processes in which microorganisms break down biodegradable material in the absence of oxygen (NNFCC Renewable Fuels and Energy Factsheet, 2013). It is used for industrial or domestic purposes to manage waste and/or to release energy.

The digestion process begins with bacterial hydrolysis of the input materials to break down insoluble organic polymers, such as carbohydrates, and make them available for other bacteria. Acidogenic bacteria then convert the sugars and amino acids into carbon dioxide, hydrogen, ammonia, and organic acids. Acetogenic bacteria then convert these resulting organic acids into acetic acid, along with additional ammonia, hydrogen, and carbon dioxide. Finally, methanogens convert these products to methane and carbon dioxide.

Anaerobic digestion is used as part of the process to treat organic and agricultural waste and sewage sludge. Anaerobic digestion can also be used to remove organic waste at waste water treatment operations. Biodegradable plastics can hold organic and agricultural waste with the use of biodegradable plastic bags. The biodegradable plastic can be a food source for the anaerobes in the digester. The biodegradable plastic and organic waste can be converted into carbon dioxide, methane, and other trace gases.

Standards are necessary to monitor the performance of biodegradable plastics in an aerobic-digestion reactor. The anaerobic decomposition takes place under high-solids (more than 30% total solids) and static nonmixed conditions. Biodegradation usually requires one to 2 weeks of dwell time. Table 12.5 lists the biodegradation standards for anaerobic digestion.

TABLE 12.5 Biodegradation Standards for Anaerobic Digestion Environment

Country	Performance Specification	Test Method	Measurement Method
Australia	None	ISO 14853	CO_2 measurement
Europe	None	ISO 14853	CO_2 measurement
Japan	None	ISO 14853	None
The United States	ASTM D7081	ASTM D5511	CO_2 measurement

5.1 US Biodegradation Standards for Anaerobic Digestion

5.1.1 Biodegradation Test Method Standard: ASTM D5511-02 (Standard Test Method for Determining Anaerobic Biodegradation of Plastic Materials Under High-Solids Anaerobic-Digestion Conditions)

5.1.1.1 SUMMARY

This test method establishes the procedures, equipment, materials, and conditions to measure the degree and rate of biodegradation of plastic materials under anaerobic thermophilic conditions in an aqueous environment. This test method is designed to produce repeatable and reproducible test results under controlled anaerobic-digestion conditions that simulate industrial conditions. The plastic test samples are exposed to an inoculum that is derived from an aerobic digester or waste water treatment operation. The anaerobic thermophilic conditions of the test are provided in an environment where temperature, methane, carbon dioxide, and oxygen levels are closely monitored and controlled. The plastic samples can be in the form of films, powders, pellets, or molded pieces and are placed in a vessel with warm inoculum with proper anaerobic bacteria. The test containers are at 50 °C (±2 °C) for 30 days. The biogas from the container is measured for CH_4, CO_2, and O_2 over the testing period. The standard for anaerobic digestion comprises a biodegradation test method and not a biodegradation specification.

5.1.1.2 PROCEDURE

The test procedures for ASTM D5511-02 involve placing 1000 g of inoculum derived from properly operating anaerobic digester that is made from pretreated household waste. The inoculum should be derived from a digester operating under >20% total solids conditions. A minimum of 12 test vessels are required for the test. The containers will be provided for the plastic sample and the following plastic samples, blank, positive control, and negative control. The testing is completed in triplicate. The plastic samples are added to each test container in quantities up to 100 g.

The positive control must obtain >70% biodegradation in 30 days. The inoculum is tested for pH, volatile fatty acids, and NH_4^+ − N ratio.

The test apparatus can include an inverted graduated cylinder or plastic column. The graduated cylinder or plastic column is inverted in a low pH fluid to avoid CO_2 loss through the dissolution in the fluid. The biogas is calculated through a pressure measurement of the inverted tubes. The biogas can also be measured with a gas chromatograph to determine the percentage of CH_4 and CO_2 in the biogas. An example of a test apparatus is provided in Figure 12.3. The testing apparatus will provide measurements of pressure over the 30 days. Through the ideal gas law, the pressure can be converted to grams of biogas. The concentration of biogas can be converted to concentrations of CO_2 and CH_4. The conversion of carbon from the plastic sample to CO_2 and CH_4 can be determined. This will result in the carbon biodegradation percentage over 30 days in a high-solids anaerobic digester.

5.1.1.3 MATERIALS

- Plastic samples
- Blank anaerobic digester inoculum
- Positive control: cellulose powder or cellulose paper
- Negative control: polyethylene film

5.1.1.4 EQUIPMENT

- Test jars
- Low pH fluid bath or other temperature control device
- Flexible tubing nonpermeable to CH_4, CO_2, and O_2
- Stoppers equipped with sampling ports
- Graduated cylinder or plastic tube

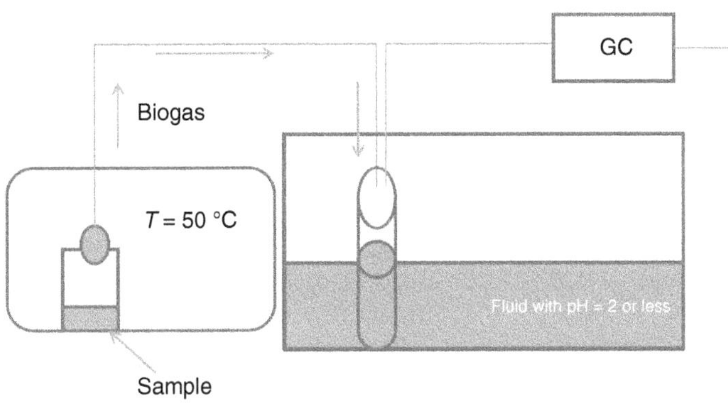

FIGURE 12.3 Experimental setup for laboratory environment for anaerobic digester test per ASTM 5511 standards. (For color version of this figure, the reader is referred to the online version of this chapter.)

- Analytical balance (±0.1 mg)
- pH meter
- Gas chromatograph

5.2 International Biodegradation Standards for Anaerobic Digestion

5.2.1 Biodegradation Test Method Standard: ISO 14853:2005 (Plastics. Determination of Ultimate Anaerobic Biodegradation of Plastic Materials in an Aqueous System. Method of Biogas production)

5.2.1.1 SUMMARY

This test method specifies a method for the determining the ultimate biodegradability of plastics by anaerobic microorganisms. The conditions described in this test method simulate the conditions of some anaerobic digesters. The plastic test samples are exposed to an inoculum that is derived from digested sludge from waste water treatment operations. The anaerobic thermophilic conditions of the test are provided in an environment where temperature, methane, carbon dioxide, and oxygen levels are closely monitored and controlled. The plastic samples can be in the form of films, powders, pellets, or molded pieces and are placed in a vessel with warm inoculum with proper anaerobic bacteria. The test containers are at 35 °C (±2 °C) for 60 days. The biogas from the container is measured for CH_4, CO_2, and O_2 over the testing period. The standard for anaerobic digestion comprises a biodegradation test method and not a biodegradation specification.

5.2.1.2 PROCEDURE

The test procedures for ISO 14853:2005 involve placing 1000 ml of inoculum derived from properly operating anaerobic digester that is made from digested sludge. The concentration of test materials is 20 to 200 mg/l of organic carbon. Approximately, 350 ml of inoculum is added to the test sample. A head space of 20% is recommended. A reference material or positive control can be cellulose, poly(hydroxybutyrate), or poly(ethylene glycol). The glass vessels have sizes of 0.1 to 1 l with gas-tight septum rated for 2000 hPa. Plastic samples are added to glass containers that include plastic samples, blank, positive control, and negative control. The testing is completed in triplicate. The positive control must obtain >70% biodegradation in 60 days. The inoculum is tested for pH, volatile fatty acids, and $NH_4^+ - N$ ratio.

Biogas measurement is achieved with a monometer or volumetric device. An example of a test apparatus is provided in Figure 12.4. The testing apparatus will provide measurements of pressure over the 60 days.

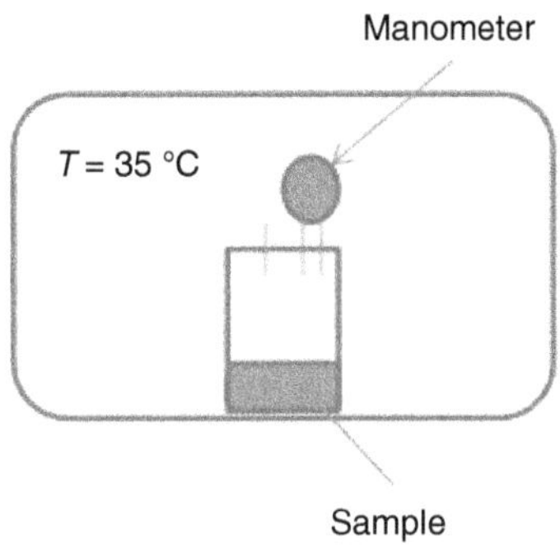

FIGURE 12.4 Experimental setup for laboratory environment for anaerobic digester test per ISO standards. (For color version of this figure, the reader is referred to the online version of this chapter.)

Through the ideal gas law, the pressure can be converted to grams of biogas. The concentration of biogas can be converted to concentrations of CO_2 and CH_4. The conversion of carbon from the plastic sample to CO_2 and CH_4 is determined. This will result in the carbon biodegradation percentage over 30 days in a high-solids anaerobic digester.

5.2.1.3 MATERIALS

- Plastic samples
- Blank anaerobic digested sludge inoculum
- Positive control: cellulose powder, poly(hydroxybutyrate), or poly (ethylene glycol)
- Negative control: polyethylene film

5.2.1.4 EQUIPMENT

- Test jars: 0.1-1 l
- Low pH fluid bath or other temperature control device
- Flexible tubing nonpermeable to CH_4, CO_2, and O_2
- Stoppers equipped with sampling ports
- Graduated cylinder or plastic tube
- Manometer
- Analytical balance (±0.1 mg)
- pH meter
- Gas chromatograph

6 ACTIVE LANDFILL

Landfills in the United States are typically built with the EPA guidelines with the use of clay linings and a landfill cap (Criteria for Solid Waste Disposal Facilities, 2013). The most common material for landfill caps is made from asphalt or concrete (Remediation Technologies Screening Matrix and Reference Guide, 2013). Landfills can operate with creation of biogas that is composed of methane, carbon dioxide, and other trace gases. Methane gas can be vented and burned or can be captured and stored for energy purposes. The carbon dioxide and other gases must be scrubbed to provide a clean methane gas without carbon dioxide or other gases. Some landfills are considered active and provide clean methane gas for energy consumption. Biodegradable plastics can hold the waste as trash bags for disposal and provide food source for the aerobic and anaerobic bacteria that are in the landfill. Standards are needed to evaluate the biodegradation of biodegradable plastics in landfills. Table 12.6 lists the biodegradation standards for an active landfill environment. The standard for active landfill comprises a biodegradation test method and not a biodegradation specification.

6.1 US Biodegradation Standards for Active Landfill

6.1.1 Biodegradation Test Method Standard: ASTM D5526-11 (Determining Anaerobic Biodegradation of Plastic Materials Under Accelerated Landfill Conditions)

6.1.1.1 SUMMARY

This test method establishes the degree and rate of anaerobic biodegradation of plastic materials under anaerobic mesophilic conditions in an accelerated landfill test environment. This test method is designed to yield percentage of carbon conversion from the plastic sample to methane or carbon dioxide biogas. This test method is designed to provide repeatable and reproducible test results under controlled anaerobic conditions that simulate high-solids (>30%) active landfill conditions. The plastic test samples are exposed to an inoculum that is derived from an aerobic digester or waste water treatment operation. The anaerobic mesophilic conditions of the test are provided in an environment where temperature, methane, carbon dioxide, and oxygen levels are closely monitored and controlled. The plastic samples can be in the form of films, powders,

TABLE 12.6 Biodegradation Standards for Active Landfill Environment

Region	Performance Specification	Test Method	Measurement Method
Australia	None	None	—
Europe	None	None	—
Japan	None	None	—
The United States	None	D5526, D7575	CO_2 measurement

pellets, or molded pieces and are placed in a vessel with cool inoculum with proper anaerobic bacteria. The test containers are at 35 °C (±2 °C) for up to 365 days. The biogas from the container is measured for CH_4, CO_2, and O_2 over the testing period.

6.1.1.2 PROCEDURE

The test procedures for ASTM D5526 involve placing 600 g of household waste with 100 g of inoculum derived from properly operating anaerobic digester that is made from pretreated household waste. The inoculum should be derived from a digester operating under >30% total solids conditions. A minimum of 27 test vessels are required for the test with a volume between 4 and 6 l. The containers will be provided for the plastic sample and the following plastic samples, blank, positive control, and negative control. The testing is completed in triplicate. The plastic samples are added to each test container in quantities between 60 and 100 g of dry weight. The mixture should contain approximately 60% dry weight of solids. The positive control must obtain >70% biodegradation at the end of the test. The inoculum is tested for pH, volatile fatty acids, and $NH_4^+ - N$ ratio.

The test apparatus can include a pressure-resistant glass vessel. The vessel should withstand a pressure of two atmospheres. The graduated cylinder or plastic column is inverted in a low pH fluid to avoid CO_2 loss through the dissolution in the fluid. The biogas is calculated through a pressure measurement of the inverted tubes. The biogas can also be measured with a gas chromatograph to determine the percentage of CH_4 and CO_2 in the biogas. An example of a test apparatus is provided in Figure 12.5. The testing apparatus will provide measurements of pressure over the 30 days. Through the ideal gas law, the pressure can be converted to grams of biogas. The concentration of biogas can be converted to concentrations of CO_2 and CH_4. The conversion of carbon from the plastic sample to CO_2 and CH_4 can be determined. This will result in the carbon biodegradation percentage over 30 days in a high-solids anaerobic digester.

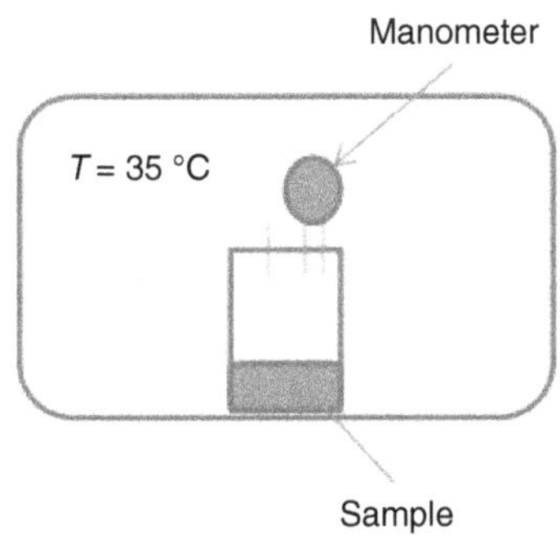

FIGURE 12.5 Experimental setup for laboratory environment for active landfill test. (For color version of this figure, the reader is referred to the online version of this chapter.)

6.1.1.3 MATERIALS

- Plastic samples
- Anaerobic digester inoculum
- Pretreated household solid organic waste
- Positive control: cellulose powder or cellulose paper
- Negative control: polyethylene film

6.1.1.4 EQUIPMENT

- Test jars
- Flexible tubing nonpermeable to CH_4, CO_2, and O_2
- Stoppers equipped with sampling ports
- Incubator
- Pressure transducer
- Analytical balance (±0.1 mg)
- pH meter
- Gas chromatograph

6.1.2 Biodegradation Test Method Standard: ASTM D7475-11 (Determining Aerobic Degradation and Anaerobic Biodegradation of Plastic Materials Under Accelerated Landfill Conditions)

6.1.2.1 SUMMARY

This two-tiered test method modifies ASTM D5526 by adding an aerobic degradation test method to the anaerobic biodegradation test method. ASTM D7475 establishes the degree of aerobic degradation and rate of anaerobic biodegradation of plastic materials under anaerobic mesophilic conditions in an accelerated landfill test environment. This test method is designed to establish mechanical tests or materials tests in addition to percentage of carbon conversion from the plastic sample to methane or carbon dioxide biogas. This test method is designed to provide repeatable and reproducible test results under controlled aerobic and anaerobic conditions that simulate high-solids (>30%) active landfill conditions. The plastic test samples are exposed to an inoculum that is derived from an aerobic digester or waste water treatment operation. The first tier of the test method is designed to estimate aerobic biodegradation through the measurement of loss of mechanical properties, e.g., tensile strength or modulus, molecular weight, or other characteristics. The second tier of the test method anaerobic is designed to provide mesophilic conditions of the test in an environment where temperature, methane, carbon dioxide, and oxygen levels are closely monitored and controlled. The plastic samples can be in the form of films, powders, pellets, or molded pieces and are placed in a vessel with cool inoculum with proper anaerobic bacteria. The test containers are at 35 °C (±2 °C) for up to 365 days. The biogas from the container is measured for CH_4, CO_2, and O_2 over the testing period.

6.1.2.2 PROCEDURE

The test procedures for the first tier of ASTM D7475-11 involve measuring baseline mechanical or chemical properties and then placing test specimens in an aerobic degradation environment. The mechanical or chemical properties are measured after a specified time. Differences in mechanical or chemical properties are recorded.

The test procedures for the second tier of ASTM D7475-11 involve placing 600 g of household waste with 100 g of inoculum derived from properly operating anaerobic digester that is made from pretreated household waste. The inoculum should be derived from a digester operating under >30% total solids conditions. A minimum of 27 test vessels are required for the test with a volume between 4 and 6 l. The containers will be provided for the plastic sample and the following plastic samples, blank, positive control, and negative control. The testing is completed in triplicate. The plastic samples are added to each test container in quantities between 60 and 100 g of dry weight. The mixture should contain approximately 60% dry weight of solids. The positive control must obtain >70% biodegradation at the end of the test. The inoculum is tested for pH, volatile fatty acids, and $NH_4^+ - N$ ratio.

The test apparatus can include a pressure-resistant glass vessel. The vessel should withstand a pressure of two atmospheres. The graduated cylinder or plastic column is inverted in a low pH fluid to avoid CO_2 loss through the dissolution in the fluid. The biogas is calculated through a pressure measurement of the inverted tubes. The biogas can also be measured with a gas chromatograph to determine the percentage of CH_4 and CO_2 in the biogas. An example of a test apparatus is provided in Figure 12.5. The testing apparatus will provide measurements of pressure over the 30 days. Through the ideal gas law, the pressure can be converted to grams of biogas. The concentration of biogas can be converted to concentrations of CO_2 and CH_4. The conversion of carbon from the plastic sample to CO_2 and CH_4 can be determined. This will result in the carbon biodegradation percentage over 30 days in a high-solids anaerobic digester.

6.1.2.3 MATERIALS

- Plastic samples
- Anaerobic digester inoculum
- Pretreated household solid organic waste
- Positive control: cellulose powder or cellulose paper
- Negative control: polyethylene film

6.1.2.4 EQUIPMENT

- Test jars
- Flexible tubing nonpermeable to CH_4, CO_2, and O_2

- Stoppers equipped with sampling ports
- Incubator
- Pressure transducer
- Analytical balance (±0.1 mg)
- pH meter
- Gas chromatograph
- Mechanical tests, e.g., tensile and impact
- Chemical tests, e.g., FTIR, DSC, GC/MS, and molecular weight

6.2 International Biodegradation Standards for Active Landfill

Currently, there are no ISO test methods that are equivalent to active landfill standard test method.

References

Bastioli, C. (2005). *Handbook of biodegradable polymers.* Shawbury: Shrewsbury, Shropshire, SY4 4NR, united Kingdom.

'BPI Certified Compostable', http://www.bpiworld.org/ [March 2013].

'Criteria for Solid Waste Disposal Facilities', Environmental Protection Agency, http://www.epa.gov/wastes/nonhaz/municipal/landfill/criteria/landbig.pdf [March 2013].

Narayan, R. (2011a). Rationale, drivers, standards, and technology for biobased materials. In M. Graziani & P. Fornasiero (Eds.), *Renewable resources and renewable energy: A global challenge* (pp. 10–16). (2nd ed.). Boca Raton: CRC Press.

Narayan, R. (2011b). Carbon footprint of bioplastics using biocarbon content analysis and life-cycle assessment. *MRS (Materials Research Society) Bulletin, 36*(09), 716–721.

Narayan, R., & Pettigrew, C. (1999). ASTM standards help define and grow a new biodegradable, plastic industry. *ASTM Standardization News*, December 1999 36.

'NNFCC Renewable Fuels and Energy Factsheet: Anaerobic Digestion', National Non-Food Crops Centre. http://www.nnfcc.co.uk/publications/nnfcc-renewable-fuels-and-energy-factsheet-anaerobic-digestion [March 2013].

'Part 503 Implementation Guidance', http://nlquery.epa.gov/epasearch/epasearch?querytext=Subpart+503-13+heavy+metals&fld=&areaname=&typeofsearch=epa&areacontacts=http%3A%2F%2Fwww.epa.gov%2Fepahome%2Fcomments.htm&areasearchurl=&result_template=epafiles_default.xsl&filter=sample4filt.hts [March 2013].

'Remediation Technologies Screening Matrix and Reference Guide', http://www.frtr.gov/matrix2/section4/4-27.html [March 2013].

Rudnik, E. (2008). *Compostable polymer materials.* Elsevier, The Boulevard, Langford Lane, Kidlington, Oxford OX5, 1GB, UK p. 101.

Title 40: Protection of the Environment', Electronic Code of Federal Regulations, http://www.ecfr.gov/cgi-bin/searchECFR [March 2013].

AS4736-2013. 'http://www.biofiba.com/documents/AS4736-2006.pdf' www.biofiba.com/documents/AS4736-2006.pdf (December 2013).

'What is the AS4736 standard for biodegradable plastics', http://www.bioplastics.org.au/faq/what-is-the-as4736-standard-for-biodegradable-plastics [March 2013].

CHAPTER

13

Effects of Formulation, Structure, and Processing on Biodegradation of Starches

Parveen Sangwan, Eustathios Petinakis, Katherine Dean

CSIRO Materials Science and Engineering, Victoria, Australia

1 INTRODUCTION

Starch, a natural polymer, is a rapidly biodegradable material that has gained immense popularity in the development of starch-based biodegradable films in various fields of applications such as food packaging, drug delivery, sutures, and agriculture mulch films. Starch presents a biodegradable, renewable, and low-cost alternative for synthetic polymers for short-term applications which require rapid degradation (Canigueral et al., 2009; Glenn, Imam, & Orts, 2011; Imam et al., 1998, 1999; Lu, Xiao, & Xu, 2009). The hydrophilic nature of starch plays an important role in initiating biodegradation process, but it also limits its technical applications. Therefore, attempts have been made to overcome this limitation by modifying starch structure, formulation, and processing techniques (Su & Cheng, 2011). These factors could significantly influence the overall performance and biodegradability of starch-based polymers.

Starch films are degradable in diverse environments such as soil, compost, marine water, and sewage; however, the rate of biodegradation may vary. It is important to assess the biodegradability of starch-polymer blends in their intended environment of disposal, as different environments are not always comparable. Each environment contains different microorganisms (in terms of species diversity and population) and different physical and chemical parameters (in terms of temperature, moisture, pH, aeration, and nutrients) that could significantly affect their rate and extent of biodegradation.

The biodegradation of starch can be significantly changed and controlled through (1) the modification of formulation, (2) the structure and type of starch itself, and (3) the way in which the starch is processed. In this chapter, some of the key factors influencing starch biodegradation are discussed. In Section 2, different methods of assessment of biodegradation are discussed. In Section 3, the effects of structure and processing of starches on biodegradation are explained in detail. In Section 4, biodegradability of starch-polymer blends is discussed. In Section 5, a review of biodegradation of starch composites is presented.

2 METHODS OF MEASURING BIODEGRADATION

To study the environmental impact of biodegradable plastics, test materials are exposed to natural environments such as soil, compost, marine water, and sewage during field trials (Breslin, 1998; Cho, Moon, Kim, Nam, & Kim, 2011; Dubey et al., 2012; Singh, Pandey, Rutot, Degee, & Dubois, 2003) or simulated natural environments in laboratory using accelerated test conditions (Abd El-Rehim, Hegazy, Ali, & Rabie, 2004; Cerruti et al., 2011; Moriana, Karlsson, & Ribes-Greus, 2010). Test specimens collected at regular intervals are examined to monitor changes in various properties such as gravimetric or molecular weight, tensile strength, chemical structure, thermal properties, and microbial growth. A series of test methods listed in the following sections are generally performed to monitor and assess the biodegradability of starch-polymer blends.

2.1 Respirometric

The method involves biodegradation of the test material by environmental microorganisms under aerobic or anaerobic conditions (Equations 13.1 and 13.2), and carbon dioxide evolved (Copinet, Bertrand, Longieras, Coma, & Couturier, 2003; Gattin, Copinet, Bertrand, & Couturier, 2002; Kijchavengkul, Auras, Rubino, Ngouajio, & Thomas Fernandez, 2006), oxygen consumed, or biogas produced is measured (Krupp & Jewell, 1992; Massardier-Nageotte, Pestre, Cruard-Pradet, & Bayard, 2006; Nayak, 2010):

$$\text{Aerobic}: \quad C_{\text{Sample}} + O_2 \rightarrow CO_2 + H_2O + C_{\text{Biomass}} + C_{\text{Residual}} \tag{13.1}$$

$$\text{Anaerobic}: \quad C_{Sample} + H_2 \rightarrow CH_4 + CO_2 + H_2O + C_{Biomass} + C_{Residual} \tag{13.2}$$

These values are then related back to the original material through compound stoichiometry and chemical structure to determine an estimate of biodegradation (Equations 13.3 and 13.4):

$$\text{Sample degradation}\,(\%) = \frac{\text{Vol}\left(CO_{2,\,\text{sample + medium}}\right) - \text{Vol}\left(CO_{2,\,\text{medium}}\right)}{\text{Theoretical vol}\left(CO_{2,\,\text{sample}}\right)} \tag{13.3}$$

$$\frac{\text{Sample degradation}\,(\%) = O_{2,\,\text{prod sample + medium}} - O_{2,\,\text{prod medium}})}{\text{Theoretical}\,O_{2,\,\text{sample}}} \tag{13.4}$$

Several renowned organizations, namely, ASTM (American Society for Testing and Materials), ISO (International Organization for Standardization), OECD (Organisation for Economic Co-operation and Development), and CEN (European Committee for Standardization), have developed standard test methods and performance standards for evaluating biodegradable plastics under accelerated conditions. A list of standard test methods that are frequently reported in literature is presented in Table 13.1.

TABLE 13.1 Standard Test Methods for Assessing Biodegradability of Plastics

Standard Test	Source of Inoculum	Measured Parameters	Expected Biodegradability of Test Material
ASTM D5338-92, ISO 14855	Aerobic compost	CO_2 evolved	No threshold (ASTM), >90%[a] (ISO)
ASTM D5988, ISO17556	Aerobic soil	CO_2 evolved	No threshold (ASTM), >90%[a] (ISO)
ASTM D5209-92	Aerobic sewage sludge	CO_2 evolved	No threshold
ASTM D5210-92	Anaerobic sewage sludge	CO_2 and CH_4 evolved	No threshold
ASTM D5511-11	Anaerobic digester	CO_2 and CH_4 evolved	No threshold
ASTM D6691-01	Aerobic marine	CO_2 evolved	No threshold
ISO9408	Aerobic aquatic	O_2 consumed	>90%[a]
ISO14852	Aerobic aquatic	CO_2 evolved	>90%[a]
ISO14852	Anaerobic aquatic	CO_2 and CH_4 evolved	>90%[a]

[a] *w/w (dry weight) in total or of the maximum degradation of a suitable reference material.*

2.2 Morphological

The physical appearance of polymers (color, shape, size, any visible cracks/holes on polymer surface, and/or microbial growth) is recorded before and after biodegradation (Muthukumar, Aravinthan, & Mukesh, 2010). Color change can also be used as an indication of biodegradation of starches. A colorimeter is commonly used for color measurements (yellowness, whiteness, light transmission, and haze).

2.3 Microscopic

Scanning electron microscopy (SEM) techniques are commonly used to investigate surface morphology of polymers and to detect growth of microorganisms on polymers during biodegradation (Dubey et al., 2012; Muthukumar et al., 2010; Pramila & Ramesh, 2011; Singh, Pandey, et al., 2003).

2.4 Gravimetric

Polymer samples are weighed before and after biodegradation and the percentage weight loss is determined (Copinet et al., 2003; Singh, Pandey, et al., 2003) as follows:

$$\%\,\text{Weight loss} = \frac{(W_{to} - W_{ts})}{W_{to}} \times 100$$

where W_{to} and W_{ts} refer to the weights of samples at time 0 (before exposure) and at specific sampling time, respectively.

This is not necessarily the best method for measuring biodegradation of starches as they typically contain a high percentage of water. This water content can increase significantly before biodegradation proceeds, showing up as a weight gain and not as a weight loss.

2.5 Physical and Morphological

Tensile tests (strength, modulus, and elongation at break) are used to determine changes in mechanical properties during biodegradation (Cerruti et al., 2011; Dubey et al., 2012). The technique of X-ray diffraction is often used to measure the degree of crystallinity in starches. Differential scanning calorimetry and thermogravimetric analysis are also used to investigate thermal transitions of polymers by measuring their glass transition and melting point temperatures and also crystallinity; all these properties are affected during biodegradation process. To assess biodegradation of starch-polymer blends in soil, thermal analysis is preferred over conventional weight loss method as it overcomes errors in test results due to adhering soil or microbial growth (Dubey et al., 2012).

2.6 Spectroscopic

Fourier transform infrared spectroscopy is routinely used to obtain qualitative information on changes in chemical structure or formation of functional groups in starches during biodegradation (Dubey et al., 2012; Jayasekara, Harding, Bowater, Christie, & Lonergan, 2003; Muthukumar et al., 2010).

2.7 Chromatographic

Gel permeation chromatography is often used to determine changes in average molecular weights (M_n and M_w) and polydispersity index (M_w/M_n) of polymer samples during biodegradation (Copinet et al., 2003; Dubey et al., 2012). High-performance liquid chromatography (Copinet et al., 2003) is used to detect the presence of monomers and oligomers formed in aqueous or gas phases during biodegradation.

2.8 Microbiological Techniques

Microbial degradation of materials is assessed by subjecting the test materials either to a wide range of environmental microorganisms found in compost, soil, marine water and freshwater, and activated sludge samples (Gattin, Copinet, Bertrand, & Couturier, 2001; Shah, Hasan, Hameed, & Ahmed, 2008) or to selected pure cultures of bacteria and fungi that are known to possess biodegradation potential (Singh, Pandey, et al., 2003). One of the following microbiological techniques is used for assessing biodegradation of starch-polymer blends by microorganisms or enzymes produced by microorganisms:

2.8.1 Microbial Growth/Clear Zone

Test specimen is disinfected, dried and placed on solid or liquid media containing no additional C source, and sprayed on with environmental samples (soil/compost/marine water) or pure cultures of bacteria and fungi. After incubation at specified test conditions, the test material is visualized for any microbial growth and/or formation of clear zone (on solid media). Numerous studies have reported susceptibility or resistance of starch-polymer blends to microbial degradation using these techniques (Geweely & Ouf, 2011; Nikazar, Safari, Bonakdarpour, & Milani, 2005; Pramila & Ramesh, 2011).

2.8.2 *Enzymatic*

The method involves incubation of sterilized film samples in a buffer solution or medium broth containing microbial or synthetic enzymes. Enzymatic activity is measured spectrophotometrically in the liquid medium and the percent weight loss over period of time is calculated. A number of studies have reported enzymatic degradation of starch-polymer blends by microbial and commercially available enzymes (Russo, Truss, & Halley, 2009; Shah et al., 2008; Singh, Pandey, et al., 2003; Swanson, Shogren, Fanta, & Imam, 1993; Vikman, Itävaara, & Poutanen, 1995).

3 BIODEGRADATION OF STARCHES

Starch is produced in the form of granules (2-100 μm in diameter) as storage energy by various plants such as potatoes, corn, wheat, cassava, and rice. It comprises two distinct microstructures, amylose and amylopectin, linked through glucosidic bonds (Figure 13.1). Amylose is the crystalline structure consisting of a linear chain of α-1,4-linked glucose units, and

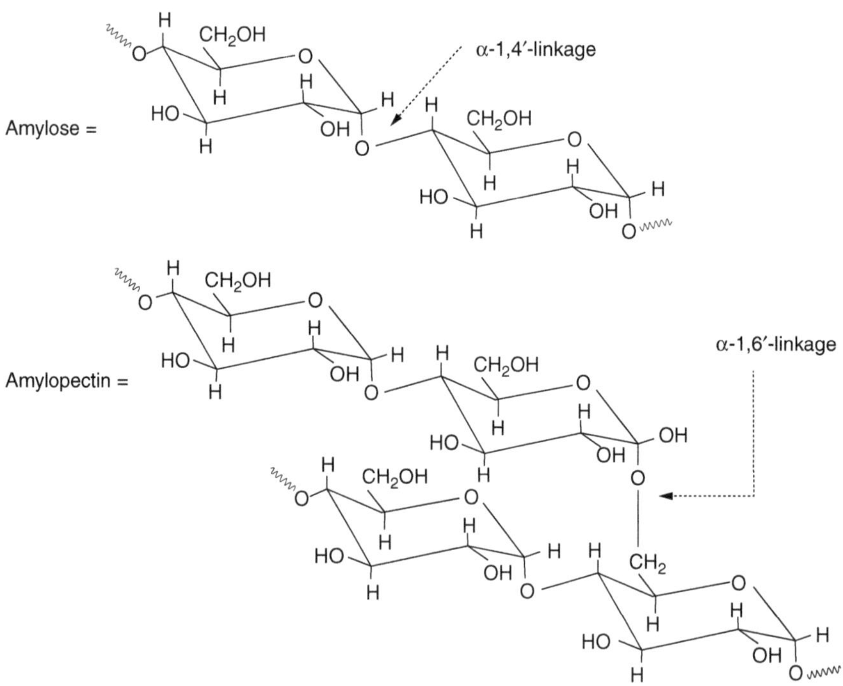

FIGURE 13.1 Molecular structure of starch.

amylopectin is the branched structure consisting of chains of α-1,4-linked glucose units cross-linked by α-1,6 bonds (Chandra & Rustgi, 1998; Flieger, Kantorova, Prell, Rezanka, & Votruba, 2003; French, 1973; Su & Cheng, 2011).

In general, the biodegradation of starch-based polymers is a result of enzymatic attack at the glucosidic linkages between the sugar units, leading to the breakdown of long-chain sugar units into oligosaccharides, disaccharides, and monosaccharides that are readily accessible to microbial or enzymatic attack (Chandra & Rustgi, 1998; Nikazar et al., 2005; Sarikaya, Higasa, Adachi, & Mikami, 2000; Swanson et al., 1993).

3.1 Effects of Starch Structure and Type on Biodegradation

The composition, morphology, and thermal and rheological properties of starch granules vary with plant source, contributing to significant differences in properties and desired functionality (Chaudhary, Miler, Torley, Sopade, & Halley, 2008; Singh, Singh, Kaur, Sodhi, & Gill, 2003; Tester, Karkalas, & Qi, 2004; Thuwall, Boldizar, & Rigdahl, 2006). Generally, starch granule is composed of approximately 25% amylose and 75% amylopectin (Jay-Lin, 2003; Rosa et al., 2008). The ratio of amylose/amylopectin content, transition temperatures, and rate of gelatinization in starch granules vary depending upon the plant source (Singh, Singh, et al., 2003). High level of crystallinity will lead to higher transition temperatures, making granules more resistant to gelatinization (Barichello, Yada, Coffin, & Stanley, 1990), thus influencing their rate of biodegradation.

Native starches are semicrystalline and hydrophilic and are readily biodegradable granules (Slade & Levine, 1993) that possess poor mechanical properties, thus restricting their use in the development of starch-based biodegradable polymers. In native starch granules, amylopectin is branched and forms the crystalline part, whereas amylose is the linear polymer associated with amorphous phase. Studies have reported that blends containing starches with linear structures are expected to biodegrade more rapidly as compared with branched starches (Chandra & Rustgi, 1998; Rosa, Lopes, & Calil, 2007). To improve functional properties, native starch granules can be modified into thermoplastic starch (TPS) using physical (milling, ultrasonication, and addition of plasticizer and other additives), chemical (esterification, etherification, grafting, cross-linking, and oxidation), or enzymatic treatments (Doane, 1992; Nabar, Narayan, & Schindler, 2006; Su & Cheng, 2011; Swanson et al., 1993). The micronization of starch granules causes size reduction, improvement in amylose content (amorphous region), compatibility with other polymers and additives, and increased susceptibility to microbial/enzymatic attack (Su & Cheng, 2011; Yang, Yu, & Ma, 2006).

A recent study has reported that structural modifications could have negative effect on the biodegradability of starch-polymer blends. Teramoto, Motoyama, Yosomiya, and Shibata (2003) observed that the biodegradability of etherified high-amylose corn starch (HACS) was significantly reduced as compared with untreated HACS samples during an activated sludge treatment. It is therefore advised to be mindful of starch modification to an extent such that it improves desired properties but does not negatively influence the overall biodegradability of starch-polymer blends.

3.2 Effects of Processing on Biodegradation

3.2.1 Granule Size and Shape

Morphological characteristics of starch granules are mainly dependent on genetic composition of the plant, but environmental factors could also influence the shape and size (Copeland, Blazek, Salman, & Tang, 2009; Haase & Plate, 1996; Teerawanichpan et al., 2008). Wheat, corn, and potato starch granules are spherical- and lenticular-shaped, angular-shaped, and cuboidal-shaped, respectively (Figure 13.2). Granule size varies with plant

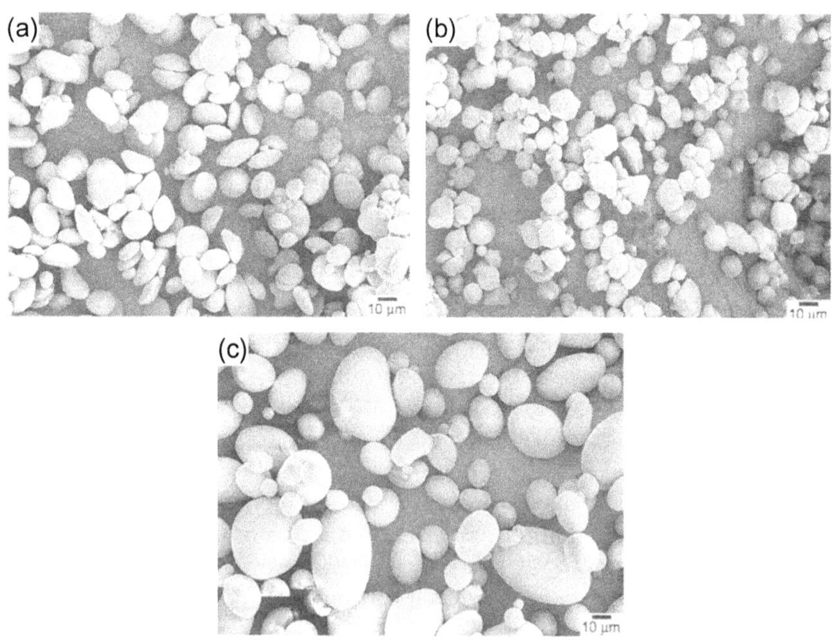

FIGURE 13.2 SEM micrographs of (a) native wheat, (b) corn, and (c) potato starches, scale bar = 10 μm, magnification 50×.

source, generally ranging from few micrometers up to approximately 110 μm (Lim, Jane, Rajagopalan, & Seib, 1992; Singh, Singh, et al., 2003; Tester et al., 2004).

Granular packing size influences the rheological properties of polymer composites that are dependent on particle-size distribution and aspect ratio (Willett, 2001; Zhou, Willett, Carriere, & Wu, 2000). The maximum packing fraction is the ratio of the density of a mass of particles to the material density:

$$\phi_{\mathrm{m}} = (m/V)/\rho$$

where ϕ_{m} is the maximum packing fraction, m and V are the mass and volume of the particles, respectively, and ρ is the material density (i.e., density of single particle).

Willett (2001) has suggested that higher maximum packing fraction values for potato and wheat as compared with corn were probably due to smooth surface and uniform shape of potato and wheat starches causing less intergranular frictions as compared with corn starch. It is generally understood that higher packing fraction values mean higher compaction, and this would affect the processability and subsequent biodegradability of starch-polymer blends (Zhou et al., 2000).

3.2.2 *Compatibilization*

Singh, Pandey, et al. (2003) studied the effects of various compatibilization techniques on the biodegradability of polycaprolactone (PCL)/starch composites. The study has reported that the use of PCL-grafted dextran copolymers as compatibilizers contributes to better dispersion of starch granules and better adhesion between starch and polymer matrix. This leads to slow rate of biodegradation during the initial period of composting, but as starch is slowly metabolized, short polyester chains diffuse into the surrounding and are metabolized by microorganisms, thus speeding up the process of biodegradation. The use of maleic anhydride (MA) as a compatibilizer has been reported to increase biodegradability of polylactic acid (PLA)/starch blends as compared with uncompatibilized PLA/starch blends at the same ratio (Jang, Shin, Lee, & Narayan, 2007). An increased biodegradability has been reported for PCL/starch composites prepared by using grafted pyromellitic anhydride as a compatibilizing agent (Avella et al., 2000).

3.2.3 *Solvent Cast*

Yavuz and Babac (2003) had also reported significant biodegradation of surface-modified PCL/starch blends prepared by solvent-casting techniques in a compost environment. Azahari, Othman, and Ismail (2011) prepared polyvinyl alcohol/corn starch blend films using solution-casting method and assessed their biodegradability by burial in soil and composting. The study found that biodegradability of PVA/CS blends increased with an increase in corn starch content.

3.2.4 *Extruded/Injection-Molded*

Native starch is converted into TPS by adding appropriate plasticizers at high temperature and under shear stress (Botana, Mollo, Eisenberg, & Torres, 2010; Liu, Han, & Dong, 2010; Swanson et al., 1993). Thermoplastic starch can then be extruded as films or be injection-molded into products. During the extrusion process, starch is subjected to high temperatures, high pressures, and high shear forces, which can cause degradation of starch granules, thus influencing functional properties of TPS resin (Chaudhary et al., 2008; O'Brien & Wang, 2009; Vansoest, Dewit, & Vliegenthart, 1996; Wiedmann & Strobel, 1991).

4 BIODEGRADATION OF STARCH-POLYMER BLENDS

4.1 Starch-Plasticizer Blends

The use of plasticizers in starch-based materials is of particular interest due to the positive effect of the processing of starch-based materials. The use of glycerol, glycol, sorbitol, formamide, and sugars as plasticizers for starch-based materials has been widely reported in literature (Liu et al., 2010; Martin & Avérous, 2001). Water and glycerol are mainly used as plasticizers in starch-based nano-biocomposites (Pandey & Singh, 2005; Tang, Zou, Xiong, & Tang, 2008). Thermoplastic starch containing glycerol (20, 30, and 40 wt%) was extruded and analyzed for its effect on material properties including enzymatic degradation (Moriana et al., 2010). It was observed that the rate of degradation of blends was proportional to an increase in the concentration of glycerol. Other additives such as sorbitol are also used (Ma & Yu, 2004; Wang, Yu, & Ma, 2007). Chivrac, Angellier-Coussy, Guillard, Pollet, and Avérous (2010) studied the effect of sorbitol and glycerol/sorbitol (Polysorb®) on plasticized starch properties and clay exfoliation process in starch nano-biocomposites. The study found that biodegradability of starch-based biocomposites was negatively influenced in the presence of sorbitol. Stein and Greene (1997) investigated the use of amino acids with differing degrees of hydrophobicity as plasticizers for starch-glycerol blends. Of the amino acids utilized, proline showed more promise due to the positive effect on mechanical properties and the glass transition temperature. Biodegradation experiments were conducted on the blends by monitoring the CO_2 evolution after activation with microorganisms (*Aspergillus niger*). The results of the biodegradation studies showed that blends prepared with amino acids exhibited greater degree of biodegradation over standard starch-glycerol blends, with proline showing again more promise due to similarities with urea and the incorporation of nitrogen groups. Shi et al. (2007) produced a novel citric acid (CA)-glycerol coplasticized TPS blend by melt blending. It was found that the new blend had some new characteristics compared with traditional glycerol-TPS in that there were partial esterification, low molecular weight, and stronger interfacial interaction between starch and citric acid. The decrease in the molecular weight could adjust the biodegradation rate of the TPS phase; however, the results of biodegradation were not reported.

4.2 Starch-Biopolymer Blends

The majority of natural polymers are hydrophilic materials since they are composed mainly of either hydroxyl or polar groups. Many synthetic biopolymers, in particular, the aliphatic polyesters, can be blended with natural materials, which can lead to a new class of biodegradable material (Averous, Fringant, & Moro, 2001; Averous, Moro, Dole, & Fringant, 2000; Tokiwa, Calabia, Ugwu, & Aiba, 2009). Studies have reported blending of starch with PLA (Ke, Sun, & Seib, 2003; Martin & Avérous, 2001; Park, Im, Kim, & Kim, 2000; Wang et al., 2007; Yu, Petinakis, Dean, Liu, & Yuan, 2011), poly-hydroxybutyrate (PHB) (Godbole, Gote, Latkar, & Chakrabarti, 2003; Innocentini-Mei, Bartoli, & Baltieri, 2003), PCL (Koenig & Huang, 1995), and many other biopolymer matrices.

Copinet et al. (2003) have reported that during the biodegradation of PLA/starch blends, starch phase is hydrolyzed and rapidly metabolized, thus enhancing the biodegradation of PLA. Shogren, Doane, Garlotta, Lawton, and Willett (2003) produced injection-molded bars composed of native corn starch, PLA, and poly(hydroxyester-ether) (PHEE), buried them in soil, and studied the effects of starch and PHEE on biodegradation rates. The blends of PLA/PHEE with native starch exhibited a rapid degree of biodegradation, compared with pure PLA and PLA with starch as a result of the softer PHEE segment, which could have facilitated higher penetration of the microbes to the inner core, leading to more rapid biodegradation. Studied biodegradation of PCL/starch films during landfill burial test and observed that smooth surface of film blends was covered with pinholes, suggesting surface deterioration due to enzymatic action of microorganisms (Figure 13.3).

Different proportions of starch were blended with poly(3-hydroxybutyrate)-*co*-3-hydroxyvalerate (PHBV) or polycaprolactone (PCL) by extrusion in a study conducted by Rosa, Rodrigues, Guedes, and Calil (2003). The biodegradability of the blends was evaluated after thermal aging. The PHBV blends showed a higher degree of biodegradation after burial in soil, compared with blends of PCL with starch. This study has reported that the biodegradation behavior of the blend with PHBV and

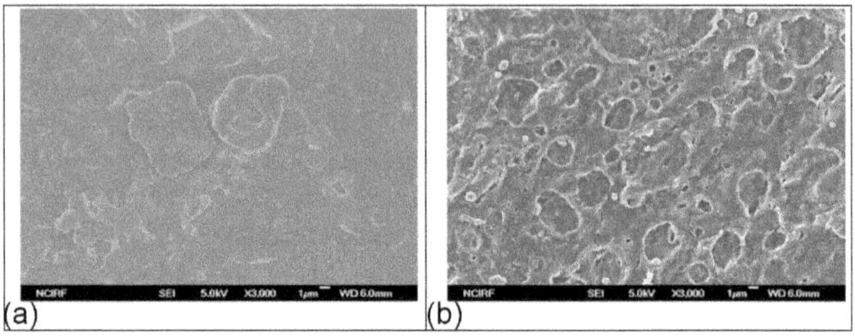

FIGURE 13.3 SEM micrographs of PCL/starch films (a) before landfill and (b) after 90 days in landfill.

25 w/w% starch was affected by thermal aging, and this was attributed to thermooxidation, which led to a breakup of the polymeric chains in the polymer backbone, producing low-molecular-weight fragments. As a result, the reduced molecular weight in the matrix would have caused significant effect on the degree of biodegradation observed in this study. Shin, Lee, Shin, Balakrishnan, and Narayan (2004) blended PCL with TPS, prepared from native corn starch and glycerol, using a twin-screw extruder. Biodegradation studies showed that the degradability of the blends increased with higher percentage of TPS and blends with 50 w/w% of TPS exhibited the most rapid degree of biodegradation. The increased degree of biodegradation was attributed to the tendency for the TPS phase to degrade, which provided a channel for further microbial attack. Imam et al. (1998) studied the biodegradability of starch/poly(3-hydroxybutyrate-*co*-3-hydroxyvalerate) blends prepared with different formulations (30% or 50% starch with or without polyethylene oxide) and observed an increase in biodegradation proportional to the amount of starch in the blends.

4.3 Starch-Polyolefin Blends

The development of degradable polyolefins has become a major topic of research in order to manage environmental problems. Blending of starch (in the form of granules or gelatinized or TPS) with low-density polyethylene renders these materials partially degradable, thus reducing the volume of nonbiodegradable plastics going to landfills. The properties of starch-polyolefin blends (tensile strength, thermal stability, and resistance to bio- and photodegradation) depend on starch content, degree of dispersion in the polymer matrix, sample morphology, interactions between components, degree of crystallinity, presence of structural defects, and preparation or processing conditions. Starch-polyolefin blends are incompatible at the molecular level and often lead to poor performance. Compatibility has been reported to be increased either by modifying the starch before blending (Thakore, Iyer, Desai, Lele, & Devi, 1999) or by adding compatibilizer (Chandra & Rustgi, 1998; Huang, Roan, Kuo, & Lu, 2005; Yoo et al., 2002). Distribution and concentration of TPS in blends influence the biodegradability of starch-polyolefin blends. Torres, Zamudio-Flores, Salgado-Delgado, and Bello-Perez (2008) have reported that uniform distribution of TPS in blends contributed to an increase in the level of biodegradation of LDPE/granular starch blends (Ishiaku, Pang, Lee, & Ishak, 2002; Yu, Gao, & Lin, 1996) as compared with oxidized starch blends. Another study has reported that PE/starch blends with more than 50 wt% TPS content lead to higher rate of biodegradation as compared with blends with relatively lower TPS content (<40 wt%) (Tena-Salcido, Rodriguez-Gonzalez, Mendez-Hernandez, & Contreras-Esquivel, 2008).

Biodegradation of starch-polyethylene films has been studied in natural environments such as soil, compost, and marine water and in the presence of purified microbial strains and enzymes (Ammala et al., 2011; Eubeler, Bernhard, & Knepper, 2010; Hakkarainen & Albertsson, 2004; Tokiwa et al., 2009). Studies have reported that the rate of biodegradation of starch-PE films was significantly enhanced as starch was rapidly metabolized by *Streptomyces* sp. (Lee, Pometto, Fratzke, & Bailey, 1991; Pometto, Lee, & Johnson, 1992), *Bacillus* sp. (Sudhakar, Doble, Murthy, & Venkatesan, 2008), and *Phanerochaete chrysosporium* (Orhan & Buyukgungor, 2000), resulting in increased biomass and changes in chemical structure. Abd El-Rehim et al. (2004) have reported that the UV exposure of LDPE/starch films prior to soil burial accelerates their rate of biodegradation (Figure 13.4).

There are a number of studies that have reported the increase in biodegradability of starch-polymer blends with an increase in starch content (Arvanitoyannis, Nakayama, & Aiba, 1998; Chandra & Rustgi, 1998; Lim et al., 1992; Zuchowska, Hlavata, Steller, Adamiak, & Meissner, 1999). Nikazar et al. (2005) analyzed starch-based LDPE samples with 5-40% starch content and observed significant increase in biodegradability of blends containing 20% or more starch content. Starch is easily susceptible to microbial attack; therefore, blends with higher starch content are more rapidly degraded, increasing porosity and surface-to-volume ratio, thus enhancing rate of biodegradation (Averous & Boquillon, 2004; Averous et al., 2001, 2000).

5 BIODEGRADATION OF STARCH COMPOSITES

5.1 Starch Microcomposites

Starch is one of the most abundant natural polymers available, which makes it a primary candidate for developing polymer composites produced entirely from renewable resources. Over the past few decades,

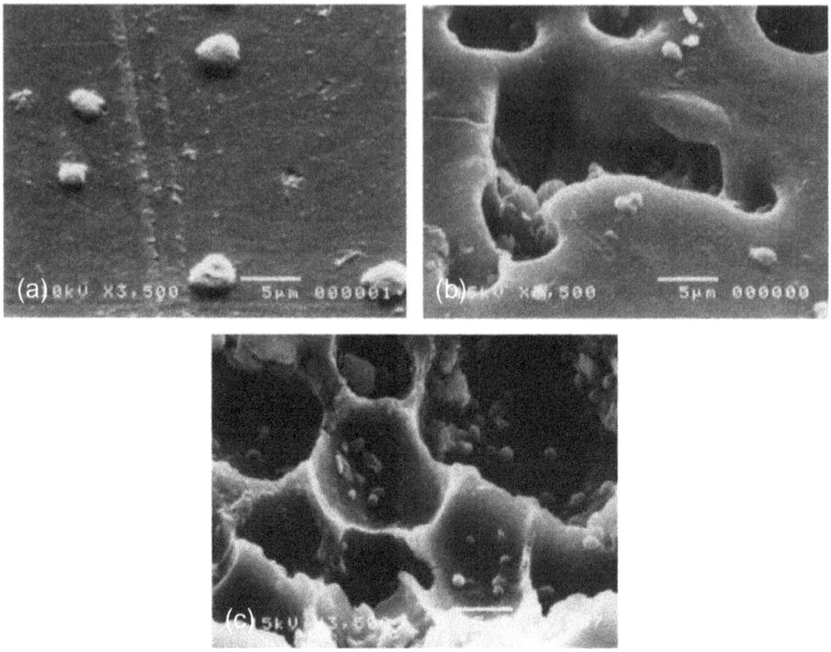

FIGURE 13.4 SEM micrographs of LDPE/starch: 75/25 (a) untreated; (b) after soil burial for 6 months; (c) after exposure to UV sunlight for 60 days followed by soil burial treatment for 6 months.

starch has been incorporated into conventional polymer matrices and, more recently, into biopolymer matrices. Starch has been incorporated as fillers into polymer matrices not only in its native form (Yu, Dean, & Li, 2006; Yu et al., 2011) but also as TPS (Bastioli, 1998), which has been blended and reinforced with natural fibers to improve mechanical properties. Iovino, Zullo, Rao, Cassar, and Gianfreda (2008) investigated the aerobic biodegradation of composites produced from PLA, TPS, and natural fiber (coir) with and without coupling agent, using ISO 14855. At the end of the incubation period, the TPS phase showed greater susceptibility to microbial attack, since the matrix showed a greater percentage of biodegradation compared with the pure matrix alone. Also, the composites with coupling agent exhibited a lower percentage of evolved CO_2 compared with the composite without coupling agent, which shows that the degree of biodegradation can be controlled or tailored according to the end use of the material.

Lu, Du, Ge, Xiao, and Meng (2006) conducted studies on the biodegradability and thermal stability of polypropylene carbonate (PPC)/starch composites. The degradability of the composites exhibited a three-stage process, which involved, first, the colonization of microorganisms; second, the degradation of the starch phase; and, finally, the degradation of the PPC component. Alvarez, Ruseckaite, and Vázquez (2006) studied the degradation of sisal fiber/Mater Bi-Y biocomposites buried in soil. Mater Bi-Y™ is a commercially available biopolymer resin based on plasticized starch and cellulose derivatives. Due to the biocompatible phases present, the composites generally showed lower sorption capacity in comparison to the water absorption intake of the matrix alone. The results of biodegradation studies indicated that the composites lost around 5% of their initial mass after the first month, but the presence of the fibers facilitated microbial growth, which induced preferential degradation of the starch phase. Wu (2008) studied the biodegradability of PLA and PLA/starch blends by encapsulating the cells of an indigenous PSB strain (*Bacillus* sp. PG01) and found that starch accelerated the biodegradation of PLA. Petinakis et al. (2010) studied the effect of hydrophilic fillers (starch and wood flour) on the biodegradation and thermal decomposition of PLA-based materials. It was found that degradation rates of both PLA/starch blends and PLA/wood flour composites were lower than that of pure cellulose but higher than that of pure PLA. The biodegradation rate was increased from about 60% to 80% when the starch content was increased from10% to 40% after 80 days.

5.2 Starch Nanocomposites

The properties of TPSs can be further enhanced by the development of nano-biocomposites (Chivrac, Pollet, Dole, & Averous, 2010; Muller, Laurindo, & Yamashita, 2012; Sinha Ray & Okamoto, 2003). The biodegradation rates of nanocomposites based on starch have also been monitored and measured by a number of different methods, and biodegradation has been shown to depend on various factors such as biodegradation environment, nanoparticle type and loading, amount of starch, and plasticizer and compatibilizer contents (Chivrac, Angellier-Coussy, et al., 2010; Pandey & Singh, 2005).

Tang et al. (2008) studied the biodegradation of several starch/polyvinyl alcohol (PVOH)/nano-silicon dioxide (nano-SiO_2) blend films in soil via gravimetric analysis. The study has reported that addition of 5 wt% nano-SiO_2 had no significant influence on the overall degree of biodegradation, but decreased the initial rate of biodegradation of the films in soil. Nayak (2010) had investigated the biodegradation of TPS/polybutylene adipate-*co*-terephthalate (TPS/PBAT) blends at starch content with two different organically modified nanoclays: Cloisite C20A and Cloisite

C30B at various clay loadings. PBAT was further modified by grafting maleic anhydride (MA) to improved interfacial adhesion with the starch and nanoclay. The biodegradability of the following film samples was evaluated in controlled compost at 58 °C: PBAT, PBAT/TPS, PBAT/TPS/C30B, and PBAT/MA-g-PBAT/TPS, with cellulose being used as the positive reference and polyethylene as the negative control in the experiment. After incubation for 130 days, only a slight increase in the rate of biodegradation was observed for the PBAT/TPS/C30B formulation. It was observed that the rate of degradation of PBAT matrix increased with the incorporation of TPS/nanoclay.

The role of α-amylase in PVA/starch nanocomposite degradation was studied by Spiridon, Popescu, Bodârlau, and Vasile (2008). Preweighed films of blends or nanocomposites were immersed in a 20 mL solution containing 6 IU α-amylase/g material. The mass loss (for both wet and dry samples) was monitored at set time intervals. The amount of nondegraded residual blends containing nanoparticles suggested the possibility to optimize the material degradability by incorporation of nanoparticles. Spiridon et al. (2008) discussed that biodegradation is mainly controlled by dissolution/erosion and enzymatic degradation of starch also occurs. The susceptibility of nanocomposite films to enzymatic degradation was shown to vary in the following order: PVA/starch/nanocore > PVA/starch/bentonite > PVA/starch/Peruvian clay, as also supported by the water adsorption capacity, which is one of the main conditions for biodegradation. Figure 13.5 illustrates the key mass loss results for the Peruvian clay series of materials.

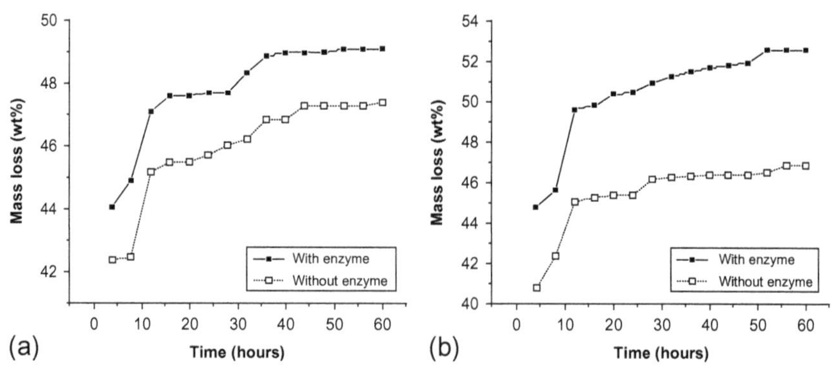

FIGURE 13.5 Mass loss vs degradation time for PVA/starch/1 wt% Peruvian clay (a) and PVA/starch/5 wt% Peruvian clay nanocomposites (b). *Reproduced from Spiridon et al. (2008).*

6 CONCLUSIONS

Starch is a natural polymer that is a biodegradable, renewable, and cost-efficient alternative to petroleum-based polymers. Its functionality has been significantly improved over the years through the modification of its structure, formulation, and processing without drastically affecting its biodegradability in the natural environment. The advancements in the field of nanotechnology promise further enhancement of the properties of TPSs and their blends, which can lead to a new class of biodegradable, biocompatible, and renewable materials.

References

Abd El-Rehim, H. A., Hegazy, E. S. A., Ali, A. M., & Rabie, A. M. (2004). Synergistic effect of combining UV-sunlight-soil burial treatment on the biodegradation rate of LDPE/starch blends. *Journal of Photochemistry and Photobiology A Chemistry, 163*, 547–556.

Alvarez, V. A., Ruseckaite, R. A., & Vázquez, A. (2006). Degradation of sisal fibre/Mater Bi-Y biocomposites buried in soil. *Polymer Degradation and Stability, 91*, 3156–3162.

Ammala, A., Bateman, S., Dean, K., Petinakis, E., Sangwan, P., Wong, S., et al. (2011). An overview of degradable and biodegradable polyolefins. *Progress in Polymer Science, 36*, 1015–1049.

Arvanitoyannis, I., Nakayama, A., & Aiba, S. (1998). Edible films made from hydroxypropyl starch and gelatin and plasticized by polyols and water. *Carbohydrate Polymers, 36*, 105–119.

Avella, M., Errico, M. E., Laurienzo, P., Martuscelli, E., Raimo, M., & Rimedio, R. (2000). Preparation and characterisation of compatibilised polycaprolactone/starch composites. *Polymer, 41*, 3875–3881.

Averous, L., & Boquillon, N. (2004). Biocomposites based on plasticized starch: Thermal and mechanical behaviours. *Carbohydrate Polymers, 56*, 111–122.

Averous, L., Fringant, C., & Moro, L. (2001). Plasticized starch-cellulose interactions in polysaccharide composites. *Polymer, 42*, 6565–6572.

Averous, L., Moro, L., Dole, P., & Fringant, C. (2000). Properties of thermoplastic blends: Starch-polycaprolactone. *Polymer, 41*, 4157–4167.

Azahari, N. A., Othman, N., & Ismail, H. (2011). Biodegradation studies of polyvinyl alcohol (PVOH)/corn starch blend film in solid and solution media. *Journal of Physical Science, 22*, 15–31.

Barichello, V., Yada, R. Y., Coffin, R. H., & Stanley, D. W. (1990). Low-temperature sweetening in susceptible and resistant potatoes—Starch structure and composition. *Journal of Food Science, 55*, 1054–1059.

Bastioli, C. (1998). Properties and applications of Mater-Bi starch-based materials. *Polymer Degradation and Stability, 59*, 263–272.

Botana, A., Mollo, M., Eisenberg, P., & Torres, S. R. M. (2010). Effect of modified montmorillonite on biodegradable PHB nanocomposites. *Applied Clay Science, 47*, 263–270.

Breslin, V. T. (1998). Degradation of starch calcium carbonate disposable packaging in a solid waste composting facility. *Journal of Environmental Polymer Degradation, 6*, 9–21.

Canigueral, N., Vilaseca, F., Mendez, J. A., Lopez, J. P., Barbera, L., Puig, J., et al. (2009). Behavior of biocomposite materials from flax strands and starch-based biopolymer. *Chemical Engineering Science, 64*, 2651–2658.

Cerruti, P., Santagata, G., D'Ayala, G. G., Ambrogi, V., Carfagna, C., Malinconico, M., et al. (2011). Effect of a natural polyphenolic extract on the properties of a biodegradable starch-based polymer. *Polymer Degradation and Stability, 96*, 839–846.

Chandra, R., & Rustgi, R. (1998). Biodegradable polymers. *Progress in Polymer Science, 23*, 1273–1335.

Chaudhary, A. L., Miler, M., Torley, P. J., Sopade, P. A., & Halley, P. J. (2008). Amylose content and chemical modification effects on the extrusion of thermoplastic starch from maize. *Carbohydrate Polymers, 74*, 907–913.

Chivrac, F., Angellier-Coussy, H., Guillard, V., Pollet, E., & Avérous, L. (2010). How does water diffuse in starch/montmorillonite nano-biocomposite materials? *Carbohydrate Polymers, 82*, 128–135.

Chivrac, F., Pollet, E., Dole, P., & Averous, L. (2010). Starch-based nano-biocomposites: Plasticizer impact on the montmorillonite exfoliation process. *Carbohydrate Polymers, 79*, 941–947.

Cho, H. S., Moon, H. S., Kim, M., Nam, K., & Kim, J. Y. (2011). Biodegradability and biodegradation rate of poly(caprolactone)-starch blend and poly(butylene succinate) biodegradable polymer under aerobic and anaerobic environment. *Waste Management, 31*, 475–480.

Copeland, L., Blazek, J., Salman, H., & Tang, M. C. M. (2009). Form and functionality of starch. *Food Hydrocolloids, 23*, 1527–1534.

Copinet, A., Bertrand, C., Longieras, A., Coma, V., & Couturier, Y. (2003). Photodegradation and biodegradation study of a starch and poly(lactic acid) coextruded material. *Journal of Polymers and the Environment, 11*, 169–179.

Doane, W. M. (1992). USDA research on starch-based biodegradable plastics. *Starch-Starke, 44*, 293–295.

Dubey, K. A., Chaudhari, C. V., Raje, N., Panickar, L., Bhardwaj, Y. K., & Sabharwal, S. (2012). Radiation-assisted morphology modification of LDPE/TPS blends: A study on starch degradation-processing-morphology correlation. *Journal of Applied Polymer Science, 124*, 3501–3510.

Eubeler, J. P., Bernhard, M., & Knepper, T. P. (2010). Environmental biodegradation of synthetic polymers II. Biodegradation of different polymer groups. *Trac-Trends in Analytical Chemistry, 29*, 84–100.

Flieger, M., Kantorova, M., Prell, A., Rezanka, T., & Votruba, J. (2003). Biodegradable plastics from renewable sources. *Folia Microbiologica, 48*, 27–44.

French, D. (1973). Chemical and physical properties of starch. *Journal of Animal Science, 37*, 1048–1061.

Gattin, R., Copinet, A., Bertrand, C., & Couturier, Y. (2001). Comparative biodegradation study of starch- and polylactic acid-based materials. *Journal of Polymers and the Environment, 9*, 11–17.

Gattin, R., Copinet, A., Bertrand, C. L., & Couturier, Y. (2002). Biodegradation study of a starch and poly(lactic acid) co-extruded material in liquid, composting and inert mineral media. *International Biodeterioration and Biodegradation, 50*, 25–31.

Geweely, N. S., & Ouf, S. A. (2011). Enhancement of fungal degradation of starch based plastic polymer by laser-induced plasma. *African Journal of Microbiology Research, 5*, 3273–3281.

Glenn, G. M., Imam, S. H., & Orts, W. J. (2011). Starch-based foam composite materials: Processing and bioproducts. *MRS Bulletin, 36*, 696–702.

Godbole, S., Gote, S., Latkar, M., & Chakrabarti, T. (2003). Preparation and characterization of biodegradable poly-3-hydroxybutyrate–starch blend films. *Bioresource Technology, 86*, 33–37.

Haase, N. U., & Plate, J. (1996). Properties of potato starch in relation to varieties and environmental factors. *Starch-Starke, 48*, 167–171.

Hakkarainen, M., & Albertsson, A. C. (2004). Environmental degradation of polyethylene. *Long-Term Properties of Polyolefins, 169*, 177–199.

Huang, C. Y., Roan, M. L., Kuo, M. C., & Lu, W. L. (2005). Effect of compatibiliser on the biodegradation and mechanical properties of high-content starch/low-density polyethylene blends. *Polymer Degradation and Stability, 90*, 95–105.

Imam, S. H., Chen, L., Gordon, S. H., Shogren, R. L., Weisleder, D., & Greene, R. V. (1998). Biodegradation of injection molded starch-poly (3-hydroxybutyrate-co-3-hydroxyvalerate) blends in a natural compost environment. *Journal of Environmental Polymer Degradation, 6*, 91–98.

Imam, S. H., Gordon, S. H., Shogren, R. L., Tosteson, T. R., Govind, N. S., & Greene, R. V. (1999). Degradation of starch-poly(beta-hydroxybutyrate-co-beta-hydroxyvalerate) bioplastic in tropical coastal waters. *Applied and Environmental Microbiology, 65*, 431–437.

Innocentini-Mei, L. H., Bartoli, J. R., & Baltieri, R. C. (2003). Mechanical and thermal properties of poly(3-hydroxybutyrate) blends with starch and starch derivatives. *Macromolecular Symposia, 197*, 77–88.

Iovino, R., Zullo, R., Rao, M. A., Cassar, L., & Gianfreda, L. (2008). Biodegradation of poly (lactic acid)/starch/coir biocomposites under controlled composting conditions. *Polymer Degradation and Stability, 93*, 147–157.

Ishiaku, U. S., Pang, K. W., Lee, W. S., & Ishak, Z. A. M. (2002). Mechanical properties and enzymic degradation of thermoplastic and granular sago starch filled poly(epsilon-caprolactone). *European Polymer Journal, 38*, 393–401.

Jang, W. Y., Shin, B. Y., Lee, T. X., & Narayan, R. (2007). Thermal properties and morphology of biodegradable PLA/starch compatibilized blends. *Journal of Industrial and Engineering Chemistry, 13*, 457–464.

Jayasekara, R., Harding, I., Bowater, I., Christie, G. B. Y., & Lonergan, G. T. (2003). Biodegradation by composting of surface modified starch and PVA blended films. *Journal of Polymers and the Environment, 11*, 49–56.

Jay-Lin, J. (2003). Starch. In Piotr Tomasik (Ed.) *Chemical and Functional Properties of Food Saccharides*. Florida, USA: Piotr Tomasik, CRC Press.

Ke, T., Sun, S. X., & Seib, P. (2003). Blending of poly(lactic acid) and starches containing varying amylose content. *Journal of Applied Polymer Science, 89*, 3639–3646.

Kijchavengkul, T., Auras, R., Rubino, M., Ngouajio, M., & Thomas Fernandez, R. (2006). Development of an automatic laboratory-scale respirometric system to measure polymer biodegradability. *Polymer Testing, 25*, 1006–1016.

Koenig, M. F., & Huang, S. J. (1995). Biodegradable blends and composites of polycaprolactone and starch derivatives. *Polymer, 36*, 1877–1882.

Krupp, L. R., & Jewell, W. J. (1992). Biodegradability of modified plastic films in controlled biological environments. *Environmental Science and Technology, 26*, 193–198.

Lee, B. T., Pometto, A. L., Fratzke, A., & Bailey, T. B. (1991). Biodegradation of degradable plastic polyethylene by phanerochaete and Streptomyces species. *Applied and Environmental Microbiology, 57*, 678–685.

Lim, S. T., Jane, J. L., Rajagopalan, S., & Seib, P. A. (1992). Effect of starch granule size on physical-properties of starch-filled polyethylene film. *Biotechnology Progress, 8*, 51–57.

Liu, H., Han, C. Y., & Dong, L. S. (2010). Study of the biodegradable poly(epsilon-caprolactone)/clay nanocomposite foams. *Journal of Applied Polymer Science, 115*, 3120–3129.

Lu, X. L., Du, F. G., Ge, X. C., Xiao, M., & Meng, Y. Z. (2006). Biodegradability and thermal stability of poly(propylene carbonate)/starch composites. *Journal of Biomedical Materials Research Part A, 77A*, 653–658.

Lu, D. R., Xiao, C. M., & Xu, S. J. (2009). Starch-based completely biodegradable polymer materials. *Express Polymer Letters, 3*, 366–375.

Ma, X., & Yu, J. (2004). The effects of plasticizers containing amide groups on the properties of thermoplastic starch. *Starch-Starke, 56*, 545–551.

Martin, O., & Avérous, L. (2001). Poly(lactic acid): Plasticization and properties of biodegradable multiphase systems. *Polymer, 42*, 6209–6219.

Massardier-Nageotte, V., Pestre, C., Cruard-Pradet, T., & Bayard, R. (2006). Aerobic and anaerobic biodegradability of polymer films and physico-chemical characterization. *Polymer Degradation and Stability, 91*, 620–627.

Moriana, R., Karlsson, S., & Ribes-Greus, A. (2010). Assessing the influence of cotton fibers on the degradation in soil of a thermoplastic starch-based biopolymer. *Polymer Composites, 31*, 2102–2111.

Muller, C. M. O., Laurindo, J. B., & Yamashita, F. (2012). Composites of thermoplastic starch and nanoclays produced by extrusion and thermopressing. *Carbohydrate Polymers, 89*, 504–510.

Muthukumar, T., Aravinthan, A., & Mukesh, D. (2010). Effect of environment on the degradation of starch and pro-oxidant blended polyolefins. *Polymer Degradation and Stability, 95*, 1988–1993.

Nabar, Y., Narayan, R., & Schindler, M. (2006). Twin-screw extrusion production and characterization of starch foam products for use in cushioning and insulation applications. *Polymer Engineering and Science, 46*, 438–451.

Nayak, S. K. (2010). Biodegradable PBAT/starch nanocomposites. *Polymer-Plastics Technology and Engineering, 49*, 1406–1418.

Nikazar, M., Safari, B., Bonakdarpour, B., & Milani, Z. (2005). Improving the biodegradability and mechanical strength of corn starch-LDPE blends through formulation modification. *Iranian Polymer Journal, 14*, 1050–1057.

O'Brien, S., & Wang, Y. J. (2009). Effects of shear and pH on starch phosphates prepared by reactive extrusion as a sustained release agent. *Carbohydrate Polymers, 77*, 464–471.

Orhan, Y., & Buyukgungor, H. (2000). Enhancement of biodegradability of disposable polyethylene in controlled biological soil. *International Biodeterioration and Biodegradation, 45*, 49–55.

Pandey, J. K., & Singh, R. P. (2005). Green nanocomposites from renewable resources: Effect of plasticizer on the structure and material properties of clay-filled starch. *Starch-Starke, 57*, 8–15.

Park, J. W., Im, S. S., Kim, S. H., & Kim, Y. H. (2000). Biodegradable polymer blends of poly (L-lactic acid) and gelatinized starch. *Polymer Engineering and Science, 40*, 2539–2550.

Petinakis, E., Liu, X., Yu, L., Way, C., Sangwan, P., Dean, K., et al. (2010). Biodegradation and thermal decomposition of poly(lactic acid)-based materials reinforced by hydrophilic fillers. *Polymer Degradation and Stability, 95*, 1704–1707.

Pometto, A. L., Lee, B., & Johnson, K. E. (1992). Production of an extracellular polyethylene-degrading enzyme(s) by Streptomyces-species. *Applied and Environmental Microbiology, 58*, 731–733.

Pramila, R., & Ramesh, K. V. (2011). Biodegradation of low density polyethylene (LDPE) by fungi isolated from marine water- a SEM analysis. *African Journal of Microbiology Research, 5*, 5013–5018.

Rosa, D. S., Carvalho, C. L., Gaboardi, F., Rezende, M. L., Tavares, M. I. B., Petro, M. S. M., et al. (2008). Evaluation of enzymatic degradation based on the quantification of glucose in thermoplastic starch and its characterization by mechanical and morphological properties and NMR measurements. *Polymer Testing, 27*, 827–834.

Rosa, D. S., Lopes, D. R., & Calil, M. R. (2007). The influence of the structure of starch on the mechanical, morphological and thermal properties of poly (epsilon-caprolactone) in starch blends. *Journal of Materials Science, 42*, 2323–2328.

Rosa, D. D., Rodrigues, T. C., Guedes, C. D. F., & Calil, M. R. (2003). Effect of thermal aging on the biodegradation of PCL, PHB-V, and their blends with starch in soil compost. *Journal of Applied Polymer Science, 89*, 3539–3546.

Russo, M. A. L., Truss, R., & Halley, P. J. (2009). The enzymatic hydrolysis of starch-based PVOH and polyol plasticised blends. *Carbohydrate Polymers, 77*, 442–448.

Sarikaya, E., Higasa, T., Adachi, M., & Mikami, B. (2000). Comparison of degradation abilities of alpha- and beta-amylases on raw starch granules. *Process Biochemistry, 35,* 711–715.

Shah, A. A., Hasan, F., Hameed, A., & Ahmed, S. (2008). Biological degradation of plastics: A comprehensive review. *Biotechnology Advances, 26,* 246–265.

Shi, R., Zhang, Z., Liu, Q., Han, Y., Zhang, L., Chen, D., et al. (2007). Characterization of citric acid/glycerol co-plasticized thermoplastic starch prepared by melt blending. *Carbohydrate Polymers, 69,* 748–755.

Shin, B. -Y., Lee, S., II, Shin, Y. -S., Balakrishnan, S., & Narayan, R. (2004). Rheological, mechanical and biodegradation studies on blends of thermoplastic starch and polycaprolactone. *Polymer Engineering and Science, 44,* 1429–1438.

Shogren, R. L., Doane, W. M., Garlotta, D., Lawton, J. W., & Willett, J. L. (2003). Biodegradation of starch/polylactic acid/poly(hydroxyester-ether) composite bars in soil. *Polymer Degradation and Stability, 79,* 405–411.

Singh, R. P., Pandey, J. K., Rutot, D., Degee, P., & Dubois, P. (2003). Biodegradation of poly (epsilon-caprolactone)/starch blends and composites in composting and culture environments: The effect of compatibilization on the inherent biodegradability of the host polymer. *Carbohydrate Research, 338,* 1759–1769.

Singh, N., Singh, J., Kaur, L., Sodhi, N. S., & Gill, B. S. (2003). Morphological, thermal and rheological properties of starches from different botanical sources. *Food Chemistry, 81,* 219–231.

Sinha Ray, S., & Okamoto, M. (2003). Biodegradable polylactide/layered silicate nanocomposites: Open a new dimension for plastics and composites. *Macromolecular Rapid Communications, 24,* 815–840.

Slade, L., & Levine, H. (1993). Water relationships in starch transitions. *Carbohydrate Polymers, 21,* 105–131.

Spiridon, I., Popescu, M. C., Bodârlau, R., & Vasile, C. (2008). Enzymatic degradation of some nanocomposites of poly(vinyl alcohol) with starch. *Polymer Degradation and Stability, 93,* 1884–1890.

Stein, T. M., & Greene, R. V. (1997). Amino acids as plasticizers for starch-based plastics. *Starch-Starke, 49,* 245–249.

Su, J. F., & Cheng, J. J. (2011). Modified methods in starch-based biodegradable films. *Advanced Materials Research, 183–185,* 1635–1641.

Sudhakar, M., Doble, M., Murthy, P. S., & Venkatesan, R. (2008). Marine microbe-mediated biodegradation of low- and high-density polyethylenes. *International Biodeterioration and Biodegradation, 61,* 203–213.

Swanson, C. L., Shogren, R. L., Fanta, G. F., & Imam, S. H. (1993). Starch-plastic materials—Preparation, physical properties, and biodegradability (a review of recent USDA research). *Journal of Polymers and the Environment, 1,* 155–166.

Tang, S., Zou, P., Xiong, H., & Tang, H. (2008). Effect of nano-SiO2 on the performance of starch/polyvinyl alcohol blend films. *Carbohydrate Polymers, 72,* 521–526.

Teerawanichpan, P., Lertpanyasampatha, M., Netrphan, S., Varavinit, S., Boonseng, O., & Narangajavana, J. (2008). Influence of cassava storage root development and environmental conditions on starch granule size distribution. *Starch-Starke, 60,* 696–705.

Tena-Salcido, C. S., Rodriguez-Gonzalez, F. J., Mendez-Hernandez, M. L., & Contreras-Esquivel, J. C. (2008). Effect of morphology on the biodegradation of thermoplastic starch in LDPE/TPS blends. *Polymer Bulletin, 60,* 677–688.

Teramoto, N., Motoyama, T., Yosomiya, R., & Shibata, M. (2003). Synthesis, thermal properties, and biodegradability of propyl-etherified starch. *European Polymer Journal, 39,* 255–261.

Tester, R. F., Karkalas, J., & Qi, X. (2004). Starch—Composition, fine structure and architecture. *Journal of Cereal Science, 39,* 151–165.

Thakore, I. M., Iyer, S., Desai, A., Lele, A., & Devi, S. (1999). Morphology, thermomechanical properties, and biodegradability of low density polyethylene/starch blends. *Journal of Applied Polymer Science, 74*, 2791–2802.

Thuwall, M., Boldizar, A., & Rigdahl, M. (2006). Extrusion processing of high amylose potato starch materials. *Carbohydrate Polymers, 65*, 441–446.

Tokiwa, Y., Calabia, B. P., Ugwu, C. U., & Aiba, S. (2009). Biodegradability of plastics. *International Journal of Molecular Sciences, 10*, 3722–3742.

Torres, A. V., Zamudio-Flores, P. B., Salgado-Delgado, R., & Bello-Perez, L. A. (2008). Biodegradation of low-density polyethylene-banana starch films. *Journal of Applied Polymer Science, 110*, 3464–3472.

Vansoest, J. J. G., Dewit, D., & Vliegenthart, J. F. G. (1996). Mechanical properties of thermoplastic waxy maize starch. *Journal of Applied Polymer Science, 61*, 1927–1937.

Vikman, M., Itävaara, M., & Poutanen, K. (1995). Measurement of the biodegradation of starch-based materials by enzymatic methods and composting. *Journal of Polymers and the Environment, 3*, 23–29.

Wang, N., Yu, J., & Ma, X. (2007). Preparation and characterization of thermoplastic starch/PLA blends by one-step reactive extrusion. *Polymer International, 56*, 1440–1447.

Wiedmann, W., & Strobel, E. (1991). Compounding of thermoplastic starch with twin-screw extruders. *Starch-Starke, 43*, 138–145.

Willett, J. L. (2001). Packing characteristics of starch granules. *Cereal Chemistry, 78*, 64–68.

Wu, C. -S. (2008). Characterizing biodegradation of PLA and PLA-g-AA/starch films using a phosphate-solubilizing Bacillus species. *Macromolecular Bioscience, 8*, 560–567.

Yang, J. H., Yu, J. G., & Ma, X. F. (2006). Preparation and properties of ethylenebisformamide plasticized potato starch (EPTPS). *Carbohydrate Polymers, 63*, 218–223.

Yavuz, H., & Babac, C. (2003). Preparation and biodegradation of starch/polycaprolactone films. *Journal of Polymers and the Environment, 11*, 107–113.

Yoo, S. I., Lee, T. Y., Yoon, J. S., Lee, I. M., Kim, M. N., & Lee, H. S. (2002). Interfacial adhesion reaction of polyethylene and starch blends using maleated polyethylene reactive compatibilizer. *Journal of Applied Polymer Science, 83*, 767–776.

Yu, L., Dean, K., & Li, L. (2006). Polymer blends and composites from renewable resources. *Progress in Polymer Science, 31*, 576–602.

Yu, J. G., Gao, J. P., & Lin, T. (1996). Biodegradable thermoplastic starch. *Journal of Applied Polymer Science, 62*, 1491–1494.

Yu, L., Petinakis, E., Dean, K., Liu, H., & Yuan, Q. (2011). Enhancing compatibilizer function by controlled distribution in hydrophobic polylactic acid/hydrophilic starch blends. *Journal of Applied Polymer Science, 119*, 2189–2195.

Zhou, G. W., Willett, J. L., Carriere, C. J., & Wu, Y. V. (2000). Effect of starch granule size on viscosity of starch-filled poly(hydroxy ester ether) composites. *Journal of Polymers and the Environment, 8*, 145–150.

Zuchowska, D., Hlavata, D., Steller, R., Adamiak, W., & Meissner, W. (1999). Physical structure of polyolefin-starch blends after ageing. *Polymer Degradation and Stability, 64*, 339–346.

PART IX

APPLICATIONS

CHAPTER

14

Starch Applications: State of Market and New Trends

Bronwyn G. Laycock, Peter J. Halley

AIBN and the School of Chemical Engineering, The University of Queensland, St Lucia, Queensland, Australia

1 INTRODUCTION

Starch is the major polysaccharide reserve material in photosynthetic plants and is stored in many organs such as seeds and swollen stems as discrete, water-insoluble granules (Halley et al., 2007). Essentially composed of two main polysaccharides, amylose and amylopectin, these granules also contain some minor components such as lipids and proteins. Ranging in size from 1 to 100 μm, they have a size, shape, and composition that depends on their botanical source (Halley et al., 2007). The processing and properties of these semicrystalline materials are therefore closely linked to starch genetics and various structure levels from granule structure to macromolecular structure and crystalline macrostructures.

Starch and its derivatives are produced commercially from a range of raw materials such as corn, wheat, pea, potato, and cassava roots (tapioca) (Gozzo & Glittenberg, 2009). Starch has a wide range of applications beyond the food industry: in the paper and board sector (for wet-end addition, size press, and surface coating and in the production of recycled paper), the pharmaceuticals sector, the industrial binder sector (including adhesives and gypsum boards), and the textile industry, where it is used to "size" and stiffen textiles and help increase mechanical strength and resistance to friction wear as well as increase moisture resistance (Association des Amidonniers et Féculiers, 2009).

Around 73 million tons of starch is produced worldwide (expressed as primary starch with 12% moisture) (Giract Starch Industry Overview, 2011). It is expected that, over time, demand for starch by nonfood industries will tend to grow faster than demand for starch by food industries.

Starch is therefore a very widely available and relatively low cost commodity that is completely biodegradable. Since the carbon in plants is derived from atmospheric CO_2, biodegradation or incineration of starch results in no net gain in CO_2. This combination makes it attractive to manufacture starch-based plastics as alternatives to the common petrochemically derived polymers such as polypropylene, polyethylene, and polystyrene (PS). For that reason, and others that will be covered later, the starch-based plastics market is currently growing strongly; this review provides an overview of the current state of this market.

2 BIOPLASTICS: BACKGROUND

2.1 Definition of a Bioplastic

Bioplastics fall into either or both of two broad categories (BioIntelligence Service, 2011):

- Bio-based plastics that are derived from renewable resources
- Biodegradable (compostable) plastics that meet standards for biodegradability and compostability

Bio-based plastics can be either biodegradable or nonbiodegradable. Similarly, biodegradable polymers can be petroleum-based. Biodegradable polymers produced in nature are renewable. Some synthetic polymers are also renewable because they are made from renewable feedstock. For example, polylactic acid is derived from agricultural feedstock. Current starch plastics fall into both categories, with some commercial blends and composites being based on nondegradable components; in addition, some products are biodegradable but in part not derived from renewable sources (such as polycaprolactone/starch blends).

2.2 Biodegradation: Definitions and Certification Environment

Biodegradable polymers are polymers that by a biological action break down to biomass, CO_2, and water in a given time period (defined by a specific biodegradation standard) and in a given environment (i.e., marine, compost, and anaerobic sludge) (Halley et al., 2007). Some industry sources have offered the term compostable in place of biodegradable. To be considered compostable, three criteria must be met: biodegradation (it has to undergo degradation by biological processes into carbon dioxide, water, and biomass at the same rate as cellulose), disintegration (the plastic must become indistinguishable in the compost), and nontoxicity (World Centric, 2004).

There are a range of standards used for certification of the biodegradability of plastics. The EN 13432 industrial standard "Proof of compostability of plastic products" is arguably the most international in scope, and compliance with this standard is required in order to claim that a product is compostable in the EU. In summary, it requires biodegradation of 90% of the materials in a lab within 180 days. EN 13432 also establishes that biodegradability shall be determined for each relevant organic component accounting for more than 1% of a product's dry weight; the total quantity of nonrelevant organic components should not exceed 5%. This means that biodegradable additives and fillers must be used to meet the standards. EN 14995 "Plastics. Evaluation of compostability. Test scheme and specifications" has the same testing scope as EN 13432 but is not as widely used. The ASTM D6400-04 "Standard Specification for Compostable Plastics" is the standard used in the regulatory framework for the United States and sets a less stringent threshold of 60% biodegradation within 180 days within commercial composting conditions, as does the ISO 17088:2008 "Specifications for compostable plastics." Similar definitions have been recognized in several countries around the world. The AS4736-2000 standard is the Australian standard, which is similar to the EN 13432 standard but has an additional requirement of a worm test. Other ASTM methods for assessment of biodegradation include D6868-03 "Standard Specification for Biodegradable Plastics Used as Coatings on Paper and Other Compostable Substrates" and the ASTM D7081-05 "Standard Specification for Non-Floating Biodegradable Plastics in the Marine Environment" and the ASTM D5511-02 "Standard Test Method for Determining Anaerobic Biodegradation of Plastic Materials Under High-Solids Anaerobic-Digestion Conditions." The ASTM D6866 method has been developed to certify the biologically derived content of bioplastics. As of 1 April 2012, an EU-wide European compostable standard was established based on EN 13432/14955 (European Bioplastics, 2012b).

Accredited institutes around the world may certify and label the biodegradability of plastics, the most significant being DIN CERTCO (Germany) and Vinçotte (Belgium). Other organizations include the European Bioplastics association (Germany), the BNPP (France), the Biodegradable Products Institute (BPI), the Australian Bioplastics Association, and the Japanese Bioplastics Association.

2.3 Sustainability and Life Cycle Assessment

There has been a recent shift of emphasis from biodegradability to sustainability as the driving force for replacement polymer materials. The Sustainable Biomaterials Collaborative based in Washington, DC, has, in its 2009 guidelines for sustainable bioplastics, defined sustainable biomaterials as those that (1) are sourced from sustainably grown and harvested croplands and forests, (2) are manufactured without hazardous inputs and impacts, (3) are healthy and safe for the environment during use, (4) are designed to be reutilized at the end of their intended use, and (5) provide living wages and do not exploit workers or communities throughout the product life cycle (Sustainable biomaterials collaborative, 2009).

Starch is one of the better studied biopolymers with respect to full life cycle assessment. Patel, Bastioli, Marini, and Würdinger (2005) found that the energy requirements for the production of starch polymer pellets were generally 25-75% below those for polyethylene. Greenhouse gas emissions were also found to be 20-80% lower. These ranges originate from the comparison of different starch/copolymer blends, different waste treatments, and different polyolefin materials used as reference. Starch pellets and starch films/bags also scored well in comparison with PE on almost all other indicators, with eutrophication being an exception, and with the effect on acidification being unclear. Many of the commercial producers have conducted detailed assessments of their materials, with a similar result. Mater-Bi® starch-based materials, for example, have been found to result in significant reduction of environmental impact, particularly with respect to carbon dioxide emissions and energy consumption, in comparison with traditional materials (Bastioli, 2005). Plantic conducted a full ISO 14040-certified life cycle assessment for their process that also found that the manufacture of their plastic trays resulted in reduced environmental impacts and used lower nonrenewable energy for its manufacturing in comparison with conventional polymers (Plantic Technologies, 2011). The results for starch polymer loose fills, on the other hand, were found to be quite variable, depending on the assumed bulk density of the fill and different approaches for quantification of the ozone depletion potential (inclusion versus exclusion of NO_x). In general, starch loose fills were found to score better than virgin EPS, but not necessarily better than the recycled EPS; the potential for these starch loose fills to result in methane generation in landfill was found to be the main issue (Patel et al., 2005).

More recent studies have found that the picture is less clear (Harding, Dennis, von Blottnitz, & Harrison, 2007; Mattila, Kujanp, Dahlbo, Soukka, & Myllymaa, 2011; Sustainability Victoria, 2007). For shopping bags in Finland, for example, the carbon footprint for recycled polyethylene bags (at −7 to 24 g CO_2 equiv./bag) was lower than for biodegradable starch-based bags (38-60 g CO_2 equiv./bag) (Mattila et al., 2011). The end-of-life destination is crucial, with landfill being a poor disposal choice for starch-based plastics (Mattila et al., 2011; Patel et al., 2005). Gironi and Piemonte (2011), in a reanalysis of the data from Patel, found that by using a "distance-to-target" method of analysis, the starch-based plastics were in most cases superior to PE in terms of overall environmental impact. By contrast, however, when they conducted a new analysis using Ecoindicator-99 methodology comparing Mater-Bi® bags (composted) with PE (recycled), it was found that the materials were roughly equivalent in terms of environmental impact, although recycling of PE was found to be a better option than compositing of starch. The advantage of the bioplastics was the use of renewable resources, but an impact on ecosystem quality was caused by the use of pesticides and fertilizers and by the consumption of land and water. In a recent review, Hottle, Bilec, and Landis (2013) also found (based on LCA studies in the ecoinvent v2.2 database) that the global warming potential of thermoplastic starch was not very different from petroleum-based plastics when end-of-life was taken into account. They also found that in some environmental impact categories, like eutrophication, ozone depletion, and noncarcinogenic human health, the impacts were higher (although the scale of these impacts may be so small such that they have little real impact).

In general, therefore, the key considerations are the amount of nonrenewable energy used in the starch plastic manufacture, the end-of-life disposal options adopted, and the longer term potential implications for land use (e.g., impacts on forests) and food crops.

3 MARKET DRIVERS FOR BIOPLASTICS

In the context of the previous discussion around biodegradation and sustainability, the market demand for bioplastics is strongly influenced by many factors, such as competitive price, possibility of widening of "fit-for-use" applications, legislative requirements, and optimization of commercial composting process. There are therefore a range of market drivers behind the surge of interest in bioplastics, including starch. These drivers have changed with time, with the early interest being due to the increasing amounts of waste and limits to landfill capacity. Today public concerns about the environment, climate change, and limited fossil fuel resources have become more important (BioIntelligence Service, 2011; European Bioplastics, 2012c; Mohan, 2010; Plackett, 2011; Platt, 2006; Sarnacke & Wildes, 2008; Shen, Haufe, & Patel, 2009; Shen, Worrell, & Patel, 2009).

Internal (industry-focused) drivers include as follows:

- Advanced technical properties, which increase product attractiveness.
- Potential cost reduction through economies of scale.

- The development of additional disposal options.
- The patent situation, the reliability of new technologies, and the pace at which technology can be developed.
- The availability of trained personnel with knowledge of the sector, collaboration with companies in the agro-industry chain, the availability of raw materials, and the usefulness of coproducts.
- Starch-based biopolymers are lower cost materials than some other biodegradable polymer types such as synthetic copolyesters and PHA. They are produced from relatively cheap agricultural feedstock and have simpler manufacturing processes compared with synthetic biopolymers.

External drivers include as follows:

- Growing consumer awareness and preference for sustainable packaging technologies, with reduction in CO_2 emissions and environmental footprint, as well as the conservation of the earth's fossil resources.
- Limited landfill capacity and concern about marine pollution, with the most prominent example being the so-called Great Pacific Garbage Patch in the central North Pacific Ocean.
- Brand owners (such as Wal-Mart) recognizing the benefits of promoting sustainable or "green" packaging.
- The price differential between biodegradable polymers and petrochemical-based plastics narrowing due to better production techniques, better material sourcing by suppliers, and higher production volumes.
- Future proofing against rising oil and plastic prices and limitation to dependence on fossil fuels.
- The price of biofeedstocks, investment risks, fiscal policy initiatives, and the availability of capital at competitive interest rates.
- Development of supporting framework conditions such as more favorable government regulations to reduce waste packaging and landfill in favor of recycling and composting. Political support is also slowly gaining ground with biodegradable packaging receiving special treatment in some countries such as Germany.
- The world biodegradable plastics industry has agreed a set of standards and certification procedures for biodegradable packaging materials, which will continue to encourage growth and possibly deter imitation.
- Composting infrastructures are being developed by local councils in major towns and cities around the world in response to the problem of packaging waste and overreliance on landfill in some countries.
- Some governments are increasing support for use of renewable plastics made from bio-based raw materials to mitigate the growing waste disposal issues especially in urban areas.

- Thermoformed starch sheets give better transparency compared with some other biodegradable polymer. The material offers good potential for home composting, which is a growing consumer trend. This will be advantageous for starch-based biopolymers over PLA, which only decomposes in a communal composting system.
- Improved functional performance, with new product and technology development by leading biodegradable polymer suppliers that have opened up new markets and potential applications.
- Developments in terms of new suppliers and production capability.

There are differences in the key drivers between different regions worldwide (BioIntelligence Service, 2011; European Bioplastics, 2012a, 2012c; Mohan, 2010; Plackett, 2011; Platt, 2006; Sarnacke & Wildes, 2008; Shen, Haufe, et al., 2009; Shen, Worrell, et al., 2009). In Western Europe, for example, the main drivers are legislation, depleting landfill capacity, pressure from retailers, growing consumer interest in sustainable plastic solutions, fossil oil and gas independence, and greenhouse gas emissions reduction (BioIntelligence Service, 2011). Governments have implemented legislation to reduce the amount of municipal waste packaging being sent to landfill. In Italy, for example, legislation banning nonbiodegradable plastic bags took effect at the beginning of 2011 (Plasteurope.com, 2011a). The Belgian government put a 300% tax on nondegradable bags in 2010, making degradable ones more economic (Schut, 2008). The European Commission Directive 94/62/EC on packaging and packaging waste set recovery and recycling targets to prevent or minimize the impact of packaging waste on the environment (Platt, 2006), with legislation introduced in 2005 to modernize the 1975 Waste Framework Directive. Additionally, municipal solid waste (including plastics) is being tested for power generation (Mooney, 2009). The Household Waste Recycling Act (2003) required all local authorities in England to provide curbside collections for all householders for a minimum of two materials by 2010 (Platt, 2006). A new EU strategy entitled "Innovating for Sustainable Growth: A Bioeconomy for Europe" is targeting to drive the transition from a fossil-based economy to a sustainable bioeconomy in Europe (European Commission, 2012). It focuses on measures to boost research and innovation to develop new bio-based technologies. There is also a recent regulation in Germany concerning bioplastic drink bottles as established in the 5th Amendment of the German Packaging Directive, which took effect beginning January 2009 (Doane, 2012).

In North America, the key drivers are the increased cost competitiveness of biodegradable polymers; growing support from authorities for addressing solid waste disposal needs at the local, state, and federal levels; growing public and industry awareness of environmental issues; concern over the depletion of petroleum-based raw materials; desire of large retailers and manufacturing companies to develop more sustainable raw material sources as well as to impact global warming; and improvements in the properties of biodegradable polymers (BioIntelligence Service, 2011; European Bioplastics, 2012c; Mohan, 2010; Plackett, 2011; Platt, 2006; Sarnacke & Wildes, 2008; Shen, Haufe, et al., 2009; Shen, Worrell, et al., 2009). In America, in contrast to Europe, the lack of a composting infrastructure limits end-of-use options (BioIntelligence Service, 2011). Plastics comprise about 12% of the 227 metric tons of municipal waste produced annually in the United States, with packaging accounting for 1/3, followed by construction and consumer products (Mooney, 2009). Recycling recovered about 30% of this waste stream, meaning that 18.2 metric tons per annum accumulates (Mooney, 2009). It is expected that when there is a large volume of compostable products (driven by their low-carbon footprint), then it will make an economic and environmental sense to compost and recycle more. In early legislation, the Marine Plastic Pollution Research and Control Act of 1987 (Public Law 100-200, 29 December 1987) required the US Navy to cease disposing of nonbiodegradable plastics at sea by the end of 1993 (Doane, 2012). A recent example of legislation is the Food Service Waste Reduction Ordinance (2007) that requires that San Francisco restaurants and food vendors no longer use PS foam as disposable food ware (Sfgov, 2007).

In Japan, bio-based renewability seems to be more important than biodegradability. The promotion of biodegradable polymers by the government and industry and increased cost competitiveness of biodegradable polymers are key factors (BioIntelligence Service, 2011). The Japanese government, for example, set a "Biomass Nippon Strategy" back in 2002 with a goal that 20% of all plastics consumed in Japan would be renewably sourced by 2020. The rising prices for petroleum and petroleum-based products have also contributed to the replacement of petroleum-based polymers with biodegradable polymers. However, Japanese consumption of biodegradable polymers has not increased as much as expected.

In China, high growth is expected in the coming years because of an increase in production capacity, higher demand for greener products, and plastic waste control legislation (BioIntelligence Service, 2011; Plasteurope.com, 2011b). Plastic pollution is turning into a serious problem, and the government has been acting to curb the use of conventional plastics over the last decade. In 2008, for instance, China banned the use of plastic bags with a thickness of less than 0.025 mm, promoting both reusable and biodegradable alternatives instead (Plasteurope.com, 2011b). At the regular state council meeting in March 2011, it was proposed that all local municipalities and 50% of the country's large- and medium-sized cities implement waste separation by 2015. In addition, environmentalists have been lobbying for reforms to mandate that all meal boxes (of which about 12 billion are used per annum in the country) be made of biodegradable materials (Plasteurope.com, 2011b).

4 THE GLOBAL BIOPLASTICS MARKET

Bioplastics development, production, and demand production is still in an early stage compared to traditional plastics and still suffers from some performance limitations and high costs (Gozzo & Glittenberg, 2009). However, it is a market that has been experiencing rapid growth: between 2003 and 2007, the average annual growth rate was close to 40% worldwide and nearly 50% in Europe (Shen, Worrell, et al., 2009). Even during the recession, the 2009 market growth was in the range of 5-10% (SRI Consulting, 2010).

Shen, Worrell, et al. (2009) estimated the global capacity of bioplastics in 2007 as 0.36 Mt, equivalent to only 0.1% of the world's paper and board production and 0.2% of the global petrochemical plastics production (Gozzo & Glittenberg, 2009; Shen, Worrell, et al., 2009). Of this total, approximately 43% or 0.17 Mt was starch-based plastics (both biodegradable and nondegradable plastics), of which 75% was located in Europe (Shen, Worrell, et al., 2009; Sustainability Victoria, 2007). Likewise, estimates by the UK group NNFCC put the global production capacity at 0.33 Mt in 2010 (Williams, 2010). Similarly, according to both the Freedonia Group and the United States consulting firm Jim Lunt & Associates, the world bioplastics market was 0.30 Mt in 2010, of which 90% was biodegradable plastics (de Guzman, 2010; Mohan, 2011). Another estimate by firm Industry Experts put capacity at 0.264 Mt in 2007 and forecast 0.476 Mt by 2011 (PRNewswire, 2012). There are, however, other estimates for current production that differ somewhat from these figures. According to BCC research, the global use of bioplastics was 0.64 Mt in 2010 and 0.85 Mt in 2011, with the European market segment accounting for around 34% of the global market in 2010 and the American market accounting for around 37.5% of the total (BCC Research, 2012). Similarly, the European Bioplastics organization put the production capacity of biodegradable plastics in 2011 at 0.486 Mt with bio-based but nonbiodegradable being 0.675 Mt (total = 1.161 Mt) (European Bioplastics, 2013). Plackett (in 2011) predicted a global bioplastics capacity for 2012 in the 0.5-1.0 Mt range, which is still a small niche in the total plastics market, which exceeded 260 billion kilograms of plastic in 2009. A summary of current bioplastics production by type (according to European Bioplastics) is given in Figure 14.1.

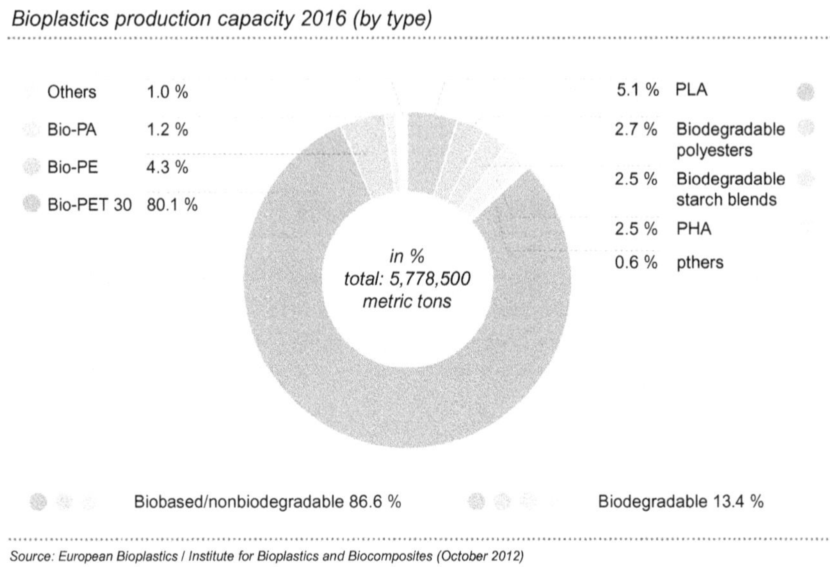

FIGURE 14.1 Biopolymers production capacity 2010 by type. (For color version of this figure, the reader is referred to the online version of this chapter.) *Reprinted with permission from European Bioplastics.*

Today in China, there are about a dozen larger corporations specialized in R&D and the production of starting raw materials for bioplastics (Plasteurope.com, 2011b). Market insiders estimate that China now has capacity to produce about 180,000 t per year of starch-based bioplastics, based mainly on corn but also tapioca (Plasteurope.com, 2011b). Because of the current product price, about 90% of China's bioplastics output is exported to North America, the EU, Japan, and Australia; by contrast, China consumes about 25 Mt of polyolefins each year (Plasteurope.com, 2011b).

There have been a number of market projections for bioplastics in general and starch plastics in particular. A product overview and market projection for emerging bio-based plastics was commissioned by the European Polysaccharide Network of Excellence (EPNOE) (http://www.epnoe.eu) and European Bioplastics (http://www.european-bioplastics.org) and completed in June 2009 (Shen, Haufe, et al., 2009; Shen, Worrell, et al., 2009). This study projected global bioplastic capacity to increase to 2.32 Mt by 2013 and to 3.45 Mt in 2020, based on company announcements. It was expected that production would grow on average 19% per year to 2020, with starch plastics, PLA, bio-based PE, polyhydroxyalkanoates (PHA), and bio-based epoxy resin expected to be the major types of bio-based plastics in the future (Shen, Haufe, et al., 2009; Shen, Worrell, et al., 2009). The announced capacity at that time for starch plastics by 2020 was 1.30 Mt, of which 40% was to consist of nondegradable starch plastics, while 60% would be biodegradable (Shen, Haufe, et al., 2009). The total maximum technical substitution potential of bio-based plastics and fibers was estimated at 240 Mt or 90% of the total consumption of plastics and fibers in 2007 (an estimate based purely on the technical properties of the bio-based plastics) (Shen, Worrell, et al., 2009). This projected capacity would still only account for 1.5% of the potential technical market (Shen, Worrell, et al., 2009).

According to the Freedonia group, global demand for biodegradable and bio-based plastics was expected to more than triple to reach 1 million metric tons (1.1 Mt) by 2015, valued at $2.9 billion (Plastics News, 2012) (although we have apparently already reached that capacity). The drivers for this were expected to be preferences for sustainable materials, improved product performance, and price parity with petroleum-based plastics. Sales of starch-based resins were expected to more than double in demand over this time (Mohan, 2011). However, the strongest growth was anticipated to be in nonbiodegradable bio-based resins, which will be fueled by the availability of commercial quantities of bio-based polyethylene from Braskem's 200,000-metric-ton-per-year plant in Brazil, which opened in late 2010 (Mohan, 2011). The group also forecast that the US demand for degradable plastics would rise 16.6 % per year to 147,000 tons in 2014 (Bioplastics Magazine, 2012).

BCC research forecast that the global biodegradable polymer market would be expected to have its largest growth over 5 years from 2012, dominated by packaging and usage in fibers, and also that the use of bioplastics will increase up to 3.7 million metric tons by 2016, a compound annual growth rate (CAGR) of 34.3% (BCC Research, 2012).

According to Germany-based Helmut Kaiser Consultancy, the global bioplastics market is growing at 20% to 30% per year and was expected to jump from 182,000 tons in 2006 to 4.5 Mt by 2015 (Mohan, 2009). Agriculture and Agri-Food Canada came up with an estimate in the same range, showing that the European bioplastics market should have reached 0.4-0.9 Mt in 2010 and 3-5 Mt in 2020 (Prevéraud, 2006).

In contrast, according to the German-based consultancy Ceresana Research (Lerner, 2012), world bioplastics demand was estimated to increase to around 1.03 Mt by 2015, with the market reaching revenues of more than $2.8 billion (€2.1 billion) in 2018, with annual growth rates of 17.8%. With a 48% share of global demand, Europe was expected to be the largest outlet for bioplastics in 2010, followed by North America and Asia-Pacific. By 2025, Helmut Kaiser Consultancy estimated that Europe would have 31%, the United States 28%, and Asia 32% of the market (Lerner, 2012).

The NNFCC predicted global annual capacity would grow more than sixfold to 2.1 Mt by 2013 (Williams, 2010), while the global market for bioplastics was forecast by the consulting firm Industry Experts to touch 1.9 million metric tons by 2017, growing by a CAGR of 22% during 2007-2017 (PRNewswire, 2012). Another projection by SRI Consulting showed slightly slower growth to just over 1.4 Mt in 2013, but the trend was still strongly positive (SRI Consulting, 2010). The SRI study projected total consumption of biodegradable polymers worldwide at an average annual growth rate of 13% from 2009 to 2014. The total consumption of biodegradable polymers in North America, Europe, and Asia was forecast to grow at an average annual rate of nearly 13% over the 5-year period from 2009 to 2014 (SRI Consulting, 2010).

A Global Markets Direct Report (USPRwire, 2009) noted that even the considerably developed market segments such as starch-based plastics were poised for higher growth due to the increasing demand and the technological advancement made in terms of their production and enhancing properties. According to this report, the total global biopolymers market by volumes was set to grow at a CAGR of 27.18% from 2008 to 2015.

In another forecast by European Bioplastics (2013), the global production capacity for bioplastics was predicted to increase to 5.8 Mt by 2016 (Figure 14.2). The composition of this production volume is also expected to change; in 2010, biodegradable materials were primarily produced, totaling around 0.4 Mt (compared to 0.3 Mt of bio-based commodity plastics). This ratio is forecast to completely change in the coming years, with the bio-based "drop-in" plastics making up 80% of forecast production capacity in 2016 (European Bioplastics, 2013). Biodegradable materials will, however, also grow substantially and will reach about 0.8 Mt by then.

Finally, according to Pira (Smithers Pira, 2010), global bioplastic packaging demand is forecast to reach 884,000 tons by 2020. A 24.9% CAGR is expected from 2010 to 2015 slowing to 18.3% in the 5 years to 2020. A new breed of bioplastics will be major drivers as packaging market demand gradually shifts from biodegradable and compostable polymers toward biopackaging based on renewable and sustainable materials.

In summary, while there are a range of predictions in terms of total capacity and growth in the bioplastics market, the forecasts are consistent that this will be an area of strong growth over the next 5-10 years. It should be noted however that it is unknown if these forecasts have taken into account the effect of the extraction of plentiful natural gas from the hydraulic fracturing process in the United States, this being a cheap feedstock for polyethylene production.

5 PROCESSING OF STARCH PLASTICS

One reason that starch plastics are particularly attractive as alternatives to petrochemical plastics is that they can be readily processed using conventional conversion technologies such as film blowing, extrusion, thermoforming, injection molding, foaming, and extrusion coating, with little or no modification. For example, starch-based Mater-Bi® films can

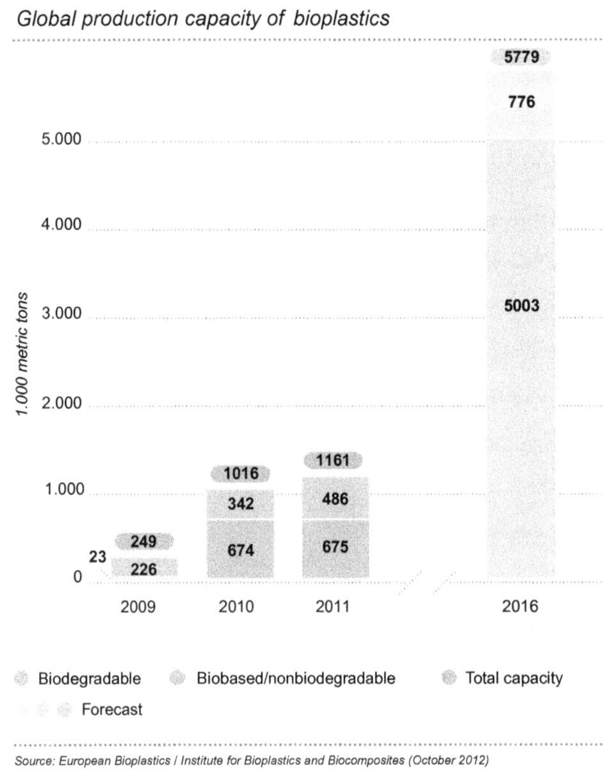

FIGURE 14.2 Global production capacity of bioplastics. (For color version of this figure, the reader is referred to the online version of this chapter.) *Reprinted with permission from European Bioplastics.*

be produced using film blowing and casting equipment traditionally used for low-density polyethylene (LDPE). The main difference is that Mater-Bi® has a lower welding temperature, so production lines with good cooling capacity are best suited for processing this film. As starch-based polymers are water-sensitive and prone to hydrolytic degradation, it is important that the materials are dried before processing (Halley et al., 2007). The melting point of native starch is higher than the thermal decomposition temperature: hence, the poor thermal stability of native starch and the need for conversion to starch-based materials with a much-improved property profile (Halley et al., 2007; Shen, Haufe, et al., 2009).

There are a range of processes used for the preparation of starch plastics from the raw material (Figure 14.3; Shen, Haufe, et al., 2009).

These can be broadly summarized as being

- partially fermented starch,
- thermoplastic starch (TPS) (or destructurized starch),
- chemically modified starch,
- starch composites.

Chemically modified, thermoplastic or even raw starch can be processed further by blending it with other polymers. Finally, one very established commercial technology is the production of starch foams for packaging. These processes are described in more details in the succeeding text.

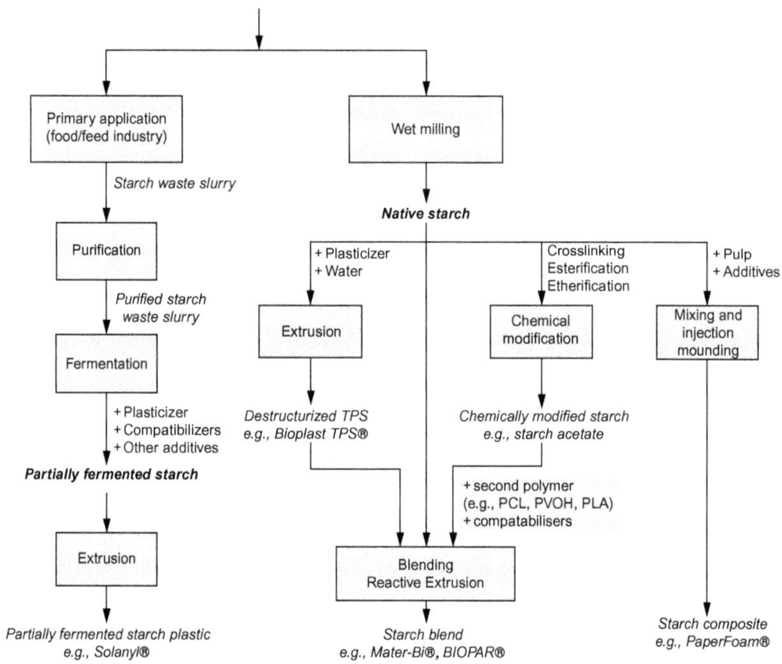

FIGURE 14.3 Starch plastic production technologies. (For color version of this figure, the reader is referred to the online version of this chapter.) *Reprinted with permission from Shen, Haufe, et al. (2009), Copyright (2009) Utrecht University.*

5.1 Partially Fermented Starch

Partially fermented starch can be produced using the Rodenburg process (Shen, Haufe, et al., 2009). In this case, the raw material is a slurry waste originating from processing of potatoes for the food industry, which consists of 72% starch, along with proteins, inorganics, fats and oils, and cellulose. After cyclone treatment to remove these impurities, the slurry is partially fermented over 2 weeks in storage silos (with some starch being naturally converted to lactic acid in this stage). This product is then dried (10% water content) and extruded (Platt, 2006; Shen, Haufe, et al., 2009) along with olein and other additives as required to make a plastic.

5.2 Thermoplastic Starch

Starch is not a real thermoplastic, but in the presence of a plasticizer (or other plasticizing additives), high temperatures (90-180 °C), and shear conditions, the crystalline structure of starch is disrupted and it forms an amorphous thermoplastic starch (TPS) (Vilpoux & Avérous, 2004). Commonly used plasticizers for TPS include water (although additional plasticizer is needed in this case), glycerol, sucrose, fructose, glucose, glycols, urea, formamide, ethanolamine, ethylene bisformamide, and amino acids (Zhang & Rempel, 2012); these ensure flexibility and mechanical properties stability of the finished product (Gozzo & Glittenberg, 2009). This destructurization process typically involves an irreversible disruption of molecular order within the granule termed gelatinization, which occurs via the following stages: (1) the loss of crystallinity of the granule, (2) an uptake of heat as the conformation of the starch is altered, (3) the hydration of the starch as accompanied by swelling of the granules, (4) a decrease in the relaxation time of the water molecules, (5) the loss of molecular (double helical) order, and (6) the leaching of the linear molecules (amylose) from ruptured granules (Gozzo & Glittenberg, 2009; Halley et al., 2007). The collapsed granules remaining (containing mostly amylopectin) are then held in a matrix of amylose. At a molecular level, there are three existing models to explain this gelatinization process: (a) water-mediated melting of starch crystallites, (b) melting process in semicrystalline polymers, and (c) the breakdown of the side-chain liquid crystalline starch structure that can occur via (i) breakage of starch-starch-OH bonds, (ii) formation of starch-solvent-OH bonds, and (iii) an unwinding helix-coil transition.

These TPS polymers are typically molecularly homogeneous (with both amylose and amylopectin dispersed uniformly throughout the material) and have superior mechanical properties and greatly enhanced processability (Shen, Haufe, et al., 2009). However, on their own, TPS polymers are extremely water-sensitive, can suffer from significant molecular weight change on extrusion (Davidson, Paton, Diosady, & Laroque, 1984; Gomez & Aguilera, 1983; Sagar & Merrill, 1995), and are thus of limited practical value. Therefore, most of the commercial research on TPSs has involved modified starches and/or blends with additives and other polymers.

5.3 Starch Blends

Starch blends are produced by mixing TPS, chemically modified starch, or sometimes even native starch with petrochemical, bio-based, or inorganic compounds into a (microscopically) homogenous material (Shen, Haufe, et al., 2009). Such blends can be tailored to produce very diverse properties (e.g., relative rigidity versus flexibility), allowing for a broad range of applications. Such starch-based blends come closest to achieving the mechanical properties of LDPE and high-density polyethylene (HDPE), as well as PS. Blending usually takes place during hot melt extrusion. Starch/aliphatic polyester blends, for example, can be prepared with concomitant destructurization of the starch resulting in an interpenetrated or partially interpenetrated structure. Blending with polycaprolactone, polyvinyl alcohol, or other biodegradable and relatively hydrophobic polymers has made it possible to effectively reduce the strong hydrophilic nature of the final polymer and improve the overall properties.

In a process of "reactive blending," starch is bonded covalently (Shen, Haufe, et al., 2009; Song, Murphy, Narayan, & Davies, 2009) and/or by van der Waals forces with other polymers. For example, reactive blending can use carboxyl, anhydride, epoxy, urethane, or oxazoline functional groups of other polymers to react with the hydroxyl or carboxyl groups of the starch. Alternatively, either the starch or the blend material can be chemically modified to improve the molecular scale mixing and thus the overall properties. An example of this is graft copolymerization, whereby synthetic monomers are coupled to the starch hydroxyl groups and then polymerized. The mechanical performance of starch blends depends on a series of parameters including blend morphology (particle size and shape and particle dispersion and distribution), interfacial adhesion, and the intrinsic characteristics of TPS (Xie, Halley, & Avérous, 2011).

The starch content in blends can be as high as 80%, depending on the target properties required. Starch blends can be completely biodegradable (such as when the polymer blend material comprises biodegradable but fossil fuel-derived polymers such as polycaprolactone) or aimed at the durable plastics market and based on nondegradable components such as PP and PUR. A recent development is the incorporation of bio-derived PP and PE in such blends.

5.4 Starch Modification

Starch modification is a process whereby some of the hydroxyl groups in the native starch structure are substituted with alternative functionalities, such as through acetylation/esterification of starch to form starch acetates, carbanilation of starch with phenyl isocyanates, addition of inorganic esters to starch to produce phosphate or nitrate starch esters, production of starch ethers, and hydroxypropylation of starches via propylene oxide modification (Halley et al., 2007). Very low levels of chemical modification can significantly reduce the hydrophilicity of the end product and lower gelatinization temperatures, reduce retrogradation (recrystallization over time), and improve flexibility of final products. Cross-linking is another form of modification in which two hydroxyl groups or neighboring starch molecules are linked chemically; this process inhibits granule swelling or gelatinization and gives increased stability to acid, heat treatment, and shear forces.

Starch modification is currently a relatively expensive and complex process and therefore not widely adopted commercially in the production of starch plastics, although research in this area is ongoing. For example, the use of new heterogeneous catalysts has the potential to make the process much cheaper (Shen, Haufe, et al., 2009). One current commercial application using modified starch is as a filler to partially substitute carbon black in tires, a technology that has been jointly developed by BMW, Goodyear, and Novamont (de Guzman, 2010).

5.5 Starch Composites

Starch composites differ from starch blends in that they are composed of native, modified, or TPS mixed with one or more constituent materials with significantly different physical or chemical properties from starch, which remain separate and distinct on a macroscopic level within the finished structure (Shen, Haufe, et al., 2009).

As an example, granular starch beads (up to 100 μm in size) can be compounded with another synthetic polymer as a filler or used as directly as a component in polymer blends. These blends can be improved through surface modification of starch using, for example, $R_1R_2R_3$-O-SiMe silane as a compatibilizing agent (Vilpoux & Avérous, 2004). Starch nanocrystals (which are about 45% crystalline) have also been prepared and have been shown to increase the tensile strength and elastic modulus of nanocomposite materials by more than twofold, depending on the matrix material (BIOP, 2012). Other starch nanocomposites with significantly improved mechanical properties have been made using starch as the matrix material and incorporating nanoparticles from other sources, such as montmorillonite clay particles (IMCD, 2011).

Extruded starch-based foam products can be made into composites to improve their strength and resiliency, by incorporating pulped cellulose fiber as a reinforcing agent. The use of baked starch-fiber composites is also common, whereby starch-based foam products reinforced with well-dispersed cellulose fibers are hot pressed in a process similar to that used for making waffles.

5.6 Starch Foams

Starch can also be softened and expanded into a foam using water as a blowing agent. This process typically involves heating the starch under pressure to superheat water absorbed into the starch matrix and then explosively releasing the pressure (BIOP, 2012). If the heating rate is quick enough, this process can also be performed under atmospheric pressure. Foamed starch is antistatic, insulating, and shock-absorbing, therefore constituting a good replacement for polystyrene foam. It is usually derived from corn, wheat, and hydroxypropylated high-amylose cornstarch, mixed with a small amount of additives like polyvinyl alcohol, glycerol, polyethylene glycol, or silicon dioxide. Starch-based foam sheets have also been made for packing and cushion applications using a similar technology to extrude starch foam through an annular die. Starch-based foams have also been produced using chemically modified starch, with significantly improved properties such as improved moisture resistance.

6 STARCH APPLICATION AREAS

Historically, loose fill foam packaging and compost bags were the principal applications for starch-based polymers. However, they are now used in a wide range of other application areas. This includes various types of bags and sacks, flexible packaging, rigid packaging such as thermoformed trays and containers, and other forms of food packaging where composting of organic waste is effective. They are also used in agriculture and horticulture for applications such as mulching film, covering film and plant pots, and some unusual application such as ties for vineyards, as well as for fishery materials. Applications are increasing in the automative industry, particularly with the development of more durable blends such as with polypropylene or polyethylene, and there is a growing market in electronics casings and recordable media (e.g., DVDs). A range of small-scale injection molding applications are also being marketed, including pencil sharpeners, golf tees, rulers, cartridges, dog chews, combs, and toys (Platt, 2006) as well as use in hygiene products including sanitary products and nappies.

Medical uses are also being explored, such as scaffolds for bone tissue engineering, because of their biocompatibility, biodegradability, and porous nature that allows blood vessel proliferation during bone growth as well as controlled release of drugs or macromolecules. Polyacrylate-modified starch microparticles have been used as protein and drug carriers, and starch/albumin microparticles have been designed for controlled protein release (Plasteurope.com, 2012). Controlled orientation injection molding has been used to develop starch-polylactide and starch-poly(ethylene-*co*-vinyl alcohol) blends for biomedical implants, with such starch materials demonstrating a promising cytocompatibility.

There are many more markets now just starting to use bioplastic materials such as building and construction, household, leisure, or fiber applications (clothing and upholstery).

7 LEADING STARCH-BASED POLYMER PRODUCERS

A summary of the main starch-based polymer producers along with reported production levels where such information could be obtained is given in Tables 14.1 and 14.2. There are some well-established firms

TABLE 14.1 Starch Bioplastics Producers (Not in the Asian Region)

Manufacturer	Polymer Type	Trade Name	Production Capacity (World) kt. p.a	Refs.
Novamont	Thermoplastic starch/polymer blends	Mater-Bi	80	Plastral (2012)
Rodenburg Biopolymers	Modified potato starch	Solanyl	65	IMCD (2011)
Cereplast	Thermoplastic starch blends	Cereplast Compostables Cereplast Hybrid	36 increasing to 86 in 2014	Cereplast (2011)
Roquette	Thermoplastic starch blends	Gaialene	25	Plasteurope.com (2012)
Stanelco, Biotec	Thermoplastic potato starch, thermoformed starch blends, and composites	Biome, Bioplast	20	Voegele (2011)
Cardia Bioplastic	Thermoplastic starch blends (biodegradable polyesters)	Cardia Compostable	11.2	Weir (2011)
	Thermoplastic starch blends (PE, polyesters, etc.)	Cardia Biohybrid		
Limagrain Cereales	Directly extruded from whole-grain cereals (no starch extraction), polymer blend	Biolice	8 (planned in 2013)	Limagrain (2012)

Continued

TABLE 14.1 Starch Bioplastics Producers (Not in the Asian Region)—cont'd

Manufacturer	Polymer Type	Trade Name	Production Capacity (World) kt. p.a	Refs.
Plantic Technologies	Thermoplastic cornstarch polymers	Plantic	7.5	Roberts (2010)
StarchTech	Foamed packaging peanuts from cornstarch	Re-NEW	5.9	StarchTech Inc. (2012)
Tecnaro	Starch blend/composites	ARBOBLEND	5	Tecnaro GmbH (2012)
BIOP Biopolymer	Biphasic starch blends (aliphatic copolyesters)	BIOPAR	~5	Shen, Worrell, et al. (2009) and Shen, Haufe, et al. (2009)
	Starch blends, starch acetate (under development)	BIOPAREN		
J C Hagen GmbH	Thermoplastic starch blends	Bioplast	Unknown	Plasteurope.com (2010)
Cerestech	Thermoplastic starch blends	Cereloy	Unknown	EON (2008)
PaperFoam	Starch composite	PaperFoam	Unknown	PaperFoam (2013)
Nvyro	Cassava starch composite	Edible Plate Company	Unknown	NVYRO (2012)
NNZ BV	Thermoplastic starch blends	Ökopack	Unknown	NNZ (2012)
Potatopak	Potato starch composites	Potatopak	Unknown	PotatoPak (2010)
Biosphere Industries Ltd.	Tapioca starch composite	Biosphere	Unknown	Biosphere® Industries, LLC (2008)
Vegemat	Starch composite	Vegeplast	Unknown	Vegeplast (2012)
Green Light Products	Thermoplastic starch blends	Green Light	Unknown	GreenLight Products Pty Ltd. (2011)
DaniMer Scientific	Thermoplastic starch blends		Unknown	DaniMer (2011)

TABLE 14.2 Starch Bioplastics Producers (Asian Region)

Manufacturer	Polymer Type	Trade Name	Production Capacity (World) kt. p.a	Refs.
Asian Producers				
Wuhan Huali Environmental-friendly Technology Co., Ltd.	Thermoplastic starch blends	PSM Plastarch Biodegradable	40	Global Market (2013)
Shanghai Disoxidation Macromolecule Materials	Thermoplastic starch blends		32	Disoxidation Macromolecule (2012)
Guangdong Shangjiu Biodegradable Plastics	Thermoplastic starch blends	Shangjiu	15	Guangdong Shangjiu Biodegradable Plastics Co., Ltd. (2012)
Beijing Macrolink Biomaterial Co.	Thermoplastic starch blends	MBM (C or T)	10	Weiku (2011)
Tianhe Environmental Technology Co., Ltd.	Thermoplastic starch blends		6	Tianhe Environmental Technology Co., Ltd. (2013)
Maribumi Starchtech	Thermoplastic starch blends	ZERO-PLASTIC resin	Unknown	Blazelead (2013)
Glycan Biotechnology Co.	Cellulose from starch	Glycan	Unknown	Glycan Biotechnology (2008)
Wuhan Greenplas	Thermoplastic starch blends	Greenplas	Unknown	Wuhan Greenplas (2012)
Greenpoly Group	Thermoplastic starch blends	Greenpoly	Unknown	Greenpoly (2005)
Japan Cornstarch	Thermoplastic starch blends	Cornpole	Unknown	Japan Cornstarch Co., Ltd. (2012)
Kingfa Science and Technology	Thermoplastic starch blends	ECOPOND FLEX-162	Unknown	Ecopond (2010)
Grace Biotech	Thermoplastic starch blends	GRABIO	Unknown	Grabio (2011)

producing starch-based polymer resins and related products, as well as a number of emerging firms, particularly in Asia. These will be discussed under the loose categories of producers of TPSs and starch blends, foams, and composites.

7.1 Profiles of Leading Thermoplastic Starch-Based Polymer Producers

7.1.1 Novamont

The Italian-based biotechnology company Novamont (www.novamont.com) is the market leader in the starch-based bioplastics industry, with an annual capacity of ~80,000 tons (Table 14.1). Created originally as a Montedison research center in 1989, Novamont acquired the worldwide patents from one of the original starch plastic development companies, Warner-Lambert, in 1997, inheriting at the same time the license agreement with Enpac, subsequently acquired by National Starch. The company further acquired the "Eastar Bio" technology from Eastman Chemical in 2004, which included all patents and technology rights but not operating facilities.

Novamont produces a range of cornstarch-based polymers under the Mater-Bi® brand name, with the fourth-generation products now in development. It holds more than 1000 patents relating to starch materials technologies, particularly in the areas of blending and additives technologies as well as TPSs (Novamont, 2013).

The Mater-Bi® product range is divided into families according to the processing technology to be used, these being film, extrusion/thermoforming, injection molding, foaming, and tire technology grades. Mater-Bi® products have a use performance that is similar to traditional plastics, with good processing properties, a wide range of mechanical properties from soft to tough and rigid, antistatic behavior, and compostability in a wide range of composting conditions (Platt, 2006). Their Z class comprises starch blended with varying levels of biodegradable polyester (polycaprolactone), plasticizers, and other additives, such as Ecoflex (which adds melt strength and flexibility) (Mooney, 2009), and is their film sheet grade. Polycaprolactone limits moisture sensitivity, boosts melt strength, and helps plasticize the starch. Their Y-class injection molded grade is blended with modified cellulose and their A class is a mixture of starch and ethylene vinyl alcohol. This latter grade is not compostable although it is slowly biodegradable and is used for rigid and expanded items (Mooney, 2009).

Recent collaborations on the part of Novamont have led to innovative product applications for starch. For example, Novamont has joined with nappy manufacturer WIP to develop a nappy with biodegradability index of greater than 80%, apparently the highest on the market (Voegele, 2011). The company has also been working closely with the Italian Association for Organic Agriculture to promote the use of biodegradable products in agriculture and catering and recycling sectors, presenting plans for joint activities to the Federal Congress in Milan (Bellini & Tomasi, 2011). Further, it is developing a new plant at Porto Torres as part of a joint venture with Eni Versalis, for the production of azelaic acid from vegetable oil in addition to a wide range of chemicals (Bellini & Tomasi, 2011). In line with this, Genomatica, Versalis and Novamont recently announced plans to produce butadiene from renewable feedstocks (Adler & Di Stefano, 2012b), an announcement that follows on from an earlier agreement between Genomatica and Novamont to produce butanediol in Adria, Italy (Adler & Di Stefano, 2012a). This monomer would be used in the manufacture of biopolyesters for blending with the Mater-Bi® bioplastic. The company is also planning to commercialize "nanostarch" particles for use in film grade Mater-Bi® resin. According to reports (Schut, 2008), addition of these particles at the 25-40% addition rate produces thinner, stronger, clearer starch films.

Novamont was recently involved in a European-wide patent dispute with German company Biotec Biologische Naturverpackungen and Sphere (one of Biotec's holding companies) (Smith, 2010), centering around three patents for reinforcing starch blends. This follows on from an earlier dispute in 2001, when a global agreement was reached involving cross-licensing arrangements (Platt, 2006). The most recent dispute was settled in June 2013.

7.1.2 Other Manufacturers

The sustainable packaging developer Stanelco, a UK-based firm, currently has two divisions: bioplastics (as Biome Bioplastics) and RF applications. Biome Bioplastics is responsible for the development and sale of the Group's new bioplastic products (http://www.biometechnologiesplc.com/), and the German company Biotec Biologische Naturverpackungen (http://www.biotec.de/), acquired in 2005 as a 50/50 joint venture with Sphere, manufactures the bioplastic resins and owns a large number of original patents for the processing of TPS. A broad range of products is produced by Biotec or sold through Biome including, among a range of others, Bioplast GF106/2, a compostable potato starch-based product that is plasticizer-free, and Bioplast TPS, a thermoplastic potato starch resin of more than 85% starch content with additional additives to make it processable. Biome Technologies is currently experiencing a recovery from difficult recent years, with recent significant increases in sales (StarchTech Inc., 2012).

The French firm Roquette (www.roquette.com), a global leader in starch, polyol, maltodextrin, etc., has just undergone a major expansion in its bioplastic production capacity, starting up a new plant for its Gaïalene range of bioplastic at its Lestrem site (Tecnaro GmbH, 2012). This is in close proximity to Europe's largest biorefinery. This product, which contains more than 50% plant-based material, is produced from starch using grafting technology.

Indiana-based Cereplast, Inc. (www.cereplast.com) produces both a compostable and "sustainable" line of TPS-based resin products, with a nominal capacity of 36,000 tons per annum. Cereplast Compostables are compostable blends of starch derived from corn, tapioca, or potatoes with "Ingeo" polylactic acid, other biopolymers (such as PHA or PHB), and natural plasticizers (soy proteins). The blending process includes a proprietary polymer modification process. Nanocomposites are also included to optimize the surface of the resin pellets. These products can be processed on blown film extruder lines as well as capable of being injection molded, thermoformed, and extruded as a coating. The WS50 line consists of a blend of starch and polyvinyl alcohol. The hybrid line (Biopropylene CP Bio-PP-50) consists of a 50/50 blend of starch plastic with polypropylene, which provides improved heat resistance and heat stability in processing. There are plans to increase the starch content in these hybrids and also to explore hybrids based on LDPE and HDPE. Following major restructuring in 2009, Cereplast is currently entering into a number of distribution agreements in Europe (such as with Mastercolor AB in Scandinavia and GAMA Plastik in Turkey) in pursuit of an aggressive growth strategy (Plasteurope.com, 2010).

BIOP Biopolymer Technologies (www.biop.eu), based in Germany, produces a biodegradable multilayered and biphasic starch-based bioplastic from potato starch—a process that it claims enables higher starch content in the final product. Both the pure granulate and blend products are produced. The starch extraction method includes a reactive extrusion process to produce polymer. The company produces a number of grades including BIOPAR®, a blend containing 60% starch-based raw material with biodegradable copolymers such as Ecoflex from BASF or Bionolle (the aromatic-aliphatic copolyester based on butanediol, adipic acid, and terephthalic acid) along with a compatibilizer (which enables the starch to be present in a bicontinuous phase with the polyester). The fully bio-based product BIOPAREN® is a blend of starch with PHA/PHB and starch acetate. Another grade (starch/PUR) is also under development (Shen, Haufe, et al., 2009; SpecialChem, 2011).

Originally incorporated in 2001, the Australian-based company, Plantic Technologies (www.plantic.com.au), launched its initial product (a rigid water-soluble chocolate box tray) in 2003. Their biodegradable starch-based polymer is produced from a high-amylose content cornstarch, modified by hydroxypropylation to make it thermoprocessable. Plantic entered into separate joint venture agreements with DuPont and the

Bemis Co. in 2007, whereby these companies distribute Plantic's starch polymer pellets, film, and thermoformable sheet internationally. Another agreement with National Starch in 2007 had the purpose of developing new starch and starch-modification technology in order to broaden the performance spectrum of the product line. In 2011, Plantic purchased the packaging company Tribond Packaging Solutions, which specializes in thermoforming and lamination technologies. The product range currently consists of thermoformed trays for applications such as chocolate trays; the new range (Plantic ecoPlastic) is a multilayer product that has a moisture-protective outer skin and up to 80% renewables content.

Cardia Bioplastics (www.cardiabioplastics.com) is an Australian-based company, established in 2002 as Biograde Limited. The production plant is located in Nanjing, China. Product lines include Cardia Compostable resins, with a variety of application grades, and Cardia Biohybrid resins, which are blends of TPS-based polymers with traditional polyolefins.

Cerestech, a spin-off company from the École Polytechnique de Montréal (www.cerestech.ca), has developed a family of TPS blends called Terraloy™ that is marketing in partnership with Teknor Apex Company. This product family consists of blends of TPS with copolymers such as biodegradable copolyester (PBAT), polylactic acid, or PHA, as well as blends of TPS with polyolefins such as polypropylene or polyethylene (EON, 2008).

The starch-based biodegradable bioplastic Biolice® is produced by Limagrain Céréales Ingrédients, owned by Groupe Limagrain (www.lci.limagrain.com)—the largest seed breeder in the European Union. In this novel process, the product is directly extruded from maize flour to produce processable granules instead of going through an initial starch extraction stage (Shen, Haufe, et al., 2009). A long breeding program was necessary to produce the specific varieties for this process.

Rodenburg Biopolymers, BV, based in the Netherlands, produces a unique thermoplastic product, Solanyl®, from a partially fermented starch product derived from potato peel (www.biopolymers.nl). Established in 1945 for the trading of plant-derived products for various industries, this company is now one of the largest starch-based bioplastics producers in Europe, with 47,000 tons per annum nominal capacity. While Solanyl® itself is biodegradable, most of the application grades currently in production are blends with synthetic or bio-based polyesters.

JC Hagen GmbH is a plastics trading organization founded in 1999, which has recently added starch-based plastics to its product line. Under the trade name of Bioplast (www.bio-plast.com), the product comprises 55-75% cornstarch along with calcium stearate, a stabilizer, a lubricant, and beeswax. This mix is compounded into a thermally processable resin.

MGP Ingredients, Inc. (www.mgpingredients.com) began operation in Kansas in 1941 and is a company dedicated to the production of products derived from natural sources; principal among these are specialty wheat proteins and starches, high-quality beverage and food grade industrial alcohol products, and plant-based bioplastics and wood-plastic composites. Their starch-based bioplastics, the Terratek® SC range, are blends of 50-65% wheat starch and polypropylene for injection molding applications that can withstand boiling temperatures.

NVYRO (a UK-based firm, nvyro.com) produces a range of dinnerware products from cassava (tapioca) starch, under the trade name "the Edible Plate Company." The production process uses thermal compression molding at a temperature of 200-240 °C, and the starch mix consists of more than 90% dry weight starch, plasticizers, emulsifier, calcium carbonate, and agar.

DaniMer, based in Bainbridge, Georgia, United States, makes biodegradable resin blends composed of starch (mainly cornstarch), lactic acid, and PLA (www.danimer.com). And NNZ bv (www.nnz.com) produces the Ökopack® range of trays, including the transparent water-soluble Ökopack®-Tray W based on starch.

Finally, Ingredion (http://www.ingredion.com), a company recently formed from the merger of Corn Products and National Starch, produces hydrophobic starches designed to improve moisture barriers and help in the production of specialty starch-based biodegradable food grade packaging that can be used in trays, foams, and functional additives. The company also developed the first starch-based loose fill in the 1980s.

7.1.3 The Asian Industry

Over recent years, there has been a surge of interest in the production of starch-based bioplastics in Asia, and in particular China, fuelled in part by government policies and programs. While this market is still in a relative state of flux, the following list includes producers that are currently active as far as it is possible to ascertain:

- Shanghai Disoxidation Macromolecule Materials Co., Ltd., based in Jiangsu (www.dmmsh.com), was certified as a new and high-technology transformation project by Shanghai Science and Technology Commission in 2006. Producing a nominal 32,000 tons per annum (Disoxidation Macromolecule, 2012), this company produces several grades of starch resin (of 50%, 62%, and 100% biodegradability).
- Wuhan Huali Environment Protection Science and Technology Co., Ltd. is a subsidiary of Hongkong-based Engrowth Investments Limited. The company markets the Plastarch Material (PSM®) brand of cornstarch-based plastic, first developed in 2001 (www.psm-hk.com). The technology involves mixing cornstarch with cellulose and enzymes under conditions that promote condensation and starch modification/cross-linking; the resulting crystals are then plasticized in a twin-screw extruder. A range of grades is available, from the bio-based but not fully

biodegradable HL-100 resins (which is a blend of starch with a small proportion of polypropylene) to the compostable HL-300 series (which is a blend of starch and high-polymerization polyester).

- Beijing Macrolink Biomaterial Co., Ltd. (MBM), located in Beijing, is part of one of China's major private corporations, the Macrolink Group. Located in Beijing, MBM produces three kinds of biodegradable resins based on starch, including MBM-C, a compostable grade, MBM-T (a blended tailored resin), and a blendable masterbatch (MBM-A) (www.mbm-bio.com).
- Guangdong Shangjiu Biodegradable Plastic Co., Ltd., also founded in 1999, produces a fully biodegradable starch-based resin in four grades for film, injection molding, sheet (and blister), and foaming applications. They hold a range of patents covering, among others, different types of starch blends (such as with polycaprolactone or polybutylene succinate plus other additives) (shangjiu.en.alibaba.com).
- Kingfa, established in 1993, is based in China and is reported to be Asia's largest company in the modified plastics industry. According to their company website, Kingfa has the capacity to produce 1,000,000 tons of modified plastics. Their product ECOPOND FLEX-162 is a fully biodegradable starch-based resin for film applications (Ecopond, 2010) but production capacity is unknown.
- For other companies, there is very limited information available; they are apparently producing and trading in starch-based resins based on their web presence and so are included for completeness. Such companies include Wuhan Greenplas (www.greenplas.com) that produces a biodegradable blend of modified starch mixed with PLA as well as a starch/LDPE blend masterbatch (60% starch) (www.greenplas.com/en/about/index.asp). Also, the Greenpoly company (based in Dongguan, China; www.greenpoly.com) is listed as producing a biodegradable resin made from natural starch or cellulose with PCL and PVA acting as a carrier resin for biodegradable resin as well as a blended resin comprising starch plus PE/PP/PS.

Away from China, there are other manufacturers in this region:

- Grace Biotech Corporation, a Taiwan-based bioplastics producer, was established in 1999, with early technology coming from scientists based in the UCLA and Stanford University in the United States. The company produces a blend of cornstarch with biodegradable polyesters, marketed as GRABIO resin (www.grabio.com.tw).
- Glycan Biotechnology Co., Ltd., also based in Taiwan, has developed a novel technology for the production of glycan (a cellulose) from starch (Glycan Biotechnology, 2008).

- The large Japanese starch manufacturer, Japan Cornstarch, markets a biodegradable bioplastic made primarily from cornstarch, named Cornpole, although information about this product is very limited (www.nihon-cornstarch.com).
- Maribumi Starchtech, in Malaysia, produces a biodegradable resin by the trade name of ZERO-PLASTIC based on 55% tapioca starch (plasticized), refined glycerol, polyvinyl alcohol, and other additives (Blazelead, 2013).

7.2 Profiles of Leading Starch-Based Foam Producers

One of the most established application areas for starch plastics is as foams for packaging. There are a number of companies that specialize in this technology. StarchTech (based in Minnesota, www.starchtech.com) is a long-established firm, founded in 1996, that produces starch-based, water-soluble, and fully biodegradable peanuts for packing applications. As part of their drive toward sustainability, they have a new system that allows manufacturers, distributors, and wholesalers to make these packing peanuts at their own site, cutting down on transport costs. They claim that one trailer load of the resin pellet material is the equivalent of approximately 23 trailer loads of packing peanuts.

The US company Storopack (originally founded in Swabian Backnang, www.storopack.us) produces an S-shaped loose fill made from corn and potato starch under the brand name PELASPAN™ BIO as well as air cushions (AIRplus BIO). The particular advantage of starch-based fills is that they are antistatic, meaning that they pack readily into voids, and are very appropriate for use with electronic components. They are also of lighter weight than the PS equivalent and the company claims that due to the particular manufacturing technology, they are produced with a harder outer surface leading to less dust.

Green Light Products in the United Kingdom produces the starch-based Eco Flo, a fully biodegradable loose fill that is claimed to prevent product migration and provide better impact properties than PS. They also manufacture air cushions and bubble film (www.greenlightproducts.co.uk).

The US-based firm FP International (www.fpintl.com) also produces a starch-based loose fill branded as Flo-Pak Bio 8.

7.3 Profiles of Leading Starch-Based Composite Producers

There are a number of producers who specialize in the production of starch-based composites. PaperFoam (www.paperfoam.com) is one of the more established brands, founded in 1998 as a packaging company. The product, a starch/cellulose fiber composite, is produced using a one-step injection molding process. Potato starch (~70%) is mixed with paper grade pulp (~20%) and other additives and then injected into a heated mold. The product is foamed by the evaporation of water. The great advantage of PaperFoam® above traditional packaging, according to the manufacturer, is the ultra-low-carbon footprint and the low weight.

Biosphere Industries (based in California and founded in 2002, www.biosphereindustries.com) produces the Biosphere® range of biodegradable rigid packaging products, comprising composites based primarily on tapioca starch and grass fibers. These are baked in a similar fashion to that of the EarthShell composite, which was composed of a blend of cellulose from paper waste, starch from potato waste, ground limestone, and water, which was effectively foamed and thermoformed in a process similar to that of making waffles. The product was used in packaging applications such as clamshells and also used for dinnerware. However, this company filed for bankruptcy in 2007.

Potatopak, based in the United Kingdom, also uses a baking technology to produce thermoformed products from potato starch (and increasingly from waste starch) (www.potatoplates.com). The process uses a high-speed thermoforming machine that presses the starch powder into a shape using a mold under heat and pressure. They produce a very broad range of injection molded products such as plates, punnets, and bowls as well as fillers, packaging, and industrial and medical products.

Tecnaro produces a composite product, ARBOBLEND®, that contains different biopolymers such as PHA, polyester, Ingeo PLA, starch, cellulose, organic additives, natural resins, or waxes and natural reinforcing fibers (www.tecnaro.de). "Plastic wood" (ARBOFORM®) is one of the primary products of this company.

Finally, the company Vegeplast produces a composite VEGEMAT® (Vegeplast, 2012) that is composed of fibers, starch, proteins, lipids, and additives.

7.4 Other Starch Applications

One key producer that utilizes starch on a large scale (Table 14.1) is Ecosynthetix, a company that produces starch-based binders under the trade name of EcoSphere Biolatex (http://www.ecosynthetix.com/).

And, while not being a starch-based material, an important synthetic but fully biodegradable copolymer often used in starch-based blends is Ecoflex, an aromatic-aliphatic copolyester based on butanediol, adipic acid, and terephthalic acid produced by BASF in Germany.

8 KEY ISSUES, BARRIERS, AND OPPORTUNITIES FOR STARCH PLASTICS

There are a number of barriers to broader market uptake of the current generation of starch plastics. These include the following:

- Higher price (production costs) in comparison with petrochemically derived (traditional) plastics.
- Lack of heat resistance and impact properties.
- The mechanical properties of the current generation of starch plastics are still in general inferior to those of petrochemical polymers—there is a need to develop a broader range of materials.
- Sensitivity to moisture and water contact and high water vapor permeability (Shen, Haufe, et al., 2009)—which means that a water barrier coating or blend is typically required.
- Starch plastics by themselves have low resistance to solvents and oil (Gironi & Piemonte, 2011; Shen, Haufe, et al., 2009), although blending can offer a solution.
- Embrittlement with age due to dehydration and free volume relaxation—starch plastics need plasticizers or to be blended to maintain properties.
- The aging and durability of thick-walled molded parts.
- Processing window sensitivity and degradation with temperature.
- There is a tendency for these plastics to stick to processing equipment; additives are often needed to increase melt strength and sometimes thermal stability or lubricity.
- Some starch plastic types, such as starch acetate, are still in a relatively early stage of development (Shen, Haufe, et al., 2009).
- The use of starch plastics in the production of durable goods or structural applications requires an in-depth study of their indoor and outdoor aging and durability properties, for which it is still difficult to find relevant data.
- The high specific gravity of TPS (1.21-1.29 g/cm^3) decreases the price competitiveness of starch plastics on a volume basis.
- Lack of government regulation (particularly in the United States) regarding waste.
- Integration of bioplastics into current end-of-life management systems (collection, sorting, recycling, etc.).
- Labeling confusion.
- Skepticism with respect to functional abilities and environmental claims.
- The risk of insufficient supply with raw materials due to competition with food, feed, bio-fuel, and other commercial end uses—not likely to be a bottleneck in the short term.

- Variability of cereal costs (particularly in light of biofuel market).
- Use of GMO plants.

Some of the opportunities include:

- Starch plastics are relatively easy to process using current processing equipment, and the processing of native starch blend-extruded with other compounds is nowadays well understood.
- The development of new, commercially attractive functional properties (such as the intrinsic antistatic properties of starch, of particular benefit in foam packaging).
- Barrier properties to gases (oxygen and carbon dioxide) are moderate to good.
- TPS and starch blend films have reasonable transparency in comparison with many other bioplastics.
- There is a good opportunity for combining biopolymers for tailor-making properties, e.g., starch-based composites with excellent functional properties are being developed (Xie et al., 2011).
- Other novel approaches to improving the properties are being developed, such as the development of starch colloids, biopolymer interpenetrating networks, and organic-inorganic hybrids (EON, 2008).
- Novel nanoscale products are a major area of development in materials research. Future applications include ultraprecise drug delivery, transparent nanofoams, nanoelectronics, coatings, and ultraselective molecular sieves (EON, 2008).

9 END-OF-LIFE ISSUES FOR STARCH PLASTICS

Starch-based biomaterials can be designed to have a range of end-of-life properties, not all being intended to be fully compostable products; for example, a high copolymer content can slow the rate of biodegradation due to the complex interaction of starch and polyester at the molecular level (Shah, Hasan, Hameed, & Ahmed, 2008), and blends with nondegradable polymers such as polypropylene are designed to be more durable. As such, the integration of such materials with waste disposal options is an important aspect of starch plastic commercialization (Figure 14.4).

According to a recent European Commission report on waste in the environment (BioIntelligence Service, 2011), the following key trends will occur over the next few years:

- "levels of *recycling*, primarily mechanical, will increase;
- levels of *energy recovery* will increase but in a more limited way than recycling levels due to the lead times associated with plant development;

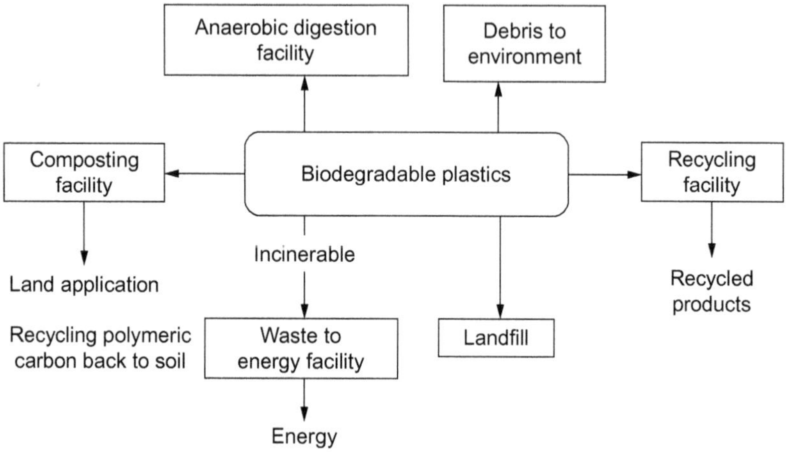

FIGURE 14.4 Integration of biodegradable plastics with waste disposal infrastructure. *Reprinted with permission from Song et al. (2009), Copyright (2009) The Royal Society.*

- proportional use of *disposal* will decrease;
- the most substantive changes in terms of volume will be seen in the treatment of *packaging* wastes. However, proportional changes will also occur in the other sectors analyzed.

Other trends likely to influence the impacts associated with generation and management of plastic wastes to 2015 are:

- a continuing upward trend in the *demand* for plastics;
- the level of *exports* of waste, in particular plastic waste for recycling and recovery, looks set to increase as overall recycling levels and volumes increase;
- the *production of plastics* will also tend to be dominated by the Asian market and particularly China;
- the *production of bioplastics*, while remaining a relatively low proportion of total plastic use, will increase rapidly;
- *waste-to-energy* (*incineration*) is set to increase, reducing the percentage of landfilling, and overall levels may decline." (BioIntelligence Service, 2011)

In the light of these trends, there are likely to be further governmental requirements for retailers to measure the sustainability of their plastic packaging and other plastic products. An independent labeling system is likely to be more broadly adopted, and the best-practice guidelines are likely to be put in place to minimize overall environmental impacts and for collection and recovery, at least in Europe at the start but with the likelihood that such measures would be more broadly adopted.

Areas of concern with end-of-life management for starch-based plastics include the fact that the end market for compost products may not be large enough yet, particularly outside Europe, to drive this technology as a viable option. The sorting of plastics is also a critical issue for proper management, and the variety of bioplastics with different degradation times (even just within the starch-based category) might complicate the process. Biodegradable starch-based plastic waste should fit within the requirements of composting or digestion (anaerobic degradation) facilities when recycled or be incinerated; disposal to landfill is a poor choice for these materials.

Other issues of concern are that due to the low volume of biopolymers and because conventional recycling facilities are not designed to accommodate bioplastics, there is a short-term risk of contamination of such plants. Over the longer term, dedicated plants will be required. Also, the biodegradability/compostability of thick-walled parts a few millimeters thick as opposed to micron thin films needs to be taken into account.

10 CONCLUSIONS AND PERSPECTIVES

Overall, the starch-based plastics market, although still in the relatively early stages of development, is very healthy and likely to experience strong growth over the coming years. There are a number of strong market drivers that are fuelling this growth that are industry based as well as governmental and societal in origin. There are now some well-established producers with significant production levels and established market penetration, as well as a number of newcomers to the field, particularly in Asia, that have the potential to be strongly competitive in this market. As new technologies are developed, so the potential application areas for these materials broaden. Further research is clearly needed in a number of areas, such as durability, processing techniques, end-of-life issues, and development of tailored formulations to broaden the application areas. A particular area of innovation is natural fiber-reinforced biocomposites and nanocomposites. A lot more research is needed to improve wettability and adhesion through suitable surface changes and to overcome problems such as high batch-to-batch variability, powder formation, fiber thermodegradation, and yellowing during compounding and processing. Additional research into customer requirements is needed, with a focus on developing formulations that can meet the specifications for a wide range of applications, biopolymer aging and weathering, processability and finishing techniques, and compostability/degradability.

Acknowledgments

P. J. H. and B. G. L. would like to acknowledge their colleagues and students in biodegradable polymers. Thanks to the CRC for the International Food Manufacture and Packaging Science, the Australian Research Council, the University of Queensland, and the Plantic Technologies for research funding.

References

Adler, I., & Di Stefano, A. (2012a). *Versalis to partner with genomatica and novamont for bio-based butadiene.* from Eni, Novamont and Genomatica, viewed 22/10/2013, http://globenewswire.com/news-release/2012/07/24/274259/263131/en/Versalis-to-Partner-With-Genomatica-and-Novamont-for-Bio-Based-Butadiene.html.

Adler, I., & Di Stefano, A. (2012b). *Novamont launches joint venture for European BDO plant using genomatica's process.* from Novamont and Genomatica, viewed 22/10/2013, http://www.genomatica.com/news/press-releases/novamont-launches-joint-venture-for-bdo-plant-in-europe/.

Association des Amidonniers et Féculiers, (2009). *The markets—Non-food.* from AAF, viewed 2 February 2012, http://www.aaf-eu.org/html/non-food.php.

Bastioli, C. (2005). Starch-based technology. In C. Bastioli (Ed.), *Handbook of biodegradable polymers* (pp. 257–286). Shrewsbury, UK: Smithers Rapra Limited.

BCC Research, (2012). *Plastics: Global markets and technologies for bioplastics.* from BCC Research, viewed 12 April 2012, http://www.bccresearch.com/report/bioplastics-markets-technologies-pls050b.html.

Bellini, A., & Tomasi, A. (2011). *Novamont to announce strategic alliance with AIAB.* from Novamont 29 November 2011, viewed 23 October 2013, http://www.novamont.com/default.asp?id=738&id_n=10909.

BioIntelligence Service, (2011). *Plastic waste in the environment—Final report.* from European Commission (DG Environment), Specific contract 07.0307/2009/545281/ETU/G2 under Framework contract ENV.G.4/FRA/2008/0112, viewed 23 October 2012, http://ec.europa.eu/environment/waste/studies/pdf/plastics.pdf.

Bioplastics Magazine, (2012). *Degradable plastics: A new study from The Freedonia Group, Inc.* viewed 12 March 2012, http://www.bioplasticsmagazine.com/en/books/degradable_plastics.php.

Biosphere® Industries, LLC, (2008). *The world's most environmentally responsible high performance packaging.* from Biosphere Industries, viewed 15 May 2012, http://www.biosphereindustries.com/.

Blazelead, (2013). *Welcome to Maribumi Starchtech Sdn Bhd.* viewed 23 October 2013, http://www.blazelead.com/Maribumi-Starchtech-Sdn-Bhd/aboutus/.

Cereplast, (2011). *Cereplast completes purchase of manufacturing plant in Italy to help serve growing demand for bioplastic materials in Europe.* from Cereplast Oct 26, 2011, viewed 23 October 2013, http://globenewswire.com/news-release/2011/10/26/459489/235957/en/Cereplast-Completes-Purchase-of-Manufacturing-Plant-in-Italy-to-Help-Serve-Growing-Demand-for-Bioplastic-Materials-in-Europe.html?print=1.

Danimer, (2011). *Danimer Scientific llc.* from Danimer, viewed 23 October 2013, http://www.danimer.com/bioplastics/.

Davidson, V. J., Paton, D., Diosady, L. L., & Laroque, G. (1984). Degradation of wheat starch in a single screw extruder: Characteristics of extruded starch polymers. *Journal of Food Science, 49*, 453–458.

de Guzman, D. (2010). *Bioplastic industry remains strong despite recession: Durability under pressure.* from icis.com, viewed 13 March 2012, http://www.icis.com/Articles/2010/03/22/9344130/bioplastic-industry-remains-strong-despite-recession.html.

Disoxidation Macromolecule, (2012). *Our factory.* from Shanghai Disoxidation Macromolecule Co., Ltd., viewed 15 May 2012, http://www.dmmsh.com/factory.html.

Doane, W. M. (2012). *New products from traditional crops; new uses for starch.* from the National Center for Agricultural Utilization Research, ARS, USDA, Peoria, IL, viewed 2 April 2012, http://science-in-farming.library4farming.org/Crops-Uses-Markets/Traditional-Crops.html.

Ecopond, (2010). *Bio-based plastics.* from Kingfa, viewed 15 May 2012, http://ecopond.com.cn/company/Bio_based_Plastics_en.html.

EON, (2008). *Cerestech introduces new family of thermoplastic starch blend resins.* from Enhanced Online News, viewed 15 May 2012, http://eon.businesswire.com/news/eon/20080401105381/en.

European Bioplastics, (2012a). *Europe and beyond.* from European Bioplastics, viewed 23 October 2013, http://en.european-bioplastics.org/market/europebeyond/.

European Bioplastics, (2012b). *First composting certification network coming soon.* European Bioplastic Bulletin, Issue 01/2012, viewed 3 April 2012, http://en.european-bioplastics.org/blog/2012/02/21/composting-certification-network/.

European Bioplastics, (2012c). *Market drivers.* from European Bioplastics, viewed 13 March 2012, http://en.european-bioplastics.org/market/market-development/market-drivers/.

European Bioplastics, (2012d). *Market development.* from European Bioplastics, October 2012, viewed 23 October 2013, http://en.european-bioplastics.org/market/market-development/.

European Bioplastics, (2013). *Market.* from European Bioplastics, viewed 23 October 2013, http://en.european-bioplastics.org/wp-content/uploads/2011/04/Global-Prod-Capacity_Total-en_klein.jpg.

European Commission, (2012). *Innovating for sustainable growth: A bioeconomy for Europe.* from the Commission to the European Parliament, the Council, the European economic and Social Committee and the Committee of the Regions, viewed 23 October 2013, http://ec.europa.eu/research/bioeconomy/pdf/201202_innovating_sustainable_growth.pdf.

Giract Starch Industry Overview, (2011). *Starches and derivatives, impact of the economic downturn. Global production and supply 2009/10–2015.* from GIRACT, Starch Italics, April/May 2011, viewed 23 October 2013, http://www.scribd.com/doc/58817825/Starch-Italics-Tenth-Edition.

Gironi, F., & Piemonte, V. (2011). Bioplastics and petroleum-based plastics: Strengths and weaknesses. *Energy Sources, Part A: Recovery, Utilization, and Environmental Effects, 33* (21), 1949–1959.

Global Market, (2013). *Wuhan Huali Environmental Technology Co., Ltd.* viewed 23 October 2013, http://ecoplastech.gmc.globalmarket.com/company.html.

Glycan Biotechnology, (2008). *Glycan.* from Glycan Biotechnology Co., Ltd., viewed 15 May 2012, http://www.glycan-biotech.com/r2.html.

Gomez, M. H., & Aguilera, J. M. (1983). Changes in the starch fraction during extrusion-cooking of corn. *Journal of Food Science, 4*, 378–381.

Gozzo, A., & Glittenberg, D. (2009). Starch: A versatile product from renewable resources for industrial applications. In R. Höfer (Ed.), *RSC green chemistry no. 4. Sustainable solutions for modern economies* (pp. 238–263). London: The Royal Society of Chemistry.

Grabio, (2011). *Grabio bioplastics for life.* viewed 15 May 2012, http://www.grabio.com.tw/.

GreenLight Products Pty Ltd., (2011). *Environmentally sustainable void fill packaging from GreenLight Products.* viewed 15 May 2012, http://www.greenlightproducts.co.uk/.

GreenPoly, (2005). *Greenpoly company.* from GreenPoly Group, viewed 15 May 2012, http://www.greenpoly.com/.

Guangdong Shangjiu Biodegradable Plastics Co., Ltd, (2012). *Biodegradable corn starch based plastic resin.* viewed 15 May 2012, http://shangjiu.en.alibaba.com/product/303770273-200428309/Biodegradable_corn_starch_based_plastic_resin.html.

Halley, P. J., Truss, R. W., Markotsis, M. G., Chaleat, C., Russo, M., Sargent, A. L., et al. (2007). A review of biodegradable thermoplastic starch polymers. In M. Celina & R. Assink (Eds.), *Polymer durability and radiation effects* (pp. 287–300). *ACS symposium series*Washington, DC: American Chemical Society.

Harding, K. G., Dennis, J. S., von Blottnitz, H., & Harrison, S. T. L. (2007). Environmental analysis of plastic production processes: Comparing petroleum-based polypropylene and polyethylene with biologically-based poly-β-hydroxybutyric acid using life cycle assessment. *Journal of Biotechnology*, *130*, 57–66.

Hottle, T. A., Bilec, M. M., & Landis, A. E. (2013). Sustainability assessments of bio-based polymers. *Polymer Degradation and Stability*, *93*, 1898–1907.

IMCD, (2011). *IMCD plastics and Rodenburg biopolymers join forces in biopolymers.* from IMCD Group H.V., viewed 13 May 2012, http://www.imcdgroup.com/wp-content/uploads/PRESS%20RELEASE_Rodenburg%20Biopolymers.pdf.

Japan Cornstarch Co., Ltd, (2012). *Biodegradable plastics created from biomass.* viewed 15 May 2012, http://www.nihon-cornstarch.com/product/bio_plastic/tabid/160/Default.aspx.

Lerner, I. (2012). *Market players are upbeat for the prospects of bioplastics.* from icis.com, viewed 12 April 2012, http://www.icis.com/Articles/2012/03/12/9539984/market-players-are-upbeat-for-the-prospects-of-bioplastics.html.

Limagrain, (2012). *Lancement du projet industriel biolice au Brésil.* from Limagrain, viewed 22/10/2012, http://www.limagrain.com/actualites/2012-2009/lancement-du-projet-industriel-biolice-au-bresil/article-410/fr.html.

Maribumi Starchtech Sdn. Bdh, (2012). *Zero-plastic resin made from starch.* from Alibaba Starch, viewed 15 May 2012, http://my1002160627.trustpass.alibaba.com/product/118312165-103311915/ZERO_PLASTIC_Resin_made_from_Starch.html.

Mattila, T., Kujanp, M., Dahlbo, H., Soukka, R., & Myllymaa, T. (2011). Uncertainty and sensitivity in the carbon footprint of shopping bags. *Journal of Industrial Ecology*, *15*, 217–227.

Mohan, A. -M. (2009). *Bioplastics market expands despite recession.* from Greener Packaging, viewed 10 April 2012, http://www.greenerpackage.com/bioplastics/bioplastics_market_expands_despite_recession.

Mohan, A. -M. (2010). *Biodegradable polymers market to grow at 13% through 2014.* from Greener Packaging, viewed 12 April 2012, http://www.greenerpackage.com/bioplastics/biodegradable_polymers_market_grow_13_through_2014.

Mohan, A. -M. (2011). *World demand for bioplastics to exceed 1 million tons in 2015.* from Greener Packaging, viewed 12 April 2012, http://www.greenerpackage.com/bioplastics/world_demand_bioplastics_exceed_1_million_tons_2015.

Mooney, B. P. (2009). The second green revolution? Production of plant-based biodegradable plastics. *Biochemistry Journal*, *418*, 219–232.

NNZ, (2012). *BioPackaging.* from NNZ, the Packaging Network, viewed 15 May 2012, http://www.nnz.nl/us/products/agricultural/bio/okopack-tray.

Novamont, (2013). *Latest milestone from Novamont's research with an extraordinary reindustrialisation potential: The fourth generation of Mater Bi®.* from Novamont 17 June 2013, http://www.novamont.com/default.asp?id=738&id_n=23319.

NVYRO, (2012). *Welcome to NVYRO—The Edible Plate Company.* from The Edible Plate Company, viewed 15 May 2012, http://nvyro.com/.

PaperFoam, (2013). *PaperFoam biobased packaging.* from PaperFoam, viewed 23 October 2013, http://www.paperfoam.com/.

Patel, M., Bastioli, C., Marini, L., & Würdinger, E. (2005). *Life-cycle assessment of bio-based polymers and natural fiber composites. In Biopolymers online.* Wiley-VCH Verlag GmbH & Co. KGaA viewed 23 March 2012, *http://onlinelibrary.wiley.com/doi/10.1002/3527600035.bpola014/full.*

Plackett, D. (2011). Introductory overview. In D. Plackett (Ed.), *Biopolymers—New materials for sustainable films and coatings* (pp. 3–15). (1st ed.). Chichester, UK: John Wiley & Sons, Ltd..

Plantic Technologies, (2011). *Sustainability: At a glance.* from Plantic Technologies, viewed 12/4/2012, http://www.plantic.com.au/Sustainability/at-a-glance.html.

Plasteurope.com, (2010). *JC Hagen; Biodegradable plastics added to portfolio.* from Plasteurope.com, viewed 13 May 2012, http://www.plasteurope.com/news/detail.asp?id=215163.

Plasteurope.com, (2011a). *Controversial ban on non-degradable plastic bags in force since 1 January/Producers face retooling.* from Plasteurope.com, viewed 2 April 2012, http://www.plasteurope.com/news/ITALY_t218185.

Plasteurope.com, (2011b). *China.* from Plasteurope.com, viewed 2 April 2012, http://www.plasteurope.com/news/CHINA_t220808.

Plasteurope.com, (2012). *Roquette.* from Plateurope.com, from Plasteurope.com, viewed 13 May 2012, http://www.plasteurope.com/news/ROQUETTE_t221654.

Plastics News, (2012). *FYI charts: World bioplastics demand.* from Plastics News, viewed 12/04/2012, http://plasticsnews.com/fyi-charts/index.html?id=24421.

Plastral, (2012). *Bioplastics.* from Plastral, viewed 23/10/2013, http://www.plastral.com.au/page_bio.php?id=21.

Platt, D. K. (2006). *Biodegradable polymers—Market report.* Shawbury, UK: Smithers Rapra Limited.

PotatoPak, (2010). *100% Biodegradable products.* from PotatoPak, viewed 15 May 2012, http://www.potatoplates.com/.

Prevéraud, J. -F. (2006). *Les bioplastiques, futur des emballages de la distribution et de l'industrie agroalimentaire?* from Industrie & Technologies, viewed 13 April 2012, www.industrie.com/it/materiaux/les-bioplastiques-futur-des-emballages-de-la-distribution-et-de-lin dustrie-agroalimentaire.

PRNewswire, (2012). *Bioplastics—Global market overview.* from IndustryExperts, viewed 10 April 2012, http://www.prnewswire.com/news-releases/bioplastics—global-market-overview-136647193.html.

Roberts, A. (2010). *BioPlastics.* from Plantic, viewed 23/10/2012, https://docs.google.com/viewer?a=v&q=cache:-hpglsEA488J:ausbiotech2010.com.au/uploads/Speaker%2520presentations/Ann%2520Roberts.pdf+Future+proofing+against+rising+oil+and+plastic+prices.&hl=en&gl=au&pid=bl&srcid=ADGEESi8WH_glAs59Cc6jz_3VLr9BKv7QuPocdY3Ip8QXTYOHRdlNXWX6h2BixcefwZTch_91Xo_i1gh2AQX8c1fPEZZOANrR59ASO-Y55WeNB3OmAAMqwtZv4RYfMQP13z20A-y0TCd&sig=AHIEtbSokQX-3XdLrNarWItzXG5vvN6cOQ.

Sagar, A. D., & Merrill, E. W. (1995). Properties of fatty-acid esters of starch. *Journal of Applied Polymer Science, 58*, 1647–1656.

Sarnacke, P., & Wildes, S. (2008). *Disposable bioplastics consumer disposables agricultural films: A market opportunity study.* from OmniTech International for the United Soybean Board, viewed 20 April 2012, http://soynewuses.org/wp-content/uploads/pdf/DisposalblePlasticsMOS.PDF.

Schut, J. H. (2008). *What's ahead for 'green' plastics: Look for more supply, more varieties, better properties.* from Plastics Technology, from Plastics Technology, viewed 5 April 2012, http://www.ptonline.com/articles/what's-ahead-for-'green'-plastics-look-for-more-supply-more-varieties-better-properties.

Sfgov, (2007). *Food service waste reduction ordinance.* from City & County of San Francisco, viewed 2 February 2012, http://www.sfgov2.org/index.aspx?page=886.

Shah, A. A., Hasan, F., Hameed, A., & Ahmed, S. (2008). Biological degradation of plastics: A comprehensive review. *Biotechnology Advances*, *26*(3), 246–265.

Shen, L., Haufe, J., & Patel, M. K. (2009). *Product overview and market projection of emerging bio-based plastics (PRO-BIP 2009).* commissioned by European Polysaccharide Network of Excellence (EPNOE, www.epnoe.eu) and European Bioplastics (www.europeanbioplastics.org), Utrecht, The Netherlands.

Shen, L., Worrell, E., & Patel, M. (2009). Present and future development in plastics from biomass. *Biofuels, Bioproducts and Biorefining*, *4*, 25–40.

Smith, C. (2010). *Novamont suffers set-back in bioplastics case.* from European Bioplastics, viewed 13 May 2012, http://www.europeanplasticsnews.com/subscriber/headlines2.html?cat=1&id=1273048901.

Smithers Pira, (2010). *PHA and bio-derived PE to drive Bioplastic packaging market to 2020.* from Smithers Pira, viewed 12 April 2012, http://www.smitherspira.com/pha-and-bio-derived-pe-to-drive-bioplastic-packaging-market-to-2020.aspx.

Song, J. H., Murphy, R. J., Narayan, R., & Davies, G. B. H. (2009). Biodegradable and compostable alternatives to conventional plastics. *Philosophical Transactions of the Royal Society Series B Biological Sciences*, *364*, 2127–2139.

SpecialChem, (2011). *BIOP biopolymer technologies receives product excellence award for BIOPAR®.* from Frost & Sullivan, viewed 23 October 2012, http://www.plastixanz.com.au/news_article.asp?NewsID=3344.

SRI Consulting, (2010). *Biodegradable polymers.* from SRI Consulting, viewed 12/03/2012, www.sriconsulting.com/CEH/Public/Reports/580.0280/.

StarchTech Inc., (2012). *Frequently asked questions.* from StarchTech Inc, viewed 13 May 2012, http://www.starchtech.com/frequently-asked-questions.html.

Sustainability Victoria, (2007). *Comparison of existing life cycle analysis of shopping bag alternatives.* from Sustainability Victoria, viewed 12/4/2012, http://www.sustainability.vic.gov.au/resources/documents/LCA_shopping_bags_full_report%5B2%5D.pdf.

Sustainable biomaterials collaborative, (2009). *Guidelines for sustainable bioplastics.* viewed 2 February 2012, http://www.sustainablebiomaterials.org/criteria.guidelines.overview.php.

Tecnaro GmbH, (2012). *Arboblend.* from Tecnaro, viewed 13 May 2012, http://www.tecnaro.de/english/arboblend.htm?section=arboblend.

Tianhe Environmental Technology Co., Ltd, (2013). *Tianhe Environmental Technology.* viewed 22 October 2013, http://www.chinatianhe.com/en/About-1.html.

USPRwire, (2009). *Global biopolymers market analysis and forecasts to 2015.* viewed 13 April 2012, http://www.usprwire.com/Detailed/Environment/Global_Biopolymers_Market_Analysis_and_Forecasts_to_2015_70377.shtml.

Vegeplast, (2012). *Vegeplast.* viewed 15 May 2012, www.vegeplast.com/en/vegemat-2/.

Vilpoux, O., & Avérous, L. (2004). Starch-based plastics. In M. P. Cereda & O. Vilpoux (Eds.), *Technology, use and potentialities of Latin American starchy tubers, Book number 3* (pp. 521–553). São Paolo, Brazil: NGO Raízes and Cargill Foundation.

Voegele, E. (2011). *Biome reports significant increase in bioplastic sales.* from Biorefining Magazine, viewed 22 October 2013, http://biomassmagazine.com/articles/7281/biome-reports-significant-increase-in-bioplastic-sales/.

Weiku, (2011). *Beijing Macrolink Biomaterial Co., Ltd.* viewed 22 October 2013, http://mbm-bio.company.weiku.com/about/.

Weir, B. (2011). *Cardia Bioplastics research and analysis.* from NNFCC, viewed 12/04/2012, http://www.stockmarketsreview.com/news/cardia_bioplastics_research_and_analysis_20110831_159940/.plastics.from StockMarketsReview.com, viewed 13 May 2012, http://www.nnfcc.co.uk/publications/nnfcc-renewable-polymers-factsheet-bioplastics.

Williams, J. (2010). *NNFCC renewable polymers factsheet: Bioplastics,* from NNFCC, viewed 12/04/2012. http://www.nnfcc.co.uk/publications/nnfcc-renewable-polymers-factsheet-bioplastics.

World Centric, (2004). *Compostable plastics.* from World Centric, viewed 10 February 2012, http://www.worldcentric.org/biocompostables/bioplastics.

Wuhan Greenplas, *New products. viewed 15 May 2012, http://www.greenplas.com/en/chanpin/index.asp?tid=4.*

Xie, F., Halley, P. J., & Avérous, L. (2011). Bio-nanocomposites based on starch. In V. Mittal (Ed.), *Nanocomposites with biodegradable polymers: Synthesis, properties, and future perspectives* (pp. 234–260). Oxford, UK: Oxford University Press.

Zhang, Y., & Rempel, C. (2012). *Retrogradation and antiplasticization of thermoplastic starch. In A. El-Sonbati (Ed.), Thermoplastic elastomers*: InTech Open Access*http://www.intechopen.com/books/thermoplastic-elastomers/retrogradation-and-antiplasticization-of-thermoplastic-starch.*

CHAPTER

15

Starch Plastic Packaging and Agriculture Applications

Gregory M. Glenn, William Orts, Syed Imam, Bor-Sen Chiou, Delilah F. Wood

USDA-ARS, Western Regional Research Center, Bioproduct Chemistry and Engineering, Albany, California, USA

1 INTRODUCTION

Paper constitutes the most important material in the United States for packaging and containers largely because of its low cost and wide availability (WPO, 2008). It is also perceived as a sustainable material because it is derived from plants and is recycled at a very high percentage (62%) (EPA, 2010). Plastic ranks as the second most used packaging material in the United States. Plastics can provide transparency, greater moisture protection, and various mechanical properties that are superior to paper packaging. Consequently, some types of plastic packaging continue to grow faster than other packaging materials (WPO, 2008). In contrast to paper, only 7% of plastic generated as waste is recycled. This explains why more plastics ultimately end up in landfills than paper or any other packaging material (EPA, 2010). Plastic processors worldwide are becoming increasingly aware that environmentally sustainable packaging has become mainstream. It can no longer be considered only a niche market that can be ignored or given token attention. Informed consumers are demanding sustainable packaging; state and local governments are mandating it; and now, even the largest retailers are building it into the foundation of their marketing strategies (Deligio, 2009; Wood, 2010).

Packaging and containers constitute a nearly $500 billion global market (WPO, 2008) and make up the largest sector (29.5%) of municipal solid waste (MSW) in the United States (EPA, 2010). Plastic packaging-container sales alone account for $130 billion worldwide (WPO, 2008). The development of renewable/sustainable plastics is perceived by the industry as a hedge against the prospect that traditional plastics will be much more costly in the future due to dramatically higher petroleum prices. The sustainability movement is further seen as a positive development for plastic processors since it will drive further innovation and a new generation of materials with properties more comparable to commodity plastics (Wood, 2010).

Until recently, the biopolymers most often discussed in reference to sustainability were starch-based polymers, cellulose, polylactic acid (PLA), and polyhydroxyalkanoate (PHA). The most recent trends have been to broaden the spectrum of sustainability to include materials and practices that move a step closer toward sustainability by simply increasing renewable content or decreasing the overall weight of traditional plastics used. Today's sustainable plastics are not necessarily biodegradable and even include polyolefins made from renewable feedstocks (Dartee, 2010). Broadening the scope of sustainability has helped balance the need for providing affordable packaging today while stimulating investment in research to develop more sustainable alternatives for tomorrow (Dartee, 2010).

Industry announcements regarding new and innovative plastic products occur on an ever more frequent basis. Coca-Cola recently announced they will begin utilizing polyethylene terephthalate (PET) bottles containing 30% renewable content from sugarcane-derived ethylene glycol. They also announced plans to convert all their plastic packaging to the new material by 2020 (Anonymous, 2011c; Whistler & Daniel, 1984). Heinz will use the same material to make 120 million bottles for their ketchup products this year. PepsiCo claims to have developed the world's first totally biobased PET bottle. It is made from biomass including switchgrass, pine bark, and corn husks. Pilot-scale production began in 2012 (Anonymous, 2011g). Other interesting new materials entering the market include a new family of resins (Panacea) containing 10-40% finely ground soy-based protein and an injection mold-grade cellulose-based resin. The cellulose-based resin is being used to make the first biodegradable tubes for toothpaste (Anonymous, 2011a, 2011b, 2011d).

Much of the focus on renewable and sustainable plastics involves the use of starch either as a feedstock or as a component. Although a stable food product, starch is also increasingly used for industrial products. Industrial products in the United States that utilize starch have grown from 13 million metric tons (MMT) in 1975 to over 160 MMT today (USDA, 2010). Starch is inexpensive, widely available, and one of the most abundant biomass products in nature (Schwartz & Whistler, 2009; Whistler, 1984). It is produced in many different plant organs including roots, leaves, seeds, and stems. Commercial starches worldwide are extracted primarily from cereal grains such as corn (82%) and wheat (8%) and tuber crops such as potato (5%) and cassava (5%) (Le Corre, Bras, & Dufresne, 2010). Millions of metric tons of starch are used annually for making nonfood products including paper sizing, adhesives, gypsum wallboard, and sizing for textiles (USDA, 2010; Whistler, 1984). Much of the recent growth in industrial use of starch can be attributed to the production of ethanol (USDA, 2010).

The hydrophilic nature of starch and its tendency to embrittle with age do not make it very suitable as a replacement for plastics. In spite of its shortcomings, starch continues to remain a viable alternative to petroleum-based plastics primarily because of its low cost and abundance and because it is a renewable material that degrades readily in composting conditions and in many landscape and aquatic environments (Law et al., 2010; Thompson et al., 2004). This chapter discusses starch as a material and addresses some of the global concerns of plastics in the environment, the early development of starch-based plastics, its properties and limitations as a material, and some of the promising technologies entering the marketplace.

2 STARCH PROPERTIES

A considerable volume of literature has been published on the properties and chemistry of starches from various sources (Schwartz & Whistler, 2009; Whistler, Bemiller, & Paschall, 1984). Starch is the principal carbohydrate for energy storage in plants and one of the most abundant plant polymers (Whistler, 1984). In photosynthetic plant organs, a fraction of the carbon fixed in plant chloroplasts is retained as starch and used at night during the respiration process (Zeeman, Kossmann, & Smith, 2010). In nonphotosynthetic organs such as roots, tubers, and seeds, sucrose is converted to starch in specialized plastids known as amyloplasts (Tyson & Ap Rees, 1988; Zeeman et al., 2010). Plant starches synthesized in amyloplasts are formed into cold water-insoluble granules (Figure 15.1) that range from a few micrometers to more than 100 μm depending on the plant source (French, 1984; Tyson & Ap Rees, 1988). Starch can be easily extracted from a variety of starch-rich crops and dried to form a white powder. Its commercial production dates back to the middle ages and has advanced to become the efficient global industry it is today (Schwartz & Whistler, 2009).

Starch is composed of two polymers consisting of D-glucose repeating units, amylose and amylopectin. Amylose is essentially a linear polymer consisting of $\alpha[1 \rightarrow 4]$-linked glucose moieties, while amylopectin is a much larger, highly branched molecule with $\alpha[1 \rightarrow 6]$ branching linkages. The amylose-amylopectin ratio within a starch granule varies with the

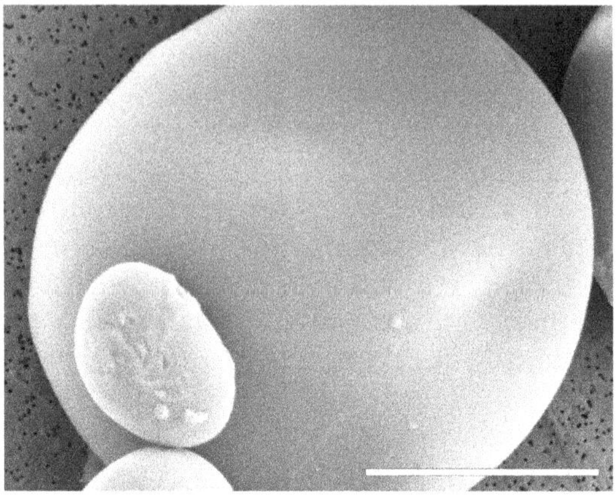

FIGURE 15.1 Scanning electron micrograph of native wheat starch granules. The granules have a bimodal size distribution consisting of small- and medium-size granules (scale bar = 10 μm) (Glenn et al., 2008).

botanical origin of the starch and can vary due to weather and growing conditions (Hoover, 2001; Morrison, Milligan, & Azudin, 1984). Normal wheat starch has 18-30% amylose content with the remainder being amylopectin (Singh, Singh, Kaur, Singh Sodhi, & Singh Gill, 2003). Starch from normal potato genotypes contains 23-31% amylose. There are now waxy genotypes for most of the major starch-rich crops, including potato, wheat, corn, and rice, that are comprised almost entirely of amylopectin (Buleon, Colonna, Planchot, & Ball, 1998; Hermansson & Svegmark, 1996). There are also genotypes with much higher than normal amylose content (Shi, Capitani, Trzasko, & Jeffcoat, 1998). These genotypes provide starches with properties markedly different from starches of normal genotypes (Shi et al., 1998).

Amylose is primarily a linear polymer composed of 500-20,000 $(1 \rightarrow 4)$ linked α-D-glucopyranosyl units (Shannon & Garwood, 1984; Whistler & Daniel, 1984). Amylose forms into a helical structure with a hydrophobic core due to the arrangement of hydrogen groups exposed in the interior of the helix (Almeida et al., 2010; Buleon et al., 1998). Amylose is able to form complexes with fatty acids, iodine, and alcohols (Almeida et al., 2010). In the case of fatty acids, the aliphatic part of the molecule is believed to reside inside, while the polar group lies outside of the helical structure (Buleon et al., 1998). There is some evidence that within the starch granule, amylose is localized in the amorphous regions (Shamekh, Forssell, Suortti, Autio, & Poutanen, 1999). However, others have found it difficult to assign a specific location for amylose and suggest that it may be randomly dispersed in the granule matrix (Atkin, Abeysekera, Cheng, & Robards, 1998; Oates, 1997). Amylose is a mobile molecule that leaches out of the granule when hydrated (Fannon & Bemiller, 1993).

Amylopectin is one of the largest known natural polymers (~10^7-10^9 Da) (Buleon et al., 1998; Singh et al., 2003) made up of α-(1→4) D-glucopyranose chains with α-(1→6) branches (Shannon & Garwood, 1984; Whistler & Daniel, 1984). About 5% of the glycosidic linkages are α-(1→6) branches. X-ray diffractometry has been used to study the crystallinity of starch granules (Hoover, 2001). Diffraction patterns that are characteristic of different plant types have been defined. Cereal starches have an A-type X-ray pattern, tuber starches generally have a B-type pattern, and legumes have mixed pattern defined as C type (Hoover, 2001; Sarko & Wu, 1978; Singh et al., 2003). Starch granules contain between 15% and 45% of crystallite material (Blanshard, 1987). Amylopectin provides the bulk of the matrix for the starch granule and it contains the crystalline fraction of the granule (Oates, 1997). The branches of the amylopectin form double helices and are arranged in crystalline domains (Sarko & Wu, 1978). The crystalline regions are believed to be hard and brittle but may be interspersed with amorphous regions of amylopectin that are assumed to have greater mobility and behave more like a rubbery polymer (Morgan, Furneaux, & Larsen, 1995). Some large starch granules have onion-like layers, suggesting that they grow by the deposition of layer upon layer (Martin & Smith, 1995) of carbohydrate. However, this structure has not been observed in small starch granules from rice or barley (Oates, 1997).

Starch granules have a relatively smooth surface (Figure 15.1) that is impervious to large molecules due to the tight packing of amylopectin molecules (French, 1984). However, the granules are hydrophilic and begin to hydrate and swell when heated in water (Fannon & Bemiller, 1993; Hermansson & Svegmark, 1996; Richardson, Kidman, Langton, & Hermansson, 2004). The behavior of starch when heated in water reveals the complex nature of the granule structure. During heating, the starch chains within the granule become more mobile as hydrogen bonding between adjacent glucose units is disrupted and water penetrates the matrix (Rodriguez-Gonzalez, Ramsay, & Favis, 2003). The increase in chain mobility allows the granule to become more elastic and to swell. At the same time, amylose and small amounts of amylopectin begin to leach from the granule matrix into the aqueous medium (Fannon & Bemiller, 1993; Ghiasi, Hoseney, & Varriano-Marston, 1982; Oates, 1997; Shamekh et al., 1999). Granule hydration may be further accelerated by the presence of small pores that reportedly span from the granule surface to the core region (Baldwin, Adler, Davies, & Melia, 1994; Huber & Bemiller, 2000; Kim & Huber, 2008).

Micrographs of starch granules from heated aqueous slurries reveal that starch granules do not simply dissolve in an "outside to inside" fashion as might be expected. Instead, they appear to dissolve from the "inside out" (Atkin et al., 1998; Glenn et al., 2008; Li, Vasanthan, Hoover, & Rossnagel, 2004) (Figure 15.2a). This pattern of starch dissolution

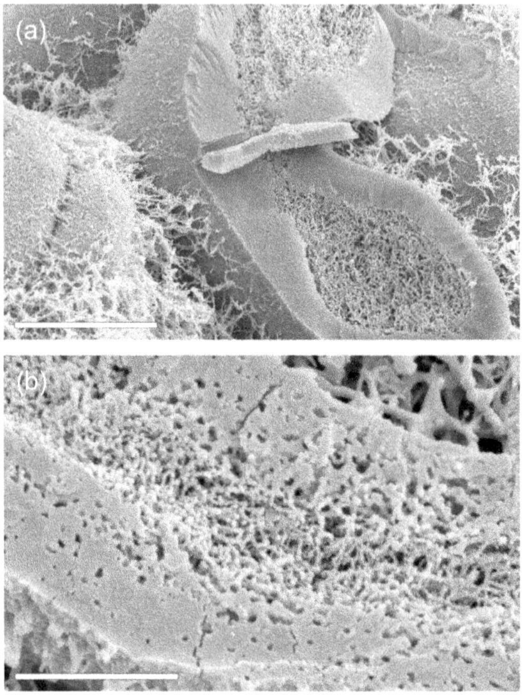

FIGURE 15.2 Scanning electron micrographs of wheat starch granules after heating to 95 °C for (a) 1 min and (b) 60 min. Note the starch granule appears to dissolve from the inside out (a). The granule remnant or envelope remains insoluble after extensive heating although it does become more porous (b) (scale bars = 5 and 2 μm, respectively) (Glenn et al., 2008).

may be due more to the insolubility of the granule surface and underlying material that has been termed collectively as the starch granule remnant or envelope. The starch granule envelope is believed to be comprised mostly of amylopectin. Water may penetrate the interior of the granule through pores or by simply penetrating the hydrated granule surface by diffusion. Swelling of the granule as it hydrates expands the volume of the granule interior that becomes filled with water-solubilized starch material (Figure 15.2a).

Aqueous slurries of starch form a viscous paste when heated. The viscosity of starch pastes increases as granules swell and hydrate. The hydration and solubilization of the granule interior creates an internal pressure that can eventually cause the envelope to rupture and release its contents into the continuous phase. Starch pastes typically increase to a peak viscosity and then drop in viscosity, presumably due to the rupture of the granule envelope (Glenn et al., 2008; Hoseney & Atwell, 1977). The granule envelopes that remain after the granules have ruptured are called "ghosts." The ghosts remain difficult to solubilize with only mild shear conditions, even after extensive cooking (Glenn et al., 2008; Hoseney & Atwell, 1977). The ghosts become more porous after extensive cooking due to extraction of the soluble fraction from the matrix (Figure 15.2b).

Gelatinization is the general term used to describe the breakdown and dissolution of the starch granule in aqueous media. The crystallinity of the granule is progressively destroyed as hydrogen bonding between adjacent glucose units is disrupted (Rodriguez-Gonzalez et al., 2003). Starch is considered fully gelatinized when the starch granules have been dissolved to the point that only starch ghosts remain in the solution (Atkin et al., 1998; Rockland, Jones, & Hahn, 1977). When a hot gelatinous starch melt is cooled, the dispersed amylose molecules reassociate in a process known as retrogradation and form elastic gels (Blaszczak, Fornal, Valverde, & Garrido, 2005; Fannon & Bemiller, 1993; Richardson et al., 2004). Starch gels consist of a three-dimensional matrix of a continuous phase of amylose molecules containing uniformly dispersed ghosts that act as filler (Figure 15.3a and b) (Morris, 1990). The heterogeneous nature of starch pastes may not be a concern in many food and industrial applications and can, in fact, even be beneficial. However, other applications may require a more homogenous melt that can be achieved only by a more complete destructuring of the starch granule using more severe processing conditions (Wittwer & Tomka, 1987). For instance, complete dissolution of the starch granule structure has been achieved in high-amylose cornstarch by processing at elevated temperatures and pressure using a batch reactor (Glenn et al., 2008). Such a destructurized starch matrix would appear as that in Figure 15.3c.

3 PLASTICS AND THE ENVIRONMENT

Much of the concern over the use of plastics today is centered around its impact on the environment. These concerns include sustainability. Nearly 6% of the world's crude oil production is used for making approximately 245 MMT of plastics globally on an annual basis (Johansson, 2009) (Table 15.1). Since plastics are mostly derived from fossil fuel feedstocks, all of the concern over the heavy dependence of the world on nonrenewable resources and the environmental impact of extracting fossil fuels applies to plastics. But while the concern with burning fossil fuels is primarily due to air pollution and the increasing concentration of greenhouse gases in the atmosphere, the challenge with plastics is the sheer volume of solid waste produced and the persistence of plastic pollution in the landscape and marine environment.

Dumping waste into oceans was common before measures were taken to curtail the practice. Still, worldwide, a significant amount of plastic

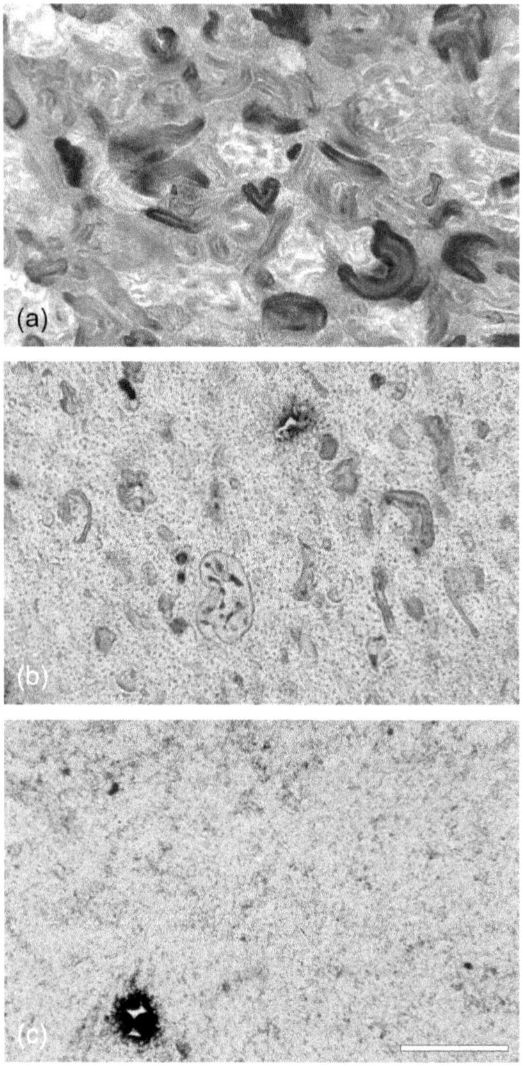

FIGURE 15.3 Light micrograph sections of starch gels stained with iodine potassium iodide. The blue is amylose and pink is amylopectin. Wheat (a), normal amylose corn (b), and high-amylose corn (c) starches (scale bar = 50 μm).

waste is believed to wash out to sea from storm drains, rivers, and waste water (Walker, Reid, Arnould, & Croxall, 1997). China, one of the world's largest plastic producers, noted that the largest source of its marine pollution was from discharging wastewater to sea. Nearly half of China's wastewater is currently dumped at sea (Zou, 2009). This practice worldwide continues to be a concern as a major source of marine pollution.

TABLE 15.1 Plastic Production in the World and United States and Municipal Solid Plastic Waste Generated in the United States in 2010 (Values in Million Metric Tons, MMT)

Global production of plastics (2009)	245[a]
U.S. production of plastics (2010)	46.7[b]
U.S. plastic waste generated—total	29.8[c]
U.S. plastic package waste generated	12.5[c]
U.S. nondurable plastic waste (plates, cups, etc.)	6.65[c]
U.S. durable plastic waste generated	10.7[c]

[a] *Johansson (2009).*
[b] *ACC (2011).*
[c] *ACC (2011), EPA (2010).*

Marine pollution can also be traced to ships and other marine businesses (Zou, 2009). Unlike many other pollutants, plastics have low density and tend to float at or near the water surface where they persist for many years (Boerger, Lattin, Moore, & Moore, 2010). The "Great Pacific Garbage Patch" also known as the North Pacific Central Gyre (NPCG) is an area about the size of the state of Texas where ocean currents gather floating waste (Boerger et al., 2010; Kaiser, 2010). The NPCG is described as the world's largest garbage dump consisting of roughly 80% plastic (Kaiser, 2010; Walker et al., 1997). Since its discovery in 1972, other gyres have been discovered in other oceans that also have a higher than normal content of plastic debris (Kaiser, 2010).

The long-term impact of plastic debris in the marine environment is of concern since birds, mammals, and fish may become entangled in plastic debris or ingest plastic particles (Boerger et al., 2010). There is now evidence that exposure from ultraviolet radiation and mechanical agitation from the movement of waves help slowly disintegrate plastics over time into microscopic debris that floats near the water surface (Kaiser, 2010). Microscopic debris can absorb chemical toxins from seawater, thus presenting an even greater threat to marine life that ingest the plastic debris (Kaiser, 2010). Recent studies have found microscopic plastic debris in plankton, thus raising fears of the accumulation of toxins and plastic debris in sea life that ultimately thrive on plankton (Kaiser, 2010; Zhang, Zhang, Feng, & Yang, 2010). The source of plastics includes both user-generated plastics as MSW and industrial waste that includes nurdles or plastic pellets used as a feedstock for plastic processing (Provencher, Gaston, Mallory, O'hara, & Gilchrist, 2010). It is believed most plastic pellets come from spills on land that are washed into waterways and eventually released into the ocean (Weiss, 2006). Because seabirds confuse the pellets for food, seabird populations are particularly susceptible to industrial plastics (Provencher et al., 2010; Weiss, 2006). The long-term impact of marine pollution from plastics is uncertain, but the issue is garnering more attention and is changing the way waste is being handled (Kaiser, 2010; Zhang et al., 2010).

4 PLASTIC USES

4.1 Agricultural Uses

4.1.1 Films

In Europe alone, an estimated 2-3 million tons of plastics is used each year in agricultural applications (Kyrikou & Briassoulis, 2007). Polyethylene films are used extensively to increase yields, extend growing seasons, reduce the usage of pesticides and herbicides, and help conserve water. About half of the agricultural plastic is film used for agricultural mulches, greenhouses, silage covers, tunnels over rows of crops, and hotboxes (Kyrikou & Briassoulis, 2007). Ideally, these films could be made of renewable polymers. Films made of starch blends were some of the first films containing renewable content tested as agricultural mulch (Otey, Mark, Mehltretter, & Russell, 1974). More recently, PLA, a polymer derived from starch fermentation, was blended with fiber to make composite films for mulch applications (Finkenstadt & Tisserat, 2010). In spite of the research efforts, petroleum-based polyolefin films continue to be the dominant material used for agricultural applications. The environmental impact of plastic films used for agricultural mulches should account for the beneficial impact of mulches including a reduction in the use of pesticides, herbicides, water, and energy. The films may be serviceable during a single growing season or for multiple years depending on the crop and the cultural practices employed (Finkenstadt & Tisserat, 2010).

4.1.2 Specialized Uses

Plastics have found uses in various niche applications in agriculture. These applications range from binders for feeds to soil amendments. One interesting application for plastics is in making controlled-release devices. Many different types of controlled-release systems have been developed for agricultural applications including hollow fiber membranes, capillaries, and polymer devices that swell and absorb the active agent (Funes, Zerba, & González Audino, 2009; Teixeira, Miller, Epstein, & GUT, 2010). These devices tend to be nonbiodegradable and may require that the spent devices be recovered, thus incurring additional labor expense.

Starch-based controlled-release devices have been made for controlling parasitic mites in honeybee colonies (Glenn et al., 2006). Reservoir-type devices were made by using a solvent displacement process to form starch gels containing an active ingredient, 2-heptanone (Erickson, Degrandi-Hoffman, Becker, Whitson, & Deeby, 2005). A heat-sealable food-grade vapor barrier film was used to envelope the starch gel and control the vapor transmission rate. The controlled-release devices were made of degradable food-grade materials that obviated the need to remove spent devices from honeybee colonies.

Starch-based microspheres (<10 μm) were also developed for mite control in honeybee colonies for delivery of less volatile natural control agents. The microspheres were loaded with 25% active ingredient yet remained a free-flowing powder. The microspheres were designed to deliver active ingredients to individual honeybees by adhering to the hairs on their thoraxes in a manner similar to pollen grains (Glenn et al., 2010).

4.2 Consumer and Industrial Products

Many of the plastic products that enter the marketplace eventually find their way to an MSW facility. The U.S. Environmental Protection Agency has kept MSW data for several decades on materials recycled or disposed of as MSW. Plastic waste began to appear in the MSW in 1970 and has grown steadily to total approximately 30 MMT (Figure 15.4). Disposal

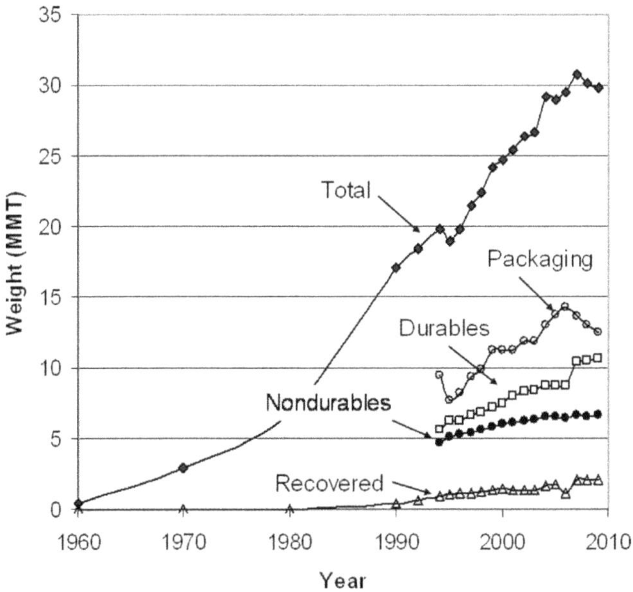

FIGURE 15.4 Plastic waste generated in the United States over the last 50 years. The data include plastic packaging waste, plastic from durable and nondurable products, and amount of recycled plastics (EPA, 2010).

of plastic waste in U.S. landfills has been a concern due, in part, to the sheer volume and the diminishing landfill capacity near urban centers (EPA, 2010). Plastic discards are categorized by the EPA as durable goods, packaging and containers, and nondurable goods (EPA, 2010) (Table 15.1).

4.2.1 *Durable Goods*

Plastic durable goods make up about 35% of the plastics deposited in landfills (Figure 15.4). Durable plastic products take advantage of the excellent chemical and moisture resistance, low density, superior mechanical properties, and low cost of plastic materials. While the public perception may be that all plastics need to be more degradable, that is not the case with this large sector of plastic products. Durability of plastics is essential in an ever-increasing number of products such as plastic auto parts, household appliances, and construction materials, including pipes for residential and industrial water and sewer lines. Some of these durable plastic goods have a service life of several decades or longer.

Assessing the environmental impact of durable plastic products can be complex since the environmental costs and benefits of such products need to be carefully examined as well (Cellura, Ardente, & Longo, 2012; Wikström & Williams, 2010; Williams, Wikstrom, & Lofgren, 2008). For instance, the environmental cost of the plastic used in manufacturing cars should be weighed against the benefit of lower gas consumption due to reduced vehicle weight. The environmental impact assessment should also consider the recycling rate, which is very low (3.8%) for plastic durable goods (EPA, 2010).

4.2.2 *Packaging and Containers*

Packaging and containers make up the largest sector (29.5%) of plastic waste in MSW (EPA, 2010) (Figure 15.4). Plastic packaging has become an integral part of the global marketplace (WPO, 2008). From Internet shopping to retail stores, packaging fills a critical role in not only protecting products during transport but also displaying products and providing consumer appeal. Changes in package design can significantly stimulate or depress product sales (Lee, Gao, & Brown, 2010). Still, packaging design has begun integrating sustainability as never before, in part, because sustainability itself has become a marketing angle. Retailers are now realizing that customers respond positively to products marketed in more sustainable or green packaging. Retailers have taken measures to require manufacturers and suppliers to implement sustainable packaging goals such as reducing the amount of primary and secondary packaging and downgauging the thickness of their plastic film to reduce the amount of plastic used to package each item (ACC, 2011; Deligio, 2009; Wood, 2010).

While efforts to minimize packaging and reduce its environmental impact must continue to be a priority, it is also important to recognize and acknowledge that packaging serves an essential role in society. Perhaps, no other packaging segment better illustrates this than the food industry. Food packaging is the largest sector of the packaging industry serving one of the largest industries in the world (WPO, 2008). The movement of much of the world's population from an agrarian to an urban society has been possible only by the development of efficient food production and distribution practices. Food packaging has served and continues to serve a critical function in facilitating food transport, preservation, sanitation, storage, and distribution from the farm to the masses throughout the world.

It can be argued that the impact of food packaging in minimizing food waste more than compensates for the resources used in the packaging itself. Estimates claim that 25-50% of the food produced globally is wasted (Mena, Adenso-Diaz, & Yurt, 2011). It may be no coincidence that the amount of food wasted based on per capita food consumption and food waste data (EPA, 2010; USDA, 2002) is much lower (12.5%) in the United States where food packaging is heavily used. Packaging that preserves foods and minimizes waste also conserves all the resources invested in bringing that product to market (Cellura et al., 2012; Wikström & Williams, 2010; Williams et al., 2008). Wasted food consumes the energy and resources used for cultivation, harvest, and transportation to market and requires additional energy for transportation to landfills. The contribution of food waste to methane production, an important greenhouse gas, also has environmental implications (Mena et al., 2011). Other tangible benefits of food packaging are improved food safety and hygiene (Chawengkijwanich & Hayata, 2008; EPA, 2010; Gemili, Yemenicioğlub, & Altınkayaa, 2009; WPO, 2008).

4.2.3 Nondurable Plastic Waste

Nondurable plastic goods consist of trash bags and single-use food service items including plates, bowls, and cups (EPA, 2010). Polystyrene (PS) is the plastic resin used mostly in plastic food service items (79%) and virtually no significant recycling is reported (EPA, 2010). The use of PS for single-use food service items is of particular concern because PS is especially persistent and difficult to degrade in the environment (Gorden, 2006). Furthermore, PS foam products are very lightweight and easily dispersed in the landscape by wind or carried into waterways by storm drains (Gorden, 2006). Much of the research aimed at using biopolymers to replace petroleum-based plastics has focused on developing bags and food service containers. Food service items comprise 3% of the plastic discarded as MSW and total about 0.9 MMT on an annual basis (EPA, 2010). Of all the plastic goods produced, plastic food service items are often considered the best candidate for replacement by renewable biopolymers. This is because materials used in food service items are typically single-use disposable items. Oftentimes, the functional service of a plate or cup amounts to less time than it takes for a hot food or beverage product to become cold. Technology developed to use starch in these products could be further developed to expand to other packaging products.

5 STARCH-BASED PLASTICS

The interest in utilizing starch as a replacement for plastics started in the 1970s and intensified in the 1980s right along with the dramatic growth in the use of plastics worldwide and the concerns about the effects of plastics on the environment (Figure 15.4). By itself, starch is a poor substitute for petroleum plastics due to its moisture sensitivity and inferior mechanical properties. However, numerous strategies have been tested resulting in commercialized technologies for incorporating starch in plastics. A survey of the more common approaches has been included in this section.

5.1 Granular Starch

The early studies on the use of starch in plastics were focused primarily on its beneficial effect as a plastic additive rather than as an ingredient for improving biodegradation or renewable content. The benefits of adding starch to polyurethane foam formulations were investigated more than 50 years ago (Dosmann & Steel, 1961). Starch was included in a mixture of an organic diisocyanate and a polyol resulting in flexible foam products with improved mechanical properties (Dosmann & Steel, 1961). A rigid urethane foam containing starch was also reported (Bennett, Otey, & Mehltretter, 1967; Otey, Westhoff, Kwolek, Mehltretter, & Rist, 1969). In these materials, starch was believed to react with -NCO groups formed by an excess of isocyanate in the reaction mixture (Otey et al., 1969).

Using starch in plastic film was reported first in 1964 (Shulman & Howarth, 1964). Starch powder was dispersed in plastic film that was intended for use in making raincoats. The starch granules increased the water vapor permeability of the film, which allowed it to "breath" and be more comfortable. Other studies investigated the use of a modified granular starch (starch xanthates) in a rubber latex matrix (Buchanan, Weislogel, Russell, & Rist, 1968). The starch xanthate was evaluated as a modifier of latex coagulation, as a reinforcing agent, and as an accelerator of vulcanization (Buchanan et al., 1968).

As concerns arose from the growing use of plastics, research exploring the use of starch to improve biodegradability began. Griffin was the first to report using starch as a filler in plastic films with the intention of improving the biodegradability of plastics (Griffin, 1973, 1974). In a later patent, Griffin claimed that adding starch to various polyolefins made them biodegradable meaning that they would ultimately break down due to the action of living organisms (Griffin, 1977). This claim was later refuted when it was shown that although the starch component could biodegrade, starch did not affect the biodegradation of the polyolefin matrix (Koenig & Huang, 1995).The Griffin process involved drying the starch (moisture content <1%) and chemically treating the granule surface to improve compatibility with the polymer matrix. Degradation of the polymer was facilitated by adding a prooxidant to the resin. Evidence for microbial degradation was based on fungal growth found in composted film samples.

Griffin was successful in extrusion-blowing film of low-density polyethylene (LDPE) containing up to 30% starch and reported acceptable thermal stability and melt-flow properties needed for processing various commodity resins. Adding starch as a filler to plastics generally increased moisture sensitivity, reduced the tensile strength, and increased the thickness of films (Griffin, 1977; Lim, Jane, Rajagopalan, & Seib, 1992). Starch granules with smaller diameters have less of a negative effect on the mechanical properties of starch-filled plastics probably due to the benefit of greater exposed surface area in smaller granules (Lim et al., 1992). Even greater exposed surface area can be obtained by forming starch nanoparticles (Le Corre et al., 2010). The nanostarch can be added in higher amounts (25-40%) compared to native starch granules yet still result in thin films with good strength and clarity (Schut, 2008).

Starch xanthates and native starch powders were evaluated as fillers in poly(vinyl chloride) (PVC) near the time of Griffin's work (Westhoff, Otey, Mehltretter, & Russell, 1974). The goal was to make the PVC more biodegradable by interspersing starch throughout the polymer matrix. The starch performed well as a filler and promoted the growth of fungi in composting studies (Westhoff et al., 1974). As mentioned earlier, however, only the starch component was biodegradable (Koenig & Huang, 1995). In contrast, using starch as a filler in a degradable synthetic polymer such as polycaprolactone provided fully biodegradable plastic composites (Koenig & Huang, 1995). Starch-filled plastic technology is still used today to incorporate starch into polyethylene and other polyolefins. Since whole starch granules are used, the amount used commercially is typically only 6-20% weight (Otey, Westhoff, & Russell, 1977; Röper & Koch, 1990). These composite materials would not fully biodegrade but would be considered more sustainable due to the renewable content provided by the starch.

5.2 Thermoplastic Starch

Native starch is not a thermoplastic material and will thermally degrade before its glass transition temperature (T_g) is reached (Dufresne, Dupeyre, & Vignon, 2000). The T_g of dry starch and its melting temperature

(230-240 °C) are both higher than the degradation temperature of starch (220 °C) (Orford, Parker, Ring, & Smith, 1989; Russell, 1987; Shogren, 1992). However, as noted in the previous section, when starch is heated in the presence of plasticizers such as water or polyols, the semicrystalline structure of the starch granule is disrupted. When this occurs, the glass transition temperature falls well below the thermal degradation temperature and the starch indeed behaves as a thermoplastic.

During the time that Griffin was investigating starch as a filler in plastics, Otey was investigating the use of starch films as a replacement for plastics (Otey et al., 1974). Using starch as a matrix material afforded the opportunity to use a much higher amount of starch in the product. Starch had long been known to gelatinize when heated in water forming a thermoplastic of sorts with a continuous matrix having good film-forming properties (Lloyd & Kirst, 1963). Interest in using starch as a nonsupported film in food and agricultural applications was deemed attractive because of its low cost, wide availability, and biodegradability. Unfortunately, the cast starch films tended to embrittle and were sensitive to moisture (Lloyd & Kirst, 1963; Otey et al., 1974). Plasticizing the starch films with glycerol or ethylene glycol improved the flexibility of the films but they had very little wet strength (Lloyd & Kirst, 1963). Cast starch-based films with improved strength could be made by blending starch with other polymers (Maxwell, 1970). Films of starch/ethylene copolymer-poly(vinyl alcohol) (EVA) blends were used to coat paper to improve strength and physical properties (Maxwell, 1970, 1971). Films made from blends of starch, glycerol, poly(vinyl alcohol) (PVA), and a cross-linking agent were then coated with a thin layer of PVC or other plastic to make cast films with good functional properties for agricultural mulch (Otey et al., 1974).

The development of thermoplastic processing technology has played an important role in achieving mass production of inexpensive plastic products. Any plastic substitute that can be processed as a "drop-in" replacement using existing plastic processing equipment has a built-in advantage of seamless integration into existing processing capacity. Otherwise, specialized equipment must be designed, manufactured, and purchased by processors, all of which can deter commercial interest.

Extrusion technology is one of the fundamental processes used in processing thermoplastics and has been used for decades to process starch-rich foods. Early concepts for extrusion technology date back to the 1700s where a piston was used to force lead through a die to make lead pipe. The process was subsequently used for extruding clay pipe, soap, and pasta. Both single- and twin-screw extruders were developed around 1870 for use in processing rubber and for making sausage (Riaz, 2000). In the mid-1930s, food companies began using extrusion to make breakfast cereals and snack foods from starch-rich cereals. During this period of time, the first extruder for thermoplastics was built by Paul Troester in Germany (Rauwendaal, 2001). In the late 1930s, Roberto Colombo and Carlo Pasquetti from Italy designed a twin-screw extruder for processing cellulose acetate (Riaz, 2000).

Initially, the extruders were heated with steam but, subsequently, were designed with electrically heated barrels and screws with longer length/diameter ratios for starch-based foods (Rauwendaal, 2001). Scientific studies and modeling work on plastic extrusion commenced during the 1950s. By 1965, the basic theories for extrusion had been developed as well as extrusion systems complete with feeders and dies (Rauwendaal, 2001). Some early studies on starch-rich foods were actually performed using extruders designed for plastics (Anderson, Conway, & Peplinski, 1970; Riaz, 2000). Studies on the effect of various extruder variables on the properties of cornstarch took place as early as 1972 (Lawton, Henderson, & Derlatka, 1972). One of the early studies on processing parameters of starch was performed using twin-screw extrusion (Mercier & Feillet, 1975). These studies were some of the first aimed at modeling starch extrusion. Subsequent work on extrusion processing of starches from a food perspective continued to provide the foundation for starch extrusion (Colonna, Doublier, Melcion, Monredon, & Mercier, 1984; Mercier, 1977; Mercier & Feillet, 1975; Wiedmann, 1987).

Today, it is widely recognized that extrusion processing can be used to completely destructurize starch and form a homogenous melt. In the food industry, extrusion is referred to as a "high-temperature (150-170 °C), short-time (20-200s)" (HTST) bioreactor that is able to gelatinize, melt, and destructurize starch granules (Colonna et al., 1984; Kokini, 1993; Lai & Kokini, 1991). Plasticized starch does indeed behave as a thermoplastic and is referred to as thermoplastic starch (TPS) (Dufresne et al., 2000). Extrusion processing melts TPS at a much lower moisture content (10-20%) than that used for conventional cooking methods (Kokini, 1993; Wiedmann, 1987).

Extrusion processing is used to compound plasticized starch and form a pelletized extrudate of TPS that can be further processed using extrusion plastic processing equipment such as injection molding or film blowing (Otey, Westhoff, & Doane, 1980; Wiedmann & Strobel, 1991). Although TPS can be processed like traditional plastics, it is not suitable for most applications because of its moisture sensitivity. For that reason, starch blends with moisture-resistant polymers were explored. Blends of starch and poly(ethylene-*co*-acrylic acid) were processed into films using plastic film blowing equipment (Otey et al., 1980; Otey, Westhoff, & Doane, 1987). The film was biodegradable and intended for use as an agricultural mulch or as a packaging material. Blends of starch and synthetic polymers such as EVA and PVA that had shown promise as films for paper or agricultural mulch (Maxwell, 1970, 1971; Otey et al., 1974) were found particularly useful when compounded by extrusion to make blends (Bastioli, 1998).

Subsequent research on blends of TPS and moisture-resistant polyesters has resulted in starch blends with adequate moisture resistance and mechanical properties (Bastioli, 1998). Other promising TPS/polymer blends are with synthetic biodegradable resins that are aliphatic polyesters and aliphatic-aromatic copolyesters. These polyesters fall within two groups, amorphous and semicrystalline. The amorphous polyesters are flexible and are comparable to LDPE. The semicrystalline polyesters are more rigid and are more comparable to PET, PS, or polypropylene (PP) (Leaversuch, 2002). The aliphatic-aromatic copolyesters are typically based on butanediol, adipic acid, and terephthalic acid. The aliphatic polyesters are a polybutylene succinate (Leaversuch, 2002). These polymers are made in modified PET facilities from petrochemical feedstocks and typically biodegrade in about 3 months. The market for these biodegradable polyesters has grown at a robust annual rate of about 30% (Leaversuch, 2002).

Another approach to improving the properties of TPS has been to form starch composite materials. Polymer composites with improved barrier, mechanical, and heat resistance properties have been reported using clay nanoparticles (Alexandre & Dubois, 2000; Park, Lee, Park, Cho, & Ha, 2003). The most commonly used nanoclays include montmorillonite, a 2:1 phyllosilicate that has a stacked platelet structure with each platelet having a thickness of approximately 1 nm and lateral dimensions on the order of micrometers. The hydrophilic interlayer between each platelet structure can be rendered more hydrophobic through a cation-exchange process with alkylammonium ions. Water vapor transmission rates of potato starch have been reduced by nearly 50% by incorporating only 5 wt% of such clays (Park et al., 2003). In addition, the dynamic elastic moduli and tensile strength were increased. The improved properties of the starch-clay composite materials are due to uniform dispersion of the clay platelets in the starch matrix (Figure 15.5). The platelets create a tortuous pathway through the starch matrix that significantly lowers the water vapor transmission rates (Chiou, Yee, Glenn, & Orts, 2005). In addition, the dispersed platelets provide more surface area for starch-nanoclay interactions that, in turn, improve the mechanical properties of the starch composites.

Cellulose microfibrils provide another nanosized material for making starch composites (Medeiros et al., 2008). In 1995, researchers in Grenoble published a notable paper where they created nanocomposites derived from cellulose whiskers dispersed in a copolymer acrylate latex film (Favier, Chanzy, & Cavaille, 1995). They showed that the dynamic storage modulus was improved more than threefold at 3-6% cellulose loadings (Favier et al., 1995). Typical cellulosic microfibrils are long crystalline

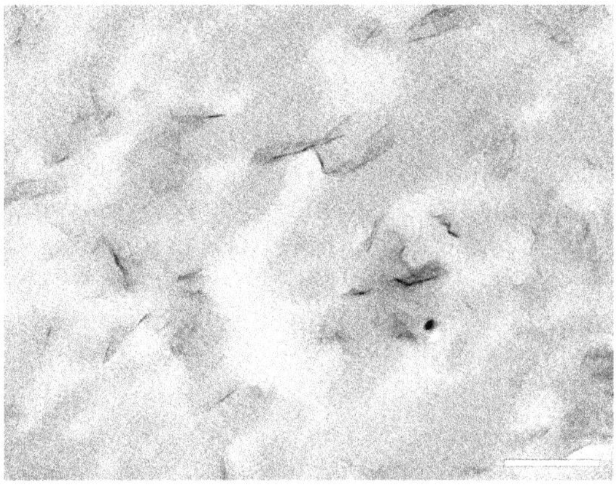

FIGURE 15.5 Transmission electron micrograph of starch-clay composite. The clay particles appear as dark lines dispersed throughout the starch matrix (scale bar = 200 nm).

"needles" ranging in size from 10 to 20 nm in width, with an average aspect ratio of 20-100 (Azizi Samir, Alloin, & Dufresne, 2005). Several groups (Dubief, Samain, & Dufresne, 1999; Favier et al., 1995) have continued work on the hypothesis that natural cellulose microfibrils can act in a fashion similar to the clay nanocomposites in reinforcing polymers. Part of the promise lies in the fact that the axial Young's modulus of the basic cellulose crystalline microfibril has been measured to be 137 GPa (Leitner, Hinterstoisser, Wastyn, Keckes, & Gindl, 2007; Wu, Henriksson, Liu, & Berglund, 2007), which is similar to Kevlar and "stronger than steel." Incorporating nanofibers in TPS increases tensile strength and modulus (Alemdar & Sain, 2008) and decreases moisture sensitivity of starch composites (Svagan, Hedenqvist, & Berglund, 2009). While research in starch nanocomposites continues to show promise, cost considerations have deterred commercialization.

5.3 Companies Marketing TPS-Based Products

The list of companies using extrusion technology to make TPS-based commercial products is ever-changing. Notwithstanding, the following list includes many of the companies currently marketing TPS-based products.

5.3.1 Novamont

One of the early innovators in developing TPS blends for commercial production is the Italian company, Novamont. Founded in 1989, Novamont has developed four different classes of materials all based on blends of TPS and synthetic polymers (Bastioli, 1998). The film-grade product is based on blends of TPS and polycaprolactone (Bastioli, Cerutti, Guanella,

Romano, & Tosin, 1995). This grade degrades in a composting environment in about 20-35 days (Bastioli, 1998). The injection molding grade is a blend of TPS and cellulose derivatives. It degrades in about 4 months and is a rigid material that may replace PS. A foaming grade is also available that contains more than 85% TPS. The foam product is used as a replacement for PS foam for loose-fill packaging (Bastioli, 1998). The final class of materials is a blend of TPS and polyethylene vinyl alcohol (EVA) (Bastioli, Bellotti, Del Giudice, & Gilli, 1993). This grade is noncompostable and degrades in about 2 years in a liquid environment (Bastioli, 1998). The current production at Novamont has exceeded 60,000 MT.

Novamont continues to develop their technology. In 1997, Novamont purchased the Warner-Lambert patents for TPS/polymer blends. They later acquired (2004) the technology from Eastman Chemical to produce the polyester (Estar Bio). This polyester is now being produced using oil from a nonfood crop and is being blended with TPS (Schut, 2008). The biobased polyester has enabled Novamont to increase the renewable polymer content of their resins to about 50% (Anonymous, 2010). They also intend to commercialize "nanostarch" particles for use in film grades of Mater-Bi. The nanostarch allows for higher renewable content in films while maintaining good strength and clarity (Schut, 2008).

5.3.2 *Cereplast*

Cereplast is located in El Segundo, CA. It was founded in 1996 and began marketing the Novamont product in North America. In 2000, Cereplast began developing their own TPS technology and started marketing Cereplast products in 2001. Cereplast recently opened a 36,000 MT facility in Indiana (Cereplast, 2011). The Cereplast products include single-use food service items including cups, bags, and cutlery (Cereplast, 2011).

5.3.3 *Ecobras*

BASF, the world's leading chemical company, has entered the starch blend market. BASF produces the biodegradable polyester, Ecoflex (polybutyrate adipate terephthalate, PBAT). In 2007, BASF aligned itself with Corn Products International and began selling a starch/PBAT blend for the Latin American market. The blend contains about 50% cornstarch and is designed for making films although it can be injection molded as well.

5.3.4 *Biotec*

Biotec GmbH in Germany has about 11,000 MT capacity (Schut, 2008). Its products include pure TPS and various blends of starch and copolyesters. It has six commercial formulations for injection molding, rigid and flexible extrusion, and foams. Many of the finished products include biodegradable carrier bags, bin liners, and refuse bags. Most sales are in Europe (Schut, 2008).

5.3.5 *Plantic*

Plantic Technologies was incorporated in Victoria, Australia, but is located today in Melbourne, Australia. Plantic acquired technology in 2001 for making high-amylose cornstarch TPS plastic sheets and trays. They found that TPS sheets could be thermoformed into trays and used to package fatty foods or products with a water activity of 35-70%. The business started making trays for a candy company. Plantic entered the global market in 2004 and recently developed multilayered polymer films with a starch film core that has improved moisture resistance, gas barrier properties, and physical properties. Plantic has announced joint ventures with several companies in recent years including DuPont for making cosmetic and food packaging and Bemis Co., Inc., Neenah, WI, to develop blown film for dry-goods packaging (Schut, 2008).

5.3.6 *Biolice*

Biolice was developed by Limagrain, a leader in the European agricultural sector. Biolice is a TPS made from cereal flour that is blended with biodegradable polyesters. Biolice is a rigid packaging material that can be thermoformed into single-use items like drink trays and cups. Films can also be made from the resin for agricultural mulch and carrier bags. The product is being marketed in France and is completely biodegradable.

5.3.7 *KTM Industries*

KTM Industries is a company located in Lansing, MI. The company uses an extrusion process similar to that used to make PS foam sheets. The process involves extruding TPS through an annular die to form a foam tube. The tube is sliced and opened flat to form sheets of starch foam that can then be used for packaging operations. The foam sheets can be cut and glued to form padding for specific packaging applications. The company also makes colored loose-fill products for children craft projects. Other companies using TPS for making starch-based loose-fill products include StarchTech, Inc., and National Starch with its Eco-Foam product.

5.4 TPS-Polyolefin Blends

5.4.1 *Cerestech*

Cerestech, Inc. was incorporated in 2001 in Montreal, Canada. The company produces blends of TPS and commodity thermoplastics in a one-step extrusion process. The process involves preparing starch/glycerol/water blends of approximately 48%, 32%, and 20%, respectively

(Rodriguez-Gonzalez et al., 2003). The starch preparation is fed into a twin-screw extruder to form a TPS melt. A second single-screw extruder is attached to the twin-screw extruder in a perpendicular position to allow a thermoplastic polyolefin such as high-density polyethylene to be melted and injected directly into the TPS melt. The melt blend is compounded further using high shear to form a blend of the two incompatible resins. Although the polyolefin and TPS form an incompatible blend, the domains of the respective polymers range from several micrometers to less than one micrometer (Rodriguez-Gonzalez et al., 2003). Blends containing up to 50% starch have been produced with excellent mechanical properties and moisture resistance (Rodriguez-Gonzalez et al., 2003).

The carbon footprint of these blends is significantly reduced compared to the neat polymer due to the starch content. A family of blends (Cereloy) based on starch and various polyolefin resins is being developed. Cerestech has granted a worldwide license to Teknor Apex to produce the blends. The blends are being sold at a similar or lower price than the neat polymer. Teknor Apex is also marketing blends of starch and recycled PP and PE to further improve the environmental profile of its products (Anonymous, 2011e).

5.5 Starch Baking Technology

A baking technology that was first developed for the food industry has been adapted for making starch-based foam food service products (Hass, Hass, & Tiefenbacher, 1996; Tiefenbacher, 1993). The starch baking process is analogous to the process used in making waffles and wafer cookies. A starch dough is first made containing gelatinized or pregelatinized starch, native starch, water, fiber, fillers, and other additives and then kneaded in a mixer for about 10 min. A predetermined amount of aqueous starch dough is placed into a preheated (150-200 °C) mold cavity that is then clamped closed. The dough quickly heats and the starch component is gelatinized forming a melt that fills the entire mold cavity. A skin forms on the upper and lower surfaces where the molten dough contacts the mold. Steam that forms during the process acts as a blowing agent and creates a foam structure in the core region of the product. Steam is allowed to vent from the mold, and within about 45-60 s, the product dries and solidifies into the desired shape. The starch-based product appears very similar to PS foam. The procedure is described in detail elsewhere (Glenn, Orts, & Nobes, 2001; Shey, Imam, Glenn, & Orts, 2006).

The properties of the baked starch foams are dependent upon various factors, including moisture content and starch type (Andersen & Hodson, 1996, 2000; Lawton, Shogren, & Tiefenbacher, 2004; Shogren, Lawton, Doane, & Tiefenbacher, 1998). Immediately following the baking process, the baked starch foams are brittle and lack flexibility but appear very similar to expanded polystyrene foam. Under dry storage conditions, starch blends and composites loose water quickly and become brittle, yielding a matrix of low modulus. Under high-moisture conditions, starch can absorb moisture, yielding a loose and flexible matrix. Thus, to obtain starch food packaging materials with acceptable properties, other substances such as fillers, compatible additives, plasticizers, and a moisture resistance film coating are generally required.

Incorporating cellulose fibers in formulations improved both the flexibility and the strength of baked starch foams (Andersen & Hodson, 1997; Glenn, Orts, & Nobes, 2001). For example, addition of softwood pulp fiber (Figure 15.6) improved flexural properties and lowered the foam density (Glenn, Orts, & Nobes, 2001; Glenn, Orts, Nobes, & Gray, 2001). Foam properties were further improved by utilizing chemically modified starches and additives such as aspen fiber, PVA, and monostearyl citrate (Lawton et al., 2004; Shogren, Lawton, Tiefenbacher, & Chen, 1998). Modified starches improved flexibility, and aspen fiber improved the strength, whereas monostearyl citrate improved water resistance. Interestingly, not all fibers help improve foam properties. Addition of corn fiber in formulations had a rather negative impact on starch foam packaging trays, as corn fiber tended to decrease the mechanical strength and increase baking time and batter volume (Cinelli, Chiellini, Lawton, & IMAM, 2006). Trays produced with a high fiber ratio in conjunction with PVA, however, showed improved water resistance.

Baked foam products with improved moisture resistance were made by incorporating PVA in the starch-based composition (Shogren, Lawton, Tiefenbacher, et al., 1998). Alternatively, protective food-grade, hydrophobic, thermostable, and degradable polyester films have been effective

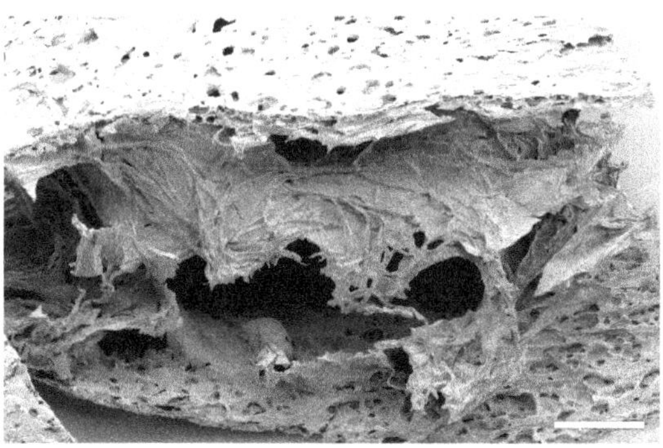

FIGURE 15.6 Scanning electron micrograph of cross section from starch/fiber baked foam composite material. A fairly dense skin was formed on the surface of the foam. The interior of the foam was porous with walls formed from the starch/fiber composite (scale bar = 0.5 mm) (Glenn, Orts, Nobes, & Gray, 2001b).

in improving moisture resistance when applied directly onto the baked product (Glenn, Orts, Nobes and Gray, 2001). More recently, natural rubber latex has been used as a moisture-resistant additive for the baked starch foams (Shey et al., 2006). Moisture resistance in starch foams improved when a small amount of latex was added in the formulation in the presence of nonionic additives. Latex also improved the flexibility of the foam product. Such approaches to improve moisture resistance, however, add to the overall cost of the product.

An Austrian company, Biopack, was the first to commercially produce starch-based baked foam trays for food packaging. Apack (Germany) also began producing starch-based food packaging made by a similar baking process. The name Apack has since changed to IBEK Verpackungshandel GmbH. This company produces over 150 million pieces a year with production facilities in China, Thailand, and Canada. Another of the early leaders in the development of baked starch containers was EarthShell Corporation located in the United States. EarthShell was granted over 130 patents in the field and was able to successfully produce commercial single-use disposable baked trays, dinner plates, and soup bowls for the fast-food industry. These products are currently being sold in the U.S. markets. More recently, EarthShell has licensed its technology to Renewable Products, Inc., located in Lebanon, MO, for manufacturing and distributing EarthShell packaging plates and bowls in the United States.

The baking technology is also being used by Biosphere Industries, LLC in Carpinteria, CA, and Durango, CO. Biosphere has a patented technology for making products to compete with cardboard and plastic foams and for baking trays and food service ware (Biosphere, 2011). They use primarily starches from cassava and potato and natural fibers in their engineered products.

5.6 Starch Fermentation Products

The use of starch in fermentation products has grown to be the largest industrial use of starch. Nearly 40% of the entire U.S. corn crop production was used for ethanol production in 2010 (USDA, 2010). Braskem, the largest petrochemical company in Latin America, has constructed a 0.2 MMT plant for making "green polyethylene" from sugarcane-based ethanol and is considering plans to build another (Murphy, 2010). Corn-based ethanol is not currently being considered as a feedstock for green polyethylene since sugarcane ethanol conversion is much more energy-efficient. However, the development and optimization of this technology could lead to future consideration (Defosse, 2009; Johansson, 2009).

Cargill Dow LLC, owner of NatureWorks, is the world's largest bioplastic manufacturer. The NatureWorks production facility is located in

Nebraska. Their Ingeo-brand PLA products are produced from the fermentation of cornstarch. At full capacity, the NatureWorks facility can produce 136,000 MT utilizing 340,000 MT of corn (BIOPOL, 2010). NatureWorks projects are at full capacity in 2-3 years and are currently enjoying a 25-30% annual growth in the market. A new plant in Thailand was recently announced that will have a production capacity similar to that of the Nebraska plant. The Thailand plant will use cassava starch and sugar from sugarcane as feedstock (Anonymous, 2011f).

PHAs are linear polyesters produced by bacteria through a sugar or lipid fermentation process using unbalanced growth conditions (Chen & Wu, 2005). More than 150 different monomers have been identified that produce polymers with a wide range in mechanical properties (Chen & Wu, 2005). The PHA is produced as an energy storage molecule in bacteria much like starch in plants. PHAs represent a family of polymers that have a considerable potential as fully degradable plastic polymers. To date, the most commercial production is from Tianan Biologic Material Co. in Ningbo, China. It has increased its capacity to more than 900 MT/year (Schut, 2008). Tianan's commercial PHA contains about 5% valerate, though developmental grades have up to 15% valerate, and even higher valerate content is being tested. Valerate adds flexibility to the polymer. Other companies producing PHAs include Metabolix (Cambridge, MA).

6 FUTURE PROSPECTS

With an ever-increasing world population, the question is not whether the global environment will be impacted by our presence but how and to what degree. The implementation of sustainable practices will help minimize our impact on the environment and conserve resources for future generations. To that end, there is a need to perpetuate the culture of environmental stewardship and sustainability that has grown stronger in recent years. Consumers and municipalities must continue to demand more sustainable packaging materials and practices. Retailers must continue to provide suppliers incentives for greater sustainability in their packaging choices. Although some of the starch-based materials and other biopolymers may not currently be cost-competitive with petroleum plastics, this may change as petroleum prices increase.

Starch is poised to establish an even stronger role in the manufacture of sustainable plastics and other bioproducts largely because it is abundant, renewable, and inexpensive. The cost and availability of starch may improve even further in the future if lignocellulose supplants cornstarch as the preferred feedstock for ethanol production in the United States. Strategies for improving the properties of starch-based plastics such as blending starch with other polymers, using starch in composite materials, and using starch as a fermentation feedstock to make other biopolymers have been successful in developing viable replacements for petroleum-based plastics. The prospects for starch in the packaging sector continue to become brighter as the market for sustainable plastics drives further innovation and development.

References

ACC. (2011). Life cycle of a plastic product. http://www.plastics.americanchemistry.com/Life-Cycle. American Chemical Council. Accessed 12.05.11.

Alemdar, A., & Sain, M. (2008). Biocomposites from wheat straw nanofibers: Morphology, thermal and mechanical properties. *Composites Science and Technology, 68*, 557–565.

Alexandre, M., & Dubois, P. (2000). Polymer-layered silicate nanocomposites: Preparation, properties and uses of a new class of materials. *Materials Science and Engineering: R: Reports, 28*, 1–63.

Almeida, M., Alves, R., Nascimbem, L., Stephani, R., Poppi, R., & De Oliveira, L. (2010). Determination of amylose content in starch using Raman spectroscopy and multivariate calibration analysis. *Analytical and Bioanalytical Chemistry, 397*, 2693–2701.

Andersen, P. J. & Hodson, S. K. (1996). Molded articles having an inorganically filled organic matrix, U.S. Patent No. 5,545,450.

Andersen, P. J. & Hodson, S. K. (1997). Starch-based compositions having uniformly dispersed fibers used to manufacture high strength articles having a fiber-reinforced starch-bound cellular matrix, U.S. Patent No. 5,679,145.

Andersen, P. J. & Hodson, S. K. (2000). Molded starch-bound containers and other articles having natural and/or synthetic polymer coatings, U.S. Patent No. 6,030,673.

Anderson, R. A., Conway, H. F., & Peplinski, A. J. (1970). Gelatinization of corn grits by roll cooking, extrusion cooking and steaming. *Starch-Stärke, 22*, 130–135.

Anonymous, (2010). *Starch-based resin adds more bio content. Plastics technology*. Cincinnati, OH: Gardner Publishing, Inc. (April issue, accessed 11/28/2011).

Anonymous, (2011a). *First biodegradable toothpaste tube. Plastics technology*. Cincinnati, OH: Gardner Publishing, Inc. (March issue, accessed 11/24/2011).

Anonymous, (2011b). *Materials: Clear biodegradable cellulosic plastic. Plastics technology*. Cincinnati, OH: Gardner Publishing, Inc. (September issue, accessed 12/3/2011).

Anonymous, (2011c). *Materials: Heinz ketchup converts to biobased bottle. Plastics technology*. Cincinnati, OH: Gardner Publishing, Inc. (March issue, accessed 11/15/2011).

Anonymous, (2011d). *Materials: Polyolefin/soy meal 'bio-composite' is new sustainable option. Plastics technology*. Cincinnati, OH: Gardner Publishing, Inc. (February issue, accessed 11/15/2011).

Anonymous, (2011e). *Materials: Starch + recycled PP = high 'green' content. Plastics technology*. Cincinnati, OH: Gardner Publishing, Inc. (April issue, accessed 112/5/2011).

Anonymous, (2011f). *NatureWorks to build world-scale PLA plant in Thailand. Plastics technology*. Cincinnati, OH: Gardner Publishing, Inc. (November issue, accessed 11/22/2011).

Anonymous, (2011g). *Pepsi develops first 100% biobased PET bottle. Plastics technology*. Cincinnati, OH: Gardner Publishing, Inc. (May issue, accessed 12/5/2011).

Atkin, N. J., Abeysekera, R. M., Cheng, S. L., & Robards, A. W. (1998). An experimentally-based predictive model for the separation of amylopectin subunits during starch gelatinization. *Carbohydrate Polymers, 36*, 173–192.

Azizi Samir, M. A. S., Alloin, F., & Dufresne, A. (2005). Review of recent research into cellulosic whiskers, their properties and their application in nanocomposite Field. *Biomacromolecules, 6*, 612–626.

Baldwin, P. M., Adler, J., Davies, M. C., & Melia, C. D. (1994). Holes in starch granules: Confocal, SEM and light microscopy studies of starch granule structure. *Starch-Stärke, 46*, 341–346.

Bastioli, C. (1998). Properties and applications of Mater-Bi starch-based materials. *Polymer Degradation and Stability, 59*, 263–272.

Bastioli, C., Bellotti, V., Del Giudice, L., & Gilli, G. (1993). Mater-Bi: Properties and biodegradability. *Journal of Polymers and the Environment, 1*, 181–191.

Bastioli, C., Cerutti, A., Guanella, I., Romano, G. C., & Tosin, M. (1995). Physical state and biodegradation behavior of starch-polycaprolactone systems. *Journal of Polymers and the Environment, 3*, 81–95.

Bennett, F. L., Otey, F. H., & Mehltretter, C. L. (1967). Rigid urethane foam extended with starch. *Journal of Cellular Plastics, 3*, 369–373.

BIOPOL. (2010). NatureWorks (PLA) doesn't reach full production capacity. http://www.biopol.free.fr/index.php/natureworks-pla-doesnt-reach-full-production-capacity/ Accessed 12.05.11.

BIOSPHERE. (2011). Biosphere Bakeware. http://www.biosphereindustries.com/index.php?option=com_content&view=category&id=14&Itemid=40, Biosphere Industries, LLC (Accessed 12/5/2011).

Blanshard, J. M. V. (1987). Starch granule structure and function. In T. Galliard (Ed.), *Starch: Properties and potential, critical reports on applied chemistry*. New York: John Wiley & Sons.

Blaszczak, W., Fornal, J., Valverde, S., & Garrido, L. (2005). Pressure-induced changes in the structure of corn starches with different amylose content. *Carbohydrate Polymers, 61*, 132–140.

Boerger, C. M., Lattin, G. L., Moore, S. L., & Moore, C. J. (2010). Plastic ingestion by planktivorous fishes in the North Pacific Central Gyre. *Marine Pollution Bulletin, 60*, 2275–2278.

Buchanan, R. A., Weislogel, O. E., Russell, C. R., & Rist, C. E. (1968). Starch in rubber. Zinc starch xanthate in latex masterbatching. *I&EC Product Research and Development, 7*, 155–158.

Buleon, A., Colonna, P., Planchot, V., & Ball, S. (1998). Starch granules: Structure and biosynthesis. *International Journal of Biological Macromolecules, 23*, 85–112.

Cellura, M., Ardente, F., & Longo, S. (2012). From the LCA of food products to the environmental assessment of protected crops districts: A case-study in the south of Italy. *Journal of Environmental Management, 93*, 194–208.

Cereplast. (2011). Cereplast's Historical Timeline. http://www.cereplast.com/company/history/ El Segundo, CA Accessed 12.04.11).

Chawengkijwanich, C., & Hayata, Y. (2008). Development of TiO2 powder-coated food packaging film and its ability to inactivate Escherichia coli in vitro and in actual tests. *International Journal of Food Microbiology, 123*, 288–292.

Chen, G. -Q., & Wu, Q. (2005). The application of polyhydroxyalkanoates as tissue engineering materials. *Biomaterials, 26*, 6565–6578.

Chiou, B. -S., Yee, E., Glenn, G. M., & Orts, W. J. (2005). Rheology of starch–clay nanocomposites. *Carbohydrate Polymers, 59*, 467–475.

Cinelli, P., Chiellini, E., Lawton, J. W., & IMAM, S. H. (2006). Foamed articles based on potato starch, corn fibers and poly(vinyl alcohol). *Polymer Degradation and Stability, 91*, 1147–1155.

Colonna, P., Doublier, J. L., Melcion, J. P., Monredon, F. D., & Mercier, C. (1984). Extrusion cooking and drum drying of wheat starch. I. Physical and macromolecular modifications. *Cereal Chemistry, 61*, 538–543.

Dartee, M. (2010). *It's time to get to know your way around bioplastics. Plastics technology*. Cincinnati, OH: Gardner Publishing, Inc. (March issue, accessed 11/22/2011).

Defosse, M. (2009). *Bioplastics: Tetra Pak tasks Braskem for bio-based HDPE. Plastics today*. Los Angeles, CA: UBM Canon (November 29 issue, accessed 11/23/2011).

Deligio, T. (2009). Designing for thinner walls, gauges. *Plastics today*. Los Angeles, CA: UBM Canon. (December 22 issue, accessed 12/7/2011). http://www.plasticstoday.com/mpw/articles/designing-thinner-walls-gauges.

Dosmann, L. & Steel, R. (1961). Flexible shock absorbing foam containing starch and method of preparing same. U.S. Patent #3,004,934.

Dubief, D., Samain, E., & Dufresne, A. (1999). Polysaccharide microcrystals reinforced amorphous poly(β-hydroxyoctanoate) nanocomposite materials. *Macromolecules, 32*, 5765–5771.

Dufresne, A., Dupeyre, D., & Vignon, M. R. (2000). Cellulose microfibrils from potato tuber cells: Processing and characterization of starch–cellulose microfibril composites. *Journal of Applied Polymer Science, 76*, 2080–2092.

EPA, (2010). *Municipal solid waste in the United States—2009 facts and figures.* Washington DC: United States Environmental Protection Agency. www.epa.gov/epawaste/nonhaz/municipal/msw99.htm.

Erickson, E. H., Degrandi-Hoffman, G., Becker, C. G., Whitson, R. S. & Deeby, T. A. (2005). Control of parasitic mites of honey bees, U.S. Patent #6,843,985.

Fannon, J. E., & Bemiller, J. N. (1993). Structure of corn starch paste and granule remnants revealed by low-temperature scanning electron microscopy after cryopreparation. *Cereal Chemistry, 69*, 456–460.

Favier, V., Chanzy, H., & Cavaille, J. Y. (1995). Polymer nanocomposites reinforced by cellulose whiskers. *Macromolecules, 28*, 6365–6367.

Finkenstadt, V. L., & Tisserat, B. (2010). Poly(lactic acid) and Osage Orange wood fiber composites for agricultural mulch films. *Industrial Crops and Products, 31*, 316–320.

French, D. (1984). Organization of starch granules. In R. L. Whistler, J. N. Bemiller, & E. F. Paschall (Eds.), *Starch: chemistry and technology*. (2nd ed.). Orlando: Academic Press Inc.

Funes, H. N., Zerba, E., & González Audino, P. (2009). Comparison of three types of traps baited with sexual pheromones for Ambrosia Beetle Megaplatypus mutatus (Coleoptera: Platypodinae) in poplar plantations. *Journal of Economic Entomology, 102*, 1546–1550.

Gemili, S., Yemenicioğlub, A., & Altınkayaa, S. A. (2009). Development of cellulose acetate based antimicrobial food packaging materials for controlled release of lysozyme. *Journal of Food Engineering, 90*, 453–462.

Ghiasi, K., Hoseney, R. C., & Varriano-Marston, E. (1982). Gelatinization of wheat starch. III. Comparison by differential scanning calorimetry and light microscopy. *Cereal Chemistry, 59*, 258–264.

Glenn, G. M., Klamczynski, A., Chiou, B. -S., Orts, W. J., Imam, S. H., & Wood, D. F. (2008). Temperature related structural changes in wheat and corn starch granules and their effects on gels and dry foam. *Starch-Stärke, 60*, 476–484.

Glenn, G. M., Klamczynski, A. P., Ludvik, C., Shey, J., Imam, S. H., Chiou, B., et al. (2006). Permeability of starch gel matrices and select films to solvent vapors. *Journal of Agricultural and Food Chemistry, 54*, 3297–3304.

Glenn, G. M., Klamczynski, A. P., Woods, D. F., Chiou, B., Orts, W. J., & Imam, S. H. (2010). Encapsulation of plant oils in porous starch microspheres. *Journal of Agricultural and Food Chemistry, 58*, 4180–4184.

Glenn, G. M., Orts, W. J., & Nobes, G. A. R. (2001). Starch, fiber and $CaCO_3$ effects on the physical properties of foams made by a baking process. *Industrial Crops and Products, 14*, 201–212.

Glenn, G. M., Orts, W. J., Nobes, G. A. R., & Gray, G. M. (2001). In situ laminating process for baked starch-based foams. *Industrial Crops and Products, 14*, 125–134.

Gorden, M. (2006). *Ban on expanded polystyrene foam takeout food packaging*. Fort Collins, CO: California Water Action. (accessed 1/12/2011). http://www.cleanwateraction.org/ca.

Griffin, G. J. L. (1973). Biodegradable fillers in thermoplastics. *Soc Div Org Coat Plast Chem, 33*, 88–90.

Griffin, G. J. L. (1974). Biodegradable fillers in thermoplastics. In *Fillers and reinforcements for Plastics* (pp. 159–170). American Chemical Society, 134.

Griffin, G. J. (1977). Synthetic resin sheet material, U.S. Patent #4,021,388.

Hass, F., Hass, J. & Tiefenbacher, K. (1996). Process of manufacturing rottable thin-walled starch-based shaped elements, U.S. Patent No. 5,576,049.

Hermansson, A. -M., & Svegmark, K. (1996). Developments in the understanding of starch functionality. *Trends in Food Science & Technology, 7*, 345–353.

Hoover, R. (2001). Composition, molecular structure, and physicochemical properties of tuber and root starches: A review. *Carbohydrate Polymers, 45*, 253–267.

Hoseney, R. C., & Atwell, W. A. (1977). Scanning electron microscopy of starch isolated from baked products. *Cereal Foods World, 22*, 56–60.

Huber, K. C., & Bemiller, J. N. (2000). Channels of maize and sorghum starch granules. *Carbohydrate Polymers, 41*, 269–276.

Johansson, J.-E. (2009). Compelling facts about plastics. PlasticsEurope Market Research Group (PEMRG).

Kaiser, J. (2010). The dirt on ocean garbage patches. *Science, 328*, 1506.

Kim, H. -S., & Huber, K. C. (2008). Channels within soft wheat starch A- and B-type granules. *Journal of Cereal Science, 48*, 159–172.

Koenig, M. F., & Huang, S. J. (1995). Biodegradable blends and composites of polycaprolactone and starch derivatives. *Polymer, 36*, 1877–1882.

Kokini, J. L. (1993). The effect of processing history on chemical changes in single- and twin-screw extruders. *Trends in Food Science & Technology, 4*, 324–329.

Kyrikou, I., & Briassoulis, D. (2007). Biodegradation of agricultural plastic films: A critical review. *Journal of Polymers and the Environment, 15*, 125–150.

Lai, L. S., & Kokini, J. L. (1991). Physicochemical changes and rheological properties of starch during extrusion (a review). *Biotechnology Progress, 7*, 251–266.

Law, K. L., Moret-Ferguson, S., Maximenko, N. A., Proskurowski, G., Peacock, E. E., Hafner, J., et al. (2010). Plastic accumulation in the North Atlantic subtropical gyre. *Science, 329*, 1185–1188.

Lawton, B. T., Henderson, G. A., & Derlatka, E. J. (1972). The effects of extruder variables on the gelatinisation of corn starch. *The Canadian Journal of Chemical Engineering, 50*, 168–172.

Lawton, J. W., Shogren, R. L., & Tiefenbacher, K. F. (2004). Aspen fiber addition improves the mechanical properties of baked cornstarch foams. *Industrial Crops and Products, 19*, 41–48.

Le Corre, D. B., Bras, J., & Dufresne, A. (2010). Starch nanoparticles: A review. *Biomacromolecules, 11*, 1139–1153.

Leaversuch, R. D. (2002). *Biodegradable polyesters: Packaging goes green. Plastics technology.* Cincinnati, OH: Gardner Publications, Inc.

Lee, J. -Y., Gao, Z., & Brown, M. G. (2010). A study of the impact of package changes on orange juice demand. *Journal of Retailing and Consumer Services, 17*, 487–491.

Leitner, J., Hinterstoisser, B., Wastyn, M., Keckes, J., & Gindl, W. (2007). Sugar beet cellulose nanofibril-reinforced composites. *Cellulose, 14*, 419–425.

Li, J. H., Vasanthan, T., Hoover, R., & Rossnagel, B. G. (2004). Starch from hull-less barley: IV. Morphological and structural changes in waxy, normal and high-amylose starch granules during heating. *Food Research International, 37*, 417–428.

Lim, S. -T., Jane, J. -L., Rajagopalan, S., & Seib, P. A. (1992). Effect of starch granule size on physical properties of starch-filled polyethylene film. *Biotechnology Progress, 8*, 51–57.

Lloyd, N. E., & Kirst, L. C. (1963). Some factors affecting the tensile strength of starch films. *Cereal Chemistry, 40*, 154–161.

Martin, C., & Smith, A. M. (1995). Starch biosynthesis. *Plant Cell, 7*, 971–985.

Maxwell, C. S. (1970). Effect of ethylene copolymer-starch blends on water resistance of paper. *Tappi, 53*, 1464–1466.

Maxwell, C. S. (1971). Ethylene copolymer-poly(vinyl alcohol) blends for treatment of paper. *Tappi, 54*, 568–570.

Medeiros, E. S., Mattoso, L. H. C., Ito, E. N., Gregorski, K. S., Robertson, G. H., Offeman, R. D., et al. (2008). Electrospun nanofibers of poly(vinyl alcohol) reinforced with cellulose nanofibrils. *Journal of Biobased Materials and Bioenergy, 2*, 231–242.

Mena, C., Adenso-Diaz, B., & Yurt, O. (2011). The causes of food waste in the suppliers-retailer interface: Evidences from the UK and Spain. *Resources, Conservation and Recycling, 55*, 648–658.

Mercier, C. (1977). Effect of extrusion-cooking on potato starch using a twin screw French extruder. *Starch-Stärke, 29*, 48–52.

Mercier, C., & Feillet, P. (1975). Modification of carbohydrate components by extrusion-cooking of cereal products. *Cereal Chemistry, 52*, 283–297.

Morgan, K. R., Furneaux, R. H., & Larsen, N. G. (1995). Solid-state NMR studies on the structure of starch granules. *Carbohydrate Research, 276*, 387–399.

Morris, V. J. (1990). Starch gelation and retrogradation. *Trends in Food Science & Technology, 1*, 2–6.

Morrison, W. R., Milligan, T. P., & Azudin, M. N. (1984). A relationship between the amylose and lipid contents of starches from diploid cereals. *Journal of Cereal Science, 2*, 257–271.

Murphy, P. (2010). *Brazil Braskem may expand green plastic on demand*. Sao Paulo, Brazil: Reuters. (March 22 issue, accessed 1/4/2012). http://www.reuters.com/article/2010/03/22/braskem-brazil-ethanol-idUSN2221095120100322.

Oates, C. G. (1997). Towards an understanding of starch granule structure and hydrolysis. *Trends in Food Science & Technology, 8*, 375–382.

Orford, P. D., Parker, R., Ring, S. G., & Smith, A. C. (1989). Effect of water as a diluent on the glass transition behaviour of malto-oligosaccharides, amylose and amylopectin. *International Journal of Biological Macromolecules, 11*, 91–96.

Otey, F. H., Mark, A. M., Mehltretter, C. L., & Russell, C. R. (1974). Starch-based film for degradable agricultural mulch. *Industrial & Engineering Chemistry Product Research and Development, 13*, 90–92.

Otey, F. H., Westhoff, R. P., & Doane, W. M. (1980). Starch-based blown films. *Industrial & Engineering Chemistry Product Research and Development, 19*, 592–595.

Otey, F. H., Westhoff, R. P., & Doane, W. M. (1987). Starch-based blown films. 2. *Industrial & Engineering Chemistry Research, 26*, 1659–1663.

Otey, F. H., Westhoff, R. P., Kwolek, W. F., Mehltretter, C. L., & Rist, C. E. (1969). Urethane plastics based on starch and starch-derived glycosides. *Product R&D, 8*, 267–274.

Otey, F. H., Westhoff, R. P., & Russell, C. R. (1977). Biodegradable films from starch and ethylene-acrylic acid copolymer. *Product R&D, 16*, 305–308.

Park, H. -M., Lee, W. -K., Park, C. -Y., Cho, W. -J., & Ha, C. -S. (2003). Environmentally friendly polymer hybrids Part I. Mechanical, thermal, and barrier properties of thermoplastic starch/clay nanocomposites. *Journal of Materials Science, 38*, 909–915.

Provencher, J. F., Gaston, A. J., Mallory, M. L., O'hara, P. D., & Gilchrist, H. G. (2010). Ingested plastic in a diving seabird, the thick-billed murre (Uria lomvia), in the eastern Canadian Arctic. *Marine Pollution Bulletin, 60*, 1406–1411.

Rauwendaal, C. (2001). *Polymer extrusion*. Cincinnati, OH: Hanser Gardner Publications.

Riaz, M. (2000). *Extruders in food applications*. Boca Raton, FL: CRC Press.

Richardson, G., Kidman, S., Langton, M., & Hermansson, A. -M. (2004). Differences in amylose aggregation and starch gel formation with emulsifiers. *Carbohydrate Polymers, 58*, 7–13.

Rockland, L. B., Jones, F. T., & Hahn, D. M. (1977). Light and scanning electron microscope studies on dry beans: Extracellular gelatinization of lima bean starch in water and a mixed salt solution. *Journal of Food Science, 42*, 1204–1207.

Rodriguez-Gonzalez, F. J., Ramsay, B. A., & Favis, B. D. (2003). High performance LDPE/thermoplastic starch blends: A sustainable alternative to pure polyethylene. *Polymer, 44*, 1517–1526.

Röper, H., & Koch, H. (1990). The role of starch in biodegradable thermoplastic materials. *Starch-Stärke, 42*, 123–130.

Russell, P. L. (1987). The ageing of gels from starches of different amylose/amylopectin content studied by differential scanning calorimetry. *Journal of Cereal Science, 6*, 147–158.

Sarko, A., & Wu, H. C. H. (1978). The crystal structures of A-, B- and C-polymorphs of amylose and starch. *Starch-Stärke, 30*, 73–78.

Schut, J. H. (2008). *What's ahead for 'green' plastics: Look for more supply, more varieties, better properties. Plastics technology*. Cincinnati, OH: Garner Publications, Inc. (February issue, accessed 1/5/2012). http://www.ptonline.com/articles/what's-ahead-for-'green'-plastics-look-for-more-supply-more-varieties-better-properties.

Schwartz, D., & Whistler, R. L. (2009). History and future of starch. In J. Bemiller & R. Whistler (Eds.), *Starch chemistry and technology*. New York, New York: Academic Press.

Shamekh, S., Forssell, P., Suortti, T., Autio, K., & Poutanen, K. (1999). Fragmentation of oat and barley starch granules during heating. *Journal of Cereal Science, 30*, 173–182.

Shannon, J. C., & Garwood, D. L. (1984). Genetics and physiology of starch development. In R. L. Whistler, J. N. Bemiller, & E. F. Paschall (Eds.), *Starch: Chemistry and technology*. (2nd ed.). New York: Academic Press.

Shey, J., Imam, S. H., Glenn, G. M., & Orts, W. J. (2006). Properties of baked starch foam with natural rubber latex. *Industrial Crops and Products, 24*, 34–40.

Shi, Y. -C., Capitani, T., Trzasko, P., & Jeffcoat, R. (1998). Molecular structure of a low-amylopectin starch and other high-amylose maize starches. *Journal of Cereal Science, 27*, 289–299.

Shogren, R. L. (1992). Effect of moisture content on the melting and subsequent physical aging of cornstarch. *Carbohydrate Polymers, 19*, 83–90.

Shogren, R. L., Lawton, J. W., Doane, W. M., & Tiefenbacher, K. F. (1998). Structure and morphology of baked starch foams. *Polymer, 39*, 6649–6655.

Shogren, R. L., Lawton, J. W., Tiefenbacher, K. F., & Chen, L. (1998). Starch–poly(vinyl alcohol) foamed articles prepared by a baking process. *Journal of Applied Polymer Science, 68*, 2129–2140.

Shulman, J. & Howarth, J. T. (1964). Waterproof plastic films of increased water vapor permeability and method of making them, U.S. Patent Office #3,137,664.

Singh, N., Singh, J., Kaur, L., Singh Sodhi, N., & Singh Gill, B. (2003). Morphological, thermal and rheological properties of starches from different botanical sources. *Food Chemistry, 81*, 219–231.

Svagan, A. J., Hedenqvist, M. S., & Berglund, L. (2009). Reduced water vapour sorption in cellulose nanocomposites with starch matrix. *Composites Science and Technology, 69*, 500–506.

Teixeira, L. A. F., Miller, J. R., Epstein, D. L., & GUT, L. J. (2010). Comparison of mating disruption and mass trapping with Pyralidae and Sesiidae moths. *Entomologia Experimentalis et Applicata, 137*, 176–183.

Thompson, R. C., Olsen, Y., Mitchell, R. P., Davis, A., Rowland, S. J., John, A. W. G., et al. (2004). Lost at sea: Where is all the plastic? *Science, 304*, 838.

Tiefenbacher, K. F. (1993). Starch-based foamed materials-use and degradation properties. *Journal of Macromolecular Science, Part A: Pure and Applied Chemistry, 30*, 727–731.

Tyson, R. H., & Ap Rees, T. (1988). Starch synthesis by isolated amyloplasts from wheat endosperm. *Planta, 175*, 33–38.

USDA, (2002). *Profiling food consumption in America. Agriculture fact book*. Washington, DC: USDA.

USDA, (2010). Agricultural yearbook. In United States: Department of Agriculture, Economic Research Service.

Walker, T. R., Reid, K., Arnould, J. P. Y., & Croxall, J. P. (1997). Marine debris surveys at Bird Island, South Georgia 1990-1995. *Marine Pollution Bulletin*, *34*, 61–65.

Weiss, K. R. (2006). *Plague of plastic chokes the seas*. Los Angeles, CA: Los Angeles Times (August 7 issue, accessed 11/22/2011).

Westhoff, R. P., Otey, F. H., Mehltretter, C. L., & Russell, C. R. (1974). Starch-filled polyvinyl chloride plastics—Preparation and evaluation. *Product R&D*, *13*, 123–125.

Whistler, R. L. (1984). History and future expectation of starch use. In R. L. Whistler, J. N. Bemiller, & E. F. Paschall (Eds.), *Starch: Chemistry and technology*. (2nd ed.). New York: Academic Press.

Whistler, R. L., Bemiller, J. N., & Paschall, E. F. (1984). *Starch: Chemistry and technology*. New York: Academic Press.

Whistler, R. L., & Daniel, J. R. (1984). Molecular structure of starch. In R. L. Whistler, J. N. Bemiller, & E. F. Paschall (Eds.), *Starch: Chemistry and technology*. (2nd ed.). New York: Academic Press.

Wiedmann, W. (1987). Extrusion cooking of starches for semi-products. *Starch-Stärke*, *39*, 352–357.

Wiedmann, W., & Strobel, E. (1991). Compounding of thermoplastic starch with twin-screw extruders. *Starch-Stärke*, *43*, 138–145.

Wikström, F., & Williams, H. (2010). Potential environmental gains from reducing food losses through development of new packaging—A life-cycle model. *Packaging Technology and Science*, *23*, 403–411.

Williams, H., Wikstrom, F., & Lofgren, M. (2008). A life cycle perspective on environmental effects of customer focused packaging development. *Journal of Cleaner Production*, *16*, 853–859.

Wittwer, F. & Tomka, I. (1987). Polymer composition for injection molding, U.S. Patent #4,673,438.

Wood, B. (2010). *Wood on plastics: Sustainable packaging: A lot more than buzzwords. Plastics technology*. Cincinnati, OH: Gardner Publishing, Inc.

WPO. (2008). Market statistics and future trends in global packaging. World Packaging Organisation/PIRA International Ltda. www.worldpackaging.org.

Wu, Q., Henriksson, M., Liu, X., & Berglund, L. A. (2007). A high strength nanocomposite based on microcrystalline cellulose and polyurethane. *Biomacromolecules*, *8*, 3687–3692.

Zeeman, S. C., Kossmann, J., & Smith, A. M. (2010). Starch: Its metabolism, evolution, and biotechnological modification in plants. *Annual Review of Plant Biology*, *61*, 209–234.

Zhang, Y., Zhang, Y. -B., Feng, Y., & Yang, Z. -J. (2010). Reduce the plastic debris: A model research on the great pacific ocean garbage patch. *Advanced Materials Research*, *113–114*, 59–63.

Zou, K. (2009). Regulation of waste dumping at sea: The Chinese practice. *Ocean & Coastal Management*, *52*, 383–389.

Index

Note: Page numbers followed by *f* indicate figures and *t* indicate tables.

C

R

www.ingramcontent.com/pod-product-compliance
Ingram Content Group UK Ltd.
Pitfield, Milton Keynes, MK11 3LW, UK
UKHW020222290926
15470UKWH00001B/333

9 780444 537300